Concepts and Strategies in Plant Sciences

Series Editor

Chittaranjan Kole, Raja Ramanna Fellow, Government of India, ICAR-National Institute for Plant Biotechnology, Pusa, Delhi, India

This book series highlights the spectacular advances in the concepts, techniques and tools in various areas of plant science. Individual volumes may cover topics like genome editing, phenotyping, molecular pharming, bioremediation, miRNA, fast-track breeding, crop evolution, IPR and farmers' rights, to name just a few. The books will demonstrate how advanced strategies in plant science can be utilized to develop and improve agriculture, ecology and the environment. The series will be of interest to students, scientists and professionals working in the fields of plant genetics, genomics, breeding, biotechnology, and in the related disciplines of plant production, improvement and protection.

Interested in editing a volume? Please contact Prof. Chittaranjan Kole, Series Editor, at ckoleorg@gmail.com

More information about this series at http://www.springer.com/series/16076

Ramesh Namdeo Pudake · Utkarsh Jain ·
Chittaranjan Kole
Editors

Biosensors in Agriculture: Recent Trends and Future Perspectives

Springer

Editors
Ramesh Namdeo Pudake
Amity Institute of Nanotechnology
Amity University Uttar Pradesh
Noida, UP, India

Utkarsh Jain
Amity Institute of Nanotechnology
Amity University Uttar Pradesh
Noida, UP, India

Chittaranjan Kole
Department of Atomic Energy
ICAR - National Institute for Plant
Biotechnology
New Delhi, Delhi, India

ISSN 2662-3188 ISSN 2662-3196 (electronic)
Concepts and Strategies in Plant Sciences
ISBN 978-3-030-66164-9 ISBN 978-3-030-66165-6 (eBook)
https://doi.org/10.1007/978-3-030-66165-6

This Springer imprint is published by the registered company Springer Nature Switzerland AG
The registered company address is: Gewerbestrasse 11, 6330 Cham, Switzerland

Preface

Agriculture and its allied industries are the backbone of global economies and need to be continuously developed to achieve higher production and sustainability. The industrial development and growing human population are among the major factors that are having adverse effects on various aspects of food, meat, and milk production. Augmentation of production but with minimum environmental impacts from agricultural inputs requires innovation and utilization of advanced technologies. The reduction in the quantity of agricultural inputs required for optimum production will benefit farmers by lowering production costs. Along with this, we also need to develop technologies which can reduce the wastage of agricultural products until it reaches the consumer. For this conventional tools will be of limited use, and technologies like precision farming with help of various sensors and electronics support have the potential to become an alternative strategy to achieve the above-mentioned goals.

A biosensor can be defined as an analytical tool that combines a biological sensing element and a physicochemical transducer; which can detect the analyte by transforming bio-molecular interactions into identifiable signal. The role of nanomaterial-based biosensors in agriculture is now a key research area of interest among scientists. They can used to monitor the problems associated with food security, water and soil quality, integrated pest management for crops and livestock. Currently the focus of research is targeted for search of novel nanomaterials for improving the effectiveness of biosensors by higher selectivity and sensitivity. Nanomaterials of different types, shapes and sizes are being used presently in the development of different sensors for precision farming, food processing industries and water quality monitoring use. The research in this area has the potential to increase and maintain the food safely for the consumption of the increasing global population. Despite potential advantages, the application and commercialization of nanosensor technology in agricultural sector is still comparably trivial. Like other technologies it also has some issues like interference from various biomolecules, portability, stability of sensing device, etc. that still need to be addressed. So, there is a need of continued research that will revolutionize the commercial utilization. In the near future the upcoming technologies such as microfluidics, robotics and artificial intelligence will further add additional features to biosensors.

The aim of this book is addressing the issue of knowledge gap by bringing together the acheievements from recent research and development pertaining to biosensors and their uses in agriculture. We strongly believe that this book will be useful to various stakeholders including students, researchers and manufacturers interested on the use of nanosensors in field of agriculture and allied sciences. This book comprises 22 chapters, authored by 68 eminent scientists, covering the diverse applications of nanosensors in a wide range of agricultural tasks.

We are indebted to the authors of the book chapters for their contributions. We thank Mrs. Zuzana Bernhart, and Mrs. Banu Dhayalan from Springer Nature for their immense help and patience in making this journey very comfortable. We also like to thank our respective organizations for the support and encouragement provided for academic endeavors. The encouragement and inspiration received from Dr. Ashok K. Chauhan (Founder President, Ritnand Balved Education Foundation, New Delhi) need special mention. We also like to thank our beloved families, friends, and colleagues for their support during all the activities done for the duration of penning of this book.

Noida, India — Dr. Ramesh Namdeo Pudake
Noida, India — Dr. Utkarsh Jain
New Delhi, India — Dr. Chittaranjan Kole

Contents

Contributors

Naina Abbineni DBT-National Institute of Animal Biotechnology, Hyderabad, India

Charles Oluwaseun Adetunji Applied Microbiology, Biotechnology and Nanotechnology Laboratory, Department of Microbiology, Edo University Iyamho, Auchi, Edo State, Nigeria

Aylin Altan Department of Food Engineering, Mersin University, Mersin, Turkey

Sapna Balayan Amity Institute of Nanotechnology, Amity University Uttar Pradesh, Noida, UP, India

Kamlesh Bali Division of Entomology, Sher-e-Kashmir University of Agricultural Sciences and Technology, Chatha, J&K, India

Shivansh Bansal Amity Institute of Nanotechnology, Amity University Uttar Pradesh, Noida, UP, India

Tinku Basu Amity Centre for Nanomedicine, Amity University Uttar Pradesh, Noida, UP, India

Sheetal K. Bharadwaj University of Amsterdam, Amsterdam, The Netherlands

Rashi Bhardwaj Amity Centre for Nanomedicine, Amity University Uttar Pradesh, Noida, UP, India

Khalid Hussain Bhat Division of Basic Sciences and Humanities, Faculty of Agriculture, Sher-e-Kashmir University of Agricultural Sciences and Technology, Kashmir, J&K, India

Basanti Brar Department of Animal Biotechnology, College of Veterinary Sciences, Lala Lajpat Rai University of Veterinary and Animal Sciences, Hisar, Haryana, India

Navakanth Vijay Challagulla Amity Institute of Nanotechnology, Amity University Uttar Pradesh, Noida, UP, India

Nidhi Chauhan Amity Institute of Nanotechnology, Amity University Uttar Pradesh, Noida, UP, India

Ruchika Chauhan Amity Centre for Nanomedicine, Amity University Uttar Pradesh, Noida, UP, India;
Department of Pharmaceutical Chemistry, College of Health Sciences, University of KwaZulu-Natal, Westville Campus, Durban, South Africa

Adil Denizli Department of Chemistry, Biochemistry Division, Hacettepe University, Ankara, Turkey

S. A. Eremin Faculty of Chemistry, M. V. Lomonosov, Moscow State University, Moscow, Russia

Martín A. Fernández-Baldo Instituto de Química de San Luis (INQUISAL), Consejo Nacional de Investigaciones Científicas Y Técnicas (CONICET), Universidad Nacional de San Luis (UNSL), San Luis, Argentina

Sonu Gandhi DBT-National Institute of Animal Biotechnology, Hyderabad, India

Mudasir Gani Division of Entomology, Faculty of Agriculture, Sher-e-Kashmir University of Agricultural Sciences and Technology, Kashmir, J&K, India

Rakesh Kumar Gupta Division of Entomology, Sher-e-Kashmir University of Agricultural Sciences and Technology, Chatha, J&K, India

Taskeena Hassan Department of Zoology, Aligarh Muslim University, Aligarh, Uttar Pradesh, India

Nicolás Hoffmann Departamento de Ingeniería Química, Centro de Excelencia En Investigación Biotecnológica, Aplicada al Medio Ambiente (CIBAMA), Universidad de La Frontera, Temuco, Chile

Vinita Hooda Department of Botany, Maharshi Dayanand University, Rohtak, Haryana, India

Beenu Jain Department of Animal Husbandry, Lucknow, Uttar Pradesh, India

Utkarsh Jain Amity Institute of Nanotechnology, Amity University Uttar Pradesh, Noida, UP, India

Monica Jaiswal Amity Institute of Nanotechnology, Amity University Uttar Pradesh, Noida, UP, India

Despina P. Kalogianni Department of Chemistry, University of Patras, Rio, Patras, Greece

Rajshekhar Karpoormath Department of Pharmaceutical Chemistry, College of Health Sciences, University of KwaZulu-Natal, Durban, South Africa

Ajit Kaushik Florida Polytechnique University, Lakeland, FL, USA

Ozcan Konur Formerly Ankara Yildirim Beyazit University, Ankara, Turkey

Ranjit Kumar Department of Chemistry, School of Engineering, University of Petroleum and Energy Studies (UPES), Bidholi, Dehradun, Uttarakhand, India

Robin Kumar Amity Institute of Nanotechnology, Amity University Uttar Pradesh, Noida, UP, India

Neelam Kushwaha Amity Institute of Nanotechnology, Amity University Uttar Pradesh, Noida, UP, India

Subhasis Mahari DBT-National Institute of Animal Biotechnology, Hyderabad, India

Ayodeji Samuel Makinde Climatic/Environmental/Telecommunication Physics Unit, Department of Physics, Edo University Iyamho, Auchi, Edo State, Nigeria

Ashish Mathur Department of Physics, School of Engineering, University of Petroleum and Energy Studies (UPES), Bidholi Campus, Dehradun, India

Neha Mazumder Amity Institute of Nanotechnology, Amity University Uttar Pradesh, Noida, UP, India

Jagjiwan Mittal Amity Institute of Nanotechnology, Amity University Uttar Pradesh, Noida, UP, India

Shalini Nagabooshanam Amity Institute of Nanotechnology, Amity University Uttar Pradesh, Noida, UP, India

Wilson Nwankwo Climatic/Environmental/Telecommunication Physics Unit, Department of Physics, Edo University Iyamho, Auchi, Edo State, Nigeria

Akinola Samson Olayinka Department of Physics, Edo University Iyamho, Auchi, Edo State, Nigeria

Sakshi Pareek Amity Institute of Nanotechnology, Amity University Uttar Pradesh, Noida, India

Sofía V. Piguillem Palacios Instituto de Química de San Luis (INQUISAL), Consejo Nacional de Investigaciones Científicas Y Técnicas (CONICET), Universidad Nacional de San Luis (UNSL), San Luis, Argentina

Gaya Prasad International Institute of Veterinary Education and Research, Rohtak, Haryana, India

Minakshi Prasad Department of Animal Biotechnology, College of Veterinary Sciences, Lala Lajpat Rai University of Veterinary and Animal Sciences, Hisar, Haryana, India

Ramesh Namdeo Pudake Amity Institute of Nanotechnology, Amity University Uttar Pradesh, Noida, UP, India

Tahira Qureshi Department of Chemistry, Biochemistry Division, Hacettepe University, Ankara, Turkey

Julio Raba Instituto de Química de San Luis (INQUISAL), Consejo Nacional de Investigaciones Científicas Y Técnicas (CONICET), Universidad Nacional de San Luis (UNSL), San Luis, Argentina

Matías Regiart Instituto de Química de San Luis (INQUISAL), Consejo Nacional de Investigaciones Científicas Y Técnicas (CONICET), Universidad Nacional de San Luis (UNSL), San Luis, Argentina

Vartika Rohatgi Amity Institute of Nanotechnology, Amity University Uttar Pradesh, Noida, Uttar Pradesh, India

Souradeep Roy Amity Institute of Nanotechnology, Amity University Uttar Pradesh, Noida, UP, India

Olga Rubilar Departamento de Ingeniería Química, Centro de Excelencia En Investigación Biotecnológica, Aplicada al Medio Ambiente (CIBAMA), Universidad de La Frontera, Temuco, Chile

Pawan Saini Central Sericultural Research and Training Institute, Central Silk Board, Ministry of Textiles, Govt. of India, Pampore, J&K, India

Kirti Saxena Amity Institute of Nanotechnology, Amity University Uttar Pradesh, Noida, UP, India

Saurabh Shakya Viral Recombination Section, HIV Dynamics and Replication Programme, National Cancer Institute, Frederick, MD, USA

Dolly Sharma Computer Science and Engineering, Shiv Nadar University, Noida, Uttar Pradesh, India

Prashant Sharma Department of Pediatrics, Division of Pediatric Hematology, Oncology & Bone Marrow Transplant, University of Wisconsin–Madison, Madison, WI, USA

Narlawar Sagar Shrikrishna DBT-National Institute of Animal Biotechnology, Hyderabad, India

Ravindra Pratap Singh Department of Biotechnology, Indira Gandhi National Tribal University (Central University), Amarkantak, Anuppur (M.P.), India

Shringika Soni Amity Institute of Nanotechnology, Amity University Uttar Pradesh, Noida, UP, India

Anuj Tewari Department of Veterinary Microbiology, College of Veterinary and Animal Sciences, Govind Ballabh Pant University of Agriculture and Technology, Pantnagar, Uttarakhand, India

Theivasanthi Thirugnanasambandan International Research Centre, Kalasalingam Academy of Research and Education (Deemed University), Krishnankoil, Tamilnadu, India

Gonzalo Tortella Departamento de Ingeniería Química, Centro de Excelencia En Investigación Biotecnológica, Aplicada al Medio Ambiente (CIBAMA), Universidad de La Frontera, Temuco, Chile

Ravi Mani Tripathi Amity Institute of Nanotechnology, Amity University Uttar Pradesh, Noida, UP, India

Deniz Türkmen Department of Chemistry, Biochemistry Division, Hacettepe University, Ankara, Turkey

Kingsley Eghonghon Ukhurebor Climatic/Environmental/Telecommunication Unit, Department of Physics, Edo University Iyamho, Auchi, Edo State, Nigeria; Informatics and CyberPhysical Systems Laboratory, Department of Computer Science, Edo University Iyamho, Auchi, Edo State, Nigeria

Shikha Wadhwa Department of Chemistry, School of Engineering, University of Petroleum and Energy Studies (UPES), Dehradun, India

Meryem Yılmaz Department of Food Engineering, Mersin University, Mersin, Turkey

Part I
Biosensors in Crop Science

Chapter 1
Recent Trends, Prospects, and Challenges of Nanobiosensors in Agriculture

Ravindra Pratap Singh

Abstract Nanotechnology (NT) is an interdisciplinary scientific approach at the nanoscale which revolutionized nanobiosensors' usage in agriculture. Agriculture is a diversified field that plays a self-sustaining role to promote the economic development of any country toward mankind. Agriculture requires nanotechnical interventions toward food processing, food safety, food quality or quality assurance, food security, disaster risk management, diagnosis, and prevention at the local and global levels. Nanobiosensors have immense potential to solve severe problems pertaining to agriculture. Nanobiotechnology (NBT) is a branch of NT at nano dimensions to create tools and techniques, and functional structures which open new applications toward agriculture and health. Agriculture is the major sector of national economy in developing countries. A recent advance in biosensor technology toward agriculture challenges is very challenging with the intervention of nanotechnology which would emerge to enhance crop productivity and sustainability and also improve real-time quality and food safety of livestock, and natural agricultural resources. In this book chapter, the commercial and futuristic nanobiosensors are highlighted to defend the challenges in the major sector of agriculture and bring out the implications of nanobiosensors design and development toward the improvement of crop productivity. Thus, nanobiosensors are playing a crucial role in sensing insecticides, herbicides, fertilizers, and pathogens.

Keywords Nanotechnology, Nanobiotechnology · Nanobiosensors · Agriculture

1.1 Introduction

Agriculture is proving to be the main source of raw materials to food industries and also is the backbone of developing countries for their economic growth and development. Nowadays, this sector is facing huge problems such as urbanization,

---

R. P. Singh (✉)
Department of Biotechnology, Indira Gandhi National Tribal University (Central University), Amarkantak, Anuppur (M.P.) Pin-484887, India
e-mail: ravindra.singh@igntu.ac.in

R. N. Pudake et al. (eds.), *Biosensors in Agriculture: Recent Trends and Future Perspectives*, Concepts and Strategies in Plant Sciences,
https://doi.org/10.1007/978-3-030-66165-6_1

harnessing bad use of cultivating soil, land, water, and runoff, and misutilization and management of pesticides and fertilizers. Natural resources are critical factors for economic growth and development phenomena pertaining to poverty and hunger of large populations locally and globally. Agriculture is the science of soil, crops, and livestock and an economic boon of any country for the source of livelihood to improve social welfare in both urban and rural areas. However, accessibility of resources such as quality of soil and water is declining for agriculture and created a huge economical loss. This continued stress on agricultural resources is increasing due to over population and high demand for food, constant increase of utilization of pesticides, insecticides, herbicides, and heavy metals in soil. All these challenges or issues can be solved by a new technology known as nanobiosensors to change agriculture-based food systems to not only improve the quality of the agricultural products but also boosts the national economy toward sustainable agriculture with quality products and its less cost. A biosensor is an analytical device that combines a biological entity with physico-chemical transducer to measure electrical signal when interacting with the desired analyte of interest. Fig. 1.1 shows a schematic representation of a biosensor.

An advanced version of biosensor known as nanobiosensor is very specific and selective for the detection of target analytes of interest such as pesticides, insecticides, herbicides, and various microbial pathogens (Singh 2017). Biosensor technology is playing a major role to utilize the physical-chemical properties of nanoscale materials which provide ultra sensitivity and performance through new signal transduction technologies. The advanced version of biosensor utilizing nanomaterial for enhancing sensitivity and performance of the system is known as nanobiosensor. Nanobiosensors are highly demanding technology at the early stage of development

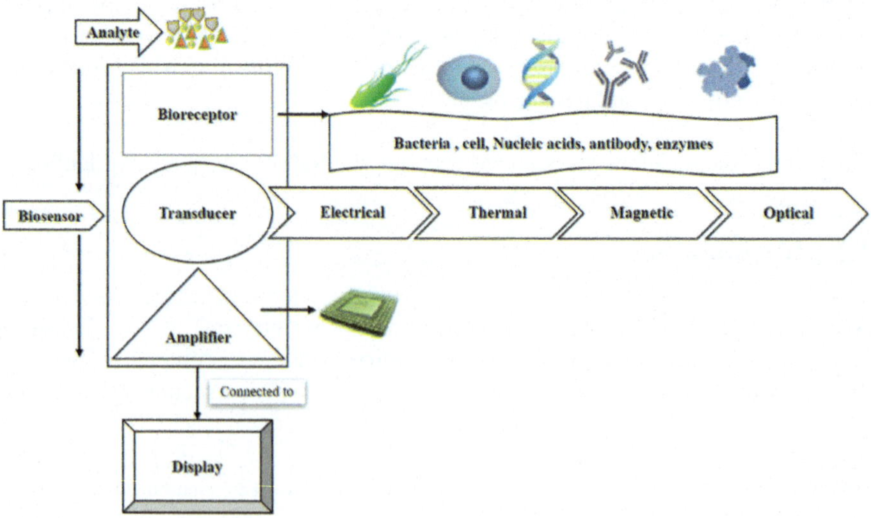

Fig. 1.1 Schematic representation of biosensor

for easy, rapid, ultrasensitive, and cost-effective systems which are very vital not only to agriculture, food, and drink, food process industries, food safety, food security but also in health care, environmental monitoring, and even for genome analysis. The nanomaterials such as metals and metal oxides nanoparticles (NPs) (Au, Ag, Cu, Co, ZnO, TiO_2, Fe_3O_4, MgO, etc.), magnetic NPs, CNT, graphene, dendrimers, polymeric, and QDs nanoparticles were used in nanobiosensors development for the detection of analytes of interests.

A nanobiosensor consists of 3 components: biological probe elements, transducer, and detector. The biological probe elements are several such as antibodies, receptors, enzymes, DNA/RNA, Peptide Nucleic Acid (PNA), Locked nucleic acid (LNA), cells, microorganisms, organelles, etc. The transducer is a physical interface to recognize signal events into a digital signal. The detector detects the signals from the transducer, then passes it to a microprocessor for amplification and analysis in terms of data, and finally displayed in the output device. Data recording and display unit consist of an amplifier, signal processor, and display that are responsible for data transferred and results displayed (Wang 2005; Singh and Choi 2010; Rai et al. 2012). The characteristic features of nanobiosensor are miniaturized, cheap, biocompatible, non-toxic, portable, specific, and stable, less reaction time, zero electrical noise, accurate, precise, and reproducible at optimum pH and temperature (Singh et al. 2014).

Nanotechnology has the potential to make an impact in the field of health, the environment, and agriculture (Singh 2017). The nanomaterials-based biosensors have offered high sensitivity, specificity, and detection time being user-friendly and detect an analyte of interest. In agriculture, nanobiosensors detect direct and indirect foodborne pathogenic microorganisms, pesticides, veterinary drugs, toxicants, contaminants, and heavy metals in foods. In addition to that nanobiosensors can also monitor soil quality, food quality, crop stress, and antibiotic resistance (Singh 2016). Keeping the above in view, nanobiosensors are helpful for healthy agriculture to enhance crops, soil, and livestock productivity which feed our populations free from any implications.

1.2 Use of Nanobiosensor in Agriculture

Nanomaterial-based biosensors are most commonly known as nanobiosensors with enhanced detection specificity, selectivity, sensitivity, and possess potential applications in clinical, environmental, and agriculture. However, in this book chapter, we are focusing on trends, prospects, and challenges of nanobiosensor in agriculture. Recent progress in the development of nanobiosensors was to improve crop health by the detection of plant pathogens, pesticides, herbicides, and soil testing. Nanoparticles are utilized in the diagnostic tool to detect plant pathogens (Singh 2019). Fig. 1.2 shows a broad spectrum of potential applications of nanobiosensors in a variety of scientific domains, and some applications in agriculture are described below.

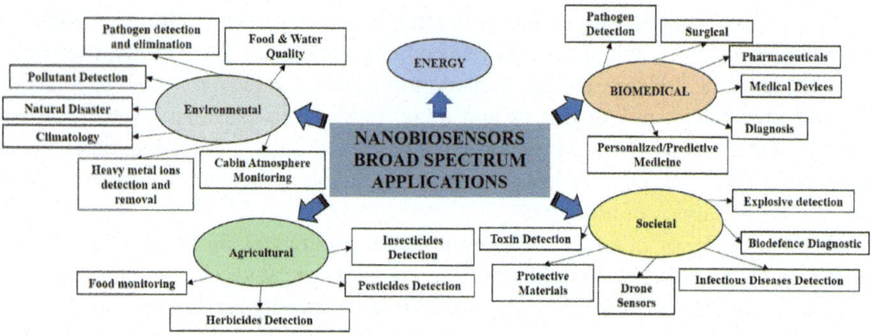

Fig. 1.2 Broad spectrum of potential applications of nanobiosensors

1.2.1 Nanosensors in pathogens detection

In developing countries consumption of foods contaminated with pathogenic foodborne bacteria is the main cause of concern posing public illness health locally and globally. In the food industry, the first line of defense against foodborne illness is laboratory-based methods to detect food pathogens with high selectivity and sensitivity before outbreaks arise. It may be possible by using nanomaterials to design nanobiosensor to detect food contaminants, pathogens, banned dyes, adulterants, antibiotics, hormones, and allergens.

Foodborne microorganism pathogens or bacteria are a serious threat to the animal's production and health to causes food poisoning, gastroenteritis, etc., due to *Salmonella sps*, *Clostriudium spp*, *Vibrio cholera*, *Escherichia coli*, and *Campylobacter spp*. Salmonellosis in humans is a global problem due to foodborne bacteria eaten contaminated meat and poultry (Nowak et al. 2007). There are no analytical devices to ascertain food samples. To overcome this problem, nanobiosensor is very promising for rapid detection in the food chain during storage, distribution, and food processing. Viswanathan et al. (2006) reported the detection of cholera toxin by an electrochemical immunosensor based on liposomal poly (3,4-ethylenedioxythiophene)-coated carbon nanotubes. Kim et al. (2013) reported the detection of Salmonella spp. in food by nanobiosensor immobilized anti-Salmonella polyclonal antibodies on streptavidin-biotin onto the quantum dot surface. Afonso et al. (2013) reported detection of Salmonella by nanobiosensor-based AuNPs in skimmed milk. Wang and Alocijia (2015) reported the detection of *Escherichia coli O157:H7* by nanobiosensor based on functionalized Fe_3O_4 NPs and AuNPs conjugated monoclonal antibodies.

Lopez et al (2009) reported fluorescent oligo capture probe onto nano-chips for detection of single nucleotide of plant pathogenic bacteria and viruses. Yao et al. (2009) reported nano-gold-based immunosensor to detect the pathogen of Karnal bunt disease in wheat. Brock et al. (2011) reported nanosensors for the detection of plant pathogens, viruses, and soil nutrients.

1.2.2 Mycotoxins detection

Nanobiosensor promises rapid detection for the trace quantity of mycotoxins. The mycotoxins are a threat to human health which influence the crops, feed, food products, especially in rainy seasons, and in turn leads to great economic losses globally. However, Masikini et al. (2015) reported the trace detection of fumonisins by nanobiosensor fabricated by using PANI-CNT doped palladium telluride quantum dots. Nanobiosensor is useful for maintaining food quality and safety for example to detect mycotoxins derived from food. There are several nanostructures materials such as QDs, nanoparticles (metal/metal oxide, polymer), nanowires, nanotubes or nanorods, and graphene, which have the ability to be functionalized and immobilized by various biomolecules such as antibodies, enzymes, DNA/RNA aptamers, receptors to detect food toxicants, adulterants, and pathogen.

Mycotoxins are fungi/molds derived toxic chemicals, i.e. natural contaminants found in foodstuffs and animal feed products which pose threat to human health and most commonly known as hepatotoxic, nephrotoxic, carcinogenic, mutagenic, for example, aflatoxins, ochratoxin, and zearalenone. Due to these causes of concern, there is an urgent need to detect or monitor by a sophisticated device known as nanobiosensor, an analytic instrument in contaminated foods and animal feeds. Parker and Tothill (2009) reported the detection of aflatoxin M1 contaminates in the milk by an electrochemical immunosensor. Xu et al. (2013) reported detection of aflatoxin B1 in peanut by an optical biosensor constructed with gold nanorod conjugated antibody. Eldin et al. (2014) reported detection of Aflatoxin B1 in peanuts, chillies, maize, and rice by immunosensor fabricated AuNPs conjugation with anti-aflatoxin B1 polyclonal antibody. Bonel et al. (2010) reported the detection of ochratoxin A by an electrochemical immunosensor constructed AuNPs conjugated polyclonal antibodies. Furthermore, Turan and Şahin (2016) reported the detection of ochratoxin A by MIP-Fe_3O_4 NPs.

1.2.3 Pesticides/ Insecticides/Herbicides detection

Pesticides are the chemicals to protect plants and animals in limited extent such as herbicide to control weeds, a fungicide to control fungus and insecticide to control insects but their application of heavy dose cause toxicity to human/animals. The traditional methods to detect pesticides are not good enough because of many disadvantages, to overcome this problem nanobiosensor is used to detect pesticides (Vimala et al. 2016; Zhao et al. 2015; McGrath et al. 2012).

Zhang et al. (2008) reported the detection of pesticide content in food by acetylcholinesterase nanobiosensor. In addition to that Norouzi et al. (2010) reported detection of monocrotophos and organophosphate pesticide by an electrochemical biosensor. Zheng et al. (2011) reported the detection of paraoxon and parathion pesticides by optical nanobiosensor using acetylcholinesterase and CdTe QDs. Furthermore, Guan et al. (2012) reported detection of dichlorvos pesticides by acetylcholinesterase biosensors. Song et al. (2015) reported the detection of carbamate pesticide by an electrochemical biosensor fabricated AuNPs/(3-mercaptopropyl)-trimethoxysilane/Au electrode sensing surface. Apart from this, Songa et al. (2009a) reported the detection of glyphosate, glufosinate herbicide in corn by biosensor using horseradish peroxidase (HRP). Further, Songa et al. (2009b) reported the detection of glyphosate herbicide and its metabolite by nanobiosensor. In another study, Songa et al. (2009c) reported detection of glyphosate herbicide by biosensor immobilized HRP on sulfonated polymer matrix. Haddaoui and Raouafi (2015) reported the detection of chlortoluron herbicide by nanobiosensor (inhibition of tyrosinase). Mousavi and Rezaei (2011) reported the detection of environmental pollutants and food contaminants by nanosmart dust. Sharon et al. (2010) reported that the intelligent system monitors and minimizes the use of pesticides and antibiotics with the help of a microelectronic circuit.

1.2.4 Veterinary drug and residues

Nanobiosensors are able to not only detect food contaminants and chemicals but also biological contaminants like veterinary drug residues in food. They are being used for the assessment and management of food safety and quality. Veterinary drugs are mostly antibiotics frequently used in farming animals to treat animal diseases and also enhance animal growth (McEwen and Fedorka-Cray 2002). However, antibiotic use in animals in some extent cause serious risks not only in animal but also in human as antibiotic resistance against microbial pathogens leads into outbreak among animals (Levy and Marshall 2004; Courvalin 2008). Furthermore, antibiotic-treated meat causes risk of resistance of flora in humans (Landers et al. 2012). Veterinary drugs are frequently present in the black market, one can procure very easily to use in poultry industry without consultation with a veterinary doctor to violate the guidelines of the veterinary drug recommendation (Idowu et al. 2010). In this context, nanobiosensors are able to detect the veterinary drug residues in food (Wu et al. 2014). In addition to that same group, i.e. Wu et al. (2015) reported to detect chloramphenicol by using aptamer-based fluorescence biosensor. In the past, Mungroo and Neethirajan (2014) had reviewed the nano-based biosensors to detect veterinary drug residues in food such as meat, milk, and poultry products. Hou et al. (2013) reported the detection of oxytetracycline by aptamer-based cantilever array sensors. Furthermore, Song et al. (2012) reported the detection of ampicillin in milk by using an aptamer biosensor.

1.3 Nanobiosensors for food safety and security

Recent challenges like food safety and security and climate changes are very important and need advanced scientific interventions such as nanotechnology to resolve human health, growth, and development. In this regard, nutritious food is very important in the diet with economical, safe, and sufficient. However, our diet is contaminated, i.e. not safe and secure so it requires food safety and food security measures. To achieve this goal advance emerging technology, i.e. nanotechnology is used as a convergent technology utilizing nanobiosensor toward food availability, food accessibility, and food utilization (Sastry et al. 2013) to detect highly toxic natural food contaminants. However, traditional methods (TLC, HPLC, LC-MS, etc.) are available but very costly instrumentation, need manpower, and complicated (Agriopoulou et al, 2020). In the recent past, Parisi et al. (2015) has reported the easiness and affordability of nanobiosensors more or less related to the crop production agriculture domain.

The nutrients not only provide nourishment but also give energy to our body to do routine activities with good health. The food at room temperature for a long time undergoes rapid bacterial growth and also chemical changes and becomes unhealthy food. Food contamination is possible because of storage or chemical changes or several bacteria and fungi that causes foodborne diseases, i.e. food poisoning globally. In addition to that, food preservation using chemicals to increase the time span of food is also a cause of illness. In this context, it is mandatory to develop a system to identify the fresh or good quality of food. A smart system like biosensor and electrical sensors can detect the freshness of food like dairy items, meat, and fruits. In addition to that pH sensor, moisture sensor, and gas sensor are also able to detect food freshness. Checking fruits and food by color and smell are possible but expensive, time consuming, and less efficient. Furthermore, nanobiosensors are able to detect vitamins, antibiotics, food spoilage, and microbial contaminants which are frequently possible.

Biological systems have used nanoscale materials such as nucleic acid, enzymes, antibodies, proteins, carbohydrates, fats, etc., whereas engineered nanomaterial use has raised potential concerns pertaining to latent toxicity risks for agriculture, environment, and biomedical or human health. The use of nanobiosensors has insufficient research pertaining to toxicity. Therefore, nanomaterials used in nanobiosensors must be characterized and tested by investigators, academics, and governmental ethical committees globally (Vamvakaki and Chaniotakis 2007).

In addition to that, nanobiosensors are monitoring soil quality, water quality, and protection against pests and microbial diseases to maintain the health of agricultural crops. The challenges related to sustainability, food safety, and security and climate changes as well are the main potential topic for the improvement of agriculture. Inbaraj and Chen (2016) reported the detection of bacterial pathogen in meat by using nanobiosensor. Detection of melamine, malathion, and foodborne contaminants in food such as milk, vegetable, and fruit are mandatory to use metal-based nanobiosensor. However, nanostructured material such as metallic nanoparticles, conjugated polymers, carbon nanotubes, and nanofibers can be utilized in variety of nanobiosensors to detect food products during early stage of manufacture, storage, and transportation if improper. In addition to that moisture, oxygen, carbon dioxide, amines, and microorganisms in food industry are important aspects to ensure safe food for the sake of human health. Toxins, pathogens, and chemicals are the main cause of concern of human/animal diseases. For example, melamine is a nitrogen-rich chemical used in food products such as in pet foods, infant milk powder foods, which is added in foods to increase protein content and achieve high price or benefit from the products. However, the high content of melamine in food products causes very detrimental cause of concern even in pet and infant death. Several methods have been established but most of the methods require manpower for the preconcentration of sample, time-taking, and using sophisticated instruments. To overcome this problem, there are increasing demands to develop a fast, simple, convenient, and sensitive method to detect melamine based on sensors as an alternative method.

Nanobiosensors have wide applications in the food industry for example QDs-based nanobiosensor is for the detection of heavy metal and organophosphate pesticides, infectious and foodborne pathogens. Nanobiosensors, e.g. electronic nose and electronic tongue are able to detect odor and taste of food products.

Adulteration of the food products such as edible oil, wine, coffee, tea, chocolate, cheese, and milk are well known. In this context food contaminants detection for food safety and security is mandatory to use nanobiosensors. Chen and Park (2016) reported identification and detection of microbial contaminants present in food and suggested the prevention and control of the outbreak.

The electrospun nanofibers have been used in various fields such as wound dressing, drug delivery, and filtration. Electrospun nanofibers' use in agriculture is in a very infancy phase. Few most common examples are reported including pheromone-loaded nanofibers, encapsulation of biocontrol agents, protective clothes, food packaging materials, nanobiosensors for pesticide detection, and filtration of beverage products (Noruzi 2016). Nehra and Singh (2015) reported the use of nanostructures in biosensing for the desired detection of analytes of interest. Nanotechnology (NT) utilization in modern agriculture is known as agrifood nanotechnology which can boost agricultural production using nanoformulated agrochemicals pesticides and fertilizers (Sekhon 2014).

1.4 Conclusion and future prospects

The agrifood nanotechnology gives benefits to our farmers through food production and food industry via food processing, preservation, and packaging. Nanomaterials-based nanosensors/nanobiosensors are being developed to detect plant pathogens and testing soil quality to improve plant health. It is well established that nanobiosensors can be used for supporting sustainable agriculture by enhancing crop productivity. Also, improved detection of food pathogens, pesticides, antibiotics, and food contaminants have to be done by using nanobiosensors toward food safety otherwise they will pose a threat to human health. Nanobiosensors in agriculture not only detect biological analyte present in agricultural food but also produce quality food to meet the local and global demands.

Acknowledgment Dr. Ravindra Pratap Singh thanks IGNTU, Amarkantak, M.P. India for providing facilities to prepare this book chapter.

References

Agriopoulou S, Stamatelopoulou E, Varzakas T (2020) Advances in analysis and detection of major mycotoxins in foods. Foods. Apr 9(4):518

Afonso AS, Perez-Lopez FRC, Mattoso LHC, Hernandez M (2013) Electrochemical detection of Salmonella using gold nanoparticles. Biosens Bioelectron 40(1):121–126

Bonel L, Vidal J, Duato P, Castillo J (2010) Ochratoxin A nanostructured electrochemical immunosensors based on polyclonal antibodies and gold nanoparticles coupled to the antigen. Anal Methods 2:335–341

Brock DA, Douglas TE, Queller DC, Strassmann JE (2011) Primitive agriculture in a social amoeba. Nature 469(7330):393–396

Chen J, Park B (2016) Recent advancements in nanobioassays and nanobiosensors for foodborne pathogenic bacteria detection. J Food Prot 79(6):1055–1069

Courvalin P (2008) Predictable and unpredictable evolution of antibiotic resistance. J Intern Med 264:4–16

Eldin TAS, Elshoky HA, Ali MA (2014) Nanobiosensor based on gold nanoparticles probe for aflatoxin B1 detection in food. Int J Curr Microbiol App Sci 3(8):219–230

Guan H, Zhang F, Yu J, Chi D (2012) The novel acetylcholinesterase biosensors based on liposome bioreactors-chitosan nanocomposites film for detection of organophosphates pesticides. Food Res Int 49(1):15–21

Haddaoui M, Raouafi N (2015) Chlortoluron-induced enzymatic activity inhibition in tyrosinase/ZnO NPs/ SPCE biosensor for the detection of ppb levels of herbicide. Sensors Actuators B Chem 219:171–178

Idowu F, Junaid K, Paul A, Gabriel O, Paul A, Sati N, Maryam M, Jarlath U (2010) Antimicrobial screening of commercial eggs and determination of tetracycline residue using two microbiological methods. Int J Poult Sci 9(10):959–962

Inbaraj BS, Chen BH (2016) Nanomaterial-based sensors for detection of foodborne bacterial pathogens and toxins as well as pork adulteration in meat products. journal of food and drug analysis 24(1): 15–28

Kim G, Park SB, Moon J, Lee S (2013) Detection of pathogenic Salmonella with nanobiosensors. Anal Methods 5:5717–5723

Landers TF, Cohen B, Wittum TE, Larson EL (2012) A review of antibiotic use in food animals: perspective, policy and potential. Public Health Reports January-February 127:1–22

Levy SB, Marshall B (2004) Antibacterial resistance worldwide: causes, challenges and responses. Nat Med 10:122–129

Lopez MM, Llop P, Olmos A, Marco-Noales E, Cambra M, Bertolini E (2009) Are molecular tools solving the challenges posed by detection of plant pathogenic bacteria and viruses? Curr. Issues Mol. Biol. 11, 13e46

Masikini M, Mailu SN, Tsegaye A et al (2015) A fumonisins immunosensor based on polyanilino-carbon nanotubes doped with palladium telluride quantum dots. Sensors 15:529–546

McEwen SA, Fedorka-Cray PJ (2002) Antimicrobial use and resistance in animals. Clin Infect Dis 34(3):93–106

McGrath TF, Elliott CT, Fodey TL (2012) Biosensors for the analysis of microbiological and chemical contaminants in food. Anal Bioanal Chem 403:75–92
Mousavi SR, Rezaei M (2011) Nanotechnology in agriculture and food production. J Appl Environ Biol Sci. 1(10):414–419
Mungroo NA, Neethirajan S (2014) Biosensors for the detection of antibiotics in poultry industry—a review. Biosensors. 4(4):472–93
Nehra A, Singh KP (2015) Current trends in nanomaterial embedded field effect transistor-based biosensor. Biosensors and Bioelectronics. 74:731–43
Norouzi P, Pirali-Hamedani M, Ganjal MR, Faridbod F (2010) A novel acetylcholinesterase biosensor for determination of monocrotophos using FFT continuous cyclic voltammetry. Int J Electrochem Sci 5:1434–1446
Noruzi M (2016) Electrospun nanofibres in agriculture and the food industry: a review. Journal of the Science of Food and Agriculture. Nov; 96(14): 4663-78
Nowak B, Müffling T, Chaunchom S, Hartung J (2007) Salmonella contamination in pigs at slaughter and on the farm: a field study using an antibody ELISA test and a PCR technique. Int J Food Microbiol 115(3):259–267
Parker CO, Tothill IE (2009) Development of an electrochemical immunosensor for aflatoxin M (1) in milk with focus on matrix interference. Biosens Bioelectron 24(8):2452–2457
Parisi C, Vigani M, Rodríguez-Cerezo E (2015) Agricultural nanotechnologies: what are the current possibilities? Nano Today. 10(2):124–7
Rai V, Acharya S, Dey N (2012) Implications of Nanobiosensors in Agriculture. J. Biomaterilas Nanobiotechnology 3:315–324
Sastry RK, Anshul S, Rao NH (2013) Nanotechnology in food processing sector-An assessment of emerging trends. Journal of Food Science and Technology. 50(5):831–41
Sekhon BS (2014) Nanotechnology in agri-food production: an overview. Nanotechnology, Science and Applications. 7:31
Sharon M, Choudhary AK, Kumar R (2010) Nanotechnology in Agricultural Diseases. J. Phytol. 2:83–92
Singh RP, Choi JW, Tiwari A, Pandey AC (2014) Functional Nanomaterials for Multifarious Nanomedicine, in Biosensors Nanotechnology (eds A. Tiwari and A. P.F. Turner), Wiley, Inc., Hoboken, NJ, USA
Singh RP (2016) Nanobiosensors: Potentiality towards Bioanalysis. J Bioanal Biomed 8:e143. https://doi.org/10.4172/1948-593X.1000e143
Singh RP (2017) Application of nanomaterials towards development of nanobiosensors and their utility in agriculture, Springer Publisher, New York, USA, Ch 14, pp 293-303 (2017). In book "Nanotechnology: An Agricultural Paradigm" Editors: Prasad, Ram, Kumar, Manoj, Kumar, Vivek (Eds.)
Singh RP (2019) Nanocomposites: Recent Trends, Developments and Applications. CRC Press, November 15, 2018 Forthcoming, Reference-552, CRC Press, 2018. pp 552 chap 2, Advances in Nanostructured Composites: Volume 1: Carbon Nanotube and Graphene Composites. 1st Edition, Mahmood Aliofkhazraei
Singh RP, Choi JW (2010) Bio-nanomaterials for versatile bio-molecules detection technology. Letter to Editors. Adv. Mat. Lett. 1(1):83–84
Song Y, Chen J, Wang LA (2015) Simple electrochemical biosensor based on AuNPs/MPS/Au electrode sensing layer for monitoring carbamate pesticides in real samples. J hazardous 304:103–109
Songa EA, Arotiba OA, Owino JH, Jahed N, Baker PG, Iwuoha EI (2009a) Electrochemical detection of glyphosate herbicide using horseradish peroxidase immobilized on sulfonated polymer matrix. Bioelectrochemistry 75(2):117–123
Songa EA, Somerset S, Waryo T, Baker PG, Iwuoha EI (2009b) Amperometric nanobiosensor for quantitative determination of glyphosate and glufosinate residues in corn samples. Pure Appl Chem 81(1):123

Songa EA, Waryo T, Jahed N, Baker PGL, Kgarebe B, Iwuoha EI (2009c) Electrochemical nanobiosensor for glyphosate herbicide and its metabolite. Electroanalysis 21(3–5):671–674
Turan E, Sahin F (2016) Molecularly imprinted biocompatible magnetic nanoparticles for specific recognition of Ochratoxin A. Sensors Actuators B Chem 227:668–676
Vamvakaki V, Chaniotakis NA (2007) Pesticide detection with a liposome-based nano-biosensor. Biosensors and Bioelectronics. 22(12):2848–53
Vimala V, Clarke SK, Urvinder Kaur S (2016) Pesticides detection using acetylcholinesterase nanobiosensor. Biosens J 5:1–4
Viswanathan S, Wu L, Huang M, Ho J (2006) Electrochemical immunosensor for cholera toxin using liposomes and poly(3,4-ethylenedioxythiophene)-coated carbon nanotubes. Anal Chem 78(4):1115–1121
Wang J (2005) Nanomaterial-based amplified transduction of biomolecular interactions. Small 1(11):1036–1043
Wang Y, Alocijia EC (2015) Gold nanoparticle-labeled biosensor for rapid and sensitive detection of bacterial pathogens. J Biol Eng 9:16
Wu Y, Tang L, Huang L, Han Z, Wang J, Pan H (2014) A low detection limit penicillin biosensor based on single graphene nanosheets preadsorbed with hematein-ionic liquids-penicillinase. Mater Sci Eng C Mater Biol Appl 1(39):92–99
Xu X, Liu X, Li Y, Ying Y (2013) A simple and rapid optical biosensor for detection of aflatoxin B1 based on competitive dispersion of gold nanorods. Biosens Bioelectron 47:361–367
Yao KS, Li SJ, Tzeng KC, Cheng TC, Chang CY, Chiu CY, Liao CY, Hsu J, Lin ZP, (2009) Fluorescence silica nanoprobe as a biomarker for rapid detection of plant pathogens. Adv. Mater. Res, 79e82, 513e516
Zhang S, Shan L, Tian Z, Zheng Y, Shi L et al (2008) Study of enzyme biosensor based on carbon nanotubes modified electrode for detection of pesticides residue. Chin Chem Lett 19:592–594
Zhao G, Wang H, Liu G (2015) Advances in biosensor-based instruments for pesticide residues rapid detection. Int J Electrochem Sci 10:9790–9807
Zheng Z, Zhoub Y, Li X, Liua S, Tangb Z (2011) Highly-sensitive organophosphorous pesticide biosensors based on nanostructured films of acetylcholinesterase and CdTe quantum dots. Biosens Bioelectron 26:3081–3085

Chapter 2
Nanostructured Platforms Integrated to Biosensors: Recent Applications in Agriculture

Sofía V. Piguillem Palacios, Nicolás Hoffmann, Matías Regiart, Olga Rubilar, Gonzalo Tortella, Julio Raba, and Martín A. Fernández-Baldo

Abstract Recently, various nanomaterials, such as nanostructured platforms, are being used for the development of biosensors. This chapter provides a brief explanation of the classification of biosensors according to their detection methods, and then it talks about the materials that are commonly used as nanostructured platforms in the manufacturing of biosensors. The field of nanotechnology continues to be a rapidly growing and developing field, and researchers are very actively inventing and discovering new nanomaterials. In this chapter, the most used sensors in the agricultural sector are cataloged and then presented in detail. The nanomaterials selected for this purpose ranges from metallic nanoparticles, magnetic nanoparticles, different types of nanomaterials derived from carbon, silicon, nanoMOF, and quantum dots. Finally, some recent works related to the subject are detailed, which have combined fast and efficient detection methods with the incorporation of nanomaterials that improve the characteristics and improve the biosensors in search of a more sensitive, selective, low cost, and reliable determination.

Keywords Nanotechnology · Nanomaterials · Platforms · Biosensors · Agriculture

S. V. Piguillem Palacios · M. Regiart · J. Raba · M. A. Fernández-Baldo (✉)
Instituto de Química de San Luis (INQUISAL), Consejo Nacional de Investigaciones Científicas Y Técnicas (CONICET), Universidad Nacional de San Luis (UNSL), D5700BWS San Luis, Argentina
e-mail: mbaldo@unsl.edu.ar

N. Hoffmann · O. Rubilar · G. Tortella
Departamento de Ingeniería Química, Centro de Excelencia En Investigación Biotecnológica, Aplicada al Medio Ambiente (CIBAMA), Universidad de La Frontera, Temuco, Chile

R. N. Pudake et al. (eds.), *Biosensors in Agriculture: Recent Trends and Future Perspectives*, Concepts and Strategies in Plant Sciences,
https://doi.org/10.1007/978-3-030-66165-6_2

2.1 Introduction

In recent years, several nanomaterials as nanostructured platforms were being used for the biosensor's development. A biosensor is a device composed of two fundamental elements: a biological receptor (i.e., some cellular macromolecules that can bind to a compound) prepared to specifically detect a substance taking advantage of the excellent specificity of biomolecular interactions, and a transducer or better called a sensor, capable of sensing the biological recognition reaction and "translates it" into a quantifiable signal (Fig. 2.1). The biosensors can be classified according to the detection method in optical, electrochemical, and piezoelectric. These biosensors have the advantage of rapid detection; the device can also be portable and, above all, sensitive; however, they are still in the laboratory research stage (Hou et al. 2018; Giorgini et al. 2018; Pohanka 2018).

A device should be called an immunosensor if the biological element is an antigen or antibody fragment (Piguillem et al. 2018; Bravo et al. 2017). They can be classified into biosensors: biocatalytic biosensors and those based on bio-affinity (Regiart et al. 2017a). The former use enzymes as recognition elements, and combine the immanent specificity of enzymes with the excellent benefits of biosensors. This type of biosensor currently attracts researchers to carry out various studies due to its enormous potential for future bioassays due to its high sensitivity and specificity. These biosensors are based on enzymes that are biocatalytic compounds, and biocatalytic

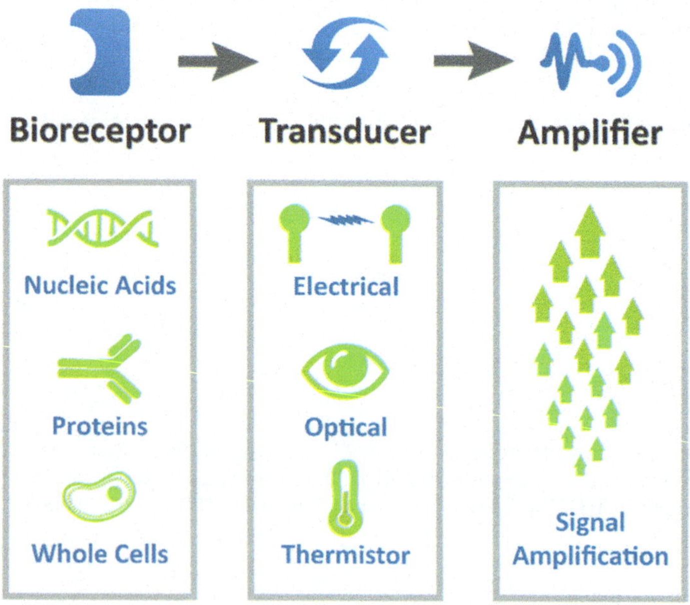

Fig. 2.1 Schematic representation of the biosensor. This device comprises three main components: a sensitive biological recognition element, a transducer, and a signal processor

reactions are generally transduced by optical and electrochemical detection methods (Zhu et al. 2019).

On the other hand, bio-affinity-based biosensors rely on specific binding proteins, nucleic acids, cells, or antibodies for biomolecular recognition. This type of biosensor is specifically developed to measure the biomolecular reaction (Campuzano et al. 2017).

Many of these biosensors are nanostructured, that is, they are modified with nanomaterials in order to obtain nanoplatforms used for biomolecules (enzymes, antigens, or antibodies) modification according to the final application (Fernández-Baldo et al. 2009). A nanomaterial is one that has at least one of its dimensions size between 0–100 nm. This makes its nanoscale properties very different from those exhibited by the same macroscale material (López-Sanz et al. 2019), and there are many techniques to determine its shape and size. There are two main approaches used for nanomaterial synthesis: bottom-up and top-down (Fig. 2.2). Bottom-up is an approach in which material elements are used at the atomic level and are then followed by self-assembly reactions between them that result in the synthesis of nanostructures (Henarava et al. 2019), while top-down is an approach where a macroscopic structure is manipulated externally until the desired nanomaterials are

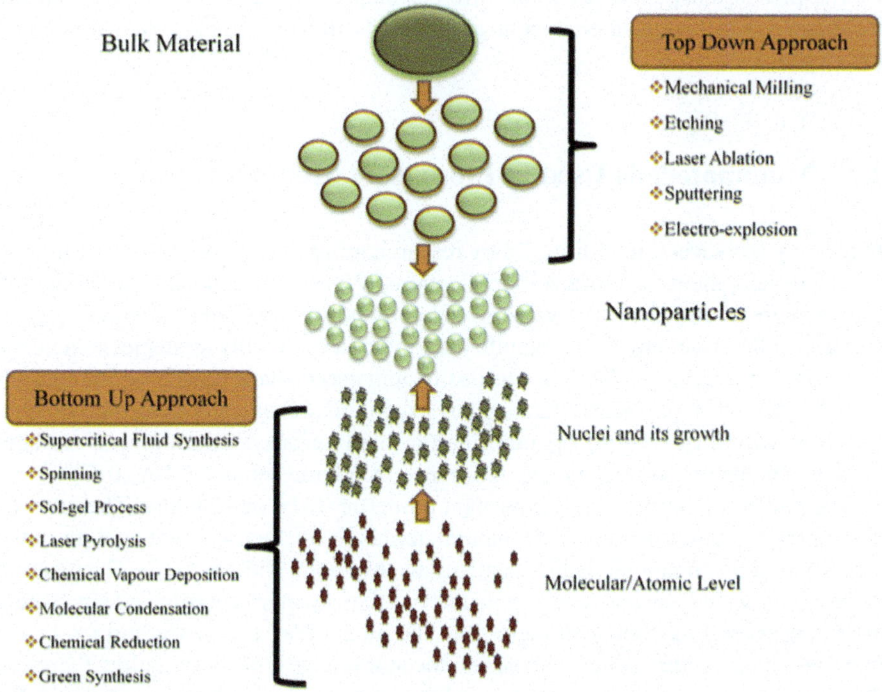

Fig. 2.2 Scheme of different methods that can be used for the nanomaterial's synthesis

obtained (Chepurov et al. 2018). That is why the first approach is known as the builder's approach and the second the sculptor's approach.

In general, the controlled synthesis of nanomaterials is performed using chemical methods such as hydrothermal sol-gel (Chukova et al. 2019), co-precipitation (Regiart et al. 2015), chemical reduction (Piguillem et al. 2018), since with these methods the shape, size, and composition of the particles can be appropriately adjusted to suit a specific catalytic activity. Another nanomaterial synthesis widely used in recent years consists in the use of electrochemical methods, where we can find methods such as nanoparticles electrodeposition (Regiart et al. 2013), dealloying (Zhu et al. 2013), and dynamic hydrogen bubble template (DHBT) (Plowman et al. 2015) that have gained so much attention. DHBT-assisted growth involves hydrogen synthesis, and the metal ions reduction/deposition on the surface of the working electrode simultaneously. The rate/size of hydrogen bubbles evolution provides a dynamic template for metal electrodeposition, creating three-dimensional nanoarchitectures porous materials. In these electrochemical methods, parameters like electrodeposition time and potential strongly influence the grain size, porosity, and density of the films (Sukeri and Bertotti 2017). Another way to produce nanomaterials is to combine methods, one of them uses ultrasound with electrochemistry, and this combination is called sonoelectrochemistry (Islam et al. 2019).

The present chapter focuses on different nanostructured platforms used in the biosensor's fabrication, their recent applications in agriculture, conclusions, and future perspectives.

2.2 Nanomaterials Used in Biosensors Methodologies

In recent years several interdisciplinary research groups are involved in the nanomaterial's development for biosensor applications. When synthesizing a nanomaterial, concerns are nucleation and crystal growth, composition without too many impurities, controlled size, interfaces, stability, and various assembly strategies to achieve low-cost, large-scale production. Various nanomaterials have been used for agricultural biosensors applications due to their exceptional characteristics, such as high (bio) compatibility, easy to operate, inertia, high surface/area ratio, and eminent thermal, chemical, and mechanical resistance (Maduraiveeran and Jin 2017).

The nanomaterials have two important properties to be outstanding. The first one is the surface area, that allows the nanomaterials to interact with the analyte in the sample strongly compared with bulk materials (Malhotra and Ali 2018). The material at the macroscale level has atoms in its interior that are more coordinated since there are many more links between them than the atoms that are on the surface of the material, in fact in the corners and edges, the atoms have less coordination; therefore they are less stable than interior atoms. If we now think at the nanoscale level, the surface of a nanomaterial becomes quite reactive since there are very few atoms that make it up, and therefore, it shows extraordinary catalytic and absorbance activity (Malhotra and Ali 2018).

The other property has to do with the quantum well, that is, two-dimensional confinement of particles. When the width or thickness of this well is similar to the de Broglie wavelength, the particles that are generally electrons, those particles that have energy, can now only take discrete energy values (Malhotra and Ali 2018). So, this quantum confinement effect produces an effect that when the particle size is too small can be compared with the wavelength of the electron, that is why if a particle has nanoscale dimensions, the limiting dimensions make the levels of energy discrete and this increases the band energy. In other words, if we have a nanoparticle whose size is similar to the Bohr excitation radius, the excitonic transition energy, the blue change in absorption, and the energy of the luminescence band gap increases, all as a result of quantum confinement (Malhotra and Ali 2018).

There are several ways to classify nanomaterials, but we will focus on their classification by size and dimensions, and according to this, there are four types of nanomaterials. Zero-dimensional Nanomaterial (0D) is when the three dimensions are at the nanoscale, such as nanoparticles (NP) of silver, gold, copper, or quantum dots. These NPs are generally spherical with a diameter of 1–50 nm, although cube and polygon shapes have also been found at 0D. One-dimensional (1D) nanomaterial is with one of its dimensions with a range between 1–100 nm, and the others can be on a macroscale. Typical examples of these are nanowires, nanofibers, nanorods, and nanotubes. Two-dimensional (2D) nanomaterials are when two of its dimensions are in the nanoscale and one at the macroscale. Such is the case of nanolayers or nanowalls. A clear example of this is graphene. In three-dimensional (3D) nanomaterials, there are no dimensions at the nanoscale, and all dimensions are at the macroscale. This category involves materials whose internal structure is nanostructured.

2.2.1 Metal Nanoparticles

Metallic NPs are often used for the development of electrochemical sensors and biosensors platforms due to interesting properties, either because synthesizing them does not present drawbacks with suitable methods, they are easy to functionalize since they can be covalently linked to functional groups, they are capable catalyze electrochemical reactions and greatly facilitate the electron transfer process. The oldest and most used noble metals in terms of nanomaterials can be found in gold, platinum, and silver (Jamkhande et al. 2019).

It can be affirmed that AuNPs have been used successfully as an effective electrocatalyst in innumerable electrochemical reactions, mainly because they are very stable and manage to recover completely in redox chemical processes. On the other hand, when using AuNP-modified electrodes, a higher signal/control (S/C), good catalytic activity, and diffusion of electroactive species are observed (Regiart et al. 2013). Due to the physical property of gold as a conductive material, its nanoparticles have surface plasmon resonance (SPR), so when they are irradiated with light of the same plasma frequency, an oscillation is generated. This phenomenon is responsible because, depending on the size of the nanoparticles, they present a colloidal suspension of different colors (Saha et al. 2012).

2.2.2 *Magnetic Nanoparticles*

There are several forms of magnetic NP depending on the metals involved in its synthesis, such as Fe_3O_4 (magnetite), c-Fe_2O_3 (maghemite), $NiFe_2O_4$ (nickel ferrite), and $CoFe_2O_4$ (cobalt ferrite). As they have paramagnetic properties, they can be manipulated by using an external magnetic field, making them extremely easy to use or incorporate into biosensing platforms, and that is why they have many applications, especially in medicine, as therapeutic drug carriers as well as labeled cell separators (Knezevic et al. 2019). Moreover, the magnetic property was used in conjunction with noble metal NPs like AuNPs, as a result giving core@shell NPs, thus the benefits of both materials can be utilized (Azharuddin et al. 2019).

2.2.3 *Carbon Nanomaterials*

If it refers to carbon, many nanomaterials can be included incorporating this element in different shapes and nanometric dimensions. There are nanotubes or nanofibers of carbon, graphene, fullerene, and carbon materials with micro/meso/macropores (Regiart et al. 2016). These materials have great application potential in agriculture (Mittal et al. 2019). The most used for the development and incorporation of biosensors are carbon nanotubes, porous carbon, and graphene.

2.2.3.1 Carbon Nanotubes

Carbon nanotubes (CNT) are tubular or cylindrical structures whose internal diameter must not exceed the nanoscale. Some roll with a single layer called single-walled or monolayer nanotubes (SWCNT), or those that roll concentrically called multiple-walled nanotubes (MWCNT). Before its use in biosensors, an acid pretreatment is performed to remove the tube caps, leaving it open and with functional oxygen groups, which makes the nanotube very versatile for later use (Yang et al. 2015).

2.2.3.2 Graphene

Graphene is a 2D structure of sp2 carbon atoms that forms a hexagonal network. The bonds between hexagons in the same plane are powerful (higher than the resistance in steel), while the bonds between the upper or lower layers are weaker, hence the allotropic property of carbon to be able to exfoliate and obtain monolayers. Its applications in electrochemical sensors are derived more specifically from its use in oxidized form (GO) or in its reduced oxide form (rGO); this is because it has excellent electrical conductivity, and its surface is easy to modify (Wang et al. 2016).

2.2.3.3 Porous Carbon

Porous carbon has a high surface area since its internal structure that contains micro pores (less than 2 nm) and meso (between 2 and 50 nm) added to the short path for the transfer of mass and electrons has particular advantages to incorporate them in electrochemical biosensors. The energy is stored in the double layer, which is why the porous carbon can be used as a structural material and as electrochemical energy storage. Without a doubt, its applications are innumerable as mostly due to its adsorption properties, which are used for water and air purification filters, or as a support for catalysts. At present, nanocasting templates are used which create precisely ordered mesoporous carbon that may include the incorporation of other atoms in your invention depending on the purpose of their use (Xu et al. 2017).

2.2.4 Silica Nanomaterials

Intelligently synthesized silica nanomaterials (SNMs) achieve a highly useful mesoporous structure for many areas of science. For example in medicine, they are used to administer medicines because they can contain, transport and release biologically active substances; or in biosensors to detect pathogens, aflatoxins in food, biomarkers of different types of cancer (Eivazzadeh-Keihan et al. 2020; Wang et al. 2015). In addition to all its advantages that make nanostructured silica a very interesting nanomaterial, it also exhibits low cytotoxicity, which is biocompatible and inexpensive (Regiart et al. 2017b). One of the methods for SNM synthesis involves high-temperature heat treatment, responsible for the formation of siloxane rings. However, a low-temperature heat treatment reduces its crystallinity, which makes it behave as a material with more exceptional biocompatibility, an objective that is very interesting in order to immobilize biomolecule's surface (Regiart et al. 2018).

2.2.5 NanoMOFs

Organometallic frameworks (MOFs) are organic–inorganic hybrid porous coordination polymers, the matrix of which is made up of metal ions bound by organic molecules. Its internal structure exhibits a large cavity that depends on the ions involved and the synthesis methods and solvents used for them. NanoMOFs are a novel class of nanomaterials whose reduction in the size of said pores or cavities improves the already existing benefits of traditional MOFs since they are currently being applied in fields such as biomedicine and agriculture (Chansi et al. 2019; Zhang et al. 2019).

2.2.6 Quantum Dots

Importantly, it is not an easy task to discern when a nanomaterial can be classified as a quantum dot (QD) since it depends not only on the size but also on the material that makes it up. QD are those nanomaterials that must be greater than 2 nm and less than 20 nm, and are produced with a limiting effect. As explained above, this depends on the size of the Bohr radius (size that varies according to the article); therefore, it depends on both criteria, the item, and the size to consider whether or not it is a true QD. Due to this energy confinement, these fluorescent nanomaterials have unique optical properties, such as CdS QD emits from blue to ultraviolet or CdSe QD that cover the entire visible spectrum (Marcelo et al. 2020).

2.3 Recent Applications of Nanostructured Biosensors in Agriculture

Agriculture is a vast monetary source for populations around the world that do all kinds of jobs related to cultivation and processing. And according to the report by The Food and Agriculture Organization of the United Nations (FAO), there are around 2 billion farmers. Nevertheless, also, a more modern definition of agriculture involves tasks that go from working on the land to reaching consumption in homes. Although in agricultural jargon there are two main stages: pre-harvest, where conditions are significant for cultivation, such as climatic conditions, minerals, nutrients, fungal contaminants such as aflatoxins, and post-harvest (Kundu et al. 2019). It is imperative to detect microbial contamination in the crop since it has a direct impact on the quality of the final product and can cause massive losses in many batches. Fortunately, there are currently various techniques to detect contamination, which generates stress in plants. Those methods are techniques to assess and analyze agriculture problems in the laboratory. In order to avoid this, biosensors have contributed to achieving new in situ methods that promise different tools for the early diagnosis of microbial contamination. In this context, the evidenced works improve the diagnostic method for some specific microorganisms that can produce plant diseases. Regiart et al. (2017a) have developed a microfluidic electrochemical immunosensor for the detection of *Xanthomonas arboricola* (XA). In this work, a monoclonal anti-XA antibody was covalently immobilized in the central channel of the chip in silica called amino-SBA-15. As is well known, standard methodologies for characterizing mesoporous materials correspond to N2 adsorption–desorption isotherm to assess its porous capacity, scanning electron microscopy (SEM) to assess its morphology, infrared spectroscopy (FTIR) to assess junctions and composition of silica groups. Once the nanomaterial was characterized, and the immobilized antibody was quantified XA using a direct sandwich immunoassay using an enzyme conjugate anti-XA labeled with the enzyme alkaline phosphatase (AP) and finally when the oxidized enzyme product is obtained the current measured in this process is directly proportional to the amount of XA in the samples from walnut. Another successful application of inmunosensors on *Botrytis cinerea* (BA) detection was carried out by Fernández-Baldo et al. (2009). BA is a phytopathogenic fungus (a fungus that uses plants as a host and its spores use soil or plant residues) that produces gray mold, which primarily attacks agricultural hosts. In this case, the researchers developed a competitive immunoassay on a previously modified spinning disk and immobilized BA antigens for plant tissue or BA monoclonal antibodies to react there. The quantification was done through a second anti-IgG-labeled horseradish peroxidase (HRP) enzyme, and when the reaction product is obtained, a direct proportion was obtained between the measured current and the amount of antigens present in the samples.

Biosensor's development has proved to be useful as a fast detection tool to prevent crop deficiencies or microbial contamination, but the biosensor's application field does not end in those aspects and is also used as pesticide detection. Organophosphate insecticides (IOPs) highly toxic chemical compounds used to combat pests and weeds (Dong et al. 2016). However, this high toxicity rate has caused its residues to generate contamination not only for the environment but also for the health of the humans who consume the crop product, as well as those who live in areas close to the treated fields (Wang et al. 2017). To avoid this, many acetylcholinesterase biosensors are used for the analysis of pesticides; however, these enzyme molecules have limited conductivity, sensitivity, stability, and biosensor performance. Below, we present two successful cases of improving the electrochemical signal through the design of nanomaterials.

First, we quote Bao et al. (2019), who developed an electrochemical acetylcholinesterase biosensor to detect organophosphate pesticides using graphene-copper oxide nanoflowers with excellent sensitivity. In this case, the main design is based on graphene, given that it has many properties mentioned above in the section on carbon materials. However, in addition to all these abilities, the advance in the development of new materials from graphene has led to the invention of three-dimensional graphene (3DG), which until now is the hardest material in the world with high porosity and lighter than air. Because acetylcholinesterase finds a favorable environment in these 3DG structures and due to the nature of this new nanomaterial, the surface area increases considerably, this biosensor has been able to detect pesticides successfully and with a low detection limit without altering the good sensitivity of the method.

As a second successful case, Vaghela et al. (2018) worked on a biosensor to detect glyphosate using gold nanoparticles. Its determination methodology was potentiometric based on a urease bioconjugate. Glyphosate (N- (phosphonomethyl) glycine) used worldwide to eliminate weeds, and for decades its use is considered highly controversial. It has been shown to be harmful to the health of humans, animals, and the environment since they cause serious diseases such as congenital disabilities and neurological effects. On the other hand, the International Agency for Research on Cancer cataloged glyphosate within the group as a possible human carcinogen. All this makes it extremely necessary to measure and detect glyphosate levels in all agricultural products, as well as in water for consumption and irrigation. The development of biosensors based on gold nanoparticles is a perfect alternative to expensive analyzes since it proposes a follow-up and monitoring of the levels of herbicide in the same place where the agricultural work is carried out, but it is also cheap (Vaghela et al. 2018). The biosensor is capable of detecting 0.5 ppm of glyphosate, which is the limit that, according to the World Health Organization, must contain drinking water. These researchers can be said to have developed a simple and inexpensive method of detecting glyphosate by urease inhibition.

2.4 Conclusions and Future Perspectives

The different types of nanomaterials used as an integrated platform for the manufacture of biosensors and their recent applications in agriculture are discussed in this chapter. Nanomaterials have shown enormous potential to immobilize biomolecules (enzymes, antigens, or antibodies) through various strategies on the surface of a biosensor, capable of effectively conserving the bioactivity of its biological receptors. Future research on new nanomaterials can be a relevant tool in the development of new biosensors for various purposes, always to improve the characteristics that make them as attractive as their low cost and time of analysis, outstanding sensitivity and selectivity, as well as its ability to be transportable.

Acknowledgements The authors thank the financing of the National University of San Luis (PROICO-1512-22/Q232), National Council for Scientific and Technical Research (PIP-11220150100004CO), National Agency for Scientific and Technological Promotion (PICT-2015-2246, PICT -2015-1575, PICT-2014-1184, PICT-2014-0375 and PICT-2013-3092) and the University of La Frontera (REDES 180003).

References

Azharuddin M, Zhu GH, Das D, Ozgur E, Uzun L, Turner AP, Patra HK (2019) A repertoire of biomedical applications of noble metal nanoparticles. Chem Commun 55(49):6964–96

Bao J, Huang T, Wang Z, Yang H, Geng X, Xu G, Samalo M, Sakinati M, Huo D, Hou C (2019) 3D graphene/copper oxide nano-flowers based acetylcholinesterase biosensors for sensitive detection of organophosphate pesticides. Sensors & Actuators: B. Chemical 279:95–101

Bravo K, Ortega FG, Messina GA, Sanz MI, Fernández Baldo MA, Raba J (2017) Integrated bio-affinity nano-platform into a microfluidic immunosensor based on monoclonal bispecific trifunctional antibodies for the electrochemical determination of epithelial cancer biomarker. Clin Chim Acta 464:64–71

Campuzano S, Yánez-Sedeño P, Pingarrón JM (2017) Electrochemical bioaffinity sensors for salivary biomarkers detection. Trends Anal Chem 86:14–24

Chansi BR, Hadwani K, Basu T (2019) Role of Metal–Organic Framework (MOF) for pesticide sensing. In: Pudake R, Chauhan N, Kole C (eds) Nanoscience for Sustainable Agriculture. Springer, Cham. https://doi.org/10.1007/978-3-319-97852-9_4

Chepurov A, Sonin V, Shcheglov D, Latyshev A, Filatov E, Yelisseyev A (2018) A highly porous surface of synthetic monocrystalline diamond: Effect of etching by Fe nanoparticles in hydrogen atmosphere. Int. J. Refract. Met. Hard Mater. 76:12–15

Chukova OV, Nedilko SA, Nedilko SG, Papadopoulos A, Slepets AA, Stratakis EI, Voitenko TA (2019) Structure and spectroscopy characterization of La1-xSmxVO4 luminescent nanoparticles synthesized co-precipitation and sol-gel methods Optical Materials 95:109248

Dong W, Zhang L, Wang C, Feng C, Shang N, Gao S, Wang C (2016) Palladium nanoparticle embedded in metal organic framework derived porous carbon: synthesis and application for efficient Suzuki-Miyaura coupling reaction. RSC Adv. 6:37118–37123

Eivazzadeh-Keihan R, Khanmohammadi Chenab K, Taheri-Ledari R, Mosafer J, Masoud Hashemi S, Mokhtarzadeh A, Maleki A, Hamblin MR (2020) Recent advances in the application of mesoporous silica-based nanomaterials for bone tissue engineering. Mater Sci Eng, C 107:110267

Fernández-Baldo MA, Messina GA, Sanz MI, Raba J (2009) Screen-printed immunosensor modified with carbon nanotubes in a continuous-flow system for the *Botrytis cinerea* determination in apple tissues. Talanta 79:681–686

Giorgini A, Avino S, Malara P, Zullo R, De Natale P, Mrkvová K, Homola J, Gagliardi G (2018) Surface-plasmon optical-heterodyne clock biosensor. Sens. Actuat. B Chem. 273:336–341

Henarava T, Leschev S, Levkina V, Rabinovich O, Zhdanok S, Chernuho A (2019) The influence of the carbon nanomaterials synthesis conditions on the content of polycyclic aromatic hydrocarbons in the waste gaseous products. Mater. Today: Proceedings 20:7–11

Hou Y, Cai G, Zheng L, Lin J (2018) A microfluidic signal-off biosensor for rapid and sensitive detection of Salmonella using magnetic separation and enzymatic catalysis. Food Control 103:186–193

Islam MH, Paul MT, Burheim OS, Pollet BG (2019) Recent developments in the sonoelectrochemical synthesis of nanomaterials. Ultrason Sonochem 59:104711

Jamkhande PG, Ghule NW, Bamer AH, Kalaskar MG (2019) Metal nanoparticles synthesis: An overview on methods of preparation, advantages and disadvantages, and applications. J. Drug Delivery Sci. Technol. 53:101174

Knezevic NZ, Gadjanski I, Durand JO (2019) Magnetic nanoarchitectures for cancer sensing, imaging and therapy. J. Mater. Chem. B 7:9–23

Kundu M, Krishnan P, Kotnala RK, Sumana G (2019) Recent developments in biosensors to combat agricultural challenges and their futures prospects. Trends Food Sci Technol 88:157–178

López-Sanz S, Guzman Bernardo FJ, Rodríguez Martín Doimeadios RC, Ríos A (2019) Analytical metrology for nanomaterials: Present achievements and future challenges. Anal Chim Acta 1059:1–15

Maduraiveeran G, Jin W (2017) Nanomaterials based electrochemical sensor and biosensor platforms for environmental applications. Trends Environ. Anal. Chem. 13:10–23

Malhotra BD, Ali MA (2018) Nanomaterials for Biosensors: Fundamentals and applications. Elsevier, Amsterdam, Netherlands, p 311

Marcelo GA, Lodeiro C, Capelo JL, Lorenzo J, Oliveira E (2020) Magnetic, fluorescent and hybrid nanoparticles: From synthesis to application in biosystems. Mater Sci Eng, C 106:110104

Mittal J, Osheen S, Gupta A, Kumar R (2019) Carbon Nanomaterials in Agriculture. In: Pudake R, Chauhan N, Kole C (eds) Nanoscience for Sustainable Agriculture. Springer, Cham. https://doi.org/10.1007/978-3-319-97852-9_7

Piguillem S, Ortega FG, Raba J, Messina GA, Fernández Baldo MA (2018) Development of a nanostructured electrochemical immunosensor applied to the early detection of invasive aspergillosis. Microchem J 139:394–400

Plowman BJ, Jones LA, Bhargava SK (2015) Building with bubbles: the formation of high surface area honeycomb-like films via hydrogen bubble templated electrodeposition. Chem Commun 51:4331–4346

Pohanka M (2018) Piezoelectric biosensor for the determination of Tumor Necrosis Factor Alpha. Talanta 178:970–973

Regiart M, Escudero LA, Aranda P, Martinez NA, Bertolino FA, Raba J (2015) Copper nanoparticles applied to the preconcentration and electrochemical determination of β-adrenergic agonist: An efficient tool for the control of meat production. Talanta 135:138–144

Regiart M, Rinaldi-Tosi M, Aranda P, Bertolino FA, Villarroel-Rocha J, Sapag K, Messina GA, Raba J, Fernández-Baldo (2017a) Development of a nanostructured immunosensor for early and *in situ* detection of *Xanthomonas arboricola* in agricultural food production. Talanta 175: 535–541.

Regiart M, Fernández Baldo MA, Villarroel Rocha J, Messina GS, Bertolino FA, Sapag K, Timperman AT, Raba J (2017b) Microfluidic immunosensor based on mesoporous silica platform and CMK-3/poly-acrylamide-*co*-methacrylate of dihydrolipoic acid modified gold electrode for cancer biomarker detection. Anal Chim Acta 963:83–92

Regiart M, Fernández O, Vicario A, Villarroel-Rocha J, Sapag K, Messina GA, Raba J, Bertolino FA (2018) Mesoporous immunosensor applied to zearalenone determination in Amaranthus cruentus seeds. Microchem J 141:388–394

Regiart M, Fernandez-Baldo MA, Spotorno VG, Bertolino FA, Raba J (2013) Ultra sensitive microfluidic immunosensor for determination of clenbuterol in bovine hair samples using electrodeposited gold nanoparticles and magnetic microparticles as bio-affinity platform. Biosens Bioelectron 41:211–217

Regiart M, Magallanes JL, Barrera D, Villarroel-Rocha J, Sapag K, Raba J, Bertolino FA (2016) An ordered mesoporous carbon modified electrochemical sensor for solid-phase microextraction and determination of triclosan in environmental samples. Sens. Actuat. B Chem. 232:765–772

Saha K, Agasti SS, Kim C, Li X, Rotello VM (2012) Gold nanoparticles in chemical and biological sensing. Chem Rev 5:2739–2779

Sukeri A, Bertotti M (2017) Electrodeposited honeycomb-like dendritic porous goldsurface: an efficient platform for enzyme-free hydrogen peroxide sensor at low overpotential. J Electroanal Chem 805:18–23

Vaghela C, Kulkarni M, Haram S, Aiyer R, Karve M (2018) A novel inhibition based biosensor using urease nanoconjugate entrapped biocomposite membrane for potentiometric glyphosate detection. Int. J. of Biological Macromolecules 108:32–40

Wang N, Lin M, Dai H, Ma H (2016) Functionalized gold nanoparticles/reduced graphene oxide nanocomposites for ultrasensitive electrochemical sensing of mercury ions based on thymine–mercury–thymine structure. Biosens Bioelectron 79:320–326

Wang Y, Wang L, Huang W, Zhang T, Hu X, Perman JA, Ma S (2017) Metal-organic framework and conducting polymer based electrochemical sensor for high performance cadmium ions detection. J. Mater. Chem. A. 5:8385–8393

Wang Y, Zhao Q, Han N, Bai L, Li J, Liu J, Che E, Hu L, Zhang Q, Jiang T, Wang S (2015) Mesoporous silica nanoparticles in drug delivery and biomedical applications. Nanomedicine: Nanotechnology, Biology and Medicine 11(2): 313–27

Xu F, Wu D, Fu R, Wei B (2017) Design and preparation of porous carbons from conjugated polymer precursors. Mater Today 20:629–656

Yang C, Denno ME, Pyakurel P, Venton BJ (2015) Recent trends in carbon nanomaterial-based electrochemical sensors for biomolecules: A review. Anal Chim Acta 887:17–37

Zhang Y, Yang L, Yan L, Wang G, Liu A (2019) Recent advances in the synthesis of spherical and nanoMOF-derived multifunctional porous carbon for nanomedicine applications. Coord Chem Rev 391:69–89

Zhu XT, Zhang LJ, Tao H, Di JW (2013) Synthesis of nanoporous gold electrode and its application in electrochemical sensor. Chin J Anal Chem 41:693–697

Zhu YC, Mei LP, Ruan YF, Zhang N, Zhao WW, Xu JJ, Chen HY (2019) Enzyme-based biosensors and their applications. InAdvances in Enzyme Technology (pp 201–223). Elsevier

Chapter 3
Advances in Nanotechnology for Bio-Sensing in Agriculture and Food

Theivasanthi Thirugnanasambandan

Abstract Biosensors are applied in biological related fields such as clinical, environmental, agricultural, and food analysis. They are used to measure the concentration of a biological analyte with the help of various transducers like electrochemical, optical, electronic, and piezoelectric biosensors. The performance of a biosensor is well defined by its stability, cost, sensitivity, and reproducibility. Most importantly, early detection of diseases becomes possible with biosensors. These analytical devices make the binding between targets and probes more specifically which is then converted to an electrical signal. Using nanomaterials in biosensors becomes more advantageous for pathogen detection particularly in fields like healthcare, food industry, and agriculture. Nanomaterials offer a low loading of bioreceptor units at reduced volumes and act as transduction elements in biosensors. Their outstanding chemical and electrical properties lead to more advanced sensor technologies like FET transistors, screen-printed electrodes, and lateral flow immunoassay. This chapter reviews the use of advanced nanomaterials like gold nanoparticles, graphene, and conducting polymers that can support these technologies. The performance of a nano biosensor is also reviewed, that is decided by the size, shape, conductivity, and morphology of the nanoparticles.

Keywords Nanomaterials · Nanocomposites · Biosensors · Soil moisture/water uptake · Pesticides · Pathogens · Phytohormones · Transgenic plants · Raman spectroscopy · Lateral flow immunoassay

Abbreviations

PCR	Polymerase chain reaction
ELISA	Enzyme linked immunosorbent assay
LED	Light emitting diode

T. Thirugnanasambandan (✉)
International Research Centre, Kalasalingam Academy of Research and Education (Deemed University), Krishnankoil 626126, Tamilnadu, India

R. N. Pudake et al. (eds.), *Biosensors in Agriculture: Recent Trends and Future Perspectives*, Concepts and Strategies in Plant Sciences,
https://doi.org/10.1007/978-3-030-66165-6_3

NPs	Nanoparticles
TiO_2	Titanium oxide
AuNPs	Gold nanoparticles
LCR	Inductance (L), capacitance (C), and resistance (R)
MEMS	Micro-electromechanical system
SERS	Surface-enhanced Raman spectroscopy
LFIA	Lateral flow immunoassay
RPA	Recombinase polymerase amplification
EIS	Electrochemical impedance spectroscopy
In_2O_3	Indium oxide
IAA	Indole-3-acetic acid

3.1 Introduction

The components of biosensors are bio-analyte (which is to be analyzed), bio-recognition elements (like enzymes, antibodies, and aptamers or nucleic acid sequences), and the transducers (like electrochemical and optical). There are two types of biosensors viz enzymatic and nonenzymatic biosensors. Nonenzymatic has become possible with the advancement of nanomaterials with its superior sensitivity. Aptamers are single stranded DNA- and RNA-based oligonucleotides used as bio-recognition elements in biosensors. They can bind to the targets with high affinity and specificity. The properties of conformation make them suitable for making label-free and portable devices for diagnostic applications (Sekhon et al. 2018).

New advanced technologies like electrochemical nanosensors, optical nanosensors, and some other nanosensors are emerging. They offer immediate detection within a short span of time. High sensitivity, low limit of detection, good selectivity (with the help of bio-probes), and expected reliability are some of the parameters that can be achieved through these kinds of sensors. These sensors can also be handled by people who are not having technical knowledge or with limited technical knowledge. Nanoscale materials such as metal nanoparticles or nanoclusters, metal oxide nanoparticles, carbon nanomaterials (carbon quantum dots, graphene, carbon nanotubes), and nanocomposites have high sensitivity. These materials integrate transduction principles (electrochemical, optical, Raman, catalysis, and superparamagnetic properties) into the nanosensors through signal amplification. Electrochemical/optical nanosensors, electronic nose/tongue, nanobarcode, and wireless nanosensors have multiplex and real-time sensing capabilities. They are applied in the food and agriculture sector for the detection of food contaminants, preservatives, microbes, pathogens antibiotics, heavy metal ions, and toxins. Also, they are utilized in the monitoring of temperature, humidity, and gas (Srivastava et al. 2018).

Conventional tools like PCR and ELISA need more time to get the results. They also need trained professionals to work with the analysis. Spectroscopic techniques like Raman spectroscopy is widely used nowadays for sensing biological molecules.

This kind of detection is in the infant stage where interpretation of the spectrum is difficult. Even though they provide more chemical information, the cost of the instrument is very high that makes it beyond the reach of common people. In agriculture, sensors are needed to measure the moisture and pH level of soil, the amount of fertilizers/nutrients applied in the soil and pesticide detection in plants.

Nano-inspired biosensors improve the quality of life through clinical, agricultural, and environmental applications. They are utilized in agriculture for applications such as detection of plant infections (fungal, viral, and bacterial), plant physiological activities (abiotic stress), metabolic content (Phytohormones), mi RNAs, genetically modified plants, and genetically encoded biosensors (Kumar and Arora 2020). Agriculture becomes smart when sensors are applied to increase crop production. Optical crop sensors evaluate crop conditions by shining the light with specific wavelengths on the crop leaves and measuring the light wavelengths reflected back to the sensors. In an NPK sensor optical transducer is used to measure the presence of Nitrogen (N), Phosphorus (P), and Potassium (K) in soil. The wavelength of LEDs is selected in such a way as to fit the absorption band of each nutrient.

Crop loss (due to environmental and pathogen-related stresses) reduction and efficient utilization of resources are some of the challenges in plant agriculture. Smart plant sensors improve plant productivity by optimizing the water and agrochemical allocation. New technologies are essential for real-time monitoring of plant physiological and developmental responses. Nanomaterials are utilized in such new technologies for the conversion of the plant chemical signals into digital information that can be monitored through electronic devices like biosensors (Giraldo et al. 2019).

Modern ICT (Information and Communication Technologies) can be applied in agriculture by IoT (Internet of things)-based smart farming system. Here the crop field is continuously monitored with the help of sensors (light, humidity, temperature, soil moisture, etc.) and automating the irrigation system. Smart farming utilizes agriculture sensors that can provide data to the farmers to monitor and optimize crops by adapting to changes in the environmental conditions. Installation of sensors can be done over weather stations, drones, and robots.

3.2 Nanoparticles for Sensing

Nanoparticles like gold NPs, Zinc oxide NPs, graphene, and carbon nanotubes possess high sensitivity. Farmers can perform field analysis in fast, accurate, and cost-effective ways using biosensors with nanoparticles. The performance of nanostructured biosensors can be improved by the functionalization of the sensor substrates and the immobilization of the bio-probes on a transducer. Since nanoparticles possess high surface-to-volume ratio, nano biosensors offer high sensitivity, fast response time, mediate fast electron-transfer kinetics, high stability, and longer lifetime (Antonacci et al. 2018).

Nano biosensors are highly helpful to the farmers. For example, nutrition sensors and soil moisture sensors are useful to measure the levels of fertilizers and water, respectively. After getting accurate information from these sensors, farmers can apply the fertilizers and water properly. Chlorophyll in plants reflects infrared light. So, cameras carried by drones are used to take near infrared images of crops (Precision Farming Market 2020). Nowadays biosensors based on aptamers have become more sensitive because of the use of nucleic acid amplification techniques. Detection of plant pathogen before appearance of disease symptoms is possible with this technology (Khater et al. 2017).

Quantum dots are semiconductor nanocrystals having a particle size from 1 to 5 nm. In agriculture, quantum dots (QDs) are used in the detection of pesticides and veterinary drug residues. Since they are available in colloidal form and are highly stable, they can easily be assembled into sensors with bio-recognition elements such as enzymes, antibodies, and aptamers. They possess very good optical properties and so signal can be received in the form of fluorescence, chemiluminescence, electrochemical luminescence, photo electrochemistry, etc.

Nanoparticles such as gold nanoparticles, magnetic nanoparticles, and quantum dots are widely used for molecular detection purposes. Gold nanoparticles are particularly well known for rapid immune diagnosis. They have unique physicochemical properties, including small-sized effect, surface plasmon resonance behavior, and catalytic effect. They react with nucleic acid through an Au-S covalent bond which is used for DNA identification. Since gold nanoparticles can produce visible color change because of their high surface area to volume ratio, they are ideal for rapid biosensing. AuNPs also have the advantage of facile surface modification for effective targeting of biomolecules.

Several parameters such as fertilizers, nutrients, herbicides, insecticides, pesticides, pathogens, moisture, and soil pH can be effectively analyzed by nanobiosensors. In addition to the analysis-related activities, these nanobiosensors support sustainable agriculture to improve crop production. Verma et al. have reported that gold nanoparticles give a simple output because of their superior optical properties. These sensors must be developed into point-of-care devices for that they need high sensitivity in complex media. With the utilization of gold nanoparticles, biosensors can be implemented in diverse environments to prevent the spread of infectious diseases (Verma et al. 2015). Silicon nanoparticles have many scopes in agricultural applications. Si-NP is utilized as the growth regulator, soil quality improving agent, pesticides, nano-carriers (delivery agents for proteins, fertilizers, nucleotides, and other chemicals) and nanosensors. Figure 3.1 exhibits its agricultural applications (Rastogi et al. 2019).

3.3 Pesticides Detection

TiO_2 nanotubes are well known for their sensing performance. In agriculture, they are used for the detection of free organophosphate pesticides, dichlofenthion, and chlorpyrifos pesticides. The nanocomposite like graphene modified with TiO_2 nanotubes

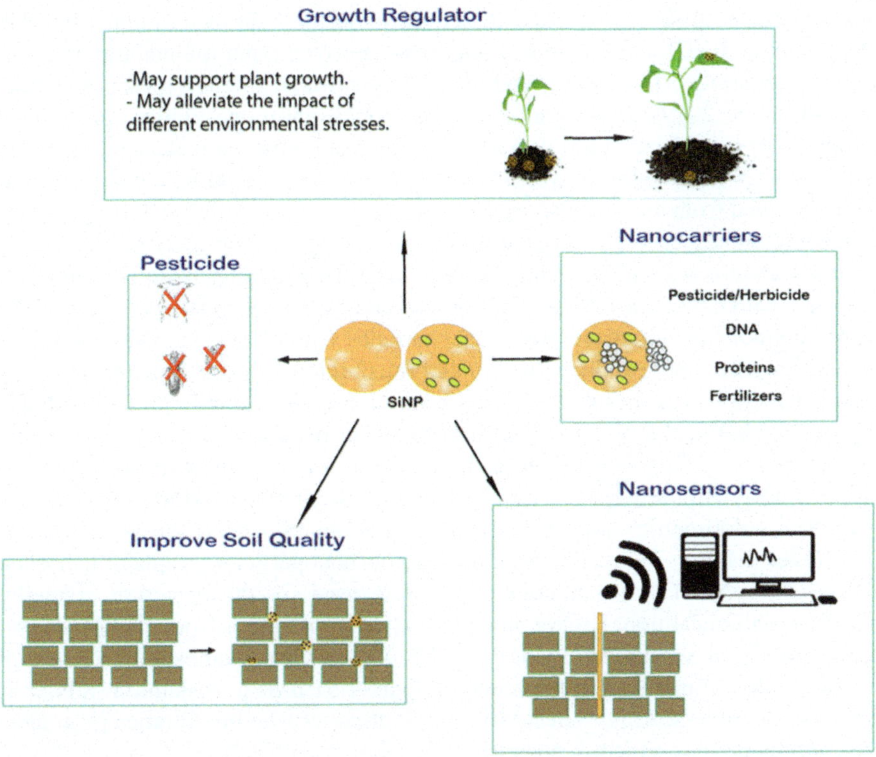

Fig. 3.1 Schematic diagram explains the various applications of silicon nanoparticles in agriculture (from Rastogi et al. 2019)

is applied for the detection of carbamate pesticides including metolcarb, carbaryl, isoprocarb, and diethofencarb with the detection limit from 2.27 to 3.26 μgL^{-1} (Wang et al. 2016). Even though the use of pesticides like pyrethorids improves crop yield in vegetable farming, the long-term consumption of fruits and vegetables (obtained from the pesticides applied farm) could lead to adverse health effects. As per the report of Yu et al., polystyrene-coated magnetic nanoparticles could detect pyrethroids at a concentration level as low as 0.02 ng g^{-1} of vegetables. The duration of the analysis is less than two hours. The nanoparticles can be reused 30 times thus making the analysis both time and cost efficient. The nanoparticles also showed strong affinity for beta-cyhalothrin, bifenthrin, fenvalerate, permethrin, and decamethrin (Yu and Yang 2017).

Simple paper-based sensors are made by incorporating quantum dots (QDs) in cellulose papers for pesticide detection. The sensing material is silicon quantum dots to detect the pesticides paraoxon and parathion by colorimetric analysis. The sensors are built by combining a green fluorescent protein with silicon quantum dots that emit red light. The paper turned yellow or green depending on the amount of toxin present (Zor et al. 2015). In paper-based techniques, while using radio frequency identification (RFID) tags, the results can be obtained in a smartphone. It is a simpler way for the farmers. There is no need for highly trained technicians for this technique (Robidillo et al. 2019). Rapid detection of organophosphorus pesticides was achieved through using gold nanoparticles and organophosphorus pesticide aptamers. The color changes from red to purple blue since the aptamers binding to the pesticides were detached from the AuNPs, resulting in aggregation of AuNPs (Bai et al. 2015).

Chemiluminescence (CL) is the production of light when a chemical reaction occurs. Luminol is a white-to-pale-yellow crystalline solid that exhibits chemiluminescence with a blue glow. Organophosphate and carbamate pesticides are detected using luminol-functionalized silver nanoparticles (Lum-Ag NP). The distinct CL response patterns are produced as "fingerprints" related to each specific pesticide after interaction with pesticides. A concentration of 24 $\mu g\ mL^{-1}$ of dimethoate, dipterex, carbaryl, chlorpyrifos, and carbofuran has been measured (He et al. 2015).

The analysis of organophosphorus pesticides was performed with Au nanoparticles and acetylcholinesterase. The color changes from claret-red to purple or even grey. The colorimetric assay gives a high reproducibility and the detection was rapid (Li et al. 2011). Plant esterase–chitosan/gold nanoparticles–graphene nanosheet (PLaE-CS/AuNPs-GNs) nanocomposites were used for the detection of methyl parathion and malathion. Highly pure plant esterase produced from plants was used as the probe. The sensing ability of PLaE-CS/AuNPs-GNs composite is as low as 50 ppt (0.19 nM) of methyl parathion and 0.5 ppb (1.51 nM) of malathion (S/N = 3). The system possessed high selectivity. Hence, there was no interference from metal ions, inorganic ions, glucose, and citric acid (Bao et al. 2015).

Glyphosate (Gly) on the surface of spinach, apple, and corn leaves was detected in situ using cysteamine-modified gold nanoparticles. Gly changed the color from red to blue on the surface of plant tissues. Gly distribution on plant tissues can facilitate the development of precision agriculture (Tu et al. 2019). Silver nanoparticles (AgNPs) were prepared by green synthesis using clove (*Syzygium aromaticum*) extract as a reducing agent. Vinclozolin (fungicide) was detected in trace amounts by UV–Vis spectroscopy with Ag NPs. Figure 3.2 shows the UV–Vis spectra and HRTEM images of the silver nanoparticles. AgNPs showed a stable LSPR band at 397 nm. The clove-AgNPs were highly selective and extremely sensitive for colorimetric sensing of vinclozolin with a lower detection limit of 21 nM (Hussain et al. 2019).

3.4 Soil Moisture Detection

It is important to measure the soil moisture to study the evaporation and transpiration in plants. There is always a change in humidity levels because of the change in air temperature and transpiration of water vapor (added by plants) to the air. The

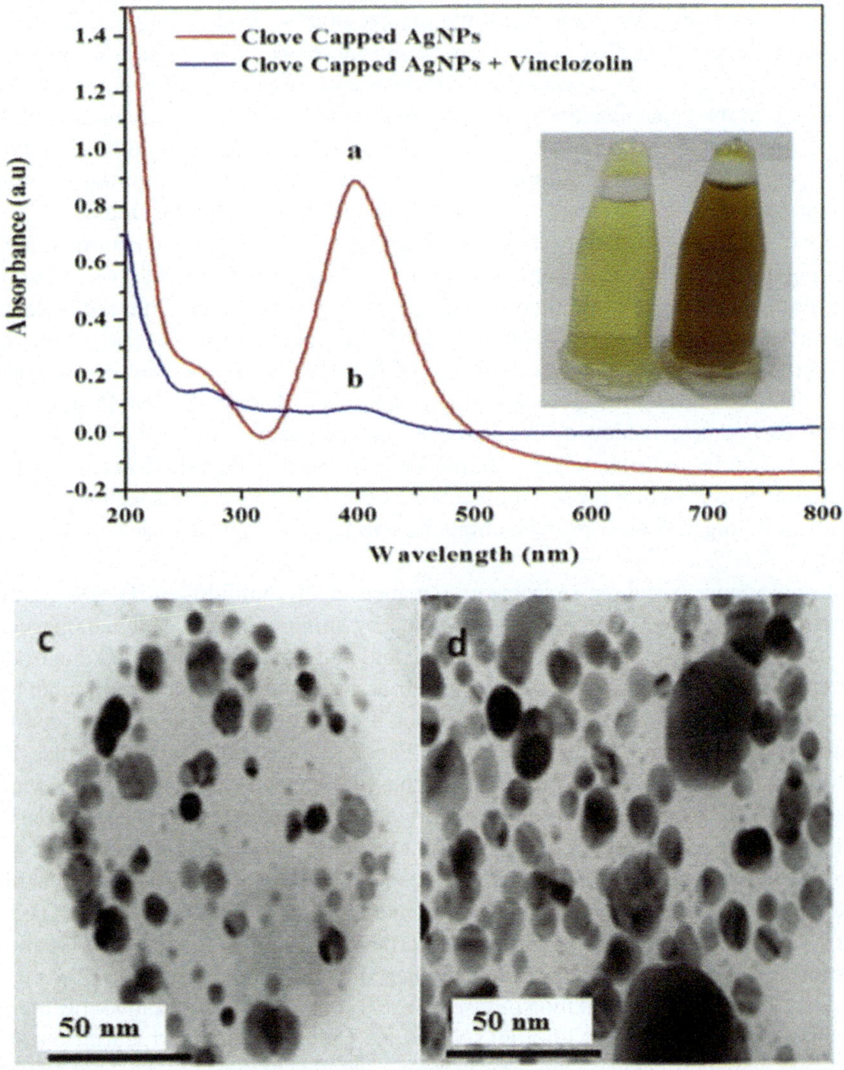

Fig. 3.2 Silver nanoparticles synthesized using aqueous solution of clove (*Syzygium aromaticum*) for vinclozolin detection. (**a** and **c**) and (**b** and **d**) exhibit the UV–Vis spectra and HRTEM images of clove-AgNPs and clove-AgNPs + VIN (16 μM), respectively. The size range = 2–20 nm and average size = 14.4 nm of AgNPs are calculated from the (**c**) (from Hussain et al. 2019)

humidity level (high or low) affects the quality of crop production, and the loss of quality reduces the selling price of crops and increases production costs which in turn reduces profits. In the case of a soil moisture sensor, the sensing material causes a change in the capacitance according to the amount of moisture in the soil which in turn is used to measure the dielectric permittivity of the surrounding medium. The applications of soil moisture sensors in agriculture include irrigation planning, climate research, solute transport studies, and soil respiration measurements.

Zinc oxide (ZnO) nanoparticles were coated as thinfilm on electrodes as the sensing material. Since ZnO nanoparticles possess oxygen vacancy dipoles, they give blue emission in the photoluminescence spectrum. So, they have enhanced dielectric properties. Moisture content from 7 to 25% was measured for wheat grains. The dielectric constant was determined using LCR meter. Because of the interaction of ZnO nanoparticles with water vapor, high sensitivity of 36.4% at 1 MHz and 97.4% at 500 Hz had been achieved. The actual sensing mechanism is the chemisorption of water molecules on activated sites of the ZnO grains. The dielectric-based sensor is becoming more popular for soil moisture measurement because this technique needs small sampling volumes, provides higher accuracy, and is of low cost (Singh et al. 2018).

Simple paper-based sensors were used to measure the humidity of 20–70% relative humidity (RH) range. The device was made by spin coating of ZnO nanoparticles on commercial printing papers as substrate and patterning interdigitated electrodes. After spin coating, the paper was annealed at 100°C for 10 min on a hotplate for binding of ZnO nanoparticles with the paper. The change in resistivity was recorded with respect to various humidity levels. Paper is a valuable substrate since it is available at low cost, is disposable, and has the capacity to absorb water vapor. Figure 3.3 exhibits the microdevice fabricated on a paper substrate and sensing activities. Figure 3.3 (a), (b), and (d) exhibit the paper substrate, gold layer, and ZnO nanoparticles deposited paper substrate, respectively. Figure 3.3 (c) and (e) exhibit the interdigitated electrodes and two fabricated devices, respectively. Figure 3.3 (f) shows the SEM image of the ZnO nanoparticles on the paper substrate. Humidity response and sensing response graph are also shown (Niarchos et al. 2017).

In another study, the soil moisture sensor was constructed using a low-cost thermo electric generator (TEG) with high sensitivity. The nanostructured thermosensitive resistors were fabricated on the same ceramic substrate of the TEG, i.e. lead sulphide (PbS) quantum dots were printed on the TEG substrate (Dias et al. 2016). Graphene oxide coated interdigitated electrode was fabricated by MEMS technique to be used as a capacitive sensor. The sensor possesses high sensitivity up to 340 and 370% for soil moisture changes from 1 to 55% for red and black soil, respectively. The diurnal temperature and salt concentration were the other parameters that can be measured based on soil conductivity. The salt concentration varies from 0 mol to 0.35 mol in the soil (Palaparthy et al. 2018).

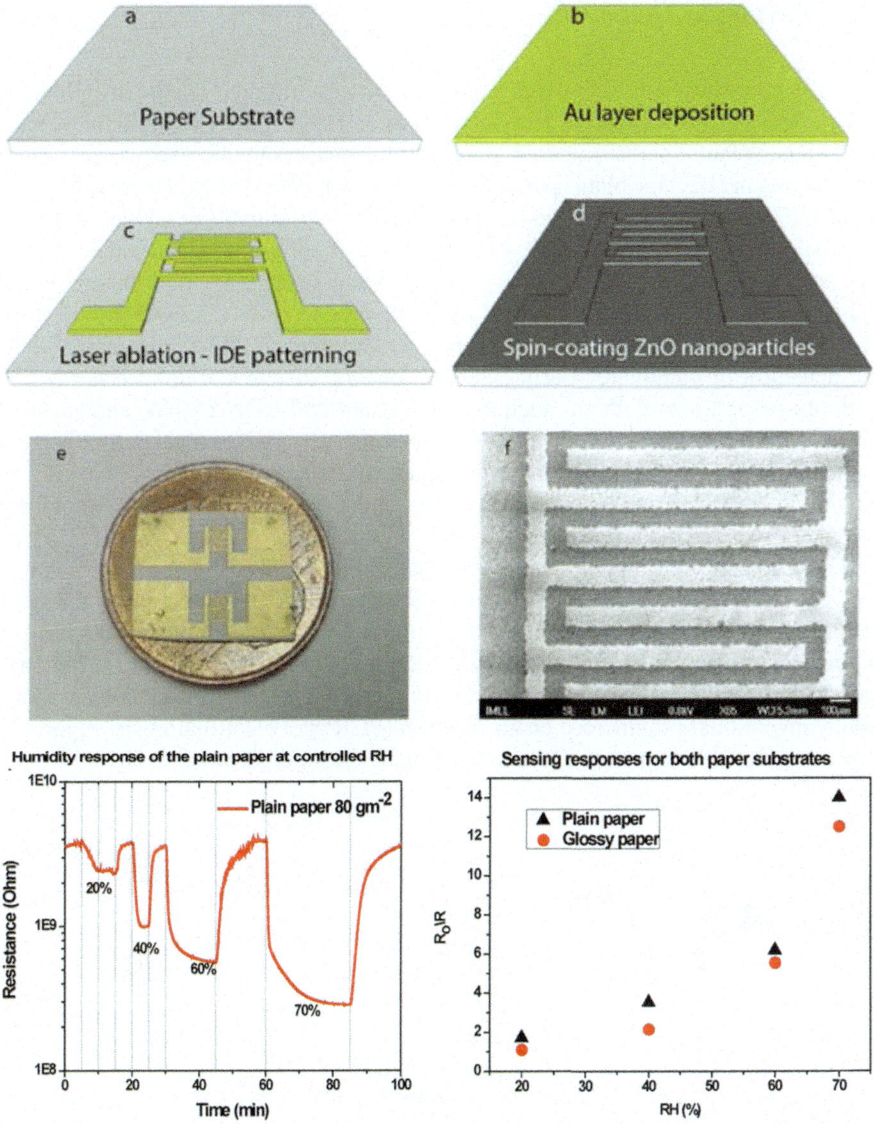

Fig. 3.3 Microdevice fabrication process. (**a**) Paper substrate. (**b**) Gold layer deposition. (**c**) Laser micromachining of the interdigitated electrodes. (**d**) Spin-coated ZnO-nanoparticle layers. (**e**) fabricated devices on paper (2 devices on a coin of 1 cent), (**f**) SEM image of the nanotextured ZnO film. Humidity response at controlled RH levels ranging from 20 to 70%, (**b**) sensing response for both papers as function of the RH (from Niarchos et al. 2017)

3.5 Pathogen Detection

Plant diseases cause more economic losses for the agricultural industry. In agriculture infections to crops arise from pathogens such as fungi, bacteria, mollicutes, parasitic higher plants, parasitic green algae, nematodes, protozoa, and viruses. These pathogens can affect the plants during growth, harvest, and postharvest processing. To attain maximum productivity and ensure agricultural sustainability, advanced disease detection and prevention in crops become more important. Conventional methods like polymerase chain reaction (PCR), immunofluorescence (IF), fluorescence in situ hybridization (FISH), enzyme-linked immunosorbent assay (ELISA), flow cytometry (FCM), and gas chromatography-mass spectrometry (GC-MS) are available for the detection of pathogens. Early detection of crop diseases is nowadays possible with bio-recognition elements such as enzyme, antibody, DNA/RNA, and bacteriophage with high selectivity (Fang and Ramasamy 2015). Nanobiosensors become more important for the proper identification of the disease and the disease-causing agent. Otherwise, disease control measures can be a waste of time and money and can lead to loss of plant production.

Early detection of infections in *Arabidopsis thaliana* by *Pseudomonas syringae* was performed through the rapid isothermal amplification of target pathogen DNA sequences by recombinase polymerase amplification (RPA) and gold nanoparticle-based electrochemical assessment with differential pulse voltammetry. This analysis takes one hour and this method is found to be 10,000 times more sensitive than conventional polymerase chain reaction (PCR)/gel electrophoresis. Figure 3.4 illustrates the electrochemical bioassay of plant pathogen (Lau et al. 2017).

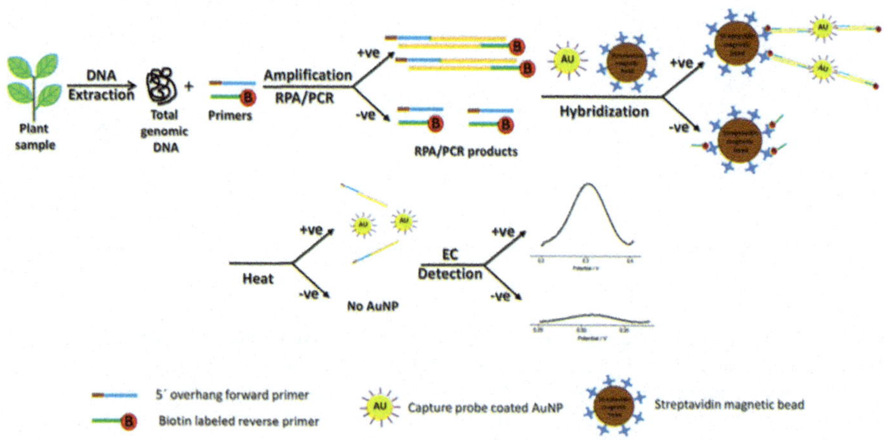

Fig. 3.4 Schematic diagram shows the electrochemical bioassay for the detection of plant pathogen DNA (from Lau et al. 2017)

Fluorescent silica nanoparticles when conjugated with antibodies are used to detect plant pathogens such as *Xanthomonas axonopodis* pv. *vesicatoria* which causes bacterial spot disease in Solanaceae plant. Yao et al., have reported that silica nanoparticles are incorporated with the organic dye, tris-2,2′-bipyridyl dichlororuthenium (II) hexahydrate (Rubpy) and conjugated with the secondary antibody of goat anti-rabbit immunoglobulin G (IgG). The fluorescence silica nano biomarker has the potential for the diagnosis of plant diseases (Yao et al. 2009). Detection of plant pathogen *Xanthomonas campestris* was performed with gold nanoparticles and bacteriophage M13 to display the receptor-binding protein from a phage. The analysis can be performed in less than one hour and can detect ~100 cells (Peng and Chen 2018). Citrus tristeza virus (CTV) causes Tristeza disease in citrus plants. The nucleic acid of CTV was measured using electrochemical sensing. A screen-printed carbon electrode was modified by electrodepositing gold nanoparticles and thiolated ssDNA was used as a probe. EIS measurements were performed in $Fe\ (CN_6)^{4-}/Fe\ (CN_6)^{3-}$ with the limit of detection 0.1–10 μM and selectivity was also good in this system (Khater et al. 2019).

3.6 Transgenic Plant Detection

Transgenic tobacco plants carrying a *Streptococcus mutans* antigen were identified using the surface resonance property of gold nanoparticles. The genomic DNA (isolated from the various parts like leaves, stems, and roots of the transgenic tobacco), as well as the biotinylated oligonucleotide probes, are immobilized onto a streptavidin (SA) sensor chip. The detection limit of this sensor was very low, i.e. 1 pM. Transgenic plant detection is shown in Fig. 3.5 (a). Figure 3.5 (b) exhibits transgenic DNA detection using SA SPR sensor and gold nanoparticles (Grześkowiak et al. 2019).

The genetically modified maize (MON 810 corn) is combating crop loss (created mainly by insects). Highly monodisperse Fe_3O_4@Au nanoparticles were used to detect them by the electrochemical sensing method. Usually, these monodisperse NPs are prepared by the thermal decomposition method. Freitas et al. have reported about the coating of Fe_3O_4@Au nanoparticles on the screen-printed electrodes. DNA covalently linked to a carboxylated self-assembled monolayer and a fluorescein isothiocyanate (FITC) signaling are used as probes. Chronoamperometric measurements were made after PCR amplification for the genoassay range from 0.25 to 2.5 nM (Freitas et al. 2016).

3.7 Raman Spectroscopy in Sensing

Surface-enhanced Raman spectroscopy (SERS) is a powerful technique for the quantitative measurement of pesticides without pre-treating the sample. Tognaccini et al.

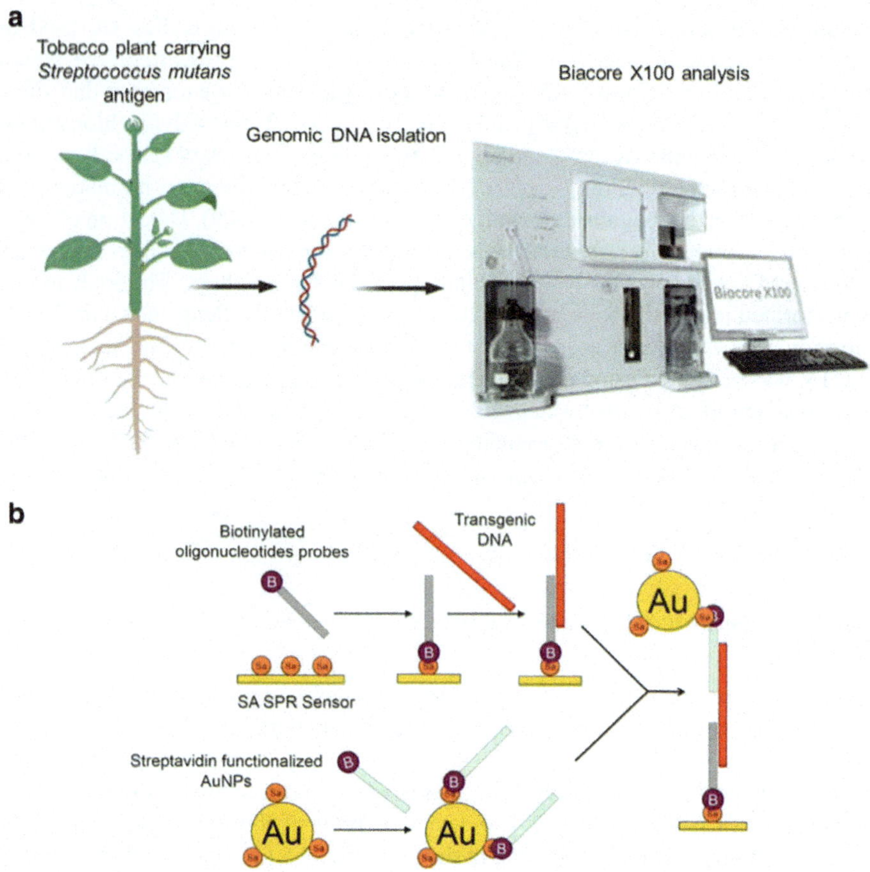

Fig. 3.5 Schematic diagram of the transgenic plant detection experimental. (**a**): a general procedure of the transgenic plant detection using gold nanoparticles (AuNPs)-based SPR biosensor, (**b**): schematic illustration of transgenic DNA detection using SA SPR sensor and AuNPs (from Grzeskowiak et al. 2019)

have reported that dimethoate (DMT) is an organophosphate insecticide applied to protect olive trees. DMT rapidly degrades than omethoate (OMT) and both disappear within two months. So, in-field determination of dimethoate becomes more important. The maximum limit of DMT permitted on olives for oil production is 3 mg Kg^{-1} and for OMT 1.5 mg Kg^{-1}. If the concentration of DMT in water is below 0.1 mg L^{-1}, such water will be sent to the sewage system (Tognaccini et al. 2019).

Simple and effective methods are essential for its in-field identification for public health and for the proper certification of organic produce. Raman scattering intensity can increase enormously with its surface coated with nanostructured silver or gold. Detection of DMT and OMT in water and on olive leaves is performed by surface-enhanced Raman spectroscopy (SERS) using portable instrumentations. Figure 3.6 shows the variation in the SERS spectra depending upon the DMT concentration

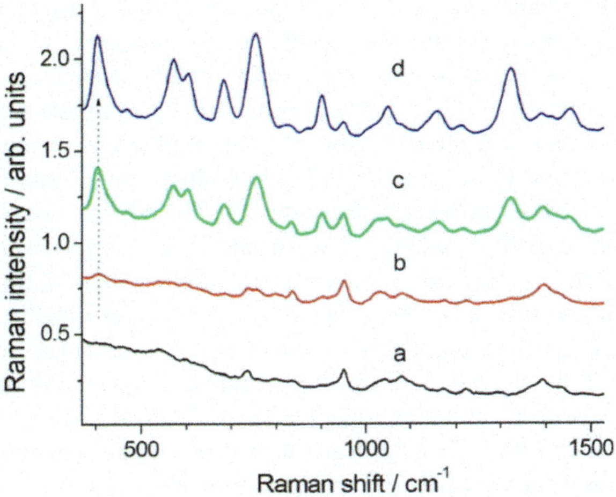

Fig. 3.6 SERS spectra related to the DMT at different concentrations. (a) 0.0 (only AgNPs in water). (**b**) 1×10^{-6}. (**c**) 5×10^{-6}. (**d**) 1×10^{-5} M. Vertical shifting in the spectra improves data readability. The arrow point at 406 cm^{-1} band can be used to differentiate in the DMT quantification (from Tognaccini et al. 2019)

(Tognaccini et al. 2019). When compared to the conventional methods like gas or liquid chromatography combined with mass spectroscopy, SERS possess ultra-sensitivity and simpler protocols for the detection of pesticide residues. Since SERS can be used for the direct detection of pesticides at trace levels in liquid samples or on the surface of solid samples. SERS provides vibrational fingerprints of molecules with non-destructive testing (Xu et al. 2017).

Trace level detection of pesticides at parts per million (ppm) or parts per billion (ppb) is possible with SERS. SERS utilizes noble metal nanostructures like gold and silver nanoparticles to increase the weak Raman signals. Chemically synthesized gold nanoparticles enhance Raman scattering in the detection of suspended pesticides (fungicides) and insecticides (neonicotinoids and organothiophosphates) with the detection limits from 0.001 to 10 parts per million (ppm).

Pesticides like phosmet and thiram on apple skin were analyzed using SERS analysis. The results showed that SERS with colloidal gold nanoparticles is a potential tool for identifying pesticides at trace levels for food safety applications. The surface of apple skin was cleaned and added to the colloidal gold nanoparticle. The suspension was then subjected to 785 nm laser excitation to observe the Raman spectral signatures of these pesticides. In this technique, pesticides less than 1 part per million can be measured. The pesticide tolerance level on apples given by the 2018 Code of Federal Regulations for the pesticides thiram, malathion, acetamiprid, and phosmet are 5 ppm, 8 ppm, 1 ppm, and 10 ppm, respectively (Dowgiallo and Guenther 2019).

In another study, silver-coated gold nanoparticles (Au@Ag NPs) were applied for the detection of pesticide residues in various fruit peels such as apple, grape, mango, pear, and peach using SERS. The Raman enhancement of Au@Ag NPs for sulfur-containing pesticides is stronger than those of bare Au and Ag NPs. The enhancement depends on the Ag shell thickness. The particle sensors can be cast onto fruit peels to measure pesticide residues like thiocarbamate and organophosphorus compounds.

The detection was highly reliable and rapid for example, 1.5 nanograms of thiram per square centimeter at apple peel was measured (Liu et al. 2012).

The extraction of targets from complex surfaces is difficult in the analysis of pesticide residues. Surface-enhanced Raman scattering active substrate is an excellent tool for the detection of target molecules. A novel substrate is constructed by decorating the commercial tape with colloidal gold nanoparticles (Au NPs) which is able to provide SERS activity. The SERS tape was pasted and peeled off on fruits and vegetables. The strong SERS signals were utilized in the detection of different pesticide residues (like parathion-methyl, thiram, and chlorpyrifos) present in the real samples having complex surfaces including green vegetable, cucumber, orange, and apple (Chen et al. 2016). The detection of the organochlorine pesticides aldrin, dieldrin, lindane, and α-endosulfan with a limit of detection reaching 10^{-8} M was achieved using high sensitivity of SERS. To improve the affinity of the pesticides with SERS substrate, the metal surface was functionalized with alkyl dithiols (Kubackova et al. 2015). Label-free detection of pathogens is possible with Raman optical spectroscopy. But it remains challenging to achieve clinically relevant speeds and accuracies due to weak Raman signal from bacterial cells.

3.8 Lateral Flow Immunoassay

Ralstonia solanacearum is a bacterium that causes potato brown rot. This pathogen was detected with Lateral flow immunoassay (LFIA) where the sensing material is gold nanoparticles conjugated with antibodies specific to *R. solanacearum*. This assay was able to detect up to 3–104 cells mL^{-1} in the potato tuber extract. Lateral flow test strips can be developed for making sensors for plants which can be highly useful for the farmers for immediate detection without a high-level scientific knowledge. Figure 3.7 shows the potato tuber extract analysis done through LFIA method (Razo et al. 2019).

Gold nanoparticles antibody conjugation, immobilization of the antibody specific to *R. solanacearum*, immobilization of protein A on the control zone and potato tuber extract with *R. solanacearum* are shown in the part (1), (2), (3), and (4) of Fig. 3.7 (a), respectively. Test zone without color band after analysis (negative result), red color band in the control zone and formation of the immune complex (immobilized specific antibodies to *R. solanacearum*, GNP conjugate with specific antibody) after completion of conventional LFIA can be seen in the (5), (6), and (7) of Fig. 3.7 (b), respectively. Figure 3.7 (c) shows the results of adding the amplification solution. After signal amplification of test zone, control zone and size enlarged gold nanoparticles are shown in the part (8), (9), and (10) of Fig. 3.7 (c), respectively (Razo et al. 2019).

Also, the rapid and in-field detection of pathogenic viruses is important in modern agricultural technologies. Potato leafroll virus was detected with a highly sensitive lateral flow immunoassay. The sensing material was gold nanoparticles with silver

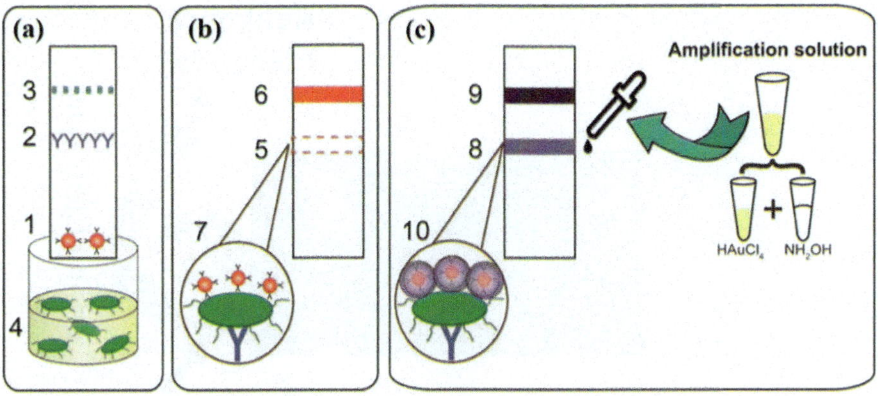

Fig. 3.7 Schematic diagram shows the LFIA analysis of pre and post signal amplification: (**a**) Insertion of the test strip into the potato tuber extract; (**b**) conventional LFIA indicates negative result; color band did not appear in the test zone; (**c**) after adding the amplification solution to the test strip color band appeared (positive result) due to the enlarged size of the gold nanoparticle (from Razo et al. 2019)

enhancement. The enhanced silver causes 15 times more sensitivity (detection limit 0.2 ng mL^{-1}; 15 min.) compared with conventional LFIA (detection limit 3 ng mL^{-1}; 10 min.) (Panferov et al. 2018).

The bacterial spot disease of stone fruits and almond is caused by the pathogen *Xanthomonas arboricola* pv. *pruni*. In one study, Lateral flow immunoassay (LFIA) was designed with polyclonal antibodies and carbon nanoparticles assembled on nitrocellulose strips. The sensitivity of the LFIA was analyzed for the plants like almond, apricot, Japanese plum, and peach using the suspensions obtained from the pure cultures of three *X. arboricola* pv. *pruni* strains and spiked leaf extracts of the plants (inoculated with this pathogen). The detection limit observed with both pure cultures and spiked samples is 10^4 CFU ml^{-1} (Lopez-Soriano et al. 2017).

3.9 Screen-Printed Electrodes

Commercial electrochemical sensors are fabricated from screen printing technology. So, miniaturized, sensitive, and portable devices can be made available lab-to-market (Hayat and Marty 2014). Figure 3.8 shows the schematic diagram of a portable sensor design using screen-printed electrode. This is a disposable sensor.

Solid-phase isothermal recombinase polymerase amplification (RPA) followed by electrochemical detection is applied for the detection of *Citrus tristeza* virus. RPA devices are limited by the need for heating sources to reach sensitive detection. This limitation can be avoided with gold nanoparticle (AuNP)-modified sensing substrates. Plant disease RPA assay for amplification of the P20 gene (387-bp) of

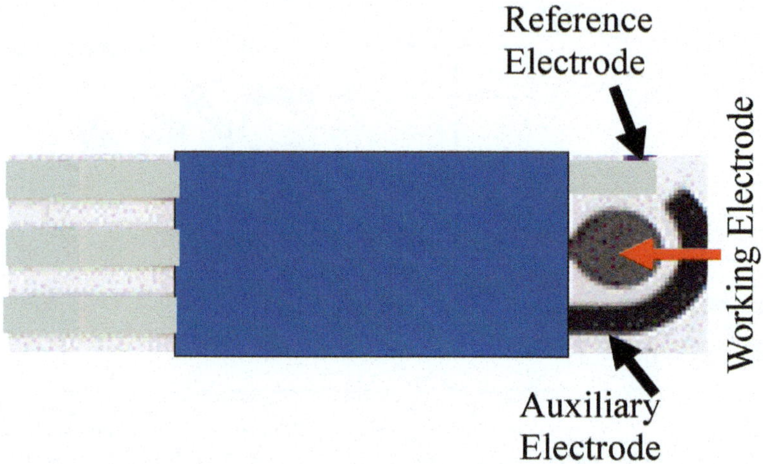

Fig. 3.8 Schematic diagram shows the screen-printed electrode. In this model, reference, working, and auxiliary electrodes are present in the same substrate. This is a disposable and portable sensor design

CTV was designed and tested by standard gel electrophoresis analysis. The Electrochemical impedance spectroscopy analysis (EIS) is performed in a Fe $(CN_6)^{4-}$/Fe $(CN_6)^{3-}$ electrolyte and a high limit of detection of up to 1000 fg μL^{-1} of nucleic acid was achieved (Khater et al. 2019).

Screen-printed electrode was modified with Ag nanoseeds and Ag nanoprisms and three different carbon substrates, graphite, graphene, and carbon nanofibers for the determination of Pb (II) and Cu (II). Figure 3.9 Voltammetric measurements of Pb (II) and Cu (II). Calibration curves are shown in the insets. The results confirm that the Ag-nanoseeds-SPCNFE has higher sensitivities. It can be applied for the simultaneous determination of Cu (II) and Pb (II) present in the natural samples (Pérez-Ràfols et al. 2017).

Nitrites are used as a preservative in dairy and meat products. They are potential carcinogens and cause detrimental health effects; if they are utilized beyond safe limits that will be harmful to public health and environment. Talbi et al. have reported that the electrochemical sensing of nitrite was performed using graphite screen-printed electrodes (GSPE) functionalized with gold nanoparticles stabilized by branched polyethyleneimine (AuNPs-PEI) with a concentration of 300 $\mu g\ mL^{-1}$. The sensor performs well in the linear range 1–10 μM and the limit of detection 1 μM (Talbi et al. 2019). Graphite screen-printed sensor developed using silver metal nanoparticle embedded chitosan composite was also employed in the quantification of nitrite at trace level (Patri et al. 2019).

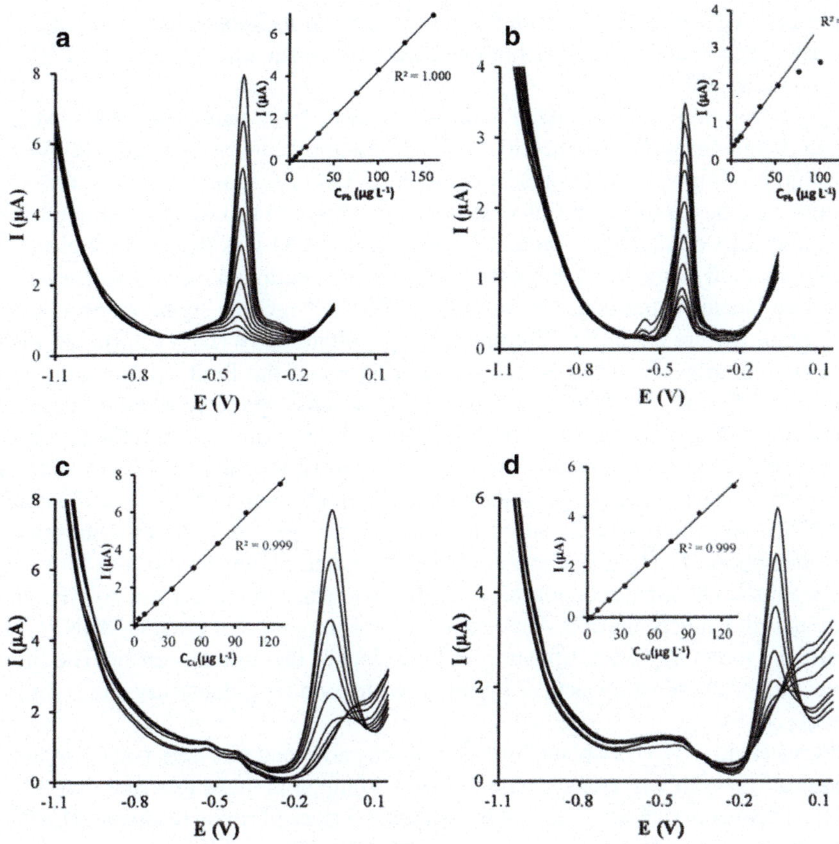

Fig. 3.9 Stripping voltammetric measurements and calibration curves (insets). (**a, b**) and (**c, d**) calibration of Pb (II) and Cu(II), respectively, in acetate buffer pH 4.5 at an E_d of −1.1 V and a t_d of 120 s. In Fig. 3.9 (**a, c**) and (**b, d**) Ag-nanoseeds-SPCNFE and Ag-nanoprisms-SPCNFE electrodes are used. SPCNFE: carbon nanofiber modified screen-printed electrode (Pérez-Ràfols et al. 2017)

3.10 Detection of Phytohormones

Salicylic acid (SA) acts as a phytohormone in plants. SA concentration changes when a plant is infected by pathogens. Electrochemical sensing of SA was performed by coating the gold electrode with copper nanoparticles. The electrocatalytic oxidation of salicylic acid in oilseed rape infected with the fungal pathogen *Sclerotinia sclerotiorum* was studied and the sensitivity was high with gold electrode modified with copper nanoparticles than the bare gold electrode (Wang et al. 2010).

$NaLuF_4$ nanoparticles are upconversion nanoparticles that can emit multicolor visible light under the excitation of 980 nm near-infrared (NIR) photons. Rhodamine-B dye in plant cells was measured effectively with a detection limit of 0.25 $\mu g\ cm^{-3}$

in plant cell. RB absorbs the green fluorescence from nanoparticles and emits red light. The phenomenon is called luminescence resonance energy transfer (LRET) (Xiaofeng et al. 2016).

Neoglycoprotein functionalized with fluorescent gold nanoclusters (AuNCs), contains biantennary N-glycan (G0) as targeting molecule, ovalbumin (OVA) as carrier/model antigen. This fluorescent gold core (G0-OVA-AuNCs) can be utilized as imaging probe for the detection of plant lectins and in vitro imaging of dendritic cells (Brzezicka et al. 2018). Indole-3-acetic acid (IAA) was detected by electrochemical method using gold nanoparticles (AuNPs) functionalized with horseradish peroxidase labeled immunoglobulins G (AuNPs-HRP-IgG) as signal amplification probe. Studies were taken by differential pulse voltammetry (DPV) in 0.1 M PBS (pH 7.4) containing Fe (CN)63 −/4 − electrolyte and the limit of detection was found to be 5.5×10^{-10} M (S/N = 3). The immunosensor possesses high selectivity toward indole-3-acetic acid and can differentiate IAA from other phytohormones, such as gibberellin, abscisic acid, and salicylic acid (Zhou et al. 2013). In another study, gold nanoparticles/polyaniline/multiwall carbon nanotubes (AuNPs/TPANI-MWCNTs) nanocomposite was used to construct an amperometric immunosensor for the detection of indole-3-acetic acid. The thiolated bionanocomposite film was made from the chemical reaction between boronic acid functionalized AuNPs and the vicinal diol functionalized AuNP labeled immunoglobulin G (IgG–AuNPs). It was utilized to get more amplification of the signal of the immunosensor. The limit of detection of the sensor was 0.97 pg mL^{-1} with good reproducibility and stability (Su et al. 2019).

Abscisic acid (ABA) is a plant hormone which participates in seed, bud dormancy, the control of organ size and stomatal closure. Wang et al. have reported that ABA from fresh leaves of rice was detected by localized surface plasmon resonance (LSPR) using aptamer-functionalized gold nanoparticles (AuNPs) without using expensive instrument and antibody. Abscisic acid involves in abiotic stress response and plays an indispensable role in plant physiological activities. The sensor shows a linear range from 5×10^{-7} M to 5×10^{-5} M with a detection limit of 0.33 μM (Wang et al. 2017).

Glassy carbon paste electrode modified with In_2O_3 nanoparticles is used for the electrochemical determination of luteolin (LU) in thyme. The analytical performance of the sensor shows a wide response to LU from 9.98×10^{-9} M to 8.84×10^{-8} M LU and a low detection limit of 1.99×10^{-10} M of LU was achieved (Ibrahim and Temerk 2015). A simple electrochemical sensor was constructed by electrodeposition of glassy carbon electrode with graphene quantum dots (GQDs) and gold nanoparticles (GNPs) for the determination of luteolin *peanut hull* samples. The nanocomposite resulted in a 16-fold increase in the oxidation current of luteolin in phosphate buffer at pH 5.0 with the detection limit of luteolin 1.0 nM (S/N = 3) (Tang et al. 2019).

3.11 Detection of Water Uptake

Direct visualization of water-conducting pathways and analysis of sap flows in xylem vessels can be done through AuNPs (gold nanoparticles) combined synchrotron X-ray imaging technique. AuNPs were used as flow tracers in the xylem vessels in the first 20–30 min without any physiological barrier. So, fluid dynamic phenomena can be visualized using gold nanoparticles in vascular plants (Hwang et al. 2014). In another study, TiO_2 nano-flowers were coated on fluorine-doped tin oxide substrate and used as a pH sensor which can measure pH range of 2–12. TiO_2 nano-flowers pH sensor had high sensitivity value, i.e. 2.7 $(\mu A)^{1/2}$/pH and a linear relationship between I_{DS} and pH (regression of 0.9991). The voltage reference and pH relationship showed the sensitivity value 46 mV/pH and a linear regression of 0.9989. The results confirm that a flower-like TiO_2 nanostructure extended gate field effect transistor pH sensor can be utilized in the detection of the pH value effectively (Yang et al. 2019) (Fig. 3.10).

3.12 Conclusion

Biosensors can help farmers to increase the crop yield by detecting various influencing factors. In agriculture, many parameters like fertilizers, herbicides, pesticides, insecticides, pathogens, moisture, and soil pH are analyzed by nanobiosensors. The challenges for the researchers in this area lie in the performance of the sensors, sampling, detection in open areas and scaling up measurements. Spectroscopic techniques like Raman spectroscopy, Fourier transform infrared spectroscopy, and fluorescence spectroscopy are the best alternatives for PCR technology and offer immediate detection. The achievement of a low limit of detection by various transduction techniques with the help of nanomaterials is discussed in this chapter. These transducers can transform plant chemical signals to digital form which can be easily measured. Wearable sensors can be produced to enhance crop productivity in order to meet the demand for food. These sensors are also useful to control the utilization of costly chemicals like pesticides and herbicides.

Slow technology transfer to the marketplace is the major drawback of the biosensors. However, biosensors technology will replace conventional analytical techniques with the help of highly sensitive nanoparticles. Nanotechnology enables the smart plant sensors for monitoring and optimizing the plant productivity and proper utilization of resources. These sensing techniques can be highly reliable for continuous monitoring of plant health which will be more helpful in modern agriculture.

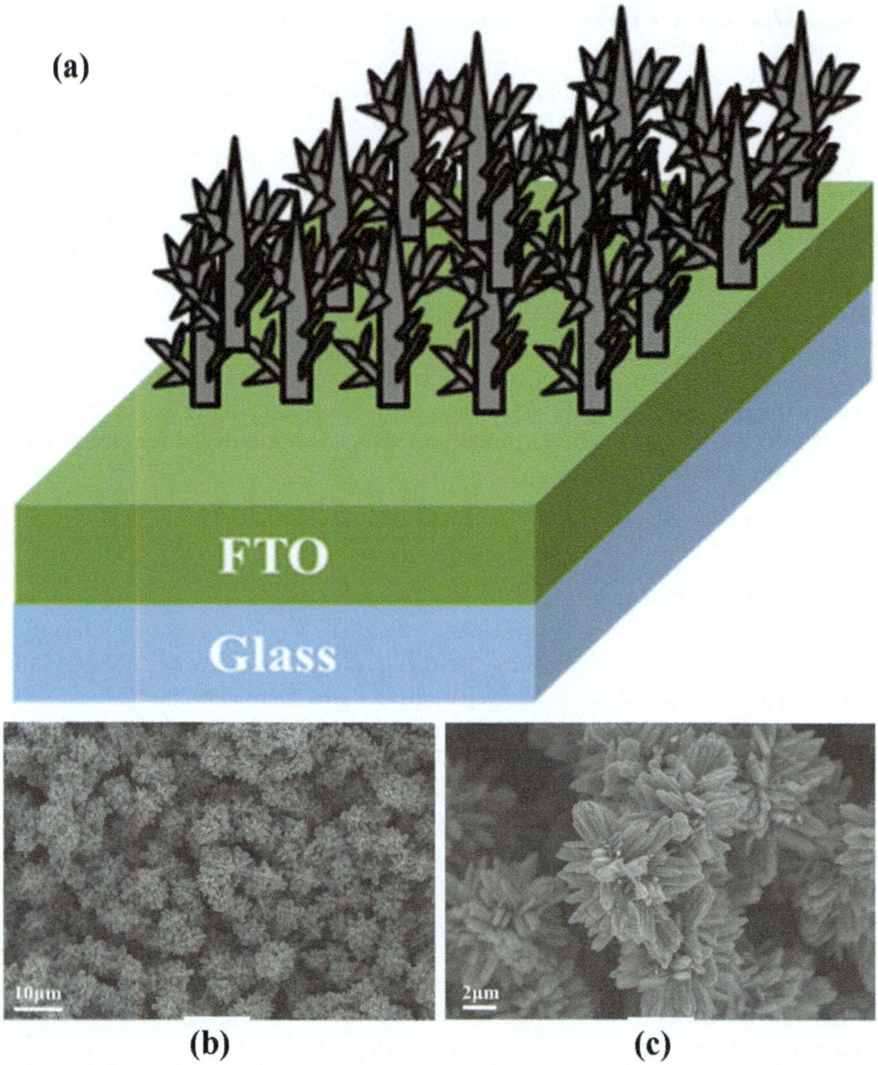

Fig. 3.10 Application of TiO_2 in the extended gate field effect transistor (EGFET) pH sensor. (**a**) Shows the schematic diagram of the TiO_2 EGFET pH sensor. (**b**) HR-SEM image of TiO_2 nano-flowers (Top view). Randomly oriented flower-like nanostructures can be seen. (**c**) The same image (TiO_2 nano-flowers) is shown with high-magnification (Yang et al. 2019)

Acknowledgements The author expresses thanks to her husband Mr. G. Sankar for his assistance in this work. Also, she acknowledges the assistance of International Research Center, Kalasalingam Academy of Research and Education (Deemed University), Krishnankoil – 626 126, Tamilnadu, (India).

References

Antonacci A, Arduini F, Moscone D, Palleschi G, Scognamiglio V (2018) Nanostructured (Bio) sensors for smart agriculture. TrAC Trends Anal Chem 98:95–103

Bai W, Zhu C, Liu J, Yan M, Yang S, Chen A (2015) Gold nanoparticle–based colorimetric aptasensor for rapid detection of six organophosphorous pesticides. Environ Toxicol Chem 34(10):2244–2249

Bao J, Hou C, Chen M, Li J, Huo D, Yang M, Luo X, Lei Y (2015) Plant esterase–chitosan/gold nanoparticles–graphene nanosheet composite-based biosensor for the ultrasensitive detection of organophosphate pesticides. J Agric Food Chem 63(47):10319–10326

Brzezicka KA, Serna S, Reichardt NC (2018) Fluorescent neoglycoprotein gold nanoclusters: Synthesis and applications in plant lectin sensing and cell imaging. Nanoscale Res Lett 13(1):360

Chen J, Huang Y, Kannan P, Zhang L, Lin Z, Zhang J, Chen T, Guo L (2016) Flexible, and adhesive surface enhance Raman scattering active tape for rapid detection of pesticide residues in fruits and vegetables. Anal Chem 88(4):2149–2155

Dias PC, Cadavid D, Ortega S, Ruiz A, França MB, Morais FJ, Ferreira EC, Cabot A (2016) Autonomous soil moisture sensor based on nanostructured thermosensitive resistors powered by an integrated thermoelectric generator. Sens Actuators, A 239:1–7

Dowgiallo AM, Guenther DA (2019) Determination of the limit of detection of multiple pesticides utilizing gold nanoparticles and surface-enhanced Raman spectroscopy. J Agric Food Chem 67(46):12642–12651

Fang Y, Ramasamy RP (2015) Current and prospective methods for plant disease detection. Biosensors 5(3):537–561

Freitas M, Sá Couto M, Barroso MF, Pereira C, de-los-Santos-Álvarez N, Miranda-Ordieres AJ, Lobo-Castañón MJ, Delerue-Matos C (2016) Highly monodisperse Fe3O4@ Au superparamagnetic nanoparticles as reproducible platform for genosensing genetically modified organisms. Acs Sensors 1(8):1044–1053

Giraldo JP, Wu H, Newkirk GM, Kruss S (2019) Nanobiotechnology approaches for engineering smart plant sensors. Nat Nanotechnol 14(6):541–553

Grześkowiak BF, Tuśnio K, Woźniak A, Szalata M, Lipiński D, Jurga S, Słomski R (2019) Transgenic plant detection using an AuNPs based SPR biosensor. Biosensors 9(4):116

Hayat A, Marty JL (2014) Disposable screen printed electrochemical sensors: Tools for environmental monitoring. Sensors 14(6):10432–10453

He Y, Xu B, Li W, Yu H (2015) Silver nanoparticle-based chemiluminescent sensor array for pesticide discrimination. J Agric Food Chem 63(11):2930–2934

Hussain M, Nafady A, Avcı A, Pehlivan E, Nisar J, Sherazi ST, Balouch A, Shah MR, Almaghrabi OA, Ul-Haq MA (2019) Biogenic silver nanoparticles for trace colorimetric sensing of enzyme disrupter fungicide vinclozolin. Nanomaterials 9(11):1604

Hwang BG, Ahn S, Lee SJ (2014) Use of gold nanoparticles to detect water uptake in vascular plants. PLoS ONE 9(12):e114902

Ibrahim H, Temerk Y (2015) Novel sensor for sensitive electrochemical determination of luteolin based on In_2O_3 nanoparticles modified glassy carbon paste electrode. Sen Actuators B: Chem 206:744–752

Khater M, de la Escosura-Muñiz A, Merkoçi A (2017) Biosensors for plant pathogen detection. Biosens Bioelectron 93:72–86

Khater M, de la Escosura-Muñiz A, Quesada-González D, Merkoçi A (2019) Electrochemical detection of plant virus using gold nanoparticle-modified electrodes. Anal Chim Acta 1046:123–131

Khater M, Escosura-Muñiz AD, Altet L, Merkoçi A (2019) In situ plant virus nucleic acid isothermal amplification detection on gold nanoparticle-modified electrodes. Anal Chem 91(7):4790–4796

Kubackova J, Fabriciova G, Miskovsky P, Jancura D, Sanchez-Cortes S (2015) Sensitive surface-enhanced Raman spectroscopy (SERS) detection of organochlorine pesticides by alkyl dithiol-functionalized metal nanoparticles-induced plasmonic hot spots. Anal Chem 87(1):663–669

Kumar V, Arora K (2020) Trends in nano-inspired biosensors for plants. Mater Sci Energy Technol 3:255–273

Lau HY, Wu H, Wee EJ, Trau M, Wang Y, Botella JR (2017) Specific and sensitive isothermal electrochemical biosensor for plant pathogen DNA detection with colloidal gold nanoparticles as probes. Sci Rep 7(1):1–7

Li H, Guo J, Ping H, Liu L, Zhang M, Guan F, Sun C, Zhang Q (2011) Visual detection of organophosphorus pesticides represented by mathamidophos using Au nanoparticles as colorimetric probe. Talanta 87:93–99

Liu B, Han G, Zhang Z, Liu R, Jiang C, Wang S, Han MY (2012) Shell thickness-dependent Raman enhancement for rapid identification and detection of pesticide residues at fruit peels. Anal Chem 84(1):255–261

Lopez-Soriano P, Noguera P, Gorris MT, Puchades R, Maquieira A, Marco-Noales E, López MM (2017) Lateral flow immunoassay for on-site detection of Xanthomonas arboricola pv. pruni in symptomatic field samples. PLoS ONE 12(4):e0176201

Niarchos G, Dubourg G, Afroudakis G, Georgopoulos M, Tsouti V, Makarona E, Crnojevic-Bengin V, Tsamis C (2017) Humidity sensing properties of paper substrates and their passivation with ZnO nanoparticles for sensor applications. Sensors 17(3):516

Palaparthy VS, Kalita H, Surya SG, Baghini MS, Aslam M (2018) Graphene oxide based soil moisture microsensor for in situ agriculture applications. Sens Actuators B: Chem 273:1660–1669

Panferov VG, Safenkova IV, Byzova NA, Varitsev YA, Zherdev AV, Dzantiev BB (2018) Silver-enhanced lateral flow immunoassay for highly-sensitive detection of potato leafroll virus. Food Agr Immunol 29(1):445–457

Patri SB, Adarakatti PS, Malingappa P (2019) Silver nanoparticles-chitosan composite embedded graphite screen-printed electrodes as a novel electrochemical platform in the measurement of trace level nitrite: application to milk powder samples. Curr Anal Chem 15(1):56–65

Peng H, Chen IA (2018) Rapid colorimetric detection of bacterial species through the capture of gold nanoparticles by chimeric phages. ACS Nano. Dec 26;13(2):1244–1252

Pérez-Ràfols C, Bastos-Arrieta J, Serrano N, Díaz-Cruz JM, Ariño C, De Pablo J, Esteban M (2017) Ag nanoparticles drop-casting modification of screen-printed electrodes for the simultaneous voltammetric determination of Cu (II) and Pb (II). Sensors 17(6):1458

Precision Farming Market (2020) Precision farming market by technology [online] Marketsandmarkets.com. Available via https://www.marketsandmarkets.com/Market-Reports/precision-farming-market-1243.html. Accessed 12 May 2020

Rastogi A, Tripathi DK, Yadav S, Chauhan DK, Živčák M, Ghorbanpour M, El-Sheery NI, Brestic M (2019) Application of silicon nanoparticles in agriculture. 3 Biotech 9(3):90

Razo SC, Panferova NA, Panferov VG, Safenkova IV, Drenova NV, Varitsev YA, Zherdev AV, Pakina EN, Dzantiev BB (2019) Enlargement of gold nanoparticles for sensitive immunochromatographic diagnostics of potato brown rot. Sensors 19(1):153

Robidillo CJ, Wandelt S, Dalangin R, Zhang L, Yu H, Meldrum A, Campbell RE, Veinot JG (2019) Ratiometric detection of nerve agents by coupling complementary properties of silicon-based quantum dots and green fluorescent protein. ACS Appl Mater Interfaces 11(36):33478–33488

Sekhon SS, Park GY, Park DY, Kim SY, Wee JH, Ahn JY, Kim YH (2018) Aptasensors for pesticide detection. Toxicol Environ Health Sci 10(5):229–236

Singh H, Kumar A, Bansod BS, Singh T, Thakur A, Singh T, Sharma J (2018). Enhanced moisture sensing properties of a nanostructured ZnO coated capacitive sensor. RSC Adv 8(7):3839–3845

Srivastava AK, Dev A, Karmakar S (2018) Nanosensors and nanobiosensors in food and agriculture. Environ Chem Lett 16(1):161–182

Su Z, Cheng Y, Li C, Xiong Y, Xiao L, Chen S, Qin X (2019) Dispersing gold nanoparticles on thiolated polyaniline-multiwalled carbon nanotubes for development of an indole-3-acetic acid amperometric immunosensor. Nanoscale Adv 9:3607–3613

Talbi M, Al-Hamry A, Teixeira PR, Bouhamed A, Azzouzi S, Paterno LG, Ali MB, Kanoun O (2019) Graphite screen printed electrodes functionalized with AuNPs-PEI for nitrite detection. In 2019 16th International Multi-Conference on Systems, Signals and Devices (SSD) March 21 (pp. 1–4). IEEE

Tang J, Huang R, Zheng S, Jiang S, Yu H, Li Z, Wang J (2019) A sensitive and selective electrochemical sensor based on graphene quantum dots/gold nanoparticles nanocomposite modified electrode for the determination of luteolin in peanut hulls. Microchem J 145:899–907

Tognaccini L, Ricci M, Gellini C, Feis A, Smulevich G, Becucci M (2019) Surface enhanced Raman spectroscopy for in-field detection of pesticides: A test on dimethoate residues in water and on olive leaves. Molecules 24(2):292

Tu Q, Yang T, Qu Y, Gao S, Zhang Z, Zhang Q, Wang Y, Wang J, He L (2019) In situ colorimetric detection of glyphosate on plant tissues using cysteamine-modified gold nanoparticles. Analyst 144(6):2017–2025

Verma MS, Rogowski JL, Jones L, Gu FX (2015) Colorimetric biosensing of pathogens using gold nanoparticles. Biotechnol Adv 33(6):666–680

Wang S, Li W, Chang K, Liu J, Guo Q, Sun H, Jiang M, Zhang H, Chen J, Hu J (2017) Localized surface plasmon resonance-based abscisic acid biosensor using aptamer-functionalized gold nanoparticles. PLoS ONE 12(9):e0185530

Wang Y, Sun C, Zhao X, Cui B, Zeng Z, Wang A, Liu G, Cui H (2016) The application of nano-TiO 2 photo semiconductors in agriculture. Nanoscale Res Lett 11(1):529

Wang Z, Wei F, Liu SY, Xu Q, Huang JY, Dong XY, Yu JH, Yang Q, Zhao YD, Chen H (2010) Electrocatalytic oxidation of phytohormone salicylic acid at copper nanoparticles-modified gold electrode and its detection in oilseed rape infected with fungal pathogen Sclerotinia sclerotiorum. Talanta 80(3):1277–1281

Xiaofeng WU, Pan HU, Shigang HU, Zenghui CH, Huanyuan YA, Zhijun TA, Zaifang XI, Yi YU, Gangtao DA, Yunxin LI (2016) Upconversion nanoparticles for differential imaging of plant cells and detection of fluorescent dyes. J Rare Earths 34(2):208–220

Xu ML, Gao Y, Han XX, Zhao B (2017) Detection of pesticide residues in food using surface-enhanced Raman spectroscopy: a review. J Agric Food Chem 65(32):6719–6726

Yang CC, Chen KY, Su YK (2019) TiO2 nano flowers based EGFET sensor for pH sensing. Coatings 9(4):251

Yao KS, Li SJ, Tzeng KC, Cheng TC, Chang CY, Chiu CY, Liao CY, Hsu JJ, Lin ZP (2009) Fluorescence silica nanoprobe as a biomarker for rapid detection of plant pathogens. Adv Mater Res 79:513–516

Yu X, Yang H (2017) Pyrethroid residue determination in organic and conventional vegetables using liquid-solid extraction coupled with magnetic solid phase extraction based on polystyrene-coated magnetic nanoparticles. Food Chem 217:303–310

Zhou Y, Xu Z, Wang M, Meng X, Yin H (2013) Electrochemical immunoassay platform for high sensitivity detection of indole-3-acetic acid. Electrochim Acta 96:66–73

Zor E, Morales-Narváez E, Zamora-Gálvez A, Bingol H, Ersoz M, Merkoçi A (2015) Graphene quantum dots-based photoluminescent sensor: a multifunctional composite for pesticide detection. ACS Appl Mater Interfaces 7(36):20272–20279

Chapter 4
Nanomaterial-Based Gas Sensors for Agriculture Sector

Robin Kumar, Monica Jaiswal, Neelam Kushwaha, Shivansh Bansal, Neha Mazumder, and Jagjiwan Mittal

Abstract Discovery of nanomaterials in the last three decades and their unique properties led the boom in their applications in various areas. Use of these materials in engineering, medical and environment also increased their attraction for the potential applications in agriculture. The current focus of agriculture research is the sustainable increase in crop production and protection. For achieving the goals, constant checking of parameters such as moisture content, soil fertility, temperature, crop nutrient capacity, pathogens, plant diseases, etc. are highly required. Use of nano sensors for monitoring these aspects are found to be very effective in increasing the healthy crop production. This chapter describes the applications of different nanomaterials in sensing of gases during different phases of agricultural practices. The objective is to evaluate the current literature using the gas sensors in agriculture and meat production, and their storage and transport applications.

Keywords Sensors · Gas sensors · Nanomaterials · Agriculture · Meat industry

4.1 Introduction

For many decades, agriculture has been mainly associated with the production of essential food crops for human survival. At present, current agriculture acknowledge not only farming but processing, distribution of crops and livestock products and their marketing. Thus, the word agriculture encompasses promotion of processing production, and distribution of agricultural products and derivatives. Agriculture plays most significant role in the economic well-being and growth for any given country. Advanced technology and ideas have helped us to be able to grow as well as store crops according to our needs, even if climate is not favourable for some crops. We can make artificial atmosphere to facilitate the growth and healthy storage of various regular as well as exotic vegetables, fruits, herbs including hybrids. Besides providing all the required conditions for favourable growth and storage of crops,

R. Kumar (✉) · M. Jaiswal · N. Kushwaha · S. Bansal · N. Mazumder · J. Mittal
Amity Institute of Nanotechnology, Amity University Uttar Pradesh, Noida, UP, India
e-mail: rkumar5@amity.edu

R. N. Pudake et al. (eds.), *Biosensors in Agriculture: Recent Trends and Future Perspectives*, Concepts and Strategies in Plant Sciences,
https://doi.org/10.1007/978-3-030-66165-6_4

several technologies combine to give expected results. Several studies are going on to improvise the agricultural standards in many aspects. Different regions of the world with varying climates require customised conditions for crop growth. These include fertilizers, pesticides and various types of sensors to keep a check on all conditions required, for example, humidity sensor, temperature sensor and gas sensors. Precision agriculture model consists sensors for collection of data, providing information to farmers by means of irrigation control model arrangement, variable-rate technology model and control system using parameter of green house (Chaudhary et al. 2011).

The demand for huge supply of food by the increasing global population motivates scientists to explore new methods and materials to accelerate agricultural production. As the natural resources such as land and water are meagre, the agricultural productivity can only be enhanced by better agronomy with the support of effectively using advanced technology (Pudake et al. 2019). Sustainable agricultural escalation is a concept designed for increasing food production from the same farmland without pernicious impact on the environment. Besides production, prime concerns of agricultural activities are food safety, storage as well as transport. With the surge in globalization of food, this suggests the importance of food quality assessment during all steps of the agri-food chain supply. Therefore, primary focus of agricultural research is on the improvement of efficiency of the crop, food processing and safety, food supplements and environmental effect on the production, storage and distribution of food (Pudake et al. 2019). Food quality assessment can be achieved qualitatively at the laboratory level by the number of methods including cell culture and instrumental analysis. These methods are time-consuming i.e. they can take several hours to days, including different pre-treatment steps of the sample as well as instrumentation. Safety of food is also vulnerable to some contaminants which cannot be noticed and sensed with standard analytical techniques e.g. plant pathogens, formaldehyde, excessive residues of pesticides (Kannan and Guo 2020). The inability of conventional techniques to assess food safety gives way to the development of novel, miniaturized, fast and handy analytical methods having low range detection limits. Nanotechnology offers one key solution for these requirements of analytical tools.

Nanotechnology consists of the power to contribute to every step of agricultural advancements by exploiting novel properties of nanomaterials. Understanding and manipulating the nanometric structure of the material provides the potential to bring significant change in the agricultural field. Due to the unique properties of nanomaterials, it helps to boost agricultural production and food processing. To use inputs more efficiently, and to head towards sustainable agriculture, a farmer needs to monitor the crop production. Therefore, a constant check on parameters such as moisture content, soil fertility, temperature, crop nutrient capacity, pathogens, plant diseases, etc. is highly required. The use of nano sensors is one way for increasing healthy crop production. Nano sensors assist in the detection of microbes, humidity and toxic pollutants at tiny levels (Kaphle et al. 2018). This is done by accurately measuring the pH of soil, nutrients, residual pesticides and moisture in the soil, examining pathogens, assessment of nitrogen uptake, etc.

4.2 Sensors

Sensor is a device which can sense a property like temperature, pressure, force and convert it into preferred output like colour, electrical and thermal signal. Measuring of these signals helps in the detecting the desired property. Occasionally, signal conditioning unit must be used with the sensor as it may not be able to analyse the obtained signal alone. Signal processing unit maintains the output voltage levels of sensor in the anticipated range with respect to the end device.

Main objective in the agriculture is to regularly monitor and adjust the environments with respect to the requirement of the crop. For this, environmental parameter sensors for measuring the temperature, humidity, amount of carbon dioxide, etc. are required (Chaudhary et al. 2011). Therefore, a network of sensors is used which consists of tiny autonomous devices known as sensor nodes in large numbers. The sensors are designed for collecting information about the climate like light, pressure, temperature, humidity, carbon dioxide, speed and direction of wind.

4.2.1 *Advantages of Nanomaterials*

The improvement of sensitivity and the increase of selectivity are quite easily possible with a decrease in components size, which in turn gives more rapid response. Nanomaterials are characterized by having unique physicochemical properties, including high electrical and thermal conductivity, extremely high surface area/volume ratio, high mechanical strength and even excellent catalytic properties. Nanomaterials can become more efficient by decreasing the sensor and sample sizes, maximizing the number of sensors and reducing the power consumption

Nanotechnology has the potential to increase the productivity in agriculture by improved techniques and sensors being identified for precise detection of pathogens and pollutants in food, convenient management of natural resources, efficient and more reliable delivery system for agro-chemicals (fertilizers and pesticides), enhanced surveillance technique for food safety and agricultural activities, processing and packaging (Pudake et al. 2019).

Nano sensors are beneficial for detecting and updating the real-time information about the product from production site to delivery site of consumers (Patel et al. 2020). Nano sensors are replacing traditional sensors as they are portable, sensitive and specific (Kaushal and Wani 2017). These sensors are extremely sensitive and can detect the desired property very minutely without using sophisticated instruments and skilled workers. Development of nano sensors exclusively for the agriculture (agro-sensors) is an emerging field which helps the farmers for monitoring the diseases and contaminants at early stage. This can prevent their economic losses by controlling the diseases and containments (Saini et al. 2017).

The ideal sensor should possess the following characteristics (Campbell and Compton 2010):

1. Specificity: Selectivity for the target species,
2. Sensitivity: Detecting changes in target species concentrations,
3. Fast response time,
4. Stability: Extended lifetime of at least several months, and
5. Miniaturization: Small size with the possibility of low-cost manufacture as well as increasing number of sensors in array.

4.2.2 *Types of Sensors*

The sensors are classified based on attributes such as stimulus, working principle, properties (attributes of the characteristic) and application.

Main type of sensors based on stimulus are listed below:

a. *Electrical Sensor*: These sensors detect the existence of electrical signals in an environmental input. e.g. metal detectors, RADAR systems.
b. *Magnetic Sensor*: These sensors notice the variation in magnetic signals (magnetic flux) in an environmental contribution.
c. *Optical Sensor*: These sensors use light to quantify the characteristics of any object. They are applied to sense objects lying outside the visible spectrum i.e. ultraviolet and infrared. Electric eye is the commonly known optical sensor. This applies a light beam for detecting the existence of an object. Fibre optic sensors use the principle of total internal refection and travelling through a glass fibre by light. This measures a various characteristic, such as temperature and strain. Motion and photo detectors are used as switches for turning on and off system of lighting.
d. *Chemical Sensor*: Chemical sensors detect the existence of a certain molecule in a location. pH, oxygen, carbon monoxide sensors are some example of chemical sensors
e. *Thermal/Radiation Sensor*: These sensor measures the change/presence of radiation or temperature of the environment.
f. *Mechanical Sensor*: These sensors notice the variation in the mechanical properties of a system or object. In this setup, strain gauge is the primary mechanical sensor. This forms the base of many mechanical sensors such as load cells, humidity sensors and pressure transducers. A strain gauge comprises of adjustable resistor which measures deformation amount in a part during use of force. Potentiometer is another mechanical sensor for measuring the linear or angular displacement.

Various types of sensors employed according to parameters that are required to measure

a. *Light Sensors*: Infrared sensor using as Transmitter or LED, IR receiver sensor for example photodiode, register depends on light
b. *Gas sensors*: Humidity sensor, sensor for toxic gases
c. *Pressure/Force/Weight Sensors*: like strain gauge, and load cells

d. *Position Sensor*: Encoder, potentiometer
e. *Hall Sensor*: To measure the variation in magnetic field
f. *Temperature Sensor*: Thermistor, thermocouple.

Among all the above sensors, present chapter describes the detailed studies on the gas sensors used in the various areas of agriculture. The objective is to summarize the current literature on usage of gas sensors in agriculture and meat production, and their storage and transport applications.

4.3 Gas Sensors

Changing climate and weather conditions for growing crops and to preserve harvested food require highly reliable sensor technologies. There are three major factors on which shelf lives of highly putrescible agricultural foodstuffs for instance fruits, vegetables, baked goods and livestock products depends (Campbell and Compton 2010). They are rate of oxidation, growth of aerobic spoilage microorganisms and attack of insects and pests. Early detection and assessment of the various gases are very important in monitoring of environment and chemical processes. For this purpose, gas sensors/electronic noses are used.

With recent development in nanotechnology, it had created a huge scope for the development of highly specific and sensitive, affordable, small sized and low power consumable portable sensors. High surface area to volume ratio and hollow structures makes nanomaterials perfect for absorption of gas molecules. Thus, gas sensors based on nanomaterials have been widely studied (Jian et al. 2020; Meng et al. 2019). Substantially smaller size, lower mass, more modest power requirements, greater sensitivity, and better specificity are the potential benefits of exploiting microelectronics and nanotechnology in sensor fabrication (Neethirajan et al. 2009). Nanomaterials such as graphene, CNT and MWCNT have been found to be promising materials for gas detection. Their special electronic properties (Mittal and Lin 2020), are strongly affected by the adsorption of gases on their surfaces. These materials have advantage of high stability and show high sensitivity and good reversibility during gas sensing, and stability. Metal, metal oxides and doped metal oxide nanoparticles (Kumar et al. 2017a, 2019) have also been studied in detail for gas sensing.

Nano sensors with functionalized and decorated nanoparticles such as SWCNTs, nanowires, nanofibers to detect different gases like NH_3, NO_2, SO_2 and VOCs, have the incredible scope in monitoring pollutants in agriculture (Wanekaya et al. 2006). For quantifying the environmental pollution, nano-smart dust consisting small wireless sensors and transponders and gas sensors can be used (Mousavi and Rezaei 2011). In this regard, some of our studies had reported graphene modified or intercalated with metal oxides for the detection of NH_3 gases (Jaiswal et al. 2020; Kumar et al. 2017a), also metal nanoparticles and modified OMS-2 material was used for detection of various gases (Kumar et al. 2016, 2017b, 2018).

4.3.1 *Type of Gas Sensors*

The gas sensors are divided into four major categories

4.3.1.1 Solid-State Gas Sensor

A solid-state sensor consists of one or more metal oxides from the transition metals, such as tin oxide, aluminium oxide, etc. A thick or thin film-chip sensor are made by the metal oxides film deposited onto a substrate with an integrated heating element to regulate the sensor temperature, since the finished sensors exhibit different gas response characteristics at different temperature ranges. In the presence of gas, the metal oxide causes the gas to dissociate into charged ions or complexes which results in the transfer of electrons. Solid-state gas sensors have numerous advantages which cause great interest in them. Some of these advantages are, small sizes, high sensitivities in detecting very low concentrations, ability in detecting a wide range of gaseous chemical compounds, possibility of online operation and low cost.

4.3.1.2 Semiconductor Gas Sensors

Semiconductor gas sensors are based on metal oxides such as SnO_2, TiO_2, InO_3, etc. and known as chemo-resistive gas sensors. The gas-sensing mechanisms is based on gas/semiconductor surface interactions such as reduction/oxidation processes, adsorption of the chemical species on the semiconductor or adsorption with surface, chemical reactions between different adsorbed chemical species, etc. These interactions occur at the grain boundaries of the polycrystalline oxide film. These surface phenomena cause changes in electrical resistance (Warneck 1999). The change in electrical resistivity could be perceived and used to detect chemical species. The basic phenomenon involved in gas detection by a film semiconducting metal oxide film using change in its electrical resistance (Jonda et al. 1996). During process, oxygen concentration change caused by the adsorption and oxidation and reduction reaction of oxidized and reduced gaseous species at the surface of metal oxides, is detected. Here, electrical conductivity of sensing material is a function of temperature and the concentration of exposed gas (Eranna et al. 2004). Figure 4.1 shows set up of device using semiconductors for gas-sensing measurements.

4.3.1.3 Optical Gas Sensors

The measurement of changes in absorption spectrum was the first method to detect the desired quantity of the gas molecules. But newest optical gas sensors use other methods including spectroscopy, surface plasmon resonance and interferometry. As shown in Fig. 4.2, a setup using optical method for gas-sensing measurements detects

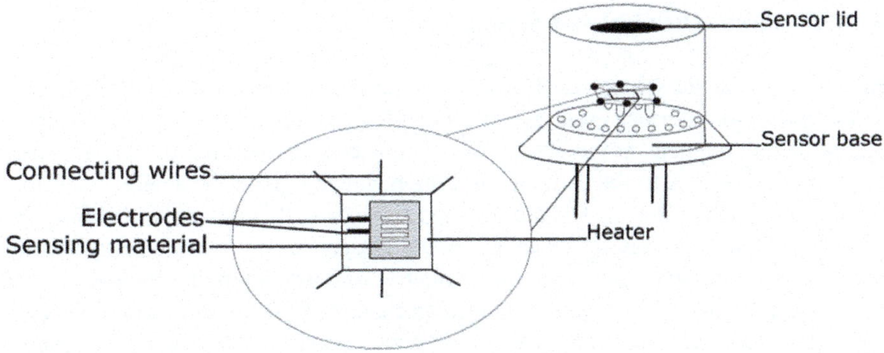

Fig. 4.1 A setup for gas sensing using semiconductors

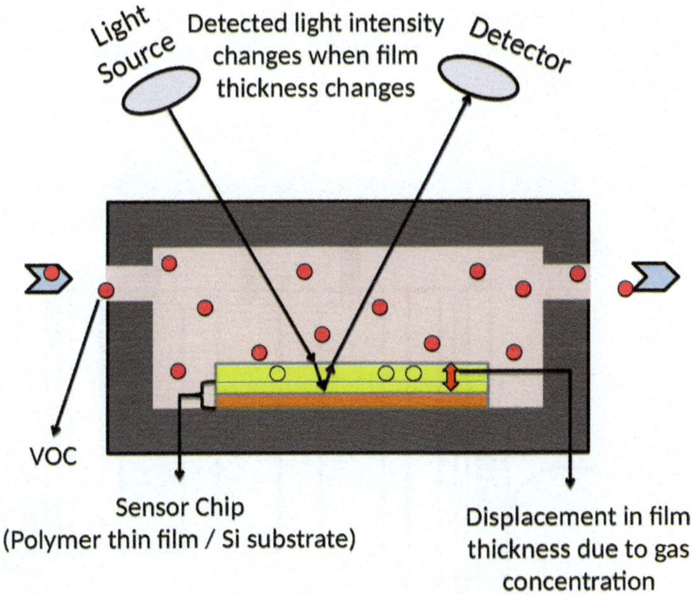

Fig. 4.2 A schematic of setup for gas sensing using optical method

changes in light intensity in sample thickness due to gas concentrations. The optical gas sensors could be used for measuring the chemical and biological quantities. Features of these sensors including safety, low power consumption, remote sensing and multiplexing of sensor arrays are advantages of theses sensors over the other sensors such as electrical gas sensors (Penza et al. 2007). The optical gas sensors are more practical than the other gas sensors.

4.3.1.4 Electrochemical Gas Sensors

Electrochemical gas sensors are based on the fact that chemical reactions at an electronic conductor interface exchange electric charge. Oxidization takes place at anode and reduction happens at the anode. This leads to the generation of current by the flow of positive ions to the cathode, and the negative ions to the anode. Therefore, reducible gases like O_2, and NO are detected at the cathode whereas oxidizable gases like CO, NO_2 and H_2S are sensed at the anode. The gas sensing using electrochemical method is either potentiometric or amperometric which depends on the electromotive force or electrical current output. Potentiometric measurements are carried out under settings of near-zero current. On the other hand, amperometric sensors are usually carried out by using high voltage by external cell to keep a nil-oxygen concentration at the surface of cathodic; here, the response of sensor is diffusion controlled. An important application of electrochemical sensors in agriculture is in the direct measurement of soil chemistry through tests such as pH or nutrient content (Li et al. 2010). Figure 4.3 shows one representative setup using electrochemical method for gas-sensing measurements.

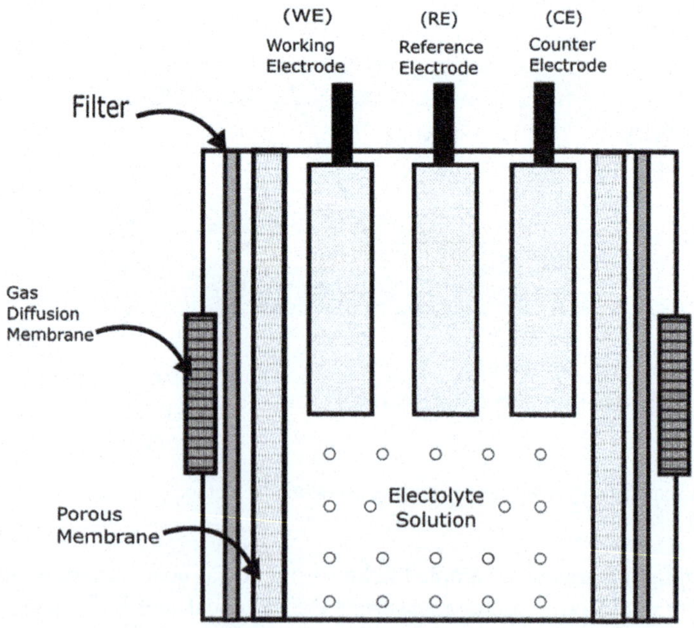

Fig. 4.3 A setup for gas sensing using electrochemical method

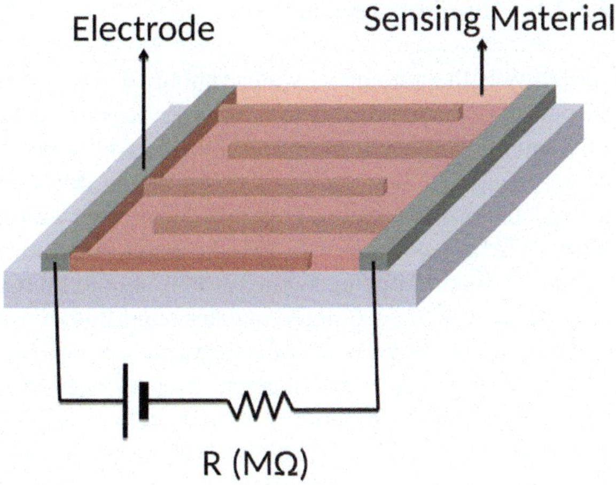

Fig. 4.4 A setup for gas sensing using electrochemical method

4.3.1.5 Chemiresistor/FET Sensors

These sensors have gained recommendable attention in the last couple of decades due to their advantages of quick response time, possible integration in manufacturing processes, miniaturization potential and parallel sensing (Bandodkar et al. 2016). As shown by a set up used for gas-sensing measurements using chemiresistor/FET in Fig. 4.4, sensor material is used between two electrodes. For chemiresistor/FET nano sensor configuration, carbon nanomaterials have been observed to be advantageous as the functional channel. Both carbon nanotubes having tubular and graphene with planar geometry uses maximum exposure of surface atoms to bind the target analyte molecules with the material of electrode. Label-free sensing of analytes with relatively high sensitivities can be carried out using carbon nanostructures (Nehra et al. 2019)

4.3.2 Development of Sensor Materials and Technologies for Gases/Vapours

In agriculture, moisture content as well as the main gases required to be detected, monitor and estimated are carbon monoxide (CO), ammonia (NH_3), oxygen (O_2), nitrogen (N_2) and volatile organic compounds (VOC). The various type of sensor materials and technologies, developed for these gases/vapours are:

4.3.2.1 Humidity Sensors

Humidity is the amount of water vapour in the air. Measurement for air humidity is carried out by the ratio of partial water vapour pressure to saturation vapour pressure and known as relative humidity (RH). Relative humidity shows the probability of occurring fog, dew, rain and precipitation. Most of the crops depend upon these natural factors for water. Therefore, sensitive, and accurate humidity sensors are needed (Srbinovska et al. 2015). As water is the most significant resource in agriculture, irrigation management systems should be installed with more accurate and reliable sensors to get ordered updates regarding moisture in soil at the roots of crops.

Widespread techniques of detections are not capable of providing profiles of accurate temperature and moisture in soil. They are also cumbersome and expensive. Recent advancements in sensor technology have led to the development of affordable, low power, multifunctional sensor nodes. Sensor nodes enable environment sensing together with data processing (Imam et al. 2015). Instrumented with a variety of sensors, such as temperature, humidity and volatile organic (VOCs) compound detection, allow monitoring of different environments. They can network with other sensor systems and exchange data with external users (Ruiz-Garcia et al. 2009).

Electric humidity sensors are divided into three groupings based on their technique for sensing:

1. Humidity sensor based on Thermal conductivity: This sensor utilizes two metallic resistors, out of which one sealed in the closed chamber and other kept in open environment. Using same electrical biasing, the resistors are heated. The resistor which is exposed cools by the humidity of environment. Therefore, in this case, thermal conductivity is a function of humidity. Thermal conductivity difference of two resistors is measure of humidity (Sohraby et al. 2007).
2. Humidity sensor based on resistivity: This sensor utilizes conductive film (may be thin or thick). This film is coated by polymers like polyvinyl alcohol (PVA). Here, number of movable ions move through the film are dependent on humidity of atmosphere. So, the change in impedance depends on relative humidity (Okcan and Akin 2007).
3. Humidity sensor based on capacitance: In this sensor, a capacitor having printed electrodes on polyimide thin films is used. Film's electric constant is related with humidity (Farahani et al. 2014).

Chang and colleagues developed a new type of antenna with an RH (relative humidity) sensing function using a modified polyimide for passive Radio Frequency Identification (RFID) sensing (Chang et al. 2007). Designed to operate at a frequency that depends on the relative humidity level, the proposed antenna is a passive device that physically and functionally combines an antenna with an RH sensor. The compactness and cost-efficiency of the antenna enables it to realize a passive tag of the RFID sensing without an additional sensor component (Pelegrí-Sebastiá et al. 2011).

4.3.2.2 Carbon Monooxide Gas Sensor

Carbon monoxide (CO), a gaseous second messenger in animals, arises in biological systems mainly during the oxidative catabolism of heme by the heme oxygenase (HO) enzymes (Raimundo Jr and Narayanaswamy 2001). Resolution of the response of CO fluxes to land use is essential since changes in the CO budget due to changes in land use could either ameliorate or exacerbate the well-known, greenhouse-reinforcing impacts of agriculture on methane and nitrogen oxides (King 2000). CO dynamics may play a significant role in strategies to mitigate greenhouse warming since CO affects the fate of methane (King 2000). Net CO consumption for a given soil type increased with decreasing organic matter content associated with forest to agriculture transitions in land use. Although interactions among soil organic matter and various microbiological, physical and chemical parameters in soils are complex, changes in organic matter at the sites described here appear to affect net CO fluxes primarily by reducing the relative rate of a biological CO production (King 2000).

Semiconducting metal-oxide sensors have been studied using metal oxides like SnO_2, ZnO, TiO_2, MoO_3 and Fe_2O_3 to produce sensors with low cost, highly sensitive and fast response time (Menil et al. 2000; Seal and Shukla 2002; Yamazoe and Miura 1995). ZnO is an interesting chemically and thermally stable n-type semiconductor with exciton binding energy and bandgap energy of 60 meV 3.37 eV, respectively, at room temperature. With these properties, ZnO nanowire in presence of noble metal elements (Pt, Pd, Au, etc.) on the surface of metal oxides can enhance the interaction of reducing gas with the absorbed oxygen on the surface (Menil et al. 2000). Recently our team had developed the Cu-doped OMS-2 and Nb-doped OMS-2 nanofibers for the detection of CO gas (Kumar et al. 2018, 2019). Table 4.1 list the response of some nanosensor material and their operating temperature studied for the sensing of different concentration of CO gas.

Table 4.1 List of some nanomaterials used for CO gas sensing

CO -Sensors	Operating temperature (K)	Response time, gas concentration	Response (%)	Reference
SnO_2 nanosheets	673	6 s, 500 ppm	1.5	Li et al. (2019), Moon et al. (2008)
Stabilized Zirconia	873	10 s, 100 ppm	90%	Liu et al. (2015)
Au-decorated SnO_2 nanobelts	523–673	30 s, 10 ppm	90%	Qian et al. (2006)
Pd-decorated In_2O_3	300	50 s, 100 ppm	85%	Lai and Chen (2012)
Cu-doped OMS2	296	60 s, 100 ppm	20%	Kumar et al. (2018)
ZnO nanowires	593	100 s, 500 ppm	60%	Choi and Kim (2012)

4.3.2.3 Ammonia Gas Sensors

Ammonia pollution also impacts species composition through soil acidification, direct toxic damage to leaves and by altering the susceptibility of plants to frost, drought and pathogens (including insect pests and invasive species) (Guthrie et al. 2018). The volatilization of ammonia (NH_3) from livestock manure has been the focus of many studies because it is a loss of nitrogen for crop production and can also have an adverse impact on the environment. Above threshold levels, NH_3 can damage vegetation. Ammonia affects freshwater ecosystems through direct agricultural run-off leading to eutrophication (accumulation of nutrients, leading to algal growth and oxygen depletion) and also has toxic effects on aquatic animals that often have thin and permeable skin surfaces (Guthrie et al. 2018).

Ammonia gas sensor based on chemically reduced graphene oxide (rGO) sheets by self-assembly technique to create conductive networks between parallel Au electrodes was studied (Wang et al. 2014). This new gas sensor showed excellent responsive repeatability to NH_3 and possess low-cost portable characteristics. Zhang and co-workers observed that upon exposure to NH_3, the conductance of the semiconducting SWNT changed dramatically (Zhang et al. 2006). An increase in the conductance by three orders of magnitude was observed within several seconds. The change is irreversible (Kong et al. 2000). When polyethyleneimine (PEI), Nafion (a polymeric perfluorinated sulfonic acid ionomer) is used for sensor studies with SWNTs for NO_2 and NH_3 sensing up to 100 ppm shows reversible response with response time of 2 min (Qi et al. 2003). Au, Pt MWNTs based sensors also studied for NH_3 sensing (Penza et al. 2007). Renganathan and his team proposed a nanocrystalline ZnO coated modified fibre optic sensor for NH_3 gas detection (Renganathan et al. 2011).

Ammonia gas sensors using metal oxides such as SnO_2, ZnO and TiO_2 are also reported. These sensors are easily to construct and shows better sensitivity towards NH_3 (Karunagaran et al. 2007; Patil and Patil 2007; Renganathan et al. 2010; Van Hieu et al. 2008; Wagh et al. 2006; Wang et al. 2001). Resistance of these sensor changes with the exposure to a particular gas and the sensors are described as electrical resistive type. But high sensitivity of these sensors works only at higher temperatures (>200 °C) (de Lacy Costello et al. 2008; Moon et al. 2004; Pourfayaz et al. 2005; Tang et al. 2006). Selectivity of these sensors is also a problem as they respond to many gases such as carbon monoxides, methanol, ammonia and methane (Wagh et al. 2006). Doping (Epifani et al. 2008; Esfandyarpour et al. 2004; Ruiz et al. 2005; Srivastava and Jain 2007), or variation in size (Lee and Park 2004; Nunes et al. 1999) or annealing (Dikovska et al. 2010; Hongsith et al. 2008; Vaezi and Sadrnezhaad 2007) of these sensors changes their physical properties so that they can work at lower temperatures with improved the gas sensitivity and selectivity.

Most frequently used detectors, for ammonia gas use conducting polymer, metal-oxide. Polymer ammonia sensors use thin film coating of polyaniline (Imamura and Yumoto 2008; Kohl 2001; Timmer et al. 2005) Nafion (Raimundo and Narayanaswamy 2001) and polypyrrole (Lähdesmäki et al. 1996). The lowest ammonia detection limit found in literature is 1 ppm and response time is ~1 min (Coutinho et al. 2007). In conducting polymer, gas detection involves an irreversible reaction between ammonia and the polymer. The irreversible nature of the reaction deteriorates the sensitivity over time when exposed to ammonia. These sensors have poor selectivity, sensitivity and are generally slower. Further, in catalytic ammonia sensors the charge carrier concentration in catalytic metal is altered by change in concentration of the gas of influence. This change in charge carries can be quantified using a field effect device. The selectivity of catalytic ammonia sensors varies with the variation of catalytic metal used in device and operating temperature.

Our group used various nanomaterials such as Sn nanoparticles (Kumar et al. 2016), metal oxides modified graphene (Jaiswal et al. 2020; Kumar et al. 2017a) and metal doped OMS-2 (Kumar et al. 2017b) for the sensing of ppm level of ammonia. Chemisorption based, MOS sensors are an economical type of gas sensors that are stable during long time exposure to NH_3 gas. These sensors can detect ammonia from small concentrations (<30 ppm) to very high concentrations (27% of air).

However, for sensing the gas in correct range, calibration of attenuated output signal must be carried out in the desired range. But this reduces accuracy and resolution at low concentrations. Another concern with chemisorption sensors is the hassle in interpretation of readings, the possible incorrect alarms, and effect of humidity on the sensor. With the variation of humidity, the sensor characteristics, and the output changes, it increases with an increase in humidity and decreases with the decrease in humidity. Even it can fall to null due to the presence of measured contaminants. Lifespan of the sensor is directly related to its exposure to NH_3. So, these are useful only in low concentration measurement applications. Table 4.2 lists the response of some sensor materials and their operating temperature studied for the sensing of different concentration of NH_3 gas.

Table 4.2 List of some graphene-based sensor materials studied for the sensing of NH_3 gas

Sensor	Response (%)	Operating temperature	Response (Sec)	Recovery (sec)	Reference
SnO_2-graphene	7	15	<60	<60	Feng et al. (2017)
Chemically rGO sheets on Au electrodes	14	27	~500	125	Wang et al. (2014)
RGO-SnO_2 10:8	9	27	~60	180	Feng et al. (2017)
SnO-graphene	21	27	15	30	Kumar et al. (2017a)
SnO_2/MWCNTs	20	27	100	210	Van Hieu et al. (2008)

4.3.2.4 Carbon Dioxide Sensors

Commonly, CO_2 gas is ejected into the food package as the excessive level of CO_2 levels in the headspace leads to reduction of the rates of metabolism, even during the existence of O_2. The reduction of carbon dioxide gas levels from the original in the headspace symbolizes the sign of leakage in the package. Besides this headspace CO_2 level may also help in spot ripeness or spoilage of agricultural products. As there is a continuous generation of CO_2 in the headspace due to growth of microorganisms and any other biological activity (Meng et al. 2014)

Severing-Haus reported the electrochemical principle based first carbon dioxide gas sensor. Using potentiometric sensors for gas detection, the potential gradient is calculated by using a reference and working electrodes (Holzinger et al. 1997). A glass pH-electrode is used to determine the pH of the thin electrolyte layer, which in last depends upon the amount of CO_2 diffused through the permeable membrane (e.g. rubber, Teflon or polyethylene). At last, the partial pressure of the gas is indirectly helping to determine the pH of the potentiometric sensor.

For ideal optical detection of carbon dioxide, the indicators are prepared by pigments (commercially available) (Lv et al. 2014; Schutting et al. 2013)

Von Bültzingslöwen and colleagues had developed an optical sol–gel-based CO_2 sensor strip using a fluorescent pH indicator. The sensor membrane comprises a ruthenium dye, filled in polymer nanobeads, and utilizing the ratiometric method, the fluorescence intensity is measured using the analyte. This sensor strip can be located inside a food package or can be incorporated by a handheld scanning device from the outside (von Bültzingslöwen et al. 2002).

4.3.2.5 Oxygen Sensors

The concentration of O_2 of the headspace impacts the shelf life and quality of agro related products which weakens by excessive O_2 quantity. Thus, O_2 gas sensors are used in agro-industry to show the presence of O_2 and measure the gas into the dissolved state (Meng et al. 2014).

As plants are immobile organisms; thus, for essential nutrients supply, plants depend on the direct environment. In the night, photosynthetic process of the plants is inactive, therefore, plants including their non green tissue like roots depends on environmental oxygen. Additionally, due to unavailability of system of O_2 distribution, plants use tissue for O_2 diffusion. As consumption rate of oxygen increases in comparison to the imparting environment O_2, the internal oxygen concentration of plant may reduce lower than ambient. According to investigation, the O_2 gas amount of cells in the tissue of plant can withstand 5 times lesser concentration of O_2 to the surface of the plant (Licausi et al. 2011; van Dongen et al. 2003). The root system can be damaged by deficiency of O_2, which restrains nutriment and ultimately growth of the plant. O_2 is a main metabolite in plants. Its deficiency induces rearrangements in metabolic and morphology of plants (Miura et al. 2011).

With the recent development of oxygen sensors, it has enhanced the understanding of oxygen distribution and consumption within the plant tissue. Still, the quantification of oxygen in the tissue of plant encircles difficulties that don't happen with the animal cells (Ast et al. 2012). Organometallic compounds have been widely used in dye structure due to their extended lifetime and intense luminescence makes them a best-suited candidate for oxygen indicator dyes in oxygen nanosensors (Ast et al. 2012). For carrying out the intracellular studies for various analytes including O_2 in macrophages, Clark and colleagues have developed sensors (Clark et al. 1998).

Gases such as oxygen, chlorine and nitrogen dioxide exhibit an intense paramagnetic effect (Kocache 1986). The gases which are attracted by the magnetic field are commonly known as paramagnetic gases. The basic principle of the thermomagnetic oxygen sensor is magnetic susceptibility. The molecular oxygen is forced to flow through an intense and heterogeneous magnetic field of the sensor, the heating of wire is used for increasing the temperature. This leads to decline of magnetic field (Uetake et al. 2000). Oxygen sensing by TiO_2, Nb_2O_5, and CeO_2 semiconductor like uses the change in resistance. The change of oxide semiconductor through the redox reaction occurs because of partial pressure of oxygen in the surroundings (Takeuchi and Igarashi 1988).

Membrane oxygen electrodes, also known as Clark electrodes, is the most commonly known oxygen electrodes. Comprising Ag as an anode, Pt as cathode and a circular ring is used as a reference electrode. Polytetrafluoroethylene/polyethylene film with a thickness of 15–25 μm is regarded as the most suitable permeable membrane for O_2, it also covers electrodes surface. KCl acts as a supporting electrolyte. This electrolyte helps in the formation of the permeable membrane for O_2 and impenetrable for all others (Nei and Compton 1996). Tracking of the dynamic change is possible with continuous measurement (Koudelka 1986). Generally, optical sensors for O_2 provides luminescence's intensity change from a probe molecule and show the change in concentration of O_2. Since molecular O_2 can reversibly quench the luminescence, this is a reversible luminescence effect (Coutinho et al. 2007).

4.3.2.6 Nitrogen Sensors

Nitrogen (N) element is considered as the most important most concerning nutrient element for maintaining healthy environment for the growing crops (Lee and Searcy 2000). Its accurate assessment in plants is a key to nutrient management for the crop. N_2O is produced and consumed by the soil microbes during nitrification and denitrification (Williams et al. 1992). The interplay between the amount of nitrogen cycling in soils and soil environmental conditions (mainly soil moisture, temperature, pH, oxygen and organic carbon concentration) governs microbial processes and thereby gaseous nitrogen production and consumption (Weitz et al. 2001). The reactive nitrogen escaping from agricultural soils, including groundwater contamination, eutrophication of freshwater and estuarine ecosystems, tropospheric pollution due to nitrogen oxides and NH_3 emissions, and accumulation of the greenhouse gas and stratospheric ozone depleting substance, nitrous oxide (Aber et al. 2003; Galloway et al. 2008; Reay et al. 2012) are the unintended adverse environmental and human health impact.

Emission growth in N_2O is also enhanced by the widespread use of nitrogenous fertilizers and manure. Need for more food production as consequence of increasing population, both agricultural land area and N_2O emissions are likely to continue to rise (Reay et al. 2012). These facts indicates that, lack of nitrogen means lower crop productivity, poor human nutrition and soil degradation (Sanchez and Swaminathan 2005), but excess N leads to environmental pollution and its concomitant threats to agricultural productivity, food security, ecosystem health, human health and economic prosperity (Zhang et al. 2015). So, to assess nitrogen stress, the requisite is to implement a real-time nitrogen sensor system.

4.3.2.7 Sensors for Volatile Organic Compounds

Plants and trees normally release volatile organic compounds (VOCs) as a by-product of everyday physiological processes. VOCs released by the plant are signature of crop and filed condition (Li et al. 2010). This emission become higher in amount and variety when plants are attacked by pests and diseases than non- attacked plants (Dudareva et al. 2006). The emission of these VOCs is not limited to the sites of injury, but also from elsewhere on the plants (Dudareva et al. 2006). Moreover, the VOC profiles emitted by plants during attack display a degree of specificity corresponding to the type of attackers. Further these profiles are markedly distinct from each other. This change in emission of VOCs can help determine the spread of disease in plants. The VOCs released by plants contribute around two-third of total emission of VOCs in environment (Guenther 1997). The range of VOCs released due to biotic and abiotic interactions were compiled earlier reports. Commonly found secondary plant volatiles are fatty acids, phenylpropanoids, amino acid volatiles and benzenoids (Dudareva et al. 2006).

The electronic nose is a highly complex device consisting of an array of various chemical sensors that are coupled to a state-of-the-art pattern-recognition program that emulates the human olfactory system. The individual sensors which have a defined electrically conductive adsorptive surfaces for VOC detection that, on interacting with volatile chemicals, display a change of electrical resistance. Even though each individual sensor responds selectively to a certain group of chemicals, they all show an overlapping response pattern called cross-selectivity. For each complex aroma, the array of sensors produces a unique response pattern called a "fingerprint" (Pinheiro et al. 2002).

VOC bouquets emitted by cucumber, pepper and tomato leaves subjected to mechanical damage or pest and disease attacks compared with undamaged control leaves. e-nose can discriminate among VOCs from undamaged leaves. It can also discriminate undamaged and artificially damaged leaves of the same plant species. In cucumber, the e-nose can discriminate among VOCs emitted from control, artificially damaged, and spider-mite-infested leaves (Laothawornkitkul et al. 2008). In another study, a conducting polymer sensor-based electronic nose was applied successfully to detect the presence and differentiate among three postharvest diseases in blueberries (Li et al. 2010).

Electronic nose (e-nose), also known as artificial olfactory, is a simulation of biological functions to identify some simple or complex odour (Gardner and Bartlett 1994). The electronic nose has derived its name because it in several aspects tries to resemble the human nose. The electronic nose is a device that detects the smell more effectively then the human sense of smell. A typical e-nose system contains a selective chemical sensor array, a signal processing and a pattern recognition subsystem is shown in Fig. 4.5. The sensors in the sensor array are sensitive to different substances, therefore an e-nose can extract the whole information of substances for identification (Wang and Li 2010).

E-nose is described as an array of chemical gas sensors with a broad and partly overlapping selectivity for measurement of volatile compounds within the headspace over a sample combined with computerized multivariate statistical data processing tools. An electronic nose consists of a mechanism for chemical detection. The electronic nose is an intelligent sensing device that uses an array of gas sensors which

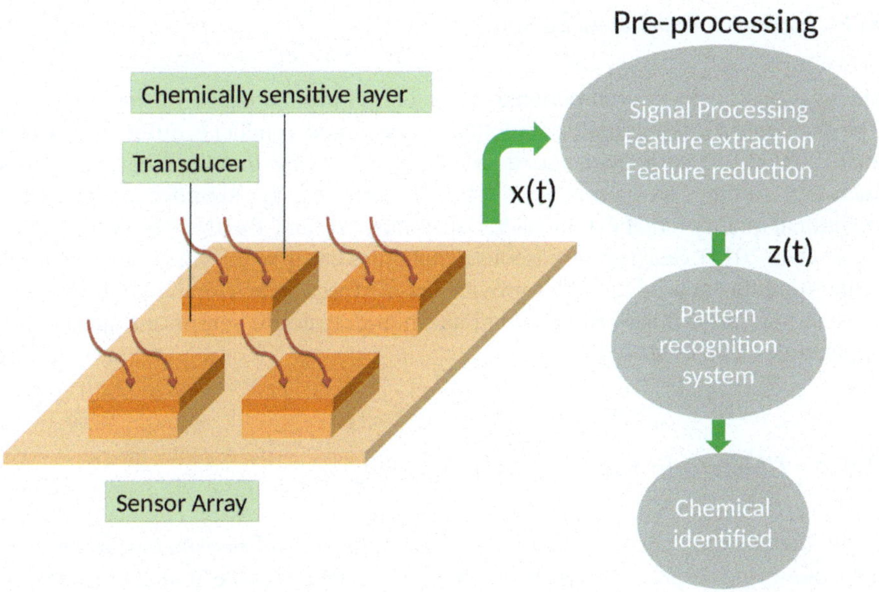

Fig. 4.5 E-nose system contains a selective chemical sensor array, a signal processing, and a pattern recognition subsystem

are overlapping selectively along with a pattern reorganization component. The electronic nose is primarily composed of gas sensor array, signal pre-processing, and pattern recognition. Gas molecules are presented in front of a sensor of an active material that converts the chemical input into an electrical signal. The response of a plurality of sensors to an odour constitutes the response spectrum of the sensor array to the odour. To achieve qualitative or quantitative analysis of the odour, the sensor signal must be properly pre-treated and processed by appropriate pattern recognition analysis (O'Farrell et al. 2007). The electronic nose is used to analyse and evaluate the odour quality according to the principle that each of the gas sensors has a response to complex component gases but different from each other.

Emission of ethylene increases with tissue senescing and ripening of fruits. A hybrid electronic nose was developed to detect the onset of biotic and abiotic stresses (Baratto et al. 2005). During the ripening of apples, trans-2-hexenal can serve as an indicator compound, because its concentration increases significantly. Accordingly, monitoring the development of this aldehyde enables monitoring of the postharvest ripening of apples (Herrmann et al. 2002). Henderson and colleagues developed an e-nose to detect stink bugs presence and to detect the damage in cotton by detecting VOCs emitted by stink bugs (Henderson et al. 2010).

4.4 Sensors for Meat Industry

Odour is one of the most important parameters for evaluating the freshness of meat. The characteristics of volatile compounds in each meat product is different therefore each meat has its own characteristic odour. Studies show that the biogenic amines, such as tyramine, tryptamine, putrescine, cadaverine, are significantly related to traditional quality indices (total bacterial counts, pH, and TVBN1) in meat products (Xiao et al. 2014). Sensing the gaseous metabolites can give information on the status of the stored meat products. This provide the platform of developing electronic nose sensing systems, which provide rapid and non-destructive sensors for meat safety applications (Shi et al. 2018).

4.4.1 Electronic Nose for Meat Quality

Main factors that are considered by consumers when choosing meat products are colour and aroma, of which the latter is a more reliable indicator of quality (Wojnowski et al. 2017). The biotic factors, ambient temperature, humidity, and transportation are causative factors of meat quality degradation (Nychas et al. 2008). The spoiled meat products consumption can trigger serious health hazards. So, it is very important to develop the mechanism to monitor and assess the quality of meat products (Wijaya et al. 2017). Volatile organic compounds (VOCs) (alcohols, aldehydes, ketones, fatty acids, esters, and sulphur compounds) also generated during meat storage and which can have an olfactory impact as well (Casaburi et al. 2015; Nychas et al. 2008).

The main applications of e-noses in meat industry are in spoilage detection of off-flavours, classification, and estimation of shelf life. There are three principal mechanisms of meat spoilage, namely microbial spoilage, lipid oxidation and autolytic enzymatic spoilage (Dave and Ghaly 2011). This can lead to a change in the composition of the volatile fraction of a meat sample, which in turn can lead to the identification of potential meat spoilage markers (Nychas et al. 2008). The main ingredients like carbohydrates, protein and fat will be decomposed by enzymes and bacteria, producing many volatile organic compounds (VOCs): the carbohydrates will be decomposed into hydrocarbons, alcohols, ketones, aldehydes and carboxylic acid gases; the protein will be decomposed into ammonia, hydrogen sulphide, ethyl mercaptan, etc. the fat will be decomposed into aldehydes and aldehyde acids. Total volatile basic nitrogen (TVB-N) content is an important reference index for evaluating pork freshness. The sensors array is chosen on the basis of their sensitivity towards the VOCs in pork meat, and they are composed of eleven different metal-oxide semiconductor (MOS) sensors positioned in a small chamber (Huang et al. 2014). The E-nose real-time responses were recorded by the ratio of conductance.

The development of Mobile e-nose for beef quality monitoring and detection has been studied (Wijaya et al. 2017). The performance of a portable quartz microbalance based electronic nose has been evaluated in monitoring aerobically packaged beef fillet spoilage at different storage temperatures. The collected signal responses could be considered as a volatile fingerprint of an active biological system, containing information for discrimination of meat samples in sensory classes corresponding to different spoilage levels (Papadopoulou et al. 2013).

Conducting polymer electronic nose was evaluated for the capability of detecting the presence of off-flavour malodorous compounds in catfish meat fillets to assess meat quality for potential merchantability. Sensors responses were measured as a percentage of electrical resistance changes to current flow in the sensors relative to baseline resistance (%ΔR/Rbase). The sorption of headspace volatiles, composed of specific VOC mixtures, to the conducting polymer sensor surfaces induce a change in the electrical resistance to current flow which is detected by a transducer to produce the output from the sensor array (Wilson et al. 2013). A study on the use of a MOS-based electronic nose prototype for spoilage classification of red meat, namely beef and mutton was carried out. Using the sensor array system, it was possible to discriminate between spoiled and unspoiled meat with a success rate of 98.87% in the case of beef, and 96.43% in the case of mutton (El Barbri et al. 2008).

4.4.2 Optical Sensor for Meat Quality Monitoring

An optoelectronic array of sensors is developed for monitoring fresh pork inside MAP. The array is based on the combination of pH indicators and selective chromogenic reagents supported on inorganic materials with diverse acidities and topologies. This approach results in a final array containing seven materials as a suitable system capable of monitoring meat spoilage of fresh pork under MAP atmosphere via simple colour changes. It is based on 16 pigments prepared by the incorporation of different dyes (pH indicators, Lewis acids, hydrogen-bonding derivatives, selective probes and natural dyes) into inorganic materials (UVM-7, silica and alumina). The colour changes of the sensor array were characteristic of chicken aging in a modified packaging atmosphere (30% CO_{2-7}0% N_2). The chromogenic array data were processed with qualitative (PCA) and quantitative (PLS) tools (Salinas et al. 2014).

4.4.3 Colorimetric Sensors

The colorimetric sensor array is designed based on two fundamental requirements. Chemo responsive pigment, which must contain centre (functional group) to interact strongly with analytes, and then this interaction centre must be strongly coupled to an intense chromophore. The first requirement implies that the interaction must not be simple physical adsorption, but rather must involve other, stronger chemical interactions (Xiao et al. 2014).

For instance, the freshness of beef degrades because microbial spoilage and biochemical reactions occur during storage. The main ingredients like protein, fat and carbohydrates are decomposed by enzymes and bacteria, producing odour: the protein will be decomposed into ammonia, hydrogen sulphide, ethyl mercaptan; the fat will be decomposed into aldehydes and aldehyde acids; the carbohydrates will be decomposed into alcohols, ketones, aldehydes and carboxylic acid gases (Smolander 2003).

Chemo responsive pigments are those pigments that change colour in either reflected or absorbed light, upon changes in their chemical environment. Image processing is used to assess the meat freshness. Images are taken from meat to identify regions of interest. Based on the colour of the meat, these regions processed to produce a RGB colour index (Dini et al. 2010; Zaragozá et al. 2012). One of the drawbacks of this method is that the relationship between colour processing and the degree of freshness cannot be determined accurately

The employment of natural pigments in colorimetric sensors is advantageous, because such sensors do not have chemical effect on the packaged meat. Xiao and colleagues developed a colorimetric sensor array based on four natural pigments, that are extracted from spinach, red radish, winter jasmine and black rice for the sensitive detection of amines generated in pork meat during its gradual abiotic decomposition (Xiao et al. 2014). Based on the investigation into the structure and properties of pigment molecules, they discover that black rice extract is the most sensitive of the extracts. This is due to the better interaction between the carbonyl and hydroxyl groups of anthocyanin molecules in the black rice extract with the amines generated during spoilage.

4.4.4 Gas Sensor for Meat Industry

Gas sensors are devices that respond reversibly and quantitatively to the presence of a gaseous analyte by changing the physical parameters of the sensor and are monitored by an external device (Kerry et al. 2006). An opto chemical CO_2 sensor (Mohebi and Marquez 2015) which uses a phosphorescent reporter dye and a colorimetric pH indicator showed robust optical responses to CO_2. The sensor is designed as a film coating to be applicable in meat packaging. The results show that the sensor is stable for at least 14 days at 4 °C and 50 days at 20 °C, however, they deteriorate within 7 days by losing colour and sensitivity to CO_2 at room temperature. Most of the well-known gas sensors such as carbon dioxide sensors are susceptible to consumption of analyte (oxygen) (Kolle et al. 1997). Due to this fact, optical oxygen sensors are superior to conventional electrochemical sensors, such as metal oxide semiconductor field effect transistors, organic conducting polymers and piezoelectric crystal sensors. Oxygen sensors are less complicated than e-noses and easy to be adopted into the meat packaging industry. These sensors are employed to compare the oxygen content in MAP and vacuum-packed beef as a non-destructive method to check the impact of oxygen content on lipid oxidation (Klimant et al. 1997). The oxygen sensors monitor the changes in oxygen levels in all samples, which have been stored at 4 °C for 15–35 days. The initial oxygen content of 1.15% in MAP and 0.07% in vacuum beef samples is reported, which increased to 1.26% in MAP and 0.55% in vacuum beef samples. The results demonstrate that the samples containing greatest levels of oxygen are the most oxidized. Moreover, it suggests that a certain level of oxygen is required for lipid oxidation and the extra oxygen has no effect on further lipid oxidation. The commercialized version of oxygen sensors is cheap and suitable for large scale production to be used in each individual meat package.

4.4.5 Meat Quality Control Technologies

As discussed above, various sensors are being used for controlling the quality of meat. Advantages and disadvantages of these sensors of these technologies are listed in Table 4.3.

Table 4.3 Meat quality control technologies

Sensor/Indicator	Advantages	Disadvantages
Time temperature indicator	Easy integration into packaging system, cheap and economical, can be checked by naked eye or measured by electronic devices	No information about quality of food, must be conditioned before use
Gas indicator	Easy integration into packaging system, cheap and economical, can be checked by naked eye or measured by electronic devices	No information about gas concentration and its chemical dye may interfere in food quality
Freshness indicator	Sensitive, can be checked by naked eye, cheap and economical, can be measured by electronic devices	Not reliable attached inside the package which may interfere in food quality
Biosensors	Easy integration into packaging system, cheap and economical, can be checked by naked eye or measured by electronic devices pathogen and microbial detection	Cannot detect low concentrated contamination, may have chemical effect on the food
Gas sensors	Sensitive, can be integrated into the packaging films, high spatial resolution, can be checked by naked eye and optical devices not affected by heat, electromagnetic and stirring	Nil
E (electronic)-nose	Accurate since uses an array of and sensors, can detect many kinds of volatile compounds, can be integrated into pattern recognition and decision-making systems	Expensive for packaging, fabrication, and commercial purpose
Fluorescence-based (oxygen)	Sensitive, concentration of can be checked by optical device, can be used in headspace or in liquid, fast and reliable	Cannot be checked by naked eye
RFID tags	Accurate, can be integrated into barcodes, wireless technology, and fast	Expensive

4.5 Sensors for Storage and Transport of Food

The agriculture and food industry have witnessed rapid and significant development in packaging technology in the last several decades, including particularly packaging materials, packaging methods, packaging equipment etc. (Moon et al. 2004). Nanosensors are far from being simply a passive, information-receiving device. They can get information from immediate and remote contexts and can analyse, record and report data (Patel et al. 2020). The part of food production where the need for the nanosensors is the most visible is food packaging and food transport (Patel et al. 2020). Food packaging prevents sensory exposure from the foods thus consumers must rely on expiry dates provided by producers based on a set of idealized assumptions about the way that the food is stored or transported. If the transport or storage conditions are violated for any period of time, the quality of food might be deteriorated which might not be known to the consumer unless the food package is opened, or even consumed (Sun and Fung 2006). Nanosensors are able to detect the presence of gasses, aromas, chemical contaminants, pathogens and even changes in environmental conditions, hence ensure that consumers purchase fresh and tasty products and reduce the frequency of food-borne illnesses which improve food safety (Patel et al. 2020).

Due to its small size this type of sensor can also be designed to detect multiple toxins and microorganisms by putting a series of cantilevers with different molecular recognition elements together in an array mounted on a single chip (Lange et al. 2002; Ziegler 2004). There are a number of nanosensors being developed that have applications in food, either for rapid detection or as nano tracers to show the history of the food product and whether it is of acceptable quality at any given time (Joyner and Kumar 2015).

Nanotechnology derived food packaging materials are the largest category of current nanotechnology applications for the food sector. Following main applications for food contact materials (FCMs):

- FCMs incorporating nanomaterials to improve packaging properties (flexibility, gas barrier properties, temperature/moisture stability).
- "Active" FCMs that incorporate nanoparticles with antimicrobial or oxygen scavenging properties.
- "Intelligent" food packaging incorporating nanosensors to monitor and report the condition of the food.
- Biodegradable polymer–nanomaterial composites (Mabeck and Malliaras 2006).

Due to very large aspect ratios, a relatively low level of nanoparticle is sufficient to change the properties of packaging materials without significant changes in density, transparency and processing characteristics (Chaudhry et al. 2008).

For example, nanocomposites have been reported to have improved properties with regard to durability (higher modulus, tensile strength, and elongation properties, rephrase) (Lei et al. 2006), temperature resistance, barrier properties (Lei et al. 2006), optical properties (mechanical, optical: The stiffness and impact strength) (Alexandre and Dubois 2000), processability due to lower viscosity (Wan et al. 2003).

Electronic noses are thought to emerge as a possibility for aroma profile analysis of wines and various food items with aroma as significant signature. The electronic nose consists of an array of gas sensors with different selectivity, a signal-collecting unit and pattern recognition software. Garcıa and colleagues, studied in same area for differentiating various types of wines electronic nose based on metal-oxide semiconductor sensors. The development of ultra-low-cost humidity sensors for large-scale packaging or storage of food and for numerous household applications has created an urgent demand for electrical-circuit free, consumer goods-integrated sensor systems (García et al. 2006). Luechinger and colleagues demonstrated the use of ultra-low-cost carbon-coated copper nanoparticles for the preparation of porous, tensile- and wetting-agent-loaded thin films for optical humidity sensing. Nanosensors based on nanoparticles have also been developed to detect the presence of moisture content inside a food packaging (Luechinger et al. 2007).

4.6 Conclusion

Studies show that nanomaterial-based gas sensors can contribute to crop production and protection in a big way. They can help in constant checking of parameters such as moisture content, soil fertility, temperature, crop nutrient capacity, pathogens, plant diseases, etc. Use of these sensors are found to be very effective in increasing the healthy crop production. Beside crops, nanomaterial-based gas sensors help in increasing the quality and quantity of meat production. These sensors also provide great help in storage and transport applications of food and meat. As a major challenge for the future, comprehensive studies on using the nano gas sensors in fields will be important for increasing the crop production for feeding the vast and increasing population.

References

Aber JD et al (2003) Is nitrogen deposition altering the nitrogen status of northeastern forests? Bioscience 53:375–389

Alexandre M, Dubois P (2000) Polymer-layered silicate nanocomposites: preparation, properties and uses of a new class of materials. Mater Sci Eng R Rep 28:1–63

Ast C, Schmälzlin E, Löhmannsröben H-G, Van Dongen JT (2012) Optical oxygen micro-and nanosensors for plant applications. Sensors 12:7015–7032

Bandodkar AJ, Jeerapan I, Wang J (2016) Wearable chemical sensors: present challenges and future prospects. Acs Sensors 1:464–482

Baratto C et al (2005) Monitoring plants health in greenhouse for space missions. Sens Actuators B: Chem 108:278–284

Campbell FW, Compton RG (2010) The use of nanoparticles in electroanalysis: an updated review. Anal Bioanal Chem 396:241–259

Casaburi A, Piombino P, Nychas G-J, Villani F, Ercolini D (2015) Bacterial populations and the volatilome associated to meat spoilage. Food Microbiol 45:83–102

Chang K, Kim Y-H, Kim Y-J, Yoon YJ (2007) Functional antenna integrated with relative humidity sensor using synthesised polyimide for passive FID sensing. Electron lett 43:259–260

Chaudhary D, Nayse S, Waghmare L (2011) Application of wireless sensor networks for greenhouse parameter control in precision agriculture. Int J Wireless Inf Networks 3:140–149

Chaudhry Q et al (2008) Applications and implications of nanotechnologies for the food sector. Food Addit Contam 25:241–258

Choi S-W, Kim SS (2012) Room temperature CO sensing of selectively grown networked ZnO nanowires by Pd nanodot functionalization. Sens Actuators B: Chem 168:8–13

Clark HA et al (1998) Subcellular optochemical nanobiosensors: probes encapsulated by biologically localised embedding (PEBBLEs). Sens Actuators B: Chem 51:12–16

Coutinho CF et al (2007) Ammonium ion sensor based on SiO2/ZrO2/phosphate-NH4 + composite for quantification of ammonium ions in natural waters. J Braz Chem Soc 18:189–194

Dave D, Ghaly AE (2011) Meat spoilage mechanisms and preservation techniques: a critical review. Am J Agric Biol Sci 6:486–510

de Lacy Costello B, Ewen R, Ratcliffe NM, Richards M (2008) Highly sensitive room temperature sensors based on the UV-LED activation of zinc oxide nanoparticles. Sens Actuators B: Chem 134:945–952

Dikovska AO, Atanasova G, Nedyalkov N, Stefanov P, Atanasov P, Karakoleva E, Andreev AT (2010) Optical sensing of ammonia using ZnO nanostructure grown on a side-polished optical-fiber. Sens Actuators B: Chem 146:331–336

Dini F, Paolesse R, Filippini D, D'Amico A, Lundström I, Di Natale C (2010) Fish freshness decay measurement with a colorimetric artificial olfactory system. Procedia Eng 5:1228–1231

Dudareva N, Negre F, Nagegowda DA, Orlova I (2006) Plant volatiles: recent advances and future perspectives. Crit Rev Plant Sci 25:417–440

El Barbri N, Llobet E, El Bari N, Correig X, Bouchikhi B (2008) Electronic nose based on metal oxide semiconductor sensors as an alternative technique for the spoilage classification of red meat. Sensors 8:142–156

Epifani M, Arbiol J, Pellicer E, Comini E, Siciliano P, Faglia G, Morante JR (2008) Synthesis and gas-sensing properties of Pd-doped SnO2 nanocrystals: A case study of a general methodology for doping metal oxide nanocrystals. Cryst Growth Des 8:1774–1778

Eranna G, Joshi B, Runthala D, Gupta R (2004) Oxide materials for development of integrated gas sensors—a comprehensive review. Crit Rev Solid State Mater Sci 29:111–188

Esfandyarpour B, Mohajerzadeh S, Famini S, Khodadadi A, Soleimani EA (2004) High sensitivity Pt-doped SnO2 gas sensors fabricated using sol–gel solution on micromachined (1 0 0) Si substrates. Sens Actuators B: Chem 100:190–194

Farahani H, Wagiran R, Hamidon MN (2014) Humidity sensors principle, mechanism, and fabrication technologies: a comprehensive review. Sensors 14:7881–7939

Feng Q, Li X, Wang J (2017) Percolation effect of reduced graphene oxide (rGO) on ammonia sensing of rGO-SnO2 composite based sensor. Sensors and Actuators B: Chemical 243:1115–1126

Galloway JN et al (2008) Transformation of the nitrogen cycle: recent trends, questions, and potential solutions. Science 320:889–892

García M, Aleixandre M, Gutiérrez J, Horrillo M (2006) Electronic nose for wine discrimination. Sens Actuators B: Chem 113:911–916

Gardner JW, Bartlett PN (1994) A brief history of electronic noses. Sens Actuators B: Chem 18:210–211

Guenther A (1997) Seasonal and spatial variations in natural volatile organic compound emissions. Ecol Appl 7:34–45

Guthrie S, Giles S, Dunkerley F, Tabaqchali H, Harshfield A, Ioppolo B, Manville C (2018) The impact of ammonia emissions from agriculture on biodiversity. The Royal Society, Santa Monica

Henderson WG, Khalilian A, Han YJ, Greene JK, Degenhardt DC (2010) Detecting stink bugs/damage in cotton utilizing a portable electronic nose. Comput Electron Agric 70:157–162
Herrmann U et al (2002) Monitoring apple flavor by use of quartz microbalances. Anal Bioanal Chem 372:611–614
Holzinger M, Maier J, Sitte W (1997) Potentiometric detection of complex gases: application to CO2. Solid State Ionics 94:217–225
Hongsith N, Viriyaworasakul C, Mangkorntong P, Mangkorntong N, Choopun S (2008) Ethanol sensor based on ZnO and Au-doped ZnO nanowires. Ceram Int 34:823–826
Huang L, Zhao J, Chen Q, Zhang Y (2014) Nondestructive measurement of total volatile basic nitrogen (TVB-N) in pork meat by integrating near infrared spectroscopy, computer vision and electronic nose techniques. Food Chem 145:228–236
Imam SA, Choudhary A, Sachan VK (2015) Design issues for wireless sensor networks and smart humidity sensors for precision agriculture: A review. In: 2015 International Conference on Soft Computing Techniques and Implementations (ICSCTI), 2015. IEEE, pp 181–187
Imamura A, Yumoto T (2008) Dynamics of fruit-body production and mycorrhiza formation of ectomycorrhizal ammonia fungi in warm temperate forests in Japan. Mycoscience 49:42–55
Jaiswal M, Kumar R, Mittal J, Jha P (2020) Synthesis of CrO_3 intercalated multilayer graphene for rapid and reversible NH_3 gas sensing. Sens Actuators B: Chem 310:127826
Jian Y, Hu W, Zhao Z, Cheng P, Haick H, Yao M, Wu W (2020) Gas sensors based on chemi-resistive hybrid functional nanomaterials. Nano-Micro Lett 12:1–43
Jonda S, Fleischer M, Meixner H (1996) Temperature control of semiconductor metal-oxide gas sensors by means of fuzzy logic. Sens Actuators B: Chem 34:396–400
Joyner JJ, Kumar DV (2015) Nanosensors and their applications in food analysis: a review. Int J Sci Technol 3:80
Kannan P, Guo L (2020) Nanosensors for food safety. Nanosensors for smart cities. Elsevier, Amsterdam, pp 339–354
Kaphle A, Navya P, Umapathi A, Daima HK (2018) Nanomaterials for agriculture, food and environment: applications, toxicity and regulation. Environ Chem Lett 16:43–58
Karunagaran B, Uthirakumar P, Chung S, Velumani S, Suh E-K (2007) TiO_2 thin film gas sensor for monitoring ammonia. Mater Charact 58:680–684
Kaushal M, Wani SP (2017) Nanosensors: frontiers in precision agriculture. Nanotechnology. Springer, Berlin, pp 279–291
Kerry J, O'grady M, Hogan S (2006) Past, current and potential utilisation of active and intelligent packaging systems for meat and muscle-based products: A review. Meat Sci 74:113–130
King GM (2000) Land use impacts on atmospheric carbon monoxide consumption by soils. Global Biogeochem Cycles 14:1161–1172
Klimant I, Kühl M, Glud RN, Holst G (1997) Optical measurement of oxygen and temperature in microscale: strategies and biological applications. Sens Actuators B: Chem 38:29–37
Kocache R (1986) The measurement of oxygen on gas mixtures. J Phys E: Sci Instrum 19:401
Kohl D (2001) Function and applications of gas sensors. J Phys D Appl Phys 34:R125
Kolle C, Gruber W, Trettnak W, Biebernik K, Dolezal C, Reininger F, O'Leary P (1997) Fast optochemical sensor for continuous monitoring of oxygen in breath-gas analysis. Sens Actuators B: Chem 38:141–149
Kong J, Franklin NR, Zhou C, Chapline MG, Peng S, Cho K, Dai H (2000) Nanotube molecular wires as chemical sensors. Science 287:622–625
Koudelka M (1986) Performance characteristics of a planar 'Clark-type'oxygen sensor. Sens Actuators 9:249–258
Kumar R, Jaiswal M, Singh O, Gupta A, Ansari MS, Mittal J (2019) Selective and reversible sensing of low concentration of carbon monoxide gas using Nb-doped OMS-2 nanofibers at room temperature. IEEE Sensors Journal 19:7201–7206
Kumar R, Kushwaha N, Mittal J (2016) Ammonia gas sensing activity of Sn nanoparticles film. Sens Lett 14:300–303

Kumar R, Kushwaha N, Mittal J (2017a) Superior, rapid and reversible sensing activity of graphene-SnO hybrid film for low concentration of ammonia at room temperature Sens Actuators B: Chem 244:243–251
Kumar R, Kushwaha N, Pandey S, Priya R, Mittal J (2017b) Superior NH_3 sensor using Ni doped K-OMS-2 nanofibers. IEEE Sens J 18:956–961
Kumar R, Mittal J, Kushwaha N, Rao BV, Pandey S, Liu C-P (2018) Room temperature carbon monoxide gas sensor using Cu doped OMS-2 nanofibers. Sens Actuators B: Chem 266:751–760
Lähdesmäki I, Lewenstam A, Ivaska A (1996) A polypyrrole-based amperometric ammonia sensor. Talanta 43:125–134
Lai H-Y, Chen C-H (2012) Highly sensitive room-temperature CO gas sensors: Pt and Pd nanoparticle-decorated In_2O_3 flower-like nanobundles. J Mater Chem 22:13204–13208
Lange D, Hagleitner C, Hierlemann A, Brand O, Baltes H (2002) Complementary metal oxide semiconductor cantilever arrays on a single chip: mass-sensitive detection of volatile organic compounds. Anal Chem 74:3084–3095
Laothawornkitkul J, Moore JP, Taylor JE, Possell M, Gibson TD, Hewitt CN, Paul ND (2008) Discrimination of plant volatile signatures by an electronic nose: a potential technology for plant pest and disease monitoring. Environ Sci Technol 42:8433–8439
Lee J-H, Park B-O (2004) Characteristics of Al-doped ZnO thin films obtained by ultrasonic spray pyrolysis: effects of Al doping and an annealing treatment. Mater Sci Eng, B 106:242–245
Lee W, Searcy S (2000) Multispectral sensor for detecting nitrogen in corn plants. In: ASAE Annual International Meeting, Midwest Express Center, Milwaukee, Wisconsin. pp 9–12
Lei S, Hoa SV, Ton-That M-T (2006) Effect of clay types on the processing and properties of polypropylene nanocomposites. Compos Sci Technol 66:1274–1279
Li G, Wang X, Yan L, Wang Y, Zhang Z, Xu J (2019) PdPt bimetal-functionalized SnO_2 nanosheets: controllable synthesis and its dual selectivity for detection of carbon monoxide and methane. ACS Appl Mater Interfaces 11:26116–26126
Li S, Simonian A, Chin BA (2010) Sensors for agriculture and the food industry. Electrochem Soc Interface 19:41–46
Licausi F, Giorgi FM, Schmälzlin E, Usadel B, Perata P, van Dongen JT, Geigenberger P (2011) HRE-type genes are regulated by growth-related changes in internal oxygen concentrations during the normal development of potato (Solanum tuberosum) tubers. Plant Cell Physiol 52:1957–1972
Liu T, Zhang X, Yuan L, Yu J (2015) A review of high-temperature electrochemical sensors based on stabilized zirconia. Solid State Ionics 283:91–102
Luechinger NA, Loher S, Athanassiou EK, Grass RN, Stark WJ (2007) Highly sensitive optical detection of humidity on polymer/metal nanoparticle hybrid films. Langmuir 23:3473–3477
Lv Y-y, Wu S-w, Li J (2014) Porphyrinic polymers for gas sensing: An overview. Curr Org Chem 18:475–488
Mabeck JT, Malliaras GG (2006) Chemical and biological sensors based on organic thin-film transistors. Anal Bioanal Chem 384:343–353
Meng X, Kim S, Puligundla P, Ko S (2014) Carbon dioxide and oxygen gas sensors-possible application for monitoring quality, freshness, and safety of agricultural and food products with emphasis on importance of analytical signals and their transformation. J Korean Soc Appl Biol Chem 57:723–733
Meng Z, Stolz RM, Mendecki L, Mirica KA (2019) Electrically-transduced chemical sensors based on two-dimensional nanomaterials. Chem Rev 119:478–598
Menil F, Coillard V, Lucat C (2000) Critical review of nitrogen monoxide sensors for exhaust gases of lean burn engines. Sens Actuators B: Chem 67:1–23
Mittal J, Lin KL (2020) Sn/SnO hybrid graphene for thermal interface material and interconnections with Sn hybrid carbon nanotubes. Mater Sci Eng, B 253:114485
Miura K, Ashikari M, Matsuoka M (2011) The role of QTLs in the breeding of high-yielding rice. Trends Plant Sci 16:319–326
Mohebi E, Marquez L (2015) Intelligent packaging in meat industry: an overview of existing solutions. J Food Sci Technol 52:3947–3964

Moon CS, Kim H-R, Auchterlonie G, Drennan J, Lee J-H (2008) Highly sensitive and fast responding CO sensor using SnO_2 nanosheets. Sens Actuators B: Chem 131:556–564

Moon WJ, Yu JH, Choi GM (2004) Selective gas detection of SnO 2-TiO 2 gas sensors. J Electroceram 13:707–713

Mousavi SR, Rezaei M (2011) Nanotechnology in agriculture and food production. J Appl Environ Biol Sci 1:414–419

Neethirajan S, Jayas D, Sadistap S (2009) Carbon dioxide (CO_2) sensors for the agri-food industry—a review. Food Bioprocess Technol 2:115–121

Nehra M, Dilbaghi N, Hassan AA, Kumar S (2019) Carbon-based nanomaterials for the development of sensitive nanosensor platforms. Advances in nanosensors for biological and environmental analysis. Elsevier, Amsterdam, pp 1–25

Nei L, Compton RG (1996) An improved Clark-type galvanic sensor for dissolved oxygen. Sens Actuators B: Chem 30:83–87

Nunes P, Malik A, Fernandes B, Fortunato E, Vilarinho P, Martins R (1999) Influence of the doping and annealing atmosphere on zinc oxide thin films deposited by spray pyrolysis. Vacuum 52:45–49

Nychas G-JE, Skandamis PN, Tassou CC, Koutsoumanis KP (2008) Meat spoilage during distribution. Meat Sci 78:77–89

O'Farrell M, Sheridan C, Lewis E, Flanagan C, Kerry J, Jackman N (2007) Online optical fiber sensor for detecting premature browning in ground beef using pattern recognition techniques and reflection spectroscopy. IEEE Sens J 7:1685–1692

Okcan B, Akin T (2007) A low-power robust humidity sensor in a standard CMOS process. IEEE Trans Electron Devices 54:3071–3078

Papadopoulou OS, Panagou EZ, Mohareb FR, Nychas G-JE (2013) Sensory and microbiological quality assessment of beef fillets using a portable electronic nose in tandem with support vector machine analysis. Food Res Int 50:241–249

Patel G, Pillai V, Bhatt P, Mohammad S (2020) Application of nanosensors in the food industry. Nanosensors for Smart Cities. Elsevier, Amstersdam, pp 355–368

Patil DR, Patil LA (2007) Ammonia sensing resistors based on Fe_2O_3 modified ZnO thick films. IEEE Sens J 7:434–439

Pelegrí-Sebastiá J, García-Breijo E, Ibanez J, Sogorb T, Laguarda-Miro N, Garrigues J (2011) Low-cost capacitive humidity sensor for application within flexible RFID labels based on microcontroller systems. IEEE Trans Instrum Meas 61:545–553

Penza M et al (2007) Enhancement of sensitivity in gas chemiresistors based on carbon nanotube surface functionalized with noble metal (Au, Pt) nanoclusters. Appl Phys Lett 90:173123

Pinheiro C, Rodrigues CM, Schäfer T, Crespo JG (2002) Monitoring the aroma production during wine–must fermentation with an electronic nose. Biotechnol Bioeng 77:632–640

Pourfayaz F, Khodadadi A, Mortazavi Y, Mohajerzadeh S (2005) CeO_2 doped SnO_2 sensor selective to ethanol in presence of CO, LPG and CH_4. Sens Actuators B: Chem 108:172–176

Pudake RN, Chauhan N, Kole C (2019) Nanoscience for sustainable agriculture. Springer International Publishing. https://doi.org/10.1007/978-3-319-97852-9

Qi P et al (2003) Toward large arrays of multiplex functionalized carbon nanotube sensors for highly sensitive and selective molecular detection. Nano Lett 3:347–351

Qian L, Wang K, Li Y, Fang H, Lu Q, Ma X (2006) CO sensor based on Au-decorated SnO_2 nanobelt. Mater Chem Phys 100:82–84

Raimundo IM Jr, Narayanaswamy R (2001) Simultaneous determination of relative humidity and ammonia in air employing an optical fibre sensor and artificial neural network. Sens Actuators B: Chem 74:60–68

Reay DS, Davidson EA, Smith KA, Smith P, Melillo JM, Dentener F, Crutzen PJ (2012) Global agriculture and nitrous oxide emissions. Nat Clim Change 2:410–416

Renganathan B, Gobi G, Sastikumar D, Srinivasan R, Bose AC (2010) Optical fiber coated with nanocrystalline tin oxide for ammonia vapour sensing. Sens Lett 8:292–296

Renganathan B, Sastikumar D, Gobi G, Yogamalar NR, Bose AC (2011) Nanocrystalline ZnO coated fiber optic sensor for ammonia gas detection. Opt Laser Technol 43:1398–1404

Ruiz-Garcia L, Lunadei L, Barreiro P, Robla I (2009) A review of wireless sensor technologies and applications in agriculture and food industry: state of the art and current trends. Sensors 9:4728–4750

Ruiz AM, Cornet A, Morante JR (2005) Performances of La–TiO_2 nanoparticles as gas sensing material. Sens Actuators B: Chem 111:7–12

Saini RK, Bagri LP, Bajpai AK (2017) Smart nanosensors for pesticide detection. New pesticides and soil sensors. Elsevier, Amsterdam, pp 519–559

Salinas Y et al (2014) A novel colorimetric sensor array for monitoring fresh pork sausages spoilage. Food Control 35:166–176

Sanchez PA, Swaminathan MS (2005) Hunger in Africa: the link between unhealthy people and unhealthy soils. The Lancet 365:442–444

Schutting S, Borisov SM, Klimant I (2013) Diketo-pyrrolo-pyrrole dyes as new colorimetric and fluorescent pH indicators for optical carbon dioxide sensors. Anal Chem 85:3271–3279

Seal S, Shukla S (2002) Nanocrystalline SnO gas sensors in view of surface reactions and modifications. Jom 54:35–38

Shi H, Zhang M, Adhikari B (2018) Advances of electronic nose and its application in fresh foods: a review. Crit Rev Food Sci Nutr 58:2700–2710

Smolander M (2003) The use of freshness indicators in packaging. In: Techniques Novel food packaging (ed) Ahvenainen, R. Woodhead Publishing, Cambridge, pp 127–143

Sohraby K, Minoli D, Znati T (2007) Wireless sensor networks: technology, protocols, and applications. John Wiley & Sons, Hoboken

Srbinovska M, Gavrovski C, Dimcev V, Krkoleva A, Borozan V (2015) Environmental parameters monitoring in precision agriculture using wireless sensor networks. J Cleaner Prod 88:297–307

Srivastava A, Jain K (2007) Study on ZnO-doped tin oxide thick film gas sensors. Mater Chem Phys 105:385–390

Sun H, Fung Y (2006) Piezoelectric quartz crystal sensor for rapid analysis of pirimicarb residues using molecularly imprinted polymers as recognition elements. Anal Chim Acta 576:67–76

Takeuchi T, Igarashi I (1988) Limiting current type oxygen sensor. Chem Sens Technol 1:79–95

Tang H, Yan M, Zhang H, Li S, Ma X, Wang M, Yang D (2006) A selective NH3 gas sensor based on Fe2O3–ZnO nanocomposites at room temperature. Sens Actuators B: Chem 114:910–915

Timmer B, Olthuis W, Van Den Berg A (2005) Ammonia sensors and their applications—a review. Sens Actuators B: Chem 107:666–677

Uetake H, Hirota N, Nakagawa J, Ikezoe Y, Kitazawa K (2000) Thermal convection control by gradient magnetic field. J Appl Phys 87:6310–6312

Vaezi M, Sadrnezhaad S (2007) Gas sensing behavior of nanostructured sensors based on tin oxide synthesized with different methods. Mater Sci Eng, B 140:73–80

van Dongen JT, Schurr U, Pfister M, Geigenberger P (2003) Phloem metabolism and function have to cope with low internal oxygen. Plant Physiol 131:1529–1543

Van Hieu N, Thuy LTB, Chien ND (2008) Highly sensitive thin film NH3 gas sensor operating at room temperature based on SnO2/MWCNTs composite. Sens Actuators B: Chem 129:888–895

von Bültzingslöwen C, McEvoy AK, McDonagh C, MacCraith BD, Klimant I, Krause C, Wolfbeis OS (2002) Sol–gel based optical carbon dioxide sensor employing dual luminophore referencing for application in food packaging technology. Analyst 127:1478–1483

Wagh M, Jain G, Patil D, Patil S, Patil L (2006) Modified zinc oxide thick film resistors as NH_3 gas sensor. Sens Actuators B: Chem 115:128–133

Wan C, Qiao X, Zhang Y, Zhang Y (2003) Effect of different clay treatment on morphology and mechanical properties of PVC-clay nanocomposites. Polym Testing 22:453–461

Wanekaya AK, Chen W, Myung NV, Mulchandani A (2006) Nanowire-based electrochemical biosensors. Electroanal Int J Devoted to Fundamental and Practical Aspects of Electroanalysis 18:533–550

Wang R, Li J (2010) Effects of precursor and sulfation on OMS-2 catalyst for oxidation of ethanol and acetaldehyde at low temperatures. Environ Sci Technol 44:4282–4287

Wang Y-D, Wu X-H, Su Q, Li Y-F, Zhou Z-L (2001) Ammonia-sensing characteristics of Pt and SiO_2 doped SnO_2 materials. Solid-State Electron 45:347–350
Wang Y et al (2014) Ammonia gas sensors based on chemically reduced graphene oxide sheets self-assembled on Au electrodes. Nanoscale Res Lett 9:251
Warneck P (1999) Chemistry of the natural atmosphere. Elsevier, Amsterdam
Weitz AM, Linder E, Frolking S, Crill P, Keller M (2001) N_2O emissions from humid tropical agricultural soils: effects of soil moisture, texture and nitrogen availability. Soil Biol Biochem 33:1077–1093
Wijaya DR, Sarno R, Zulaika E, Sabila SI (2017) Development of mobile electronic nose for beef quality monitoring. Procedia Comput Sci 124:728–735
Williams E, Hutchinson G, Fehsenfeld F (1992) NO_x and N_2O emissions from soil. Global Biogeochem Cycles 6:351–388
Wilson AD, Oberle CS, Oberle DF (2013) Detection of off-flavor in catfish using a conducting polymer electronic-nose technology. Sensors 13:15968–15984
Wojnowski W, Majchrzak T, Dymerski T, Gębicki J, Namieśnik J (2017) Electronic noses: powerful tools in meat quality assessment. Meat Sci 131:119–131
Xiao Y et al (2014) Determination of the freshness of beef strip loins (M. longissimus lumborum) using electronic nose. Food Anal Methods 7:1612–1618
Yamazoe N, Miura N (1995) Development of gas sensors for environmental protection. IEEE Trans Compon Packag Manuf Technol. Part A 18:252–256
Zaragozá P et al (2012) Fish freshness decay measurement with a colorimetric array. Procedia Eng 47:1362–1365
Zhang T, Nix MB, Yoo BY, Deshusses MA, Myung NV (2006) Electrochemically functionalized single-walled carbon nanotube gas sensor. Electroanalysis: An Int J Devoted to Fundam Pract Aspects of Electroanalysis 18:1153–1158
Zhang X, Davidson EA, Mauzerall DL, Searchinger TD, Dumas P, Shen Y (2015) Managing nitrogen for sustainable development. Nature 528:51–59
Ziegler C (2004) Cantilever-based biosensors. Anal Bioanal Chem 379:946–959

Chapter 5
Volatile Organic Compounds (VOCs) Sensors for Stress Management in Crops

Vartika Rohatgi, Navakanth Vijay Challagulla, and Ramesh Namdeo Pudake

Abstract Due to current environmental challenges, crop production all over the world is in urgent need for new sustainable solutions. Nanotechnology has a potential to provide answers to various problems associated with productivity and food security. The agronomics importance of volatile organic compounds (VOCs) from plant is well established in processes associated with plant defence against the biotic and abiotic stresses. They have been playing important role in enhanced stress resistance through maintaining reactive oxygen species (ROS) homeostasis, regulator of plant growth and development, and as a potential antimicrobial agent. Current limitations associated with its real-time detection in open field is hampering its better exploitation in future sustainable agriculture. Nano-based sensors are being developed as emerging devices for early detection of VOCs as plant stress signal molecules/reporters for real-time monitoring of plant health. This chapter analyses the current research done in this field for improving our understanding of plant stress communication. The development of novel tools for precision agriculture through VOCs sensors is also envisioned.

Keywords Volatile organic compounds · Plant stress · Sensors · Precision agriculture · Non-invasive diagnostics

5.1 Introduction

Almost every single sector is influenced directly or indirectly by agriculture. The increasing population has put a lot of pressure on fulfilling agricultural demands. So, it is very important to meet the challenges faced by the agriculture sector. Moreover,

V. Rohatgi · N. V. Challagulla · R. N. Pudake (✉)
Amity Institute of Nanotechnology, Amity University Uttar Pradesh, Noida, UP 201313, India
e-mail: rnpudake@amity.edu

R. N. Pudake et al. (eds.), *Biosensors in Agriculture: Recent Trends and Future Perspectives*, Concepts and Strategies in Plant Sciences,
https://doi.org/10.1007/978-3-030-66165-6_5

plant diseases and environmental stresses can cause huge economic losses in the agriculture sector by reducing crop yield (Chakraborty and Newton 2011; Velásquez et al. 2018). It is very essential to manage and look after plant health with sustainable methods. New innovating strategies for plant protection should decrease the losses in food production. Rapid technologies with low cost and less time consuming need to be developed for plant protection.

Volatile organic compounds (VOCs) are chemical substances produced and emitted by plants and other organisms in gaseous form. Plants emit these volatile organic compounds through flowers, leaves, roots, or other tissues with complex processes. The emission of VOCs is significantly affected by the surrounding plant atmosphere. Every part of the plant or tissue releases a different kind of volatile organic compound. Particularly leaves have many specialized cells that produce a variety of organic compounds. During the process of biosynthetic conversion of solar energy to chemical energy, multiple organic compounds are produced. Few of these smaller organic molecules are used to build various important biological macromolecules like lipids, carbohydrates, proteins, or even nucleic acids. These macromolecules have generally lower volatile nature due to their massive molecular weight. While there are intermediates of these biochemical processes that form various metabolites of primary and secondary nature. Then there are few lower than 350 Daltons compounds having functional groups containing non-polar and polar nature. This polarity also affects the volatility of the compound. Thus, bringing us to the compounds that have a low boiling point (high vapour pressure) and are easily vapourized.

These released VOCs from plants can interact with other organisms or neighbouring plants, during signalling and protecting plant from biotic or abiotic stress that are affecting our crops globally (Fig. 5.1) (Brilli et al. 2019). VOCs released from plants are involved in different functions like attracting pollinators, defending against herbivore insects and pathogens, and serve as signals to other plants. These molecules also help plant in production of defence proteins, metabolites, and promotion of defence related gene expression, thereby improving tolerance to stress. There are hundreds of these (such) volatile compounds and their compositions that can be correlated with the biological information of the plant and their processes. Plant volatile organic compounds consist of isoprene, mono-, sesquiterpenes, cis-jasmone (CJ), α-pinene, limonene, γ-terpinene. The release of VOCs constitutes a significant amount of reactive carbon released by plants into the atmosphere. The most common type of VOCs is terpenes, that is present in resin, aromatic plant tissues like flower. It helps in communication between plants and insects, and may be attractants or repellents (Brilli et al. 2019; Maleknia et al. 2007).

As discussed above, the emission rate of plant VOCs largely depends on the environmental conditions or due to stresses from bio-agents. In this chapter, we have summarized the roles of the VOCs in plants and the role of nanosensors in their detection and monitoring of plant health.

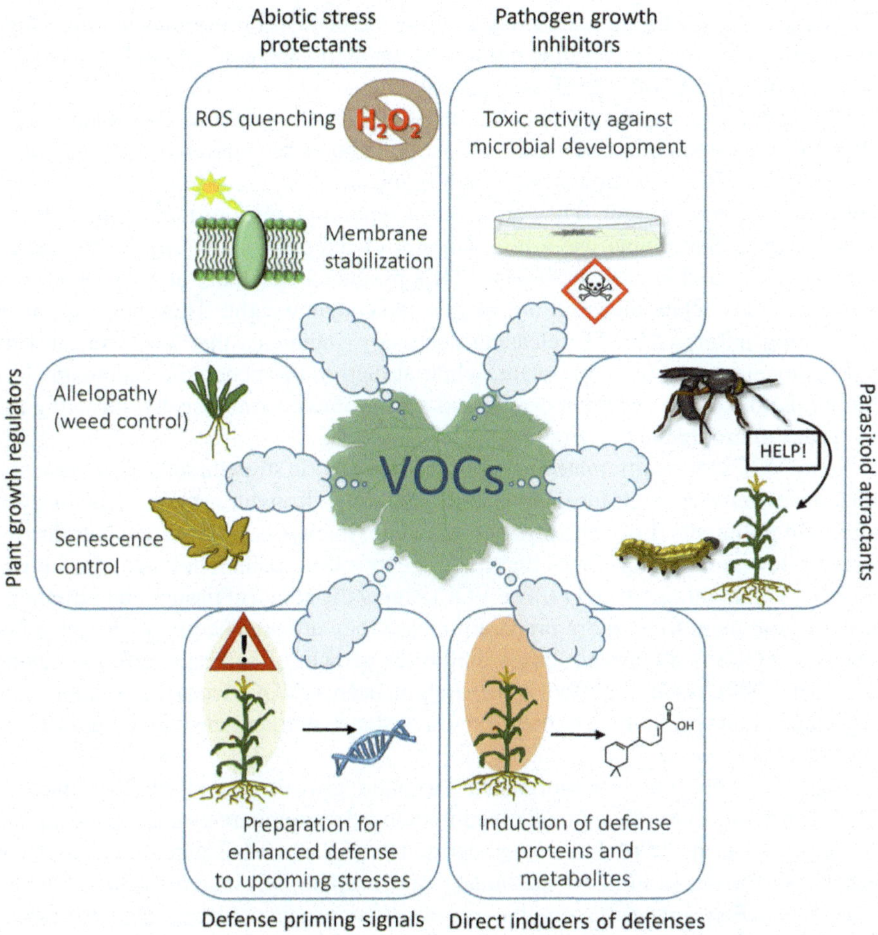

Fig. 5.1 Schematics showing the possible role of VOCs in various plant processes ranging from plant growth regulator, abiotic stress tolerance, inhibition of pathogen growth, herbivore repellence, and activation of plant defence mechanism. [Reprinted with permission from (Brilli et al. 2019) Copyright (2019) Brilli, Loreto and Baccelli]

5.2 Role of VOCs in Plant Health

Organisms have a keen way of sensing the environment around them or coordinating with other organisms that are present in their vicinity. The organisms perform this by employing series of signalling pathways. These signals allow the organism to have proper growth, homeostasis, and control over the organism's behaviour. Most of the organisms sense environment via recognition of certain molecules that are often proteins or metabolites secreted by organisms in the vicinity. Since VOCs are known for their volatile nature and can easily permeate through soil and fluids, therefore, they serve as the ideal candidate as a signalling molecule for mid-range to long-range communications for intercellular and for the whole organism (Bednarek et al. 2010; Bonfante and Anca 2009; Brilli et al. 2019).

Biological VOCs have a few common properties (Schulz and Dickschat 2007). These VOCs often comprise esters, alcohols, terpenoids, derivatives of fatty acid, aldehydes, thiols. They show high lipophilicity, and often have high vapor pressure and low molecular weight. They are easily evaporated at room temperature. It has been recorded that plants use a high amount of fixed CO_2 (20%) just to produce VOCs (Schulz and Dickschat 2007). VOCs produced by plants may be secondary or specialized metabolites and are of low molecular weight. They help the plant by creating an aroma or by releasing necessary chemicals that steers the insects and pathogens away from the plant, while attracting the beneficial organisms like the pollinators or the seed dispersers, and also promote symbiosis by attracting the necessary microorganisms (Fig. 5.1).

Some VOCs also participate in moderating the abiotic stresses such as better resistance to extreme temperatures, high light intensity, droughts, saline soils, or even some soil pollutants. Ethylene, methyl salicylate, methyl jasmonate (or their derivative) are few volatile hormones that help in controlling plant's defence mechanism (Robert-Seilaniantz et al. 2011). The VOCs signal the adjacent tissues and ultimately to the whole plant to be more prepared in case they are attacked by pathogens. For instance, VOCs from tomato leaves inhibit the growth of pathogenic fungi (Zhang and Chen 2009). A similar effect is seen in lima bean as VOCs from the neighbouring resistance plants primed the expression of resistance marker genes in the uninfected plant (Yi et al. 2010).

Also the symbiotic microorganism or plant growth promoting rhizobacteria (PGPR) release VOCs that are involved in induced systemic resistance against plant diseases (Han et al. 2006). It has been reported that 2R and 3R-Butanediol from plant-beneficial rhizobacterium, *Pseudomonas chlororaphis* O6, increases the tobacco's resistance in *Erwinia carotovora,* but was not effective against *P. syringae*, pv. *Tabaci* (Han et al. 2006). 2R, 3R-butanediol was also found to be effective in improving drought tolerance in *Arabidopsis thaliana* (Cho et al. 2008). It has been reported that PGPR strain *Paenibacillus polymyxa* E681, increases the growth and induced systemic resistance of *A. thaliana,* and the presence of tridecane, a C13 hydrocarbon was responsible for both (Lee et al. 2012). In another study, it was found that *Burkholderia cepacia* (rhizobacteria) produce VOCs that can increase the growth rate of *A. thaliana* while the identity of the specific compound is still unknown (Vespermann et al. 2007). But certain microbes may release compounds that reduce the productivity or growth of the plants. Few of those microbes include bacteria such as *Pseudomonas*, *Stenotrophomonas*, *Serratia*, *Burkholderia*, and *Chromobacterium* (Bailly and Weisskopf 2012; Kai and Piechulla 2009). Few bacteria in soil can hinder the defensive response of plants by using VOC as an effector (Blom et al. 2011).

These findings suggest that plant VOCs have great agronomic potential in developing some modern and eco-friendly solutions for plant stress management and increased crop production. The current research in this area is about designing and validating a VOC sensing platform, and many innovative solutions have been developed to use VOCs in crop stress management. The sole objective behind the recent studies is to detect plant volatility that helps in monitoring the plant's health and diseases which decreases the agricultural losses.

5.3 Limitations of Current Methods VOCs Detection

A conventional gas chromatography-mass spectrometry (GS-MS) plays a major role in the analysis of VOCs in plants (Borges et al. 2018; Chang et al. 2014; Cui et al. 2018) that are exposed to various stresses. The characteristic features of analysis of VOC using MS include detection limits, automated and detailed structural characterization of different types of compounds (Maleknia et al. 2007). However, this routine technique is costly and time consuming, and not suitable for real-time monitoring in field conditions. So, to take full advantage of the real-time VOC pattern, generated in plant species for various stress management, there is a need to develop convenient, easy to operate, real-time detection devices.

The real-time analytical techniques can be used to measure the VOCs to provide information about the health of plants at any given time. Thus, evaluating and measuring the VOCs is an engaging avenue as a non-invasive and quick approach in supervising the structure and physiological health of plants (Materić et al. 2015). It is possible to collect the sample of plant VOC by in situ approach from whole plant or organs (like fruits, flowers, fruits) or precisely from the removed or isolated plant tissues.

VOC patterns emitted by plants are extremely complex and thus it becomes challenging to directly correlate its production with a definite cause. There are four main aspects of this analysis that are necessarily and meticulously considered before we can understand changes in VOC emission pattern and then correlate them to possible reasons. First and foremost is the recognition of the variations in VOC due to the biological state of the plant and attributing it to a cause as many factors like biotic and abiotic stress may cause VOC change at the same time. Also, the age of the plant parts may also show the differential pattern of VOC release, like older foliage is expected to be less metabolically active than the young foliage resulting in more production of VOCs. So, it is very important that we should consider all these aspects while designing the experiments, so that correlation of VOC production can be accurately correlated to the factor responsible. The second most important aspect is VOC sampling, as for practical field use purpose the sampling needs to be done through non-invasive tools. The use of improper sampling techniques can introduce large errors. Also, a sampling technique can be bias towards certain compounds, which may or may not be essential in that specific plant system. The third essential aspect is to select an appropriate analytical chemistry technique used in the VOC study. The technique chosen should enable the identification of sampled chemical differences. It helps in identifying the biomarker VOCs of interest. Finally, the information collected during the analysis needs to be processed, interpreted, and correlated to the set of conditions being studied through various mathematical models/algorithms (Aksenov et al. 2013). If the issues discussed above are not addressed suitably may lead to negative results or some cause may be ignored during the establishment cause–effect relationship between VOCs and stress.

5.4 Nanosensors for Detection of Plant VOCs

For the site-specific crop management needs the knowledge of variability in various field attributes like soil and plant characteristics. The agriculture industry in many countries is now heavily dependent on sensor technology that can provide precise data for crop health. Nanotechnology has the potential to provide a promising solution for existing problems in agriculture. The nanosensors can provide high specificity and selectivity to the components that are related to plant protection. It is a rapid technology with low cost and less time-consuming. Agricultural industries have recognized the benefits of these technologies in determining the presence of specific gas mixtures or to find the specific source or even to monitor the release of specific gases during the storage and production period. It helps in monitoring plant health too. This will ultimately increase productivity and improves the quality of food production. The sensor technology has evolved into different spectrums which can help agriculture that ranges from aerial-based remote sensing, environment-based greenhouse sensor, electrochemical sensor, e-nose (electronic nose) sensors, optical sensors, and many others (Li et al. 2010). The various nanomaterial-based tools like e-nose that mimic the functions of the human nose can be used as a real-time approach to detect VOCs. Other tools like plant wearable and nanoparticle sensors coupled with data acquisition systems and analytical algorithms are widely being used in recent decades to monitor plant health. Few of the studies are reviewed in the following sections of this chapter.

5.4.1 Use of E-Nose for VOCs Analysis

E-nose has been playing a major role in real-time monitoring of crops, as they gather information on plant's health, the interaction of plants with animal/insects, pesticide, and other factors influencing the plants. The different principle and mechanism of Electronic Aroma Detection (EAD) has led to the development of e-nose sensor for diverse application in agriculture. Analysis of animal and plant-based VOCs can be performed on the simple to complex mixture of organic compounds that are volatilized from living tissues or processed products of plant or animal cells. Most often the e-nose devices are used to identify the unique mixture of VOCs in the samples derived from plant, animal, or their products. They are not generally used to detect the specific compound in the sample. With the help of e-nose, one can find the age, product consistency, freshness of the fruit, ripeness, quality of the product, purity, and one can even estimate the shelf life. These are highly beneficial when evaluating batches of wine, and find which can be considered vintage and which is recently produced. In wines, it also plays a critical role in determining the uniformity of the wine, aroma, age, taste, etc., which are important characteristics to decide its market value (Wilson 2013).

Olfactory senses of human are able to detect food degradation at much later stages and that time the food is not suitable for eating. This degradation can be due to fungal or bacterial growth that produces a stale smell as a by-product of their metabolic activity. Therefore, it needs to be detected in early stage to preserve the edible nature of the food, or else the nutritional as well as the economic value of the food deteriorates. E-nose sensors are used to constantly check the quality as well as the freshness of the agriculture products. E-nose alarms us by letting us know the deterioration of food and the presence of contaminants near the product that may spoil it (Wilson 2013). The e-nose can have different sensor platform that contributes to EAD, and few of the most commonly used are, conducting polymer-based (CP), complementary metal-oxide-semiconductor (CMOS), metal oxide semiconductor (MOS), quartz crystal microbalance (QCM), nano-electrochemical, and surface or bulk acoustic wave-based sensors. The advantages and limitations of these various e-nose sensor platforms have been reviewed earlier (Wilson and Baietto 2009).

From the literature, it was found that MOS-based sensors are the most used sensors as they have low production cost, response time is fast, require little time to recovery, and their reproducibility result is high (de Julián Fernández et al. 2008; Ge et al. 2007; Itoh et al. 2008; Kim 2009; Zhu et al. 2006). But these MOS-based sensors are not too selective with the types of VOCs with similar structures and so, cannot be used indoors to monitor air quality. MOS has shown rather poor distinguishing ability towards pollutants such as benzene from toluene, xylene, or formaldehyde. By using a mixture of multiple metal oxides-based semiconductor sensors showed a differential selectivity towards different gases at defined operating conditions (Xu et al. 2007). This shows that sensor selection should be done appropriately depending on the conditions in which the sensor has to operate. The other thing to keep in mind while selecting the sensor is to find the range in which the sensor works and that whether the sensor shows cross-sensitivity and how much of zero-error the sensor has, as this may affect the analyte recognition capabilities of it (Xu et al. 2007).

The concept of nano e-nose comes from the olfactory sensors that are present in the biological organisms and so, they are known to be acting as biomimicking the systems that are present in nature (Harun et al. 2009). Some of the key important details captured by e-noses related to plant health are rate of production of certain VOCs upon catching an infection, or determination of plant physiological processes with change in conditions of the field, weather/humidity, light, surrounding temperature (Komaraiah et al. 2004; Su 2008). A multisensor array with 20 different MOS sensor was used to monitor cell culture of *Morinda citrifolia* and *Nicotiana tabacum* that are cultivated under batch conditions. The data generated was able to predict the biomass concentration and secondary metabolite quantity in cell culture (Komaraiah et al. 2004). In another study, VOCs emitted by tomato plant under aphid attack was measured by e-nose and able to correlate the VOCs generated to the extent of damage (Cui et al. 2019).

In order to study the possibility of using Metal-OXide (MOX) gas sensors in water management in tomato and maize, Fabbri and colleagues have analysed in situ data gathered from portable sensing units, and compared this with meteo-sat data and relevant information related to irrigations, rainfalls, or pesticide-based treatments in the farm. The results had confirmed the correlation between the VOCs emissions and the hydric/metabolic status of the plants (Fabbri et al. 2020).

Several e-nose platforms are readily available that can be used to monitor plant VOCs as indicators related to stress. By using them one can easily monitor the plant's growth and health. An e-nose system, Alpha MOS Fox 3000 was used for the detection of a Citrus Tristeza Virus (CTV) in Khasi Mandarin Orange with a precision of 95.30% (Hazarika et al. 2020). Wireless sensor networks (WSNs) are an ideal solution to simultaneously and remotely observing many plants. The total VOCs produced by European privet (*Ligustrum jonandrum*) that was kept under different conditions was monitored by Sensirion device (SGP30) containing metal oxide semiconductor sensors. The networked nodes confirmed the measurable correlation of VOCs with environmental conditions that are prevailing near the plant (Catini et al. 2019).

5.4.2 Plant Wearable Sensors for Monitoring Plant Stress

Recent studies are focusing on one of the future applications of nanosensor networks that can obtain real-time information with a non-invasive method from difficult-to-reach areas. Due to the small dimensions of nanosensors, they can be embedded in small gadgets and have the ability to execute auxiliary in-network processing. These properties open horizons of its applications in the fields of agricultural, industrial, and biomedical settings; that are not possible by the standard network of the sensors. Micro-electro-mechanical systems (MEMS) technology is rapidly advancing in recent days, and the number of implantable and wearable diagnostic devices based on conductive nanomaterials are being manufactured for monitoring human health issues. This technology has great potential in long term and on-demand measuring plant VOC and other biomarkers and can be a new tool for crop health diagnostics in near future. Here, a flexible sensor that is very thin and lightweight nanosensors can be directly attached to plant tissues such as leaves for continuous observation of VOCs that are induced by stresses.

In one study, in situ synthesis of monolithically integrated electronic devices based on single-walled carbon nanotube (SWCNT) was achieved. This flexible device was then precisely placed onto various bio surfaces. In one experiment, it was placed on the leaf surface of lucky bamboo (*Dracaena sanderiana* cv. Virens) plant to detect the simulants of sarin nerve agent (Lee et al. 2014). Recently, a vapour-printing technique was used to prepare a plant tissue mounted polymer electrode that allowed long-term plant health monitoring using bioimpedance measurements.

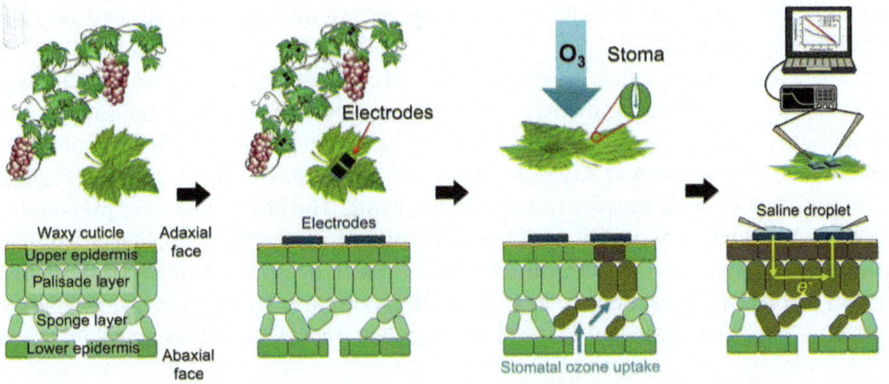

Fig. 5.2 Schematic showing the printing of ProDOT electrode and impedance-based detection of ozone damage in plants. Reprinted with permission from (Kim et al. 2020) Copyright (2020) AAAS

With this plant mounted electrode, they were able to monitor drought and photo-damage. The monomer 3,4-propylenedioxythiophene (ProDOT) was directly polymerized in a specially designed hot wall reactor that was used to print electrode on 14 plant species. Oxidative polymerization mediated by $FeCl_3$ progressed inside the reactor tube where monomer and oxidant vapours met. They also found that, printed polymer doesn't significantly influence the plant growth pattern (Kim et al. 2019). The same team has also used these polymer patches on the plant leaves to monitor the ozone damage to grapes and apple plant (Kim et al. 2020) (Fig. 5.2). The plant wearable sensor technology has the capability to real-time monitoring of plant health remotely with wireless network (Coppedè et al. 2017; Koman et al. 2017; Nassar et al. 2018). These studies have not used VOCs to monitor plant health, but the results will be helpful in designing plant wearable VOC sensors. The smart and portable electronic devices will be available in future with specificity in detection of plant molecules due to the development of flexible and biocompatible materials, and low power requirement.

5.4.3 Nanoparticle-Based Smart Sensors for Detection of Plant Stress

5.4.3.1 Electrochemical Sensors

As compared to other techniques, there are unique advantages in using electrochemical sensing for VOC profiling. Selectivity for multi-species of VOCs, real-time monitoring with rapid detection, and portability suitable for field applications are some of them. They can be used in real field conditions and can be connected wirelessly or on some moving vehicles. There are some studies with above said goals to monitor the key chemicals released by plants during infection or other stresses, and they are reviewed below.

The gold nanoparticles modified screen-printed carbon electrode (SPCE) was developed for detecting methyl salicylate in soybean plants (Umasankar and Ramasamy 2013). Methyl salicylate is a vital plant volatile and is involved in pathogen and pest attack in soybean. The AuNP–SPCE has shown to be highly stable, and sensitive towards the methyl salicylate sensing. The electrodes were prepared in a way so that they can easily detect methyl salicylate even in the existence of other compounds. This electrochemical sensor was able to detect the hydrolysis of methyl salicylate and the oxidation of negative species, correlate the characteristic signals to quantify the volatile present in the soybean pod extract (Umasankar and Ramasamy 2013).

In another study, NPs of metal oxide namely TiO_2 and SnO_2 with screen-printed carbon (SP) electrodes have been used to develop a potential device for sensing *p*-ethylguaiacol volatiles in strawberry fruits and plants. *p*-ethylguaiacol is one of the VOCs released from fruits and plants when it is infected with *Phytophthora cactorum*—a pathogenic fungus. The electrochemical analysis with the electrode developed in this study had shown high sensitivity (174–188 μA cm^{-2} mM^{-1}) and low detection limit (35–62 nM) for target analyte with high repeatability. The authors had suggested that their findings support the metal oxides as a cheaper alternative to expensive gold or platinum electrodes for amperometric sensor applications (Fang et al. 2014a, b).

In later studies, the same group had developed a bi-enzyme (alcohol oxidase and horseradish peroxidase) modified CNT sensor for selective and sensitive determination of methyl salicylate (Fang et al. 2016). On this novel electrode, the hydrolysis of methyl salicylate produced methanol. Then the two enzymes came into action, where alcohol oxidase produced formaldehyde while simultaneously reducing oxygen into H_2O_2. This H_2O_2 was further reduced to water by horseradish peroxidase, that resulted in an amperometric signal via direct electron transfer. A rapid method developed by them can be used for crop disease detection based on VOC production.

5.4.3.2 Optical Sensors

The optical nanosensors have the potential to detect the beginning of stress in plant with the help of nIR imaging devices. They can be used as new and rapid monitoring tools for plant health and under stressed growth conditions. H_2O_2 and NO signalling pathways have been reported to play a very important role in plant defence responses (Giraldo et al. 2015; Lew et al. 2020a). The functionalized SWCNT has been reported to produce unique optical signals and was used in ratiometric sensing platform to detect H_2O_2 and NO in living plant tissue. Recently new approach had been reported to detect how plants respond or communicate to stress such as injury, infection, and light damage by using sensors made of carbon nanotubes (Lew et al. 2020b). This sensor tries to track the hydrogen peroxide which is used by the plants for repairing damage and defend themselves from predators. The plants used hydrogen peroxide for sending signals to leaf cells. This new sensor with SWCNTs is exposed on the leaves of the plants. The SWCNTs were wrapped in the single stranded $(GT)_{15}$ oligonucleotides, and it was found to be specific to H_2O_2. The binding of H_2O_2 to functionalized SWCNTs resulted in quenching of the fluorescence intensity, making it a suitable probe for active sensor. The hydrogen peroxide signals used by this sensor helped in identifying and tracking light, heat, and wound stress responsive changes in 6 plant species. The functionalized SWCNTs were infiltered in test plant species, and the images were taken before and after the stress treatment. Ratiometric sensor responses were correlated with H_2O_2 responses. This advancement in technology can help the farmers in monitoring crop as the device was species-independent and capable of real-time, spatial and temporal biochemical measurements in plants (Lew et al. 2020a, b). Similarly, near-infrared (nIR) fluorescent SWCNTs functionalized with DNA aptamer that attaches to hemin (HeAptDNA-SWCNT) were designed for developing an optical sensor for H_2O_2 detection (Fig. 5.3). These kinds of optical nanobiotechnology sensors allow the detection of both environmental and pathogen related stresses, will increase our understanding of plant physiology, and provide a sustainable long-term solution in crop management (Wu et al. 2020).

Localized surface plasmon resonance (LSPR) phenomenon has been used as a valuable tool in biosensor, that can detect the collective electron charge oscillations

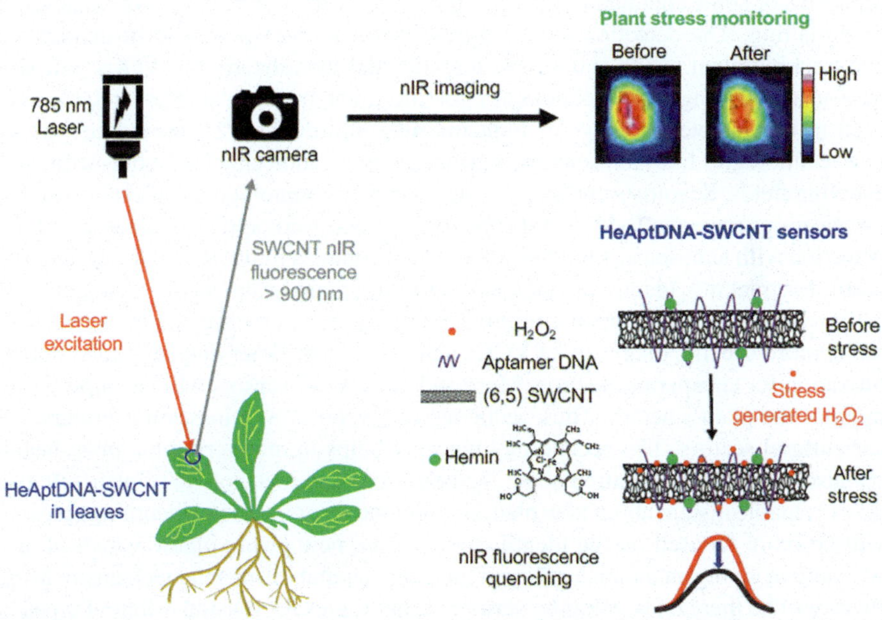

Fig. 5.3 In planta monitoring of H_2O_2 by HeAptDNA-SWCNT sensor. The binding of analyte resulted into quenching of nIR fluorescence of HeAptDNA-SWCNT, that was remotely measured by nIR camera. Reprinted with permission from (Wu et al. 2020) Copyright (2020) American Chemical Society

at surfaces of NPs of noble metal (Au, Pt) by converting changes in refractive index (RI) into spectral shifts. In one study, LSPR sensor coated with gold nanoparticles (AuNPs) doped in molecularly imprinted sol-gel (MISG) used for determination of plant volatile. The gold nanoparticles provide more specificity, stability, and sensitivity to the LSPR sensor. A five-channel AuNPs@MISG LSPR sensor showed more effective results as 12.33 times higher than the sensor without AuNPs under optimum conditions. The size of AuNP was 30 nm and the amount of AuNPs doped in the MISG was 20 μL. This technology helps in the detection and identification of VOCs which ultimately provides various benefits in the agricultural sector (Shang et al. 2018). The results of the above studies have demonstrated that optical sensors combined with other electronic devices can be used for plant VOC detection and identification, which may become a useful technology for future agricultural applications.

5.5 Conclusion and Future Perspective

The volatile organic compounds (VOCs) emitted by a plant are a notable and vital part of the plant's life cycle and are introspective of the physiological status of the plant. The release of VOCs may occur as a side effect of continuous metabolic processes or some chemicals could be emitted for a specific reason, such as defence or signalling. The detection and identification of these VOCs help in monitoring and tracking plant health status. The tracking and identification of VOCs will also help in developing new technologies for the plant betterment. Researchers have tried to use nanotechnology for reducing agriculture losses by providing various new technologies like nanosensors. Agriculture and nanotechnology both are always evolving fields. Scientists are trying to develop new technologies that can be used to protect crops for better food security and reduce the various agricultural losses. VOCs detection with nanosensors has the potential to become a future tool in the agriculture sector for tracking, identification, and monitoring plant health status remotely.

Few of the most anticipated improvements by incorporating nanomaterials will be in, more specific diagnosis of VOCs and sensitivity, faster detection rate, miniaturization for cheaper production. More and more VOC sensors will be employed to monitor the plant's health. Thus, reducing a large portion of losses for farmers and agricultural sectors. This technology not only helps in monitoring the plant health but also provides information about the behaviour and metabolism of plants. It also gives valuable insight about how plants communicate or interact. Upcoming research will be more focused on the identification of the new plant volatile which in turn will help in giving an understanding of the plant's metabolism and mechanism. More focus will be there on developing various technologies for sensing different types of plant stresses which will help us in understanding their behaviour. This will provide us with more understanding of the pathogen infestation over the crop and allow us to take actions before the crop is destroyed. In future, it is expected that new sensor technology will be developed that is capable of tracking various kinds of plants VOCs with low price and take less time in sensing.

References

Aksenov AA et al. (2013) Volatile organic compounds (VOCs) for noninvasive plant diagnostics. In: Pest management with natural products. American Chemical Society, Washington, pp 73–95
Bailly A, Weisskopf L (2012) The modulating effect of bacterial volatiles on plant growth: current knowledge and future challenges. Plant Signali Behav 7:79–85
Bednarek P, Kwon C, Schulze-Lefert P (2010) Not a peripheral issue: secretion in plant–microbe interactions. Curr Opin Plant Biol 13:378–387
Blom D, Fabbri C, Eberl L, Weisskopf L (2011) Volatile-mediated killing of Arabidopsis thaliana by bacteria is mainly due to hydrogen cyanide. Appl Environ Microbiol 77:1000–1008
Bonfante P, Anca I-A (2009) Plants, mycorrhizal fungi, and bacteria: a network of interactions. Annu Rev Microbiol 63:363–383
Borges EdO, Martins CB, da Silva RR, Zarbin PH (2018) Terpenoids dominate the bouquet of volatile organic compounds produced by *Passiflora edulis* in response to herbivory by *Heliconius erato phyllis* (Lepidoptera: Nymphalidae). Arthropod-Plant Interact 12:123–131
Brilli F, Loreto F, Baccelli I (2019) Exploiting plant volatile organic compounds (VOCs) in agriculture to improve sustainable defense strategies and productivity of crops. Front Plant Sci 10:264
Catini A et al (2019) Development of a sensor node for remote monitoring of plants. Sens 19:4865
Chakraborty S, Newton AC (2011) Climate change, plant diseases and food security: an overview. Plant Pathol 60:2–14
Chang KP, Zakaria A, Nasir AA, Yusuf N, Thriumani R, Shakaff AYM, Adom AH (2014) Analysis and feasibility study of plant disease using e-nose. In: 2014 IEEE International Conference on Control System, Computing and Engineering (ICCSCE 2014), IEEE, pp 58–63
Cho SM et al. (2008) 2R, 3R-butanediol, a bacterial volatile produced by *Pseudomonas chlororaphis* O6, is involved in induction of systemic tolerance to drought in *Arabidopsis thaliana*. Mol Plant Microbe Interact 21:1067–1075
Coppedè N et al. (2017) An in vivo biosensing, biomimetic electrochemical transistor with applications in plant science and precision farming. Sci Rep 7:1–9
Cui S, Inocente EAA, Acosta N, Keener H, Zhu H, Ling PP (2019) Development of fast E-nose system for early-stage diagnosis of aphid-stressed tomato plants. Sens 19:3480
Cui S, Ling P, Zhu H, Keener HM (2018) Plant pest detection using an artificial nose system: a review. Sens 18:378
de Julián Fernández C et al (2008) Surface plasmon resonance optical gas sensing of nanostructured ZnO films. Sens Actuators, B 130:531–537
Fabbri B et al (2020) Correlation of gaseous emissions to water stress in tomato and maize crops: from field to laboratory and back. Sens Actuators, B: Chem 303:127227
Fang Y, Umasankar Y, Ramasamy RP (2014a) Electrochemical detection of p-ethylguaiacol, a fungi infected fruit volatile using metal oxide nanoparticles. Analyst 139:3804–3810
Fang Y, Umasankar Y, Ramasamy RP (2014b) Plant volatile sensor: enzymatic transducer for selective and sensitive determination of methyl salicyalte. In: Meeting abstracts,MA2014-01 414
Fang Y, Umasankar Y, Ramasamy RP (2016) A novel bi-enzyme electrochemical biosensor for selective and sensitive determination of methyl salicylate. Biosens Bioelectron 81:39–45
Ge C, Xie C, Cai S (2007) Preparation and gas-sensing properties of Ce-doped ZnO thin-film sensors by dip-coating. Mater Sci Eng, B 137:53–58
Giraldo JP et al (2015) A ratiometric sensor using single chirality near-infrared fluorescent carbon nanotubes: Application to in vivo monitoring. Small 11:3973–3984
Han SH et al. (2006) GacS-dependent production of 2R, 3R-butanediol by Pseudomonas chlororaphis O6 is a major determinant for eliciting systemic resistance against Erwinia carotovora but not against Pseudomonas syringae pv. tabaci in tobacco. Mol Plant Microbe Interact 19:924–930
Harun FC, Covington J, Gardner J (2009) Portable e-Mucosa system: mimicking the biological olfactory. Procedia Chem 1:991–994

Hazarika S, Choudhury R, Montazer B, Medhi S, Goswami MP, Sarma U (2020) Detection of Citrus Tristeza virus in Mandarin Orange using a customdeveloped electronic nose system. IEEE Transactions on Instrumentation and Measurement, 69(11):9010–9018

Itoh T, Wang J, Matsubara I, Shin W, Izu N, Nishibori M, Murayama N (2008) VOCs sensing properties of layered organic–inorganic hybrid thin films: MoO3 with various interlayer organic components. Mater Lett 62:3021–3023

Kai M, Piechulla B (2009) Plant growth promotion due to rhizobacterial volatiles–An effect of CO_2? FEBS Lett 583:3473–3477

Kim JJ, Allison LK, Andrew TL (2019) Vapor-printed polymer electrodes for long-term, on-demand health monitoring. Sci Adv 5:eaaw0463

Kim JJ, Fan R, Allison LK, Andrew TL (2020) On-site identification of ozone damage in fruiting plants using vapor-deposited conducting polymer tattoos. Sci Adv 6:eabc3296

Kim YS (2009) Thermal treatment effects on the material and gas-sensing properties of room-temperature tungsten oxide nanorod sensors. Sens Actuators, B: Chem 137:297–304

Koman VB, Lew TT, Wong MH, Kwak S-Y, Giraldo JP, Strano MS (2017) Persistent drought monitoring using a microfluidic-printed electro-mechanical sensor of stomata in planta. Lab Chip 17:4015–4024

Komaraiah P et al (2004) Growth behavior in plant cell cultures based on emissions detected by a multisensor array. Biotechnol Prog 20:1245–1250

Lee B, Farag MA, Park HB, Kloepper JW, Lee SH, Ryu C-M (2012) Induced resistance by a long-chain bacterial volatile: elicitation of plant systemic defense by a C13 volatile produced by Paenibacillus polymyxa. PloS one 7:e48744

Lee K et al. (2014) In-situ synthesis of carbon nanotube–graphite electronic devices and their integrations onto surfaces of live plants and insects. Nano Lett 14:2647–2654

Lew TTS, Koman VB, Gordiichuk P, Park M, Strano MS (2020a) The emergence of plant nanobionics and living plants as technology. Adv Mater Technol 5:1900657

Lew TTS et al. (2020b) Real-time detection of wound-induced H_2O_2 signalling waves in plants with optical nanosensors. Nature Plants 6:404–415

Li S, Simonian A, Chin BA (2010) Sensors for agriculture and the food industry. Electrochem Soc Interface 19:41

Maleknia SD, Bell TL, Adams MA (2007) PTR-MS analysis of reference and plant-emitted volatile organic compounds. Int J Mass Spectrom 262:203–210

Materić D, Bruhn D, Turner C, Morgan G, Mason N, Gauci V (2015) Methods in plant foliar volatile organic compounds research. Appl Plant Sci 3:1500044

Nassar JM, Khan SM, Villalva DR, Nour MM, Almuslem AS, Hussain MM (2018) Compliant plant wearables for localized microclimate and plant growth monitoring. npj Flex Electron 2:1–12

Robert-Seilaniantz A, Grant M, Jones JD (2011) Hormone crosstalk in plant disease and defense: more than just jasmonate-salicylate antagonism. Annu Rev Phytopathol 49:317–343

Schulz S, Dickschat JS (2007) Bacterial volatiles: the smell of small organisms. Natural Prod Rep 24:814–842

Shang L, Liu C, Chen B, Hayashi K (2018) Plant biomarker recognition by molecular imprinting based localized surface plasmon resonance sensor array: Performance improvement by enhanced hotspot of au nanostructure. ACS Sens 3:1531–1538

Su WW (2008) Bioreactor engineering for recombinant protein production using plant cell suspension culture. In: Gupta SD, Ibaraki Y (eds) Plant Tissue Culture Engineering. Focus on Biotechnology, vol 6. Springer, Dordrecht.

Umasankar Y, Ramasamy RP (2013) Highly sensitive electrochemical detection of methyl salicylate using electroactive gold nanoparticles. Analyst 138:6623–6631

Velásquez AC, Castroverde CDM, He SY (2018) Plant–pathogen warfare under changing climate conditions. Curr Biol 28:R619–R634

Vespermann A, Kai M, Piechulla B (2007) Rhizobacterial volatiles affect the growth of fungi and *Arabidopsis thaliana*. Appl Environ Microbiol 73:5639–5641

Wilson AD (2013) Diverse applications of electronic-nose technologies in agriculture and forestry. Sens 13:2295–2348
Wilson AD, Baietto M (2009) Applications and advances in electronic-nose technologies. Sens 9:5099–5148
Wu H et al. (2020) Monitoring plant health with near-infrared fluorescent H_2O_2 Nanosensors. Nano Lett 20:2432–2442
Xu J, Jia X, Lou X, Xi G, Han J, Gao Q (2007) Selective detection of HCHO gas using mixed oxides of ZnO/$ZnSnO_3$. Sens and Actuators B: Chem 120:694–699
Yi H, Ryu C, Heil M (2010) Sweet smells prepare plants for future stress: Airborne induction of plant disease immunity. Plant signal behav 5:528–531
Zhang P, Chen K (2009) Age-dependent variations of volatile emissions and inhibitory activity toward *Botrytis cinerea* and *Fusarium oxysporum* in tomato leaves treated with chitosan oligosaccharide. J Plant Biol 52:332–339
Zhu B, Xie C, Wu J, Zeng D, Wang A, Zhao X (2006) Influence of Sb, In and Bi dopants on the response of ZnO thick films to VOCs. Mater Chem Phys 96:459–465

Chapter 6
Current Trends of Plasmonic Nanosensors Use in Agriculture

Tahira Qureshi, Deniz Türkmen, and Adil Denizli

Abstract Plasmonic sensors based on the surface plasmon resonance (SPR) have shown prosperous growth in recent years due to their flexibility in designing it at smaller scale devices. The sensing techniques predominantly include surface-enhanced spectroscopic sensors such as surface-enhanced Raman scattering (SERS), surface-enhanced fluorescence (SEF) and surface-enhanced infrared absorption (SEIR), and SPR sensors which already have good commercial options. This chapter overviews and discusses the plasmonic nanosensors advancement and the most relevant applications in the field of agriculture. We focus on the effects that distinguish plasmonic nanosensors and give them their particular conduct for agrotechnology. We also evaluated the main applications of plasmonic nanosensors developed within the last five years for agriculture sector benefit.

Keywords Plasmonic nanosensors · Surface plasmon resonance · Surface-enhanced Raman scattering · Agricultural sensors · Nanotechnology in agriculture

6.1 Introduction

Agriculture sector plays a major role in the ineradicable development and financial flourishing of a country. In current times with growing populations, the agriculture sector also requires fast and advance approaches for its productive outcome. The advance agricultural productivity has some adverse effects on soil quality, water contamination, hypertrophication of water bodies and imbalance generated in waterways consequently disturbing the environment (Bezemer and Headey 2008; Milder et al. 2014). In recent times, food and nutritional safety measures are monitored by authorities and published for consumers knowledge and betterment. The modified agricultural product directly impacts the consumers health, climate changes,

T. Qureshi · D. Türkmen · A. Denizli (✉)
Department of Chemistry, Biochemistry Division, Hacettepe University, Ankara, Turkey
e-mail: denizli@hacettepe.edu.tr

R. N. Pudake et al. (eds.), *Biosensors in Agriculture: Recent Trends and Future Perspectives*, Concepts and Strategies in Plant Sciences,
https://doi.org/10.1007/978-3-030-66165-6_6

energy, ecosystem processes, natural resources, etc. Therefore, sustainable agricultural reinforcing helps in the practical approach toward better environment outcomes. Advancement in agriculture sector for sustainable production and efficient yield is also required for the improvement of agrochemicals (pesticides, fertilizers, etc.) too (Hobbs et al. 2008; Pretty 1995). Although, all agrochemicals are to some extent source of contamination in water bodies or residues on yielded crops. This issue leads to adverse effect on human health as well as on environment, which needs to resolve by smart and advance management (Almasri and Kaluarachchi 2004; Burney and Ramanathan 2014; Smith and Gilbertson 2018). It is sure that for the sustainable growth of agriculture also needed innovative technologies such as nanotechnology. This budding technology controls several exclusive electronic associations, plasmonic and optical properties which are related with the quantum confinement effects, the alteration of the electronic energy levels that may appear due to the surface area in relation to volume ratio. The merge of nanotechnology in agrotechnology has offered an advancement and beginning of new prospects in traditional agriculture practices. The nanotechnology approach has improved irrigation and fertilizer utilization, and enhanced food production and processing, packaging and storage. In addition, the nanotechnology-based sensing gained enormous momentum and provided broad-spectrum application in the field of agriculture (Kumari and Yadav 2014; Mousavi and Rezaei 2011; Prasad et al. 2014).

Scientific inventions in agriculture have been the main requirement to ease worldwide load of population growth, climate change and limited food availability. Sustainable agriculture has the power of strengthening a notion targeted at increasing yield of crops from the same agricultural fields without harmful impacts on environment. The utilization of nanotechnology approach in agriculture sector is an efficient tool for sustainable agrotechnology. Nanotechnology is efficiently helping in plant protection, reduces nutrient deficiency, helps fast pest detection and enriches yields through improved nutrient management (Poddar et al. 2018; Pudake et al. 2019).

The wide spectrum applications of nanotechnology in agriculture contain (1) soil efficiency enhancement such as use of nanoclays and nanozeolites which improves the water-holding ability of soil; (2) crop production enhancement by use of nanoparticles (NPs) in insecticides and pesticides; (3) crop improvement by using nanomaterials, which enabled the delivery of genetic material for crop development; (4) precision farming achieved by utilizing nanosensors for soil analysis, water management and transmission, pesticide, nutrient deficiency, nanomagnets for removal of soil contaminants. The precision farming in agriculture sector has helped in enormous need for rapid, consistent and cost-effective system for monitoring, detection and analysis of agricultural products. Precision farming applies nanosensors and nano-based smart delivery systems which help out farmers to employ agricultural resources such as water, nutrients and agrochemicals to increase crop productivity with improved fertilization supervision and reducing inputs (Chen and Yada 2011; Chhipa and Joshi 2016; Ditta 2012; Handford et al. 2014; Sekhon 2014).

Real-time monitoring of crop progress and farm condition such as moisture content, soil fertility, temperature, crop nutrient capacity, pathogens, plant diseases, etc., can be done through the advancement of nanotechnology. Utilization of nanosensors to accurately measure the soil condition (pH, nutrients, residual pesticides and soil moisture), checking of pathogens helps farmers to utilize inputs more proficiently. The current advancement in plasmon-based nanosensors has significantly overcome the shortcomings of traditional optical sensors, in terms of sensitivity, adjustability, photostability and in vivo applicability, by utilizing nanostructured materials. The aim of this chapter is to summarize the most relevant contributions in the agriculture sector of nanosensors under the plasmonic system (Rameshaiah et al. 2015; Sanchez-Cortes et al. 2001; Yang et al. 2014; Zeng et al. 2014).

6.2 Plasmonic Nanosensors Employed Techniques in Agriculture

Plasmonic sensors are availing SPR technique more often, which have shown prosperous development in recent years. In the agriculture field the sensing techniques predominantly comprise: surface-enhanced spectroscopic sensors such as surface-enhanced Raman scattering (SERS), surface-enhanced fluorescence (SEF) and surface-enhanced infrared absorption (SEIR), and SPR sensors.

6.2.1 Surface Plasmon Resonance (SPR)

Plasmonic tools and systems created on bottom-up synthesized or top-down fabricated metal nanostructures can be used to enhance light-matter interactions at the nanoscale in physical, chemical, and biological systems. Plasmonic is one main area of nano-optics. Due to its ability to generate nanoscale hot spots which are close to the size of bioparticles, it has been applied in biosensing mostly with enhanced sensitivity for refractive index (RI) changes and enhanced light/matter interactions. Plasmonic systems work under three modes such as propagating surface plasmon polariton (SPP) modes at extended continuous metal-dielectric interfaces, localized surface plasmon (LSP) modes at confined nanoscale metal-dielectric interfaces and delocalized Bloch plasmonic modes (or lattice plasmon modes) in periodic metal-dielectric nanostructures (Fig. 6.1). In some reports by connecting different modes of plasmonic building blocks into a composite plasmonic system, it is possible for their associated elementary plasmonic modes to strongly interact with each other and consequently produce multiple new hybridized modes at different resonant wavelengths with spatial mode overlap. From these plasmonic modes LSP is of our interest, which plays a vital role in the analyte detection or monitoring at the molecular level. LSPR mode is primarily used for plasmonics assembled on nanomaterials (Homola and Piliarik 2006; Piliarik and Homola 2009; Simsek 2010).

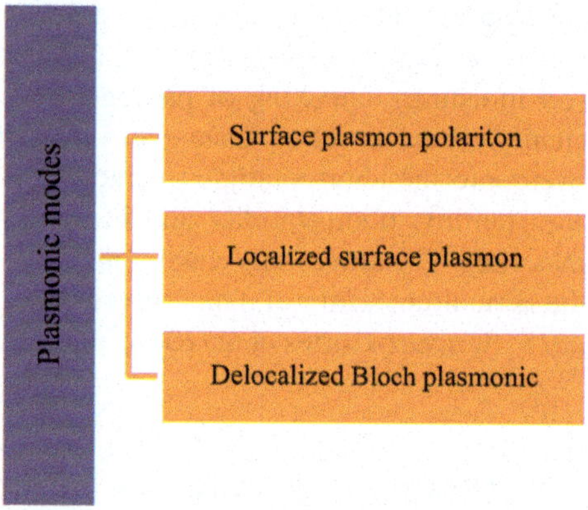

Fig. 6.1 Basic plasmonic system modes

LSPR is based on metallic nanoparticles, and the plasmonic nanosensors using it appeared as a strong tool for biosensing applications. Many detection schemes have been developed and the field is rapidly growing to incorporate new methodologies and applications. Propagating SPR is performed on a noble metal thin film, which shows even the slightest change in refractive index near the metal surface and produces a resonance angle shift. Hence, SPR sensors have been used practically as biosensors and chemical sensors in many scientific reports. However, LSPR of a noble metal nanoparticle (NP) can be utilized as sensors too because of their shift in resonant wavelength in response to a refractive index change. LSPR sensors basically calculate an extinction (absorption + scattering) peak shift or change in extinction energy, it simply works via configuration without a prism which makes it cost-efficient and appropriate for miniaturize setup. This characteristic helps in real-time and on-site environmental, food, and medical investigations. LSPR sensors also have some shortcomings like for bulk refractive index changes its response is less sensitive as compare to SPR sensors. SPR is sensitive to refractive index changes in the close vicinity of the NP surface due to a closely confined optical near field (Hong et al. 2012; Sari et al. 2018).

Propagating SPR is performed on a noble metal thin film, which shows even slightest change in refractive index near the metal surface and produces a resonance angle shift. Hence, SPR sensors have been used practically as biosensors and chemical sensors in many scientific reports. However, LSPR of a noble metal nanoparticle (NP) can be utilized as sensors too because of their shift in resonant wavelength in response to a refractive index change. LSPR sensors basically calculate an extinction (absorption + scattering) peak shift or change in extinction energy, it simply works via configuration without a prism which makes it cost-efficient and appropriate for miniaturize setup. This characteristic helps in real-time and on-site environmental, food and medical investigations. LSPR sensors also have some shortcomings like for bulk refractive index changes its response is less sensitive as compare to SPR sensors. SPR is sensitive to refractive index changes in the close vicinity of the NP surface due to a closely confined optical near field (Gupta and Kondoh 2007; Sherry et al. 2005).

Guo and coworkers have reviewed recent improvements for LSPR and described them in schematic form as follows: (1) refractive index nanosensors, (2) colorimetric sensing based on plasmon coupling and (3) amplified sensitivity based on nanoparticle growth (Fig. 6.2). They suggested to impact the local dielectric properties surrounding plasmonic nanoparticles that are motivated by several factors such as nanoparticle characteristics, amplification schemes and plasmon coupling. They also give distinct approaches for plasmonic nanosensors response enhancement with outcome prediction (Table 6.1) (Guo et al. 2015a).

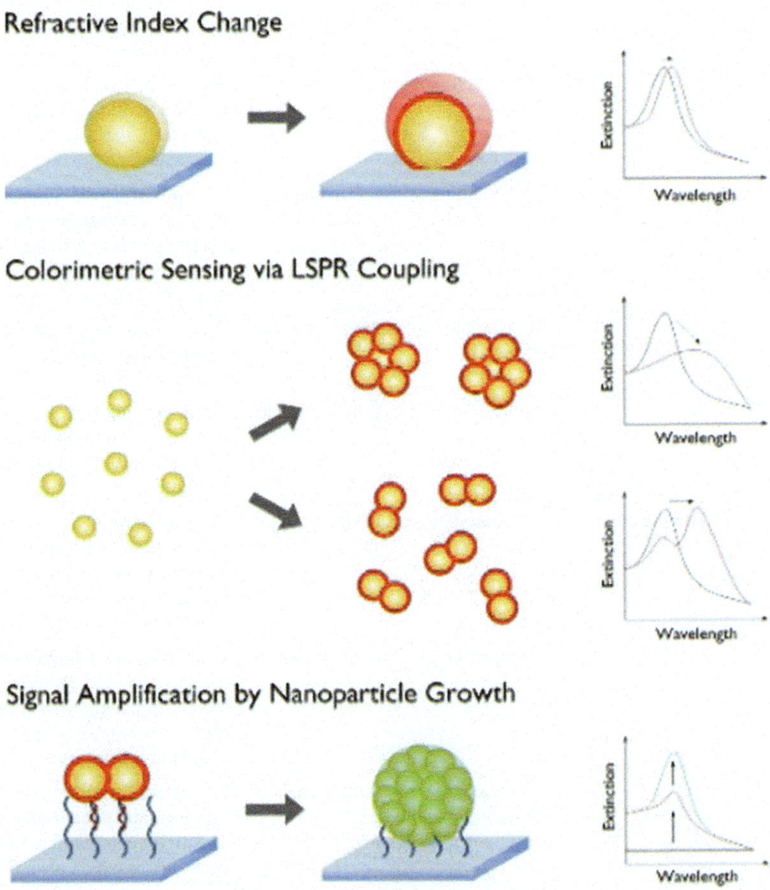

Fig. 6.2 Illustration of different strategies to improve the sensitivity of plasmonic nanosensors. Reprinted with permission (Gupta and Kondoh 2007)

Table 6.1 Strategies used for enhancing the sensitivity of plasmonic nanosensors (Reprinted with permission from Guo et al. 2015a)

Category	Strategies for enhancing sensitivity	Comments
Refractive index-based sensing	Composition	Ag is superior to Au. However, the high reactivity of Ag nanoparticles must be considered
	Size	Small size nanoparticles have high absorption ratios, while big size nanoparticles have high scattering ratios
	Shape	Anisotropic nanostructures exhibit higher refractive index sensitivity than spherical nanoparticles
	Increased molecular mass	An antibody or a nanoparticle labelled antibody could effectively increase the molecular mass of a target
	3D assembled nanostructures	3D assembly may enhance the intensity of sensing signal but the interparticle distance should be carefully controlled to avoid peak broadening
	Plasmonic-molecular resonance coupling	An effective way to detect small-molecular-weight chromophores
	Sensing with single nanoparticles	Single nanoparticle sensing greatly improves the molecular detection limit and the spatial resolution
Sensing based on plasmon coupling	Increased aggregate size	Enzyme-amplified gold nanoparticle aggregation may increase the size of aggregates, enhancing the detection sensitivity
	Minimized interparticle distance	Decreasing the interparticle distance may greatly enhance the plasmon coupling so that effectively improves the detection sensitivity
	To detect the scattering signals of plasmon coupling	The detection of scattering signal offers greatly enhanced detection sensitivity compared with detection based on absorbance
Sensing based on nanoparticle growth	Scanometric array detection coupled with silver enhancement	The silver enhancement significantly improves the detection sensitivity

(continued)

Table 6.1 (continued)

Category	Strategies for enhancing sensitivity	Comments
	Biocatalytic enlargement of AuNP	Any compounds involved in the biocatalytic reaction can be detected

6.2.2 *Surface-Enhanced Raman Scattering*

Raman spectroscopy in early times was considered as weak scattering process, till the advancement of Surface Enhanced Raman Scattering (SERS) in the early 1970s. Here, LSPs have coupled with Raman scattered light, which generates the intense signal in response on the molecular level (Chang 2013). The SERS boosted signal has resulted due to the Raman signals acquired from a given number of molecules in the presence of metal nanostructure, in this size and morphology of the nanostructures play a major role in the signals response of the analyte. The SERS system is competent to detect and differentiate in a broad range of analytes within a minute time scale and with negligible interference from environmental contaminants (Meng et al. 2014; Song et al. 2015; Zhu et al. 2016).

Also, in the study of pathogens, the SERS-based detection for slow growing bacteria is an effective tool as compared to basic laboratory testing which on average takes a few weeks for completion of analysis. Although the response obtained by SERS on quantification of bacteria shows lesser molecular specificity. SERS-based analysis does not have direct application for pure bacterial culture but for clinical samples can be detected directly. The multi-disciplinary applications of SERS have been enhanced via advanced instrumentation of RAMAN along with nanotechnology innovation makes it a valuable tool (Cheng et al. 2010; Jarvis et al. 2006).

6.2.3 *Colorimetric Nanosensors*

Plasmonic colorimetric sensors have appeared as a predominant device in chemical and biological sensing devices as compare to LSPR because of its lack of response in the visible range. The most prominent plasmonic colorimetric sensor has an efficient working response at plasmon coupling anticipated by aggregation of nanoparticles. The colorimetric nanosensors have applied to detect a particular analyte on color changing, which can be observed by naked eyes or portable optical detectors for quantitative measurement (Tang and Li 2017; Xiao et al. 2014).

6.2.4 *Miscellaneous Coupled Techniques*

The sensitivity improvement has been achieved by coupling between plasmonic-based materials detection of various approaches. For instance, LSPR combined with SERS allows sensitive molecular identification by overcoming LSPR limitations. The electromagnetic improvement in LSPR mode of nanoparticles stimulates the enhanced Raman excitation and emission in SERS mode for sensitive identification of analyte (Camden et al. 2008; Singh et al. 2009). Although, these multidisciplinary approaches have broad room for investigating in detail to conclude its characteristics. There is a need for improvement required to make beneficial research outcomes at cost-effective and reliable analysis. Enhancing the detection limit of plasmonic nanosensing facilitating a variety of sample matrices (water, soil, serum, etc.) will allow the detection of compounds that were previously not detected.

6.3 Application of Plasmonic Nanosensors in Agriculture

Nanotechnology has broadened the unconventional agricultural practices by producing safer and better quality of agricultural products, additionally has helped in generating new agricultural products, improved storage options and provided better treatment options for its directly associated field such as water and soil quality. Agro-nanotechnology has helped in various domains, like for better yield of crops with nanomaterial-based fertilizers, pesticides, herbicides, advanced genetic material agent for transfer, nutrient improvement of soils or crops and sensors for directly or indirectly linked compounds. To investigate and further improve in agro-nanotechnology, both prospects are required specific and rapid analysis in these domains. To achieve accurate, reliable and fast analysis in this area plasmonic nanosensors are playing a key role. With the minimum requirement of sample volume, real time analysis and sophistication of nanomaterials improve the outcome. Various applications of these plasmonic nanosensors have reported, which are shown in Table 6.2 from 2015 to 2019.

Kubackova and coworkers have reported the detection of the organochlorine pesticides (OPs) aldrin, dieldrin, lindane, and α-endosulfan by using SERS. To overcome the problem associated with the binding ability of these pesticides, they functionalized the metal surfaces with alkyl dithiols. By doing this they have achieved increased nanoparticle linkages and created interparticle junctions that can act as sensitive hot spots that were needed for SERS enhancement. Also, they were able to create suitable nanogaps between the metal nanoparticles, which helped in assembly and detection of pesticides. In experiments the metal surface was treated with dithiol, which forms aliphatic chain layers on the metallic surface and the interparticle cavity, this makes the reported system efficient for OPs detection (Kubackova et al. 2015). Thiram (tetramethylthiuram disulfide) has an important status among dithiocarbamate (DTC) pesticide due to its frequent usage in fungal affected crops, vegetables, seeds, etc.

Table 6.2 Plasmonic nanosensors utilized in agriculture (2015–2019)

Nanosensor/Sensor	Analyte	Linear range	Detection limit	Analytical technique	Ref.
Au Nanostars	Thiram	10^{-5}–10^{-10} M	1×10^{-10} M	SERS	Zhu et al. (2018)
Au NPs and Ag NPs colloidal suspension	(a) aldrin, (b) endosulfan, (c) lindane, and (d) dieldrin	(a) 0.1×10^{-7}–5×10^{-7} M (b) 0.1×10^{-7}–10×10^{-7} M (c) 0.2×10^{-6}–7×10^{-6} M (d) 0.1×10^{-7}–10×10^{-7} M	(a) 45 μg L^{-1}, (b) 170 μg L^{-1} (c) 1028 μg L^{-1}, (d) 316 μg L^{-1}	SERS	Kubackova et al. (2015)
(a) Au@Ag core–shell NCs (b) Au@Ag NBs	Thiram	1×10^{-6}–1×10^{-10} M	(a) 100 pM (b) 80 pM	SERS	Guo et al. (2015)
Au nanoisland films	Thiram	5–250 ppb	20 fM	SERS	Khlebtsov et al. (2015)
AgNP/TiO_2/SiO	Parathion	NA	10^{-5} M	RS whispering-gallery modes	Zhang et al. (2015)
Imprinted film	Parathion methyl	10^{-13}–10^{-10} mol/L	10^{-13} mol L^{-1}	SPR	Tan et al. (2015)
Ag-NC@PE	(a) Thiram, (b)4- Polychlorinated biphenyl, (c) Methyl parathion	10 nM to 1-pM	(a) 10 nM (b) 1 μM (c) 10 nM	SERS	Zhou et al. (2016)
Ag Nanoslits	Imidacloprid	10 ppm–10 ppb	1 ppb	SPR	Lee et al. (2016)

(continued)

Table 6.2 (continued)

Nanosensor/Sensor	Analyte	Linear range	Detection limit	Analytical technique	Ref.
AuNPs/DW	(a) Thiram (b) Carbaryl	10^{-4} M–10^{-8} M	10^{-7} M	SERS	Wei et al. (2016)
L-cys-Ag NPs	(a) Cypermethrin (b) Monosultap	0–100 ppm	100 ppm	Colorimetric sensing	Kodir et al. (2016)
MIP/PHEMA NPs	Atrazine	0.5–15 ng mL^{-1}	0.7134 ng mL^{-1}	SPR	Yılmaz et al. (2017)
Au nanopillars	Thiram	10^{-9} M–10^{-5} M	62 pg/mL	LSPR	Rippa et al. (2017)
3D MoS_2-NS@Ag-NP/MoS_2-NS@Au-seeds	Thiram	10 ppb–1 ppm	42 nM	SERS	Liang et al. (2017)
Au NPs	(a) Parathion-methyl (b) chlorpyrifos	10^{-5}–10^{-10} M	(a) 1×10^{-3} M (b) 1×10^{-4} M	SERS	Wang et al. (2017)
Au NPs	Boric acid	0.01–100 mM	0.01 mM	LSPR	Morsin et al. (2017)
Ag-Au NPs	(a) Ethion, (b) Fenthion, (c) Malathion, (d) Parathion (e) Paraoxon-ethyl	NA	(a) 7.5×10^{-4} (b) 8.3×10^{-4} (c) 3.6×10^{-3} (d) 6.3×10^{-3} (e) 0	Colorimetric sensing	Dissanayake et al. (2019)
Au nano-chestnuts	Thiram	10^{-4} M–10^{-12} M	1×10^{-14} M	SERS	Geng et al. (2018)
Fe_3O_4@SiO2@ZnO@Ag	Thiram	10^{-5}–10^{-9} M.	1×10^{-9} M	SERS	Han et al. (2019)
AuAg nanochains	Thiram	10^{-3}–10^{-7} M	1×10^{-7} M	SERS	Jiao et al. (2019)

(continued)

Table 6.2 (continued)

Nanosensor/Sensor	Analyte	Linear range	Detection limit	Analytical technique	Ref.
AuNPs	Acetamiprid	0–5.0 μM	0.5 μM	Colorimetric sensing	Tian et al. (2016)

*Nanoparticles = NPs; Nanocubes = NCs; Nanocuboids = NBs; PE = Polyethylene; DW = dragonfly wing; L-cysteine = L-cys

Furthermore, its contamination in edible stuff such as fruits and vegetables is often reported. Zhu and coworkers detected Thiram from apple peels by SERS technique. They synthesized gold nano stars via the seed growth method. The linear range was observed from 10^{-5} to 10^{-10} M, and the method detection limit was 1×10^{-10} M. However, the detection limit of thiram on apple peels was obtained as 0.24 ng/cm^2 (Zhu et al. 2018). In another report, thiram was detected by SERS technique via plasmonic core-shell nanoparticle, with respect to change in dimension of analyte. They had reported the synthesis of monodisperse Au@Ag nanocubes (NCs) and Au@Ag nanocuboids (NBs). The detection limits were observed for NCs and NBs of thiram as 100 pM and 80 pM, respectively. Khlebtsov et al. (2015) used a wet chemical approach for fabrication of gold nanoisland films (NIFs) to detect thiram from apple peels. For synthesis, they used a uniform seeding method by using glass or silicon substrate to achieve manageable growth of gold nanoisland films. The reported SERS-based NIFs detected thiram from apple peels was found 5 ng/cm^2 (Guo et al. 2015b; Khlebtsov et al. 2015).

Parathion is a broad-spectrum organophosphorus insecticide. Parathion is classified in Restricted Use Pesticide (RUP) due to its toxicity. Zhang et al. (2015) synthesized twisted tubular AgNP/TiO2/SiO multiple nanomembranes to detect parathion. They pumped the pesticide parathion solution in the nanocavity by the capillary effect. The obtained detection limit was 10^{-5} M parathion. Another study had reported the development of an imprinted films (MIF) SPR chip for parathion methyl (PM) detection. The MIF was prepared by thermally initiated polymerization reactions (Tan et al. 2015; Zhang et al. 2015). Zhou et al. also detected PM along with Thiram and 4-polychlorinated biphenyl contamination by SERS by in situ approach. They synthesized Ag-nanocuboids (NC), which was placed on transparent polyethylene (PE) film to obtain Ag-NC@PE composite film. The flexible transparent AgNCs@PE composite film was used to detect thiram, polychlorinated biphenyl and methyl parathion from orange peels by in situ SERS method (Zhou et al. 2016). Imidacloprid is a well-renowned and widely used neonicotinoid insecticide, it is a broad-spectrum neonicotinoid insecticide. Imidacloprid is required to be detected from various matrices due to its extensive use. Lee et al. stated a portable sensing platform for label-free detection of imidacloprid pesticides. In this study, SPR biochips were used for simultaneous detection of four different concentrations of imidacloprid pesticides. The visual detection limit was found around 1 ppb (Lee et al. 2016). Wei et al. reported detecting the pesticide residues by using Au nanoparticles/dragonfly wing (AuNPs/DW) arrays as SERS substrate. They have functionalized SERS substrate for thiram and carbaryl, and the result has shown that the detection limit of both pesticides was obtained at 10^{-7} M (Wei et al. 2016). Kodir et al. have synthesized colorimetric sensor based on L-cysteine modified silver nanoparticles. In this study, Ag NPs were modified by L-cysteine (L-cys-AgNPs). The monosultap and cypermethrin were detected by L-cys-AgNPs substrate (Kodir et al. 2016). Atrazine is widely used in corn production due to its effectiveness, application flexibility and cost-effectiveness. After long-term exposure disrupts regular hormone function, initiating birth defects, reproductive tumors and weight loss in mothers and embryos. Atrazine may cause skin sensitization and allergy. Yilmaz et al. synthesized affinity-based SPR sensor

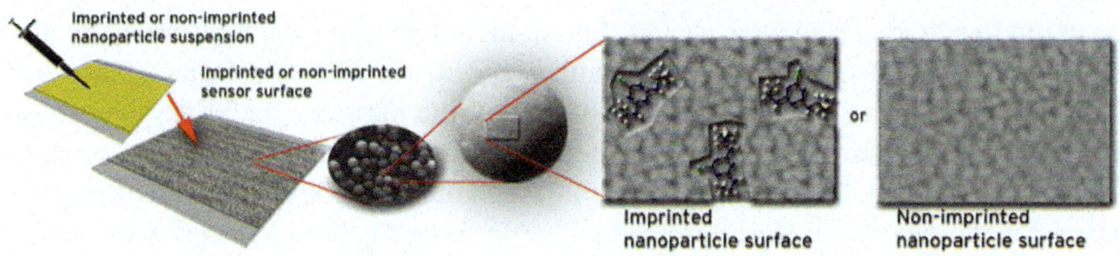

Fig. 6.3 Preparation of SPR sensor (Kodir et al. 2016)

system by molecular imprinting. Figure 6.3 shows the schematic representation of the SPR sensor preparation process. They combined SPR and MIP techniques for selective sensing and the sensor had long shelf-life too. The LOD and LOQ were determined as 0.7134 ng mL^{-1} and 2.378 ng mL^{-1}, respectively (Yılmaz et al. 2017).

Rippa and group mates analyzed the characteristics of the LSPR of aperiodic nanostructures with the arrangement in Thue-Morse (ThMo) NPs as a function of both their shapes and interparticle distances, comparing experimental and numerical results. They obtained a detection limit of 260 pM (62 pg mL^{-1}) of thiram pesticide was achieved using an LSPR nanosensor based on an optimized ThMo pattern with triangular NPs employing a phase-sensitive setup to increase the figure-of-merit (FOM) of the sensor (Rippa et al. 2017). In one more report, thiram was detected by uniform and well-dispersed 3D MoS2-NS@Ag-NP hybrids, which were fabricated by interfacial synthesis via a seeded growth route from MoS2-NS@Au seeds, which provide well-defined and simple platforms to attain a composite nanostructure (Liang et al. 2017). Wang et al. proposed sandwich configuration, which was accomplished by post-assembling a flexible transparent gel tape embedded with plasmonic nanoparticles onto an Au film decorated with desired analyte molecules. The main characteristic response obtained for pesticides, indicating that a post-assembly strategy is an effective approach for on-site detection of pesticide molecules (Wang et al. 2017). Morsin and the team studied the LSPR response of Au nanoplates to the presence of boric acid in water has been investigated. This study has shown that the LSPR of Au nanoplates is sensitive toward the presence of boric acid in concentrations as low as 0.01 mM (Morsin et al. 2017). Dissanayake et al. demonstrate the use of plasmonic Ag, Au, and bimetallic Ag-Au nanoparticles to detect and distinguish between organophosphorus (OP) pesticides. All pesticides showed nearly the same trend in their sedimentation with the plasmonic NPs (Dissanayake et al. 2019). Geng et al. fabricated gold nanochestnuts (GNCs) as ultra-sensitive SERS substrate for detecting trace pesticide residues and have been developed based on anodic aluminum oxide (AAO) templates. Then SERS measurement of thiram indicates a remarkable SERS-active sensitivity of the as-prepared GNCs with a detection limit of thiram up to 10^{-14} M (Geng et al. 2018). Han et al. also detected thiram by synthesized, $Fe_3O_4/SiO_2/ZnO/Ag$ (FSZA), which consisted of Fe_3O_4 core with strong magnetic field response and an intermediate SiO_2 layer as an electronic barrier to keep the stability of magnetite particles and outer ZnO and Ag as the effective layers for detecting pesticide (Han et al. 2019). Jiao et al. detected pesticide residues on agricultural products using SERS. They synthesized worm-like AuAg nanochains with highly interconnected ultrafine (~6.2 nm) bimetallic particles for SERS nanosensor via laser-assisted strategy. The SERS detection limit of thiram molecules on apple surfaces was about 10^{-7} M (Jiao et al. 2019). Tian et al. developed a colorimetric assay for the detection of acetamiprid. The flanking nucleotides were truncated to generation a 37-mer and 25-mer aptamers used as recognition elements in the sensing assays. The sensitivity was enhanced by removing flanking nucleotides outside of the binding region. This enhancement was attributed to complete dissociation of shorter target-binding aptamer without excess bases adhered to AuNPs (Tian et al. 2016).

6.4 Summary

In summary, a plasmon nanosensor has basically responded in two manners in accordance with substrates: (1) individual nanoparticles and their random aggregates, or (2) chip-based two-dimensional (2D) or three-dimensional (3D) nano-array patterns. The accessibility of fabrication methods along with the invention and knowledge of plasmonic modes have played role in frequent use of plasmon approach for research. Innovative approaches of fabrication process have made it easy in comparison to previous conventional photolithography. Although, nano-arrays fabrication at nano level is cost-efficient and possesses elasticity for adjusting optical characteristics. There are yet technological hurdles for the fabrication of large-area, long-range-ordered periodic nano-array models. The optical probes like colorimetric have performed the qualitative detection of pesticide residues, heavy metal and some hazardous organic molecules in various edible crops, fruits and vegetables by using plasmonic nanomaterials substrate. With plasmonic nanosensors colorimetric technique performs sensitive and rapid analysis. Above all plasmonic nanosensors facilitate for on-site monitoring of the pollutants, either from agricultural soil, crops or water assessment. From this point forward, the expansion of various morphological plasmonic nanosensors will increase the sensitivity and rapid detection of pesticides residues, and also their delivery mechanisms for enhancement of the yield. Further, the NPs clustered evenly on the surfaces like papers or chips is broadly emerging trend too. The progress of plasmonic nanosensors have the potential for inexpensive, competent, durable, and eco-friendly substrates with the potential in the direction of onsite applications. Although, to employ agro-nanotechnology has critical aspect too, the approach of synthesis needs to be improved such as biocompatible and ecofriendly in nature to gain better outcomes.

References

Almasri MN, Kaluarachchi JJ (2004) Assessment and management of long-term nitrate pollution of ground water in agriculture-dominated watersheds. J Hydrol 295:225–245

Bezemer D, Headey D (2008) Agriculture, development, and urban bias. World Dev 36:1342–1364

Burney J, Ramanathan V (2014) Recent climate and air pollution impacts on Indian agriculture. Proc Natl Acad Sci 111:16319–16324

Camden JP, Dieringer JA, Zhao J, Van Duyne RP (2008) Controlled plasmonic nanostructures for surface-enhanced spectroscopy and sensing. Acc Chem Res 41:1653–1661. https://doi.org/10.1021/ar800041s

Chang R (2013) Surface enhanced Raman scattering. Springer Nature Switzerland AG

Chen H, Yada R (2011) Nanotechnologies in agriculture: new tools for sustainable development. Trends Food Sci Technol 22:585–594

Cheng I-F, Lin C-C, Lin D-Y, Chang H-C (2010) A dielectrophoretic chip with a roughened metal surface for on-chip surface-enhanced Raman scattering analysis of bacteria. Biomicrofluidics 4:034104

Chhipa H, Joshi P (2016) Nanofertilisers, nanopesticides and nanosensors in agriculture. In: Nanoscience in food and agriculture 1. Springer, pp 247–282

Dissanayake NM, Arachchilage JS, Samuels TA, Obare SO (2019) Highly sensitive plasmonic metal nanoparticle-based sensors for the detection of organophosphorus pesticides. Talanta 200:218–227. https://doi.org/10.1016/j.talanta.2019.03.042

Ditta A (2012) How helpful is nanotechnology in agriculture? Adv Nat Sci Nanosci Nanotechnol 3:033002

Geng F, Zhao H, Fu Q, Mi Y, Miao L, Dong Y, Wu M, Lei Y (2018) Gold nanochestnut arrays as ultra-sensitive SERS substrate for detecting trace pesticide residue. Nanotechnology 29:295502. https://doi.org/10.1088/1361-6528/aac22b

Guo L, Jackman JA, Yang H-H, Chen P, Cho N-J, Kim D-H (2015) Strategies for enhancing the sensitivity of plasmonic nanosensors. Nano Today 10:213–239. https://doi.org/10.1016/j.nantod.2015.02.007

Guo P, Sikdar D, Huang X, Si KJ, Xiong W, Gong S, Wei Yap L, Premarante M and Cheng W (2015b) Plasmonic core-shell nanoparticles for SERS detection of the pesticide thiram: size- and shape-dependent Raman enhancement. Nanoscale 7:2862–2868. https://doi.org/10.1039/c4nr06429a

Gupta G, Kondoh J (2007) Tuning and sensitivity enhancement of surface plasmon resonance sensor. Sens Actuators B Chem 2:381–388. https://doi.org/10.1016/j.snb.2006.06.005

Han D, Yao J, Quan Y, Gao M, Yang J (2019) Plasmon-coupled charge transfer in FSZA core-shell microspheres with high SERS activity and pesticide detection. Sci Rep 9:13876. https://doi.org/10.1038/s41598-019-50374-y

Handford CE, Dean M, Henchion M, Spence M, Elliott CT, Campbell K (2014) Implications of nanotechnology for the agri-food industry: opportunities, benefits and risks. Trends in Food Sci Technol 40:226–241

Hobbs PR, Sayre K, Gupta R (2008) The role of conservation agriculture in sustainable agriculture. Philos Trans R Soc B: Biol Sci 363:543–555

Homola J, Piliarik M (2006) Surface plasmon resonance (SPR) sensors. In: Surface plasmon resonance based sensors. Springer, pp 45–67

Hong Y, Huh Y-M, Yoon DS, Yang J (2012) Nanobiosensors based on localized surface plasmon resonance for biomarker detection. J Nanomater 2012:759830. https://doi.org/10.1155/2012/759830

Jarvis RM, Brooker A, Goodacre R (2006) Surface-enhanced Raman scattering for the rapid discrimination of bacteria. Faraday Discuss 132:281–292

Jiao A, Dong X, Zhang H, Xu L, Tian Y, Liu X, Chen M (2019) Construction of pure worm-like AuAg nanochains for ultrasensitive SERS detection of pesticide residues on apple surfaces. Spectrochim Acta Part A Mol Biomol Spectrosc 209:241–247. https://doi.org/10.1016/j.saa.2018.10.051

Khlebtsov BN, Khanadeev VA, Panfilova EV, Bratashov DN, Khlebtsov NG (2015) Gold nanoisland films as reproducible SERS substrates for highly sensitive detection of fungicides. ACS Appl Mater Interfaces 7:6518–6529. https://doi.org/10.1021/acsami.5b01652

Kodir A, Imawan C, Permana IS, Handayani W (2016) Pesticide colorimetric sensor based on silver nanoparticles modified by L-cysteine. In: 2016 International Seminar on Sensors, Instrumentation, Measurement and Metrology (ISSIMM), 10–11 Aug, pp 43–47

Kubackova J, Fabriciova G, Miskovsky P, Jancura D, Sanchez-Cortes S (2015) Sensitive surface-enhanced Raman spectroscopy (SERS) detection of organochlorine pesticides by alkyl dithiol-functionalized metal nanoparticles-induced plasmonic hot spots. Anal Chem 87:663–669. https://doi.org/10.1021/ac503672f

Kumari A, Yadav SK (2014) Nanotechnology in agri-food sector. Crit Rev Food Sci Nutr 54:975–984. https://doi.org/10.1080/10408398.2011.621095

Lee K-L, You M-L, Tsai C-H, Lin E-H, Hsieh S-Y, Ho M-H, Hsu J-C, Wei P-K (2016) Nanoplasmonic biochips for rapid label-free detection of imidacloprid pesticides with a smartphone. Biosens Bioelectron 75:88–95. https://doi.org/10.1016/j.bios.2015.08.010

Liang X, Wang Y-S, You T-T, Zhang X-J, Yang N, Wang G-S, Yin P-G (2017) Interfacial synthesis of a three-dimensional hierarchical MoS 2-NS@ Ag-NP nanocomposite as a SERS nanosensor for ultrasensitive thiram detection. Nanoscale 9:8879–8888

Meng X, Kim S, Puligundla P, Ko S (2014) Carbon dioxide and oxygen gas sensors-possible application for monitoring quality, freshness, and safety of agricultural and food products with emphasis on importance of analytical signals and their transformation. J Korean Soc for Appl Biol Chem 57:723–733

Milder JC, Hart AK, Dobie P, Minai J, Zaleski C (2014) Integrated landscape initiatives for African agriculture, development, and conservation: a region-wide assessment. World Dev 54:68–80

Morsin M, Mat Salleh M, Ali Umar A, Sahdan MZ (2017) Gold nanoplates for a localized surface plasmon resonance-based boric acid sensor. Sensors (Basel) 17(5):947. https://doi.org/10.3390/s17050947

Mousavi SR, Rezaei M (2011) Nanotechnology in agriculture and food production. J Appl Environ Biol Sci 1:414–419

Piliarik M, Homola J (2009) Surface plasmon resonance (SPR) sensors: approaching their limits? Opt Express 17:16505–16517

Poddar K, Vijayan J, Ray S, Adak T (2018) Chapter 10—Nanotechnology for sustainable agriculture. In: Singh RL, Mondal S (eds) Biotechnology for sustainable agriculture. Woodhead Publishing, pp 281–303. https://doi.org/10.1016/B978-0-12-812160-3.00010-6

Prasad R, Kumar V, Prasad KS (2014) Nanotechnology in sustainable agriculture: present concerns and future aspects. Afr J Biotech 13:705–713

Pretty JN (1995) Participatory learning for sustainable agriculture. World Dev 23:1247–1263

Pudake RN, Chauhan N, Kole C (2019) Nanoscience for sustainable agriculture. Springer International Publishing

Rameshaiah G, Pallavi J, Shabnam S (2015) Nano fertilizers and nano sensors—an attempt for developing smart agriculture. Int J Eng Res Gen Sci 3:314–320

Rippa M, Castagna R, Pannico M, Musto P, Tkachenko V, Zhou J, Petti L (2017) Engineered plasmonic Thue-Morse nanostructures for LSPR detection of the pesticide Thiram. Nanophotonics 6:1083–1092

Sanchez-Cortes S, Domingo C, Garcia-Ramos J, Aznarez J (2001) Surface-enhanced vibrational study (SEIR and SERS) of dithiocarbamate pesticides on gold films. Langmuir 17:1157–1162

Sari E, Üzek R, Duman M, Denizli A (2018) Detection of ciprofloxacin through surface plasmon resonance nanosensor with specific recognition sites. J Biomater Sci Polym Ed 29:1302–1318

Sekhon BS (2014) Nanotechnology in agri-food production: an overview. Nanotechnol Sci Appl 7:31–53. https://doi.org/10.2147/NSA.S39406

Sherry LJ, Chang SH, Schatz GC, Van Duyne RP, Wiley BJ, and Xia Y (2005) Localized surface plasmon resonance spectroscopy of single silver nanocubes. Nano Lett 10:2034–2038. https://doi.org/10.1021/nl0515753

Simsek E (2010) Full analytical model for obtaining surface plasmon resonance modes. Opt Express 18:1722–1733

Singh NK, Alqudami A, Annapoorni S, Sharma V, Muralidhar K (2009) Enhanced bio-molecular sensing capability of LSPR, SPR-ATR coupled technique. In: AIP Conference Proceedings. vol 1. American Institute of Physics, pp 331–338

Smith AM, Gilbertson LM (2018) Rational ligand design to improve agrochemical delivery efficiency and advance agriculture sustainability. ACS Sustainable Chem Eng 6:13599–13610

Song W, Ji W, Vantasin S, Tanabe I, Zhao B, Ozaki Y (2015) Fabrication of a highly sensitive surface-enhanced Raman scattering substrate for monitoring the catalytic degradation of organic pollutants. J Mater Chem A 3:13556–13562

Tan Y, Ahmad I, Wei T-X (2015) Detection of parathion methyl using a surface plasmon resonance sensor combined with molecularly imprinted films. Chin Chem Lett 26:797–800. https://doi.org/10.1016/j.cclet.2015.04.001

Tang L, Li J (2017) Plasmon-based colorimetric nanosensors for ultrasensitive molecular diagnostics. ACS Sens 2:857–875

Tian Y, Wang Y, Sheng Z, Li T, Li X (2016) A colorimetric detection method of pesticide acetamiprid by fine-tuning aptamer length. Anal Biochem 513:87–92. https://doi.org/10.1016/j.ab.2016.09.004

Wang X, Zhu X, Chen Y, Zheng M, Xiang Q, Tang Z, Zhang G, Duan H (2017) Sensitive surface-enhanced Raman scattering detection using on-demand postassembled particle-on-film structure. ACS Appl Mater Interfaces 9:31102–31110. https://doi.org/10.1021/acsami.7b08818

Wei Y, Zhu Y-y, Wang M-l (2016) A facile surface-enhanced Raman spectroscopy detection of pesticide residues with Au nanoparticles/dragonfly wing arrays. Optik 127:10735–10739. https://doi.org/10.1016/j.ijleo.2016.08.118

Xiao Y, Jiaojiao J, Guohua H, Fangyuan Y, Minmin W, Jie H, Xiaoguo Y, Shanggui D (2014) Determination of the freshness of beef strip loins (*M. longissimus lumborum*) using electronic nose. Food Anal Methods 7:1612–1618. https://doi.org/10.1007/s12161-014-9796-8

Yang S, Wu T, Zhao X, Li X, Tan W (2014) The optical property of core-shell nanosensors and detection of atrazine based on localized surface plasmon resonance (LSPR) sensing. Sensors 14:13273–13284

Yılmaz E, Özgür E, Bereli N, Türkmen D, Denizli A (2017) Plastic antibody based surface plasmon resonance nanosensors for selective atrazine detection. Mater Sci Eng, C 73:603–610

Zeng S, Baillargeat D, Ho H-P, Yong K-T (2014) Nanomaterials enhanced surface plasmon resonance for biological and chemical sensing applications. Chem Soc Rev 43:3426–3452

Zhang J, Li J, Tang S, Fang Y, Wang J, Huang G, Liu R, Zheng L, Cui X, Mei Y (2015) Whispering-gallery nanocavity plasmon-enhanced Raman spectroscopy. Sci Rep 5:15012. https://doi.org/10.1038/srep15012

Zhou N, Meng G, Huang Z, Ke Y, Zhou Q, Hu X (2016) A flexible transparent Ag-NC@PE film as a cut-and-paste SERS substrate for rapid in situ detection of organic pollutants. Analyst 141:5864–5869. https://doi.org/10.1039/c6an00807k

Zhu C, Meng G, Zheng P, Huang Q, Li Z, Hu X, Wang X, Huang Z, Li F, Wu N (2016) A hierarchically ordered array of silver-nanorod bundles for surface-enhanced Raman scattering detection of phenolic pollutants. Adv Mater 28:4871–4876. https://doi.org/10.1002/adma.201506251

Zhu J, Liu M-J, Li J-J, Li X, Zhao J-W (2018) Multi-branched gold nanostars with fractal structure for SERS detection of the pesticide thiram. Spectrochim Acta Part A Mol Biomol Spectrosc 189:586–593. https://doi.org/10.1016/j.saa.2017.08.074

Chapter 7
Relevance of Biosensor in Climate Smart Organic Agriculture and Their Role in Environmental Sustainability: What Has Been Done and What We Need to Do?

Kingsley Eghonghon Ukhurebor and Charles Oluwaseun Adetunji

Abstract The application of climate smart agriculture (CSA) procedure has been recognized as a significant part in offering resolutions to the challenging issues of climate change (CC) and its extenuation for agricultural and environmental sustainability (AES). Moreover, it has been highlighted that several agricultural activities in several countries could be linked to the high amount of greenhouse gas emissions. The application of CSA technologies would play a significant role in the aspect of capacity building, skill and managerial future adoptions as well as intelligent management of natural resources for AES. Numerous biosensors and biosensing technologies (BSTs) like nanoparticles (NPs) and polymers along with their enormous benefits are presently being applied globally for resolving most of the challenging issues in AES. Nevertheless, it is important to integrate multi-faceted procedures for AES, so as to effectively apply biosensors that could conceivably be employed for various applications particularly in the area of CSA for AES. Hence, this chapter provides an assessment of BSTs and their wide-ranging benefits in CSA along with their role in AES. The restrictions confronted by some of the projecting BSTs particularly in relationship with CSA will be emphasized, and as well offer suitable recommendations for the improvement in AES.

Keywords Sustainability · Food · Environment · Biosensors · Agriculture

K. E. Ukhurebor (✉)
Climatic/Environmental/Telecommunication Unit, Department of Physics, Edo University Iyamho, P.M.B. 04, Auchi 312101, Edo State, Nigeria
e-mail: ukhurebor.kingsley@edouniversity.edu.ng

C. O. Adetunji
Applied Microbiology, Biotechnology and Nanotechnology Laboratory, Department of Microbiology, Edo University Iyamho, P.M.B. 04, Auchi 312101, Edo State, Nigeria

R. N. Pudake et al. (eds.), *Biosensors in Agriculture: Recent Trends and Future Perspectives*, Concepts and Strategies in Plant Sciences,
https://doi.org/10.1007/978-3-030-66165-6_7

7.1 Introduction

Agriculture is a domineering and steady part for the survival of human life. Engaging in activities of agricultural has hypothetically and incessantly been of huge assistance in producing food, as well as the production of several natural resources for both human consumption and industrial benefits (Garnett 2013; Karimi et al. 2018). The agricultural segment as described by the Food and Agriculture Organization of the United Nations (FAO) is faced with several problems such as failures in the market system and barriers in the trading system, uneven and futile socio-economic strategies, insufficient information, availability of finance and infrastructures, pressure at a result of an upsurge in the population and insufficiency resources, agronomic practices unsustainability and dilapidation in the environment, etc. (FAO 2015). These complications are further challenged by the effects of CC, since agriculture is mainly dependent on the variables or parameters of weather (Field et al. 2014; FAO 2015; Akrofi-Atitianti et al. 2018; Ukhurebor and Abiodun 2018). According to Akrofi-Atitianti et al. (2018), the agricultural segment in Africa and most of the other developing countries, are the greatest vulnerable segments to these complications that are resulting from CC.

The problem of food safety and security (FSS) along with the shortage in AES techniques as well as the problem of CC have great connexion (Karimi et al. 2018). Therefore, there is a need to lay more emphasis on their synergetic influence. In view of this, the United Nations Framework Convention on Climate Change (UNFCCC); as well as the Intergovernmental Panel on Climate Change (IPCC) have continuously emphasized the significance of AES, as they continually placed great precedence on AES (Field et al. 2014; FAO 2015). Both Karimi et al. (2018) and Ukhurebor et al. (2020a) reported that the effects of CC on AES are still insulated with some reservations and uncertainties. Hence, CC is expected to hostilely affect AES as well as other aspects of human doings universally. This could be as a result of the resultant effects of changes in most climatic variables or parameters such as temperature, rainfall, carbon dioxide pollination, etc. (Ukhurebor and Abiodun 2018; Herndon and Whiteside 2019; Nwankwo and Ukhurebor 2019; Ukhurebor and Nwankwo 2020; Nwankwo et al. 2020a, b). Accordingly, climate adaptation procedures are eventually critical for the mitigation and extenuation of these incessant climatic activities particularly in our atmospheric environment (Ukhurebor and Azi 2019; Ukhurebor et al. 2019; Ukhurebor and Umukoro 2018).

As reported by Abegunde et al. (2019), CSA has been identified as sustainable technique that is necessary for resolving both CC mitigation and AES, respectively. According to them the following are ways by which agricultural actions could contribute suggestively to CC mitigation:

- The prevention of more deforestation and adaptation or/and modification of the wetlands.
- The strengthening and escalation in the storage of carbon content of the soil and environment.

- The decrease and avoidance of present and impending surges in greenhouse gases (GHGs) emissions.

Hence, genuine efforts should be gear towards the lessening of GHG emissions, especially from agricultural activities. As reported by Fanen and Olalekan (2014), some of the utmost significant produce from agriculture have the budding of shading the gaps between innovative agricultural produces and some of these innovative produces have the potential for increasing contributions and management of abridged GHS emission opportunities.

CSA has some exceptional opportunities in handling the issues of food safety, adaptation, and control determinations (Fanen and Olalekan 2014; Abegunde et al. 2019). Undoubtedly, CSA is a reliable alternative that could be of assistance in the management of food insecurity problems, which are allegedly been instigated by CC (Fanen and Olalekan 2014; Abegunde et al. 2019). However, most unindustrialized countries have comprehended that some of these perceptions of CSA which have been recommended as a possible solution for solving the prevailing issues resulting from CC are by some means not too appropriate in their backgrounds due to some environmental unpredictability (Fanen and Olalekan 2014; Karimi et al. 2018; Abegunde et al. 2019). In addition, agriculture occupies a fundamental aspect in the assuagement of poverty and in enacting some core objectionable effects that CC is anticipated to have on quite a lot of regions universally.

Reportedly, initial achievement in CSA has been recognized as an essential means of capacity development along with skill and guide for imminent prospects (Abegunde et al. 2019). Nevertheless, it is advisable to have an appropriate meaning of "smart system (SS) or smart product" before reconnoitering to CSA practice. Abegunde et al. (2019) describe a SS as "that which facilitates the interface of a system with persons/users and is able to acclimate the framework of the user without compelling the user to acclimate to it". According to Karimi et al. (2018) and Abegunde et al. (2019), a SS will possibly encompass the following features:

- Competence or ability to work together with other mechanisms and devices.
- Flexibility in acquiring and improving the compatibility among its operational environment.
- Self-adequacy, specifies that SS can work without interference from the operator.
- Competence in networking with a person by means of a natural or ordinary interface.
- Multi-determined, specifies that a sole product of a SS is proficient in implementing several roles.
- Personality, specifies that a SS is capable of dynamic and accomplishing the structures of reliable performance.
- Reactivity, this specifies that a SS could respond to its surroundings in a distinct technique.

A SS has been identified as a crucial method in sensing, acting, and performing optimum operational control for routine or product eminence via what is known as "smart sensing techniques". The development of biosensors has significantly contributed to the improvement of both SS and CSA. Biosensors technology has been identified as a potential technique that could lead to increase in agricultural efficiency and the improvement in food production. These techniques could be utilized for controlling and management of immediate diagnosis of infectious diseases, improvement of plants capacity for nutrients fascination, the enhancement of the capability for animal production, etc. (Fanen and Olalekan 2014).

In addressing the influence of agricultural actions to these complications of ensuing from CC as a result of the modifications in the climate scheme, CSA is increasingly being endorsed in several regions globally, to contribution in the incorporation of the economy, societal and ecological extents of development sustainability in the development of the three main aspects; "sustainably increasing agricultural productivity and revenues; acclimating and building resilience to climate change and; reducing and/or removing GHGs emissions relative to conventional practices" (FAO 2015).

Moreover, the application of CSA has not yet been completely espoused in several developing regions, particularly in Africa. It is however alleged that the inadequate understanding on how to efficiently implement the adaptation procedures faced by those involved in agricultural actions across these regions are some of the limitations (Antwi-Agyei et al. 2014; Akrofi-Atitianti et al. 2018; Ukhurebor 2020). Nevertheless, several industrialized regions that are alleged to have the sufficient understanding that has started adopting and applying CSA, still have enormous arrangement to be put in place for its enhancement and development. Hence, biosensor in CSA are now been incorporated, and this would absolutely play an important role in AES.

This chapter will therefore attempt to present an appraisal of BSTs for the development of CSA drawn from previous studies as well as their role in AES as it relates to food security and CC, in the area of BST, for the enhancement and development of the agricultural sector. The constraints and restrictions that are confronted with most of the projecting procedures particularly as it recounts to CSA will be emphasized. It is believed that this be of great assistance in proffering beneficial recommendations for imminent research studies for the improvement and development of AES.

7.2 Benefits of Biosensors in CSA

According to Ukhurebor (2020), biosensors are known as investigative devices that combined biological ingredients and transducers for the detection of samples such as medications, metabolites, pollutants, infectious load, control variables, etc. They accomplish the task via the translation of "biochemical reactions" into computable physiochemical signals, which then quantify the quantity of sample that are employed for detecting the analyte concentration (Borgmann et al. 2011; Arora 2013; Vigneshvar et al. 2016; Salek-Maghsoudi et al. 2018). Biosensors have been reported to have numerous applications in the various aspects of human endeavours such as processing and safety of food, medical detection, and diagnosis, security and defence, AES, etc. (El-Said et al. 2020; Ukhurebor 2020).

There are numerous categories of biosensors employed for environmental sciences as it recounts to water, soil, and air in the area of CSA. These biosensors rely on detecting transducers or rudiments. Since the first discovering of biosensors which was known as the glucose one in 1956, by Clark, L.C Jnr, that came into public interest in 1975 (Turner et al. 1987), there are presently numerous biosensors that are now established for several commercial tenacities. These present-day biosensors have a widespread range of benefits that proposes supplementary sensitive, profligate, specific, perceptible, and multiplicative consequences as against the preceding ones (Borgmann et al. 2011; Arora 2013; El-Said et al. 2020; Ukhurebor 2020).

At the moment, with the development and improvement of nanotechnology, advanced NPs are now being developed and their advanced structures as well as their numerous applications that are present in biosensors (Rodriguez-Mozaz et al. 2005; El-Said et al. 2020; Ukhurebor 2020). NPs-built biosensors, comprise the combination of molecular engineering, biotechnology, chemistry, environmental science, physics, and material science. These diverse fields of study have been of boundless assistance in evolving the comprehension and specificity of "bioparticle/biomolecule" detection, the capability of sensing or operating particles (molecules/atoms), analysis and diagnosis of pathogenic, biomolecular recognition, along with the controlling and management of agricultural and the environmental actions (Rodriguez-Mozaz et al. 2005; El-Said et al. 2020). The benefits of the innumerable biosensors such as NPs, microorganisms built biosensors, and polymers for environmental and agricultural dealings, have helped in reducing the amount of substances spread, reduction in the losses of nutritional substances for insemination, and increase in the yields through pest and nutrient (Antonacci et al. 2016).

Biosensors are generally characterized into two groups which are instituted on sensing mechanisms and transduction procedures. The detecting/sensing mechanisms comprise "enzymes, antibodies (immunosensors), micro-organisms (cell biosensors), biological tissues and organelles". Whereas, the transduction approaches are centered on the physiochemical difference ensuing from the sensing mechanisms (Ukhurebor 2020). Consequently, unlike transducer biosensors could be either piezoelectric, electrochemical, calorimetric, or optical (Borgmann et al. 2011; Reyes De Corcuera and Cavalieri 2003). Reyes De Corcuera and Cavalieri (2003), reported that some of the utmost categories of piezoelectric transducer biosensors are acoustic and ultrasonic; some of the utmost categories of electrochemical transducer biosensors are amperometric, conductometric, and potentiometric; whereas, some of the utmost categories of optical transducer biosensors are absorbance, fluorescence, and chemiluminescence.

Arora, (2013) also reported that biosensors could be characterized on the basis of the period or order that they were established. Consequently, in these groupings we could identify; first-generation biosensors (FGBs), second-generation biosensors (SGBs), and third-generation biosensors (TGBs).

The first-generation biosensors are the biosensors that are reportedly the modest procedure concerning the unwavering discovery of either upsurge of an enzymatically produced product or decline of a substratum of redox enzymes employing ordinary mediator for transferring particles or electrons. Examples of FGBs are the glucose biosensor that makes use of the enzyme glucose oxidase and oxygen for detecting a decline in the range of oxygen or upsurge in hydrogen peroxide conforming to the range of the glucose. The second-generation biosensors are biosensors which make use of non-ordinary redox mediators such as ferricyanide, quinones, and ferrocene for transferring particles or electrons, which upsurges the sensitivity and reproducibility. Examples of SGBs are self-monitoring amperometric glucose biosensors. The third-generation biosensors are those that fall wherein the redox enzymes, controlled on the surface electrode, in a manner that unwavering particles or electrons transference are imaginable among the transducer and enzyme.

At the moment, in relation to the development, and improvement of nanotechnology, state-of-the-art NPs are presently being developed and their advanced structures along with their benefits occur in biosensors (Rodriguez-Mozaz et al. 2005; Ukhurebor 2020). Implausible enhancement has been accomplished together with the expertise involved and the uses of biosensors by means of innovative strategies, encompassing electrochemistry, nanotechnology, and bioelectronics (Borgmann et al. 2011; Clark and Lyons 1962; Fracchiolla et al. 2013; Turner 2013; Vigneshvar et al. 2016; El-Said et al. 2020; Ukhurebor 2020).

The development of biosensors as a dominant and ground-breaking analytical mechanism (that entail biological identifying constituents with numerous benefits) has indisputably adopted foremost reputation in several fields of study. Its exploitation has achieved some momentous benefits in the field of biomedicine, pharmacology, environmental science, food safety, and dealings (as well as agriculture in general). The development of biosensors has resulted in the growth and advancement of precise and dominant analytical mechanisms via biological identifying constituent as biosensors (Vigneshvar et al. 2016; Salek-Maghsoudi et al. 2018; Ukhurebor 2020).

According to Ukhurebor (2020) and Turner (2013), the methodological procedures that are employed in biosensor mechanisms are based on the label-assembled and label-unlimited discovery. Label-assembled discovery is mostly hooked on the obvious structures of label compounds for detection targeting. However, these groups of biosensors are reliable; but are however consistently encompass in bringing together obvious sensing mechanisms, invented with the controlled target protein. Whereas, label-unlimited procedure tolerates the recognition of the target substances that are not characterized or difficult in tagging (Sang et al. 2015; Citartan et al. 2013; Ukhurebor 2020). Contemporary interdisciplinary methods of electronics mechanisms and biotechnology brought about the evolution and advancement of label-unlimited biosensors for numerous discovery procedures with several benefits in environmental and medical sciences.

Also, BSTs are been applied in AES. According to Ukhurebor (2020) and Verma and Bhardwaj (2015), this is an additional beneficial aspect in which BST is presently gaining ground. These BSTs in AES will indisputably be of assistance in the instantaneous identification of pesticidal deposits that would have resulted in some averted health menaces in form of CSA (Long et al. 2013a; Fanen and Olalekan 2014; Marrazza 2014; Verma and Bhardwaj 2015; Verma and Bhardwaj 2015; Antonacci et al. 2016; Adesina et al. 2017; Neethirajan et al. 2018; Ukhurebor 2020).

Verma and Bhardwaj (2015) reported that the traditional or conventional approach, such as high-performance liquid chromatography, capillary electrophoresis, and mass spectrometry are effective in the inquiry of environmental pesticides. Yet, there are some limitations such as difficulty, time-intense procedures, need of high-end mechanisms, and operational proficiencies.

In light of all these, even if it is supposed that the modest biosensors have enormous benefits; yet, it is somehow hard to formulate integrated biosensors which can investigate numerous groups of pesticides. Therefore, stable enzyme-built biosensors have been developed for understanding the physiological (such as physical and biological) effect ensuing from the use of pesticides in the environment, food safekeeping, and quality monitoring and management (Marrazza 2014; Ukhurebor 2020; Verma and Bhardwaj 2015). Pundir and Chauhan (2012) reported that acetylcholinesterase inhibition-built biosensors that been invented still require some advancement to effectively deal with these limitations. For quite some time now, in driving for instantaneous analysis, this technique has received extreme enhancement with further contemporary developments in acetylcholinesterase inhibition-built biosensors together with immobilization mechanisms and other diverse procedures for their fabrication (Vigneshvar et al. 2016; Ukhurebor 2020).

Similarly, piezoelectric biosensors have been developed for identifying the organophosphate and carbamate on the environmental effect of pesticides (Marrazza 2014). According to Senthilkumaran (2015), organochlorine pesticides are known for disturbing the ecology of pesticides, like Endosulfan which could cause considerable environmental damages. Undoubtedly, organochlorine pesticides have been described to be one of the utmost causes of modification in the reproductive system of some fishes (Senthilkumaran 2015). Consequently, the development of biosensors for sensing, detecting, or noticing the aquatic ecology will have further consequences due to biomagnification (Senthilkumaran 2015; Vigneshvar et al. 2016). So, to deal with these issues, the electrochemical biosensors have qualified revolution with instantaneous advances in its fabrication and the use of ingredients such as NPs (Turner 2013; Senthilkumaran 2015; Verma and Bhardwaj 2015).

Hence, it is of enormous importance to place discrete prominence in selecting or collecting receptors for biosensor improvement, as well as the application of the various transduction actions and instantaneous screening methods for the use of biosensor in agricultural events such as the production of food and food safety along with dealing with environmental issues. To support this, biosensor manufacturing seems to be dynamic and the enhancements in this aspect have been categorically explained by numerous scholars and scientists.

7.3 Advantages and Technological Comparison of Some Biosensors

The different kinds of biosensors have their applications and machinery variance. Hence, this section will discuss the technological comparison and specificity as well as the detection extent, linear extent, investigation time, cost, and compactness of some of these biosensors.

Advancements in electrochemical sensors with substantial approaches concentrating on discovery extent, investigation time, and compactness afforded huge-scale user markets for low-cost biosensors designed for glucose and prenatal period assessments via anti-human chorionic gonadotropin immobilization strip with lateral-flow technique (Turner 2013; Ukhurebor 2020). According to Vigneshvar et al. (2016), the control of analytes by means of polymers and nanoparticles/nanomaterials are vital constituents for the improvement of the sensitivity and detection extent. As a result of this, lateral-flow technique permits the direct conveyance of samples to anticipated spots so as to generate precise interactions as an alternative to random interactions. Some of the above-mentioned biosensors have applied this technique; it has also given prospects for bio-fabrication by means of either contact or non-contact-built modeling (Vigneshvar et al. 2016).

The use of NPs such as gold, silver, and silicon-built bio-fabrication produced innovative procedures. Also, coating/covering of polymers on these NPs brought enormous advancements in contact-built electrochemical sensing. The foremost benefit of these categories of electrochemical sensors is sensitivity and specific nature with the real-time investigation. Though, there are some limitations such as the recovering capability or long-term procedure of polymers or/and other constituents; hitherto, the inexpensive nature of such electrochemical sensors make them more affordable (Vigneshvar et al. 2016).

Single or sole analyte discovery by means of contact-built sensing has great benefits, for instance, real-time quantification of particles with lofty specificity. To support this, Fluorescence Resonance Energy Transfer (FRET), bioluminescent resonance energy transmission, fluorescent-built as well as surface plasmon resonance-built transducers were developed for improving the sensitive and specific nature in terms of sole or single molecule/particle discovery (Dias et al. 2014; Vigneshvar et al. 2016). These procedures also have some limitations in terms of multiple analyte discovery as a result of signal emission intersection/overlap; hitherto, resonance energy transmission procedures habitually established for numerous analyte discoveries, that are extremely rewarding in medication diagnosis as a result of transformation in bio-markers between patients and interrelated infections (Vigneshvar et al. 2016).

The application of microcantilevers or nanocantilevers by way of transducers in electrochemical sensor bio-fabrication is similarly having widespread potential in numerous analytes discovery (Vigneshvar et al. 2016). Also, non-contact-built sensors by means of three-dimensional (3D) bioprinting applying inkjet or laser direct inscription afforded improved consequences. However, the expensive and customizable capability has great implications in these procedures.

Remarkably, some of these substantial biosensors have been joint with electrochemical sensing for precise purposes. Most of the prominent real-time and sensitive, as well as portable amperometric electrochemical biosensors, have been established for infection/disease diagnosis by means of the fluid content of the body (body fluids). Generally, electrochemical biosensors combined with bio-fabrication have truncated discovery extent for sole or single analyte discovery specific nature by means of real-time investigation and at an inexpensive price in view of the portable nature of the device (Kim et al. 2015a; Vigneshvar et al. 2016; El-Said et al. 2020).

Optic-built biosensors are also now leading technology in biosensing that has to do with fiber-optic chemistry. Single or sole particle/molecule discovery or detention, such as Deoxyribonucleic acid (DNA) or peptide, is done appropriately by means of hydrogel-built cross-linking as a result of the benefit of having great packing capability and hydrophilic nature. More so, optical biosensors which had widespread applications in biomedical and forensic science for DNA measurement and quantification was developed (Kwon and Bard 2012). According to Vigneshvar et al. (2016), the combination of organic or biological constituents, for example, "enzyme/substrate, antibody/antigen, and nucleic acids", resulted in a great version of optical BST. Also, it is now been conceivable to integrate microbes, other living organism cells, and tissue segments in the biosensing mechanisms.

The advancements in molecular optoelectronics brought forth optical biometric recognition systems. Combined optics machinery gave room for the integration of both inert and dynamic optical mechanisms on the similar substrates for evolving abated or curtailed dense sensing strategies by means of fabrication of numerous sensors on a sole chip. In this situation, top class polymers brought forth hybrid systems for optical BSTs. In effect, optic-built BSTs were enhanced as a result of recent innovations in surface morphology investigation by means of advanced electron and atomic force microscopy.

Nevertheless, the discovery extent of optical biosensors at no time would have been near femto-level in view of the instrumentation cost and device non-portability. As a result of this, recent optic machinery that has to do with nano-mechanical biosensors built on microcantilevers and surface resonance machinery assisted in the provision for advanced DNA chips at best for instantaneous precise and delicate investigation (Scheller et al. 2014; Ukhurebor 2020; Wang et al. 2015; El-Said et al. 2020; Zhang et al. 2015).

Basically, the benefits of optical biosensors comprise of a fast instant investigation by means of resistance of signal to electrical/magnetic intrusion in addition with the budding for spectrum of data (Vigneshvar et al. 2016). However, the main disadvantage is the expensive nature as a result of certain instrumentation necessities. Further procedural complications include complexity in immobilization particularly for bio-fabrication in addition with the need to disparagingly resolved decontaminate the environment in order to take complete benefit of optical biosensors (Vigneshvar et al. 2016).

Bio-fabrication of mechanical devices needs improved outcomes for mass-built biosensors. Irrefutably, electrochemical and optical biosensors together make one of the furthermost of this machinery for devising higher biosensors (Vigneshvar et al. 2016; El-Said et al. 2020). According to Arlett et al. (2011), the major developments in micro-fabrication and nano-fabrication machineries made it possible to produce mechanical devices with nano-sized stirring fragments. The capability to produce such devices, by means of semiconductor processing measures are linked to biophysics and bioengineering moralities in the direction of the advancement of applied micro-electromechanical and nano-electromechanical biosensors which were produced in huge quantities (Arlett et al. 2011; Vigneshvar et al. 2016).

Glass, silicon as well as quartz constituents have been reportedly applied whichever after tagging with fluorescence or gold NPs efficaciously. Although they have superior precision in terms of sole particle discovery, cost-effectiveness, multiple productions is hardly achievable (Shen et al. 2014; Peng et al. 2014; Vigneshvar et al. 2016). Several challenges still are however still noticed in mass-built sensors especially in aspects of improved capture agents to produce at nanoscale by means of microelectronic fabrication for top-throughput investigation. As such, the great application possibilities of semiconductor materials and quantum dot machineries are worth mentioning (Vigneshvar et al. 2016).

However, there are presently technologies/machineries in biosensors that are capable of simultaneously executing real-time and quantitative analyzes for bulky arrays. It is the use of micro-scale and nanoscale cantilevers fabrication that has made this achievable (Dias et al. 2014; Vigneshvar et al. 2016).

Further vital mechanical uprising in biosensors machinery is the detection of genetically encoded or synthetic fluorescent biosensors for the analysis of molecular devices by means of biological or organic procedures (Randriamampita and Lellouch 2014; Oldach and Zhang 2014; Kunzelmann et al. 2014). However, these categories sensors have great vision for sole or single particle discovery with precise analyte quantification, the procedure and investigative preparation and discovery are hard, and similarly need advanced instrumentation (Vigneshvar et al. 2016).

Biomaterials/bioparticles and microbial fuel-built biosensors also have discrepancy as a result of the highly sensitive and selective nature. Yet, multiple productions and genetic engineering procedures for developing microbial strain, need intricate measures and are expensive. But, the foremost benefit of microbial biosensors is the capability to employ them as a means for bioremediation, that have a superior consequence for agriculture as well as environmental management and monitoring. Nevertheless, the advancement of such improved genetic strain needs to be sternly analyzed by means of appropriate governing rules and regulations as well as ethical clearance and production cost management. Hence, Scheller et al. (2014) reported that the advancement of diverse micro-biosensor and nano-biosensor frameworks that as to do with combined technologies that utilize electrochemical or optic bio-electronic principles through the grouping of bioparticles or biological constituents, polymers and NPs are needed so as to attain exceedingly delicate and miniaturized mechanisms.

As the quest and need for applying biosensors for swift investigation with cost-efficiency, entail bio-fabrication that now pave way for the identification of cellular to complete instinctive action with a discovery extent of top precision for sole particles. According to Dias et al. (2014), biosensors ought to be designed to function under multiplex and complex circumstances. In that condition, both "three-dimensional (3D) and two-dimensional (2D) discovery" are needed by means of elegant transducers for detecting and measuring minor analytes of interest. With this in mind, some breakthrough discoveries have been achieved with contact or non-contact-built modeling at diverse stages.

We now have biosensors that are developed for discovering additional robust recovering long-term usage. These diagnostic biosensors are developed by means of therapeutics, which assist both the medical staff/team and their patients for additional assimilative consideration of infections and the treatment procedures (El-Said et al. 2020). Hence, Fracchiolla et al. (2013) reported that fluorescence resonance energy transfer-built biosensor provides an exceptional analytical technique for evaluating the effectiveness of "imatinib" management and handling of continuing or chronic myeloid leukemia. The present use of affibodies, aptamers, peptide arrays as well as imprinted molecular polymers are conventional illustrations for potential research methods in this aspect (Verma and Bhardwaj 2015; Vigneshvar et al. 2016; Citartan et al. 2013; El-Said et al. 2020; Abe et al. 2014).

However, little achievement has also been reported with limited possible particles for advanced therapeutic, antimicrobial, and medication delivery. According to Grabowska et al. (2014), the development in this aspect resulted in the discovery of electrochemical biosensors as consistent and dependable diagnostic means for pathogen discovery of avian influenza virus in intricate conditions. However, Mazzei et al. (2014) reported that an additional new report showed possible benefits of affinity-built biosensors in medication for sports and doping monitored investigation. Also, Bandodkar and Wang (2014), reported that there is now a multiplicity of even or wearable electrochemical biosensors for instantaneous non-hostile transmission of electrolytes and metabolites in the fluid content as indicators or determinants of a wearer's medical situations. According to Lawal and Adeloju (2012), there are also additional attractive applications in the assessment of meat and fish value by means of hypoxanthine biosensors through fabrication. According to Kumar and Rani (2013), the advancement in biosensors for the discovery of biological or organic warfare agents such as virus and bacteria as well as contaminants, are now sometimes use for various means of biosensors such as nucleic acid, electrochemical, piezoelectric and optical, which now have massive benefits in military (defence, protection, safety and security) and medical science.

The alliance of NPs with polymers of different categories of biosensors has assisted in the provision of hybrid devices that have now improved the use for the above-mentioned applications (Vigneshvar et al. 2016; Citartan et al. 2013; El-Said et al. 2020; Turner 2013). Moreover, scientific advancement has been of great assistance in the development of microbial biosensors by means of biological or organic synthesis approaches that are now contributing to energy and environmental management, monitoring, and sustainability (Sun et al. 2015; Vigneshvar et al. 2016; El-Said et al. 2020).

There are presently several other reports that have shown the significance of using microbial fuel cells for evolving techniques for the treatment/remediation of water as power sources for environmental sensors (Vigneshvar et al. 2016).

7.4 Application of Biosensors for the Detection of Pesticides Available in the Environment

It has been observed that toxic and poisonous herbicides, insecticides, and pesticides are usually applied in industries that are agriculturally based (Adetunji et al. 2018a, b, c, 2019, 2020; Alavanja et al. 2004). The application of pesticides has been documented for improving the increase in agricultural crops. But it has been observed that there are several health and environmental hazards that are involved in whenever pesticides are applied in an uncontrolled manner (Ukhurebor et al. 2020b). This can lead to the incidence of cancer, allergies, and several breathing challenges (Criswell et al. 2013; Givaudan et al. 2014). The application of high-performance liquid chromatography and gas chromatography have been discovered that not to be economical and requires great expertise as well as consume lot of times.

Therefore, the application of a biosensor might be an alternative technique that could determine the level of pesticides available in the environment.

Organophosphorus has been identified as a pesticide that is commonly utilized in most parts of the globe irrespective of its negative effect on the ecosystem and human health when applied. Hence, it is very important to develop a technique that could be used for the effective determination of these pesticides. The utilization of enzymatic biosensors has been identified in this regard which might be linked to the following facts such as their accuracy, stability, and their sensitivity. The utilization of piezoelectric, optical, electrochemical, and calorimetric have been manufactured based on the enzyme inhibition to determine the level of pesticides (Arduini and Amine 2014). Moreover, ureases and cholinesterases, organophosphorus hydrolase, are some of the different types of enzymes that have been utilized in the fabrication of biosensors for effective detection of pesticides (Zhang et al. 2014). Some of the typical examples of these biosensors which are enzyme based includes a screen-printed electrode, cetylcholinesterase, butyryl cholinesterase, organophosphorus hydrolyase, organophosphorus while others were multi-walled carbon nanotubes, screen-printed electrode, mesoporous carbon/carbon black, reduced graphene oxide and carbon nanotubes, respectively (Yada 2015).

7.5 Utilization of Biosensors for the Detection of Heavy Metals in the Environment

It has been discovered that the presence of heavy metals in the environment have several detrimental effects on animal and human health. They could build up in organism which might later result in metabolic changes in animals, most especially those that feeds on the industrial area and highly polluted environment. These might result in adverse effects through a decrease in hormonal action, malfunctioning of principal organs, cardiovascular and respiratory challenges, high level of infertility, and irritation and eventual death.

The movement of heavy metal has been identified majorly from the industrial environment to constitute a heavy challenge to human health. Some of them have been discovered to be non-biodegradable in nature such as lead, mercury, and cadmium. Most of these heavy metals are derived from lead-acid batteries, vehicle emissions, and chemical fertilizers (Gammoudi et al. 2014). Therefore, there is a need to protect the environment and human health from constant exposure to these heavy metals. However, it has been discovered that the detection of these heavy metals using chromatography, and spectrometry are time consuming, non-economical and has a greater demand for well knowledgeable experts (Bagal-Kestwal et al. 2008).

Therefore, the application of inexpensive, portable biosensors for the detection of these heavy metals will go a long way in mitigating against all these aforementioned challenges. The application of microbial sensors has been identified for the identification of heavy metals ions when applied at very low cost. A typical example of this includes microbial fluorescence-based biosensor devices (Amaro et al. 2014). This was design based on the application of reporter genes that have a greater tendency to react whenever there is a biochemical reaction between inducer molecules and cellular reporters. Kim et al. (2015b) stated the application of microbial biosensors and chemostat-like microfluidic platform that could enhance the identification of molecular analyte through the uses of a chip. Moreover, for effective determination of heavy metal ion the application of DNA optical biosensor together with evanescent wave evaluation could be applied in the identification of heavy metal in an in situ detection (Long et al. 2013b). He et al. (2013) fabricated a luminescent G-quadruplex-based sensing device that could be identify sub-nanomolar Pb^{2+} ions in liquefied solutions (He et al. 2013). Also, the detection of heavy metal ion present in the environment have been carried out using several techniques such as inhibition-based biosensors, amperometric and electrochemical sensors (Amine et al. 2014; Wang et al. 2013; Ghica et al. 2013; Sbartai et al. 2012; Gammoudi et al. 2014).

It has been discovered that graphene oxide is eco-friendly, sensitive and it is cost feasible. The application of graphene oxide-based fluorescent sensors has been employed for the detection of Hg2+ ions while graphene oxide plays a crucial role of fluorophore. It was observed that the molecular recognition probe joins to the Hg^{2+} ions through a DNA aptamer (Li et al. 2013).

Furthermore, β-Carotene available in the palm kernels could be utilized as a source of biological reporter most especially in the biosensor that are biopolymer-based utilized for detection of heavy metals such as aluminum (Wong and Wong 2015). Also, the application of an on-chip label-free sensing device was fabricated for high-throughput identification of doses as minimal as five parts per billion of heavy metal such as cadmium-chelate conjugates (Yan et al. 2016).

7.6 Application of Biosensor in CSA

According to Neethirajan et al. (2018) and Ukhurebor (2020), some of the utmost challenges confronting agriculture in relation to food protection and sustainability are contain directly or indirectly on following three fundamental aspects:

- NPs and their benefit for agricultural sustainability issues.
- Energy sustainability issues.
- Industrialization of maintainable mechanism issues.

Nanotechnology (NT) is one of the utmost benefits in the monitoring and management of AES. It has some valuable features and applicability (Ukhurebor 2020; Neethirajan et al. 2018). Prasad et al. (2017) reported that NT has the required potentials to advance food security and eminence, develop and improve the capability absorption of the nutritional content of the soil, decrease agricultural contributions and ensure the possibilities in the shrunk mechanism quantifications.

Reportedly, NT has been applied efficiently for/in the following; "precision in farming machinery (agricultural precision), smart feed management, food waste management, production of agro-chemicals/agro-materials such as nano-pesticides, nano-herbicides and nano-fertilizers, labelling and packaging of agricultural products and multiple other agricultural fields" (Neethirajan et al. 2018; Ukhurebor 2020).

Also, Neethirajan et al. (2018) reported that the use of NT for agricultural dealings in relation to food sustainability is expected to result in some issues in the forthcoming years. This affirmed the work of Dasgupta et al. (2015), that nevertheless the usefulness from the present combination of NPs and animated charcoal for the improvement of antimicrobial features, food grade nano-emulsion (used for fruit juice), integrated nano-microbials employed as water sterilizers, active nutraceutical nano-delivery and improved plant extracts conjugated via nano-packaging could have some defects also.

A key prominence in NT is in its usefulness in agricultural precision also known as precision in farming machinery (Ukhurebor 2020), such that plant excerpts from its core parts like leaves, flowers, stems and roots, from several species have been efficiently combined into NPs (Duhan et al. 2017; Dasgupta et al. 2015).

NPs have the potential to advance green synthesis (GS) in a one-step via ion along with metal diminishing consequences. This according to Ukhurebor (2020) and Prasad et al. (2017), is favourable for use at room temperature, easy-use, adaptable, and scalable as well as ecological sustainability. During GS, co-enzymes and soluble metabolites such as phenolic composites, alkaloids and terpenoids are absolutely condensed to NPs. Essentially, NPs are known as magic bullets ensuing in enhanced plant growth, location-specific delivery of nutritional values, and improved resistance of plants to contagions or diseases.

According to Ukhurebor (2020), one of the greatest considerable issues with NT is in the advancement of dependable risk-benefit assessments via standardized evaluation and measures. Hence, the formation of dependable and standardized measures in NPs quantifications, arrangements, and evaluation of their environmental consequence on living creatures as well as the participation of all pertinent stakeholders such as agriculturalists, food industrialists, etc. in a critical debate on the way further would be of great assistance (Fraceto et al. 2016; Prasad et al. 2017; Ukhurebor 2020). The ensuing issues in sustainable energy could be efficiently taken care of via the application of organic/biological resolutions. Consequently, Adesina et al. (2017), reported that one of the foremost benefits in this regard has been established in applied organic/biological for energy generation.

Also, biofuels could be dumped, generated, rehabilitated, and changed in electricity (bio-electricity) to meaningfully moderate the cost in fabricating or generating solar electricity. This according to Dasgupta et al. (2015), could be proficient via leveraging by means of the ingestion of H_2 or electron lashing carbon fixing metabolism, for the simplification of the their combination via the use of photovoltaics efficiency in a process recognized as electro-photosynthesis (Dasgupta et al. 2015; Ukhurebor 2020).

According to Neethirajan et al. (2018), hydrogen-built electrosynthesis is one of the furthermost efficacious bioengineering energy creation systems. It displays some excellent physiognomies like; top effective bioenergy storage ability for electrical energy of approximately 80%, extensive length transportability with minimum energy forfeit, hydrogen oxidation in microbes involving Nicotinamide Adenine Dinucleotide (NAD^+; $C_{21}H_{27}N_7O_{14}P_2$) reduction of the potential inconsistency and affordability due to the lesser cell-protein necessities of hydrogen oxidation (Neethirajan et al. 2018). The transmission of electrons according to Neethirajan et al. (2018) could extracellularly mediate electro-synthesis proficiently. This could be carried out via reproducible through a nanostructured surface to abridge the of generation or production of biofilm. It could cavort the need for prolonged surface area and advance the transmission of hydrogen electrons (Neethirajan et al. 2018).

Advanced organic/biological energy generation according to Neethirajan et al. (2018), can be meaningfully enhanced via the development of some other technologies. Such state-of-the-art technologies are gene engineering, whole genome engineering, protein engineering, and biosensing. These state-of-the-art technologies can interestedly improve the development and generation of biofuel. Reportedly, since it is one of the utmost prominent benefits of advanced biological science in the aspect of energy sustainability; biofuel supposedly is to act as one of the utmost sited measures for capturing and storage of solar energy with little costs (Neethirajan et al. 2018).

Presently, some of the challenges confronting the production, advancement, and generation of biofuel according to Neethirajan et al. (2018) and Ukhurebor (2020) are; the energy generation effectiveness and scale, competence investigation in cell self-assembly as well as duplication monitoring and management, and antagonistic environmental consequences. Expectedly, it is projected that in the future, biofuel will develop, advance, and encompass sources of conventional energy for reutilization and replication for the generation of constituents of energy and for the improvement and advancement of hybrid energy photosynthesis (Neethirajan et al. 2018). Nevertheless, some of the commercialization/industrialization of sustainable technologies in the agronomy aspect is enduring via some essential prominences such as; biosensor commercialization/industrialization, sensing technology commercialization/industrialization and intelligent agricultural/food commercialization/industrialization, such as CSA (Ukhurebor 2020).

In the commercialization/industrialization of biosensors, some significant aspects for the purpose of its commercialization/industrialization are; modest sample pretreatment, bioreceptor steadiness, multi-detecting/multi-sensing features, impoverishment/miniaturization, quicker testing period, wireless accessibility, and affordability (Bahadır and Sezgintürk 2015; Ukhurebor 2020). Some of the present-day technologically advanced biosensors industries for food safety, security, monitoring, management, and quality control, their merchants as well as their anticipated analytes are summarily enumerated in Table 7.1 as revised from Neethirajan et al. (2018).

Antonacci et al. (2016) and Neethirajan et al. (2018) also accounted for some of the utmost possessions of commercialized available recognized biosensors establishments as their simple structure, reduced sizes, and ideal potentials for point of care benefits. They targeted food configuration, progression control, monitoring, and management in addition to food security such as allergens, pathogens, toxins,

Table 7.1 Some technologically advanced biosensors industries for food safety, security, monitoring, management and quality

Industries	Anticipated analytes
Analox Instruments; in the UK and USA	Ethanol, methanol, glucose, lactate, glycerol
Biacore AB; in Sweden	Vitamins that are soluble in water, veterinary deposits (chemical) and mycotoxins
Biomerieux; in France	Microorganisms
Biosensores S.L.; in Spain	Microorganisms, biochemical oxygen demand
Biosentec; in France	Alcohol, allergen, acids, sulphites
Biotech-IgG; in Sweden	Allergens, Vitamins, Microorganisms
Eurofins; in Luxembourg	Microorganisms, drug residue
Gwent Sensors; in UK	Glucose, lactate, ammonia, pyruvate
Nova Biomedical; in USA	Bio profile chemistry analyser
Tectronik, in Italy	Ethanol, glucose, malate, D-lactate, L-lactate, fructose
Trace Analytics; in Germany	Glucose, L-glutamate, L-glutamine
Yellow Springs Instrument; in USA	Glucose, sucrose, galactose, ethanol, lactose, L-lactate, L-glutamate, H_2O_2, glutamine, choline

pollutants/contaminants, and additives. Neethirajan et al. (2018), stated that some of the utmost establishments for food quality biosensors is principally from the following metabolites like glucose, sucrose, glycerol, cholesterol, creatinine, alcohol, methanol, lactate, lactose, glutamate, malate, and ascorbic acid.

Bahadır and Sezgintürk (2015) reported that likened to the earlier recognized biosensors in scientific laboratories, the contemporary biosensors which are habitually commercialized are faraway scarcer indicatory of the abridged accomplishment extents in agronomy-associated biosensor development. The issues burdening biosensor advancement in the agronomy sector are considerable impairments. Such considerable impairments according to Neethirajan et al. (2018) are mass production, sensor lifetime, component integration, and handling practicability.

The motivations behind these limitations are that the utmost technologies employed in present and imminent agronomy-associated biosensor technologies are in their early phases and they include NT, agriculture/food material science, biomimetic chemistry, and microengineering (Neethirajan et al. 2018; Ukhurebor 2020). According to Neethirajan et al. (2018), these basic issues can be of assistance for the purpose of imminent biosensors industries as well as its safety to human comfort, which infers that it is those with inadequate or no human comfort consequence will have their scientific improvement in the near future (Ukhurebor 2020). The scientific improvement of intelligent agriculture/food industry stipulates crucial requirements in innovative and operative measures to guarantee AES vis-à-vis food security, protection, safety, and quality, as well as to reduce production techniques and diminishing losses in agronomy (Vanderroost et al. 2014; Ukhurebor 2020).

7.7 Conclusion and Future Contribution to Knowledge

CSA purposes excellent prospects for dealing with the issues of AES vis-à-vis food security, protection, safety, and quality in addition to enabling the adaptation and extenuation succours for AES. Consequently, CSA has been of enormous assistance in this aspect, particularly to several industrialized nations. The implementation of CSA as a proficient and instantaneous CC response is tremendously critical for the development of capacity and accomplishing food protection and safety as well as AES universally. In unindustrialized nations particularly those of Sub-Sahara Africa, considering the vulnerability to CC, their considerable reliance on agriculture for livelihoods and the vital role agricultural dealings play in their pecuniary sector; they would largely benefit from CSA. However, there is a need for modification procedures in inspiring the acceptance and improvement of CSA.

The minor agricultural section in some of these unindustrialized nations is characterized by various inhabitant. Accordingly, a solitary uniform technique would hardly be appropriate in advancing CSA dealings among these sets of agriculturalists. The consequence ensuing from this is that methods in supporting CSA execution should factor in explicit and specific communal dealings as an addition for mainstreaming procedures universally. Therefore, all relevant stakeholders must anticipate applying modalities that could accommodate the various structures and physiognomies of CSA, and avoid the possible issues that could otherwise arise. Furthermore, since CSA development in most unindustrialized regions of the world relies on the preparedness of those involved in agricultural dealings; consequently, it is necessary for all relevant stakeholders to comprehend and recognize the multi-dimensional CC issues and the succeeding self-mobilization for developing and implementing approaches that will react to the issues at suitable scales.

Irrefutably, despite the several benefits of biosensors and BSTs such as polymers, NPs and microorganisms built biosensors in resolving some of the problems in AES; there is still the necessity to expressively integrate multi-faceted approaches in evolving biosensors that could hypothetically be employed for various applications in CSA for AES. Consequently, it is recommended that suitable combination of BSTs along with bio-fabrication with non-natural/synthetic biology approaches via the application of either/both electrochemical, optical, bio-electronic moralities would be critical for the effective enhancement of wide-ranging and dominant biosensors for present-day and forthcoming involvement to knowledge in the aspect of BSTs in CSA for AES.

References

Abe K, Yoshida W, Ikebukuro K (2014) Electrochemical biosensors using aptamers for theranostics. Adv Biochem Eng Biotechnol 140:183–202

Abegunde VO, Sibanda M, Obi A (2019) The dynamics of climate change adaptation in Sub-Saharan Africa: a review of climate-smart agriculture among small-scale farmers. Climate 7:132

Adesina O, Anzai IA, Avalos JL, Barstow B (2017) Embracing biological solutions to the sustainable energy challenge. Chem 2:20–51

Adetunji CO, Oloke JK, Phazang P, Sarin NB (2020) Influence of eco-friendly phytotoxic metabolites from lasiodiplodia pseudotheobromae C1136 on physiological, biochemical, and ultrastructural changes on tested weeds. Environ Sci Pollut Res. https://doi.org/10.1007/s11356-020-07677-9

Adetunji CO, Adejumo IO, Oloke JK, Akpor OB (2018a) Production of phytotoxic metabolites with bioherbicidal activities from lasiodiplodia pseudotheobromae produced on different agricultural wastes using solid-state fermentation. Iran J Sci Technol Trans Sci 42(3):1163–1175

Adetunji CO, Oloke JK, Mishra P, Oluyori AP, Jolly RS, Bello OM (2018b) Mellein, a dihydroisocoumarin with bioherbicidal Activity from a new strain of lasiodiplodia pseudotheobromae C1136. Beni-Suef Univ J Basic Appl Sci. https://doi.org/10.1016/j.bjbas.2018.06.001

Adetunji CO, Oloke JK, Prasad G (2018c) Effect of carbon-to-nitrogen ratio on eco-friendly mycoherbicide activity from lasiodiplodia pseudotheobromae C1136 for sustainable weeds management in organic agriculture. Environ Dev Sustain 1–14

Adetunji CO, Oloke JK, Bello OM, Pradeep M, Jolly RS (2019) Isolation, structural elucidation and bioherbicidal activity of an eco-friendly bioactive 2-(hydroxymethyl) phenol, from pseudomonas aeruginosa (C1501) and its ecotoxicological evaluation on soil. Environ Technol Innov 13:304–317

Akrofi-Atitianti F, Speranza CI, Louis Bockel L, Asare R (2018) Assessing climate smart agriculture and its determinants of practice in Ghana: a case of the cocoa production system. Land 7:30

Alavanja MCR, Dosemeci M, Samanic C, Lubin J, Lynch CF, Knott C, Barker J, Hoppin JA, Sandler DP, Coble J, Thomas K, Blair A (2004) Pesticides and lung cancer risk in the agricultural health study cohort. Am J Epidemiol 160(9):876–885

Amaro F, Turkewitz AP, Martin-Gonzalez A, Gutierrez JC (2014) Functional GFP metallothionein fusion protein from Tetrahymena thermophila: a potential whole-cell biosensor for monitoring heavy metal pollution and a cell model to study metallothionein overproduction effects. Biometals 27(1):195–205

Amine A, El Harrad L, Arduini F, Moscone D, Palleschi G (2014) Analytical aspects of enzyme reversible inhibition. Talanta 118:368–374

Antonacci A, Arduini F, Moscone D, Palleschi G, Scognamiglio V (2016) Commercially available (bio)sensors in the agrifood sector. In: Comprehensive analytical Chemistry. Elsevier: Amsterdam, the Netherlands, pp 74, 315–340

Antwi-Agyei P, Dougill AJ, Stringer LC (2014) Barriers to climate change adaptation in Sub-Sahara Africa: evidence from Northeast Ghana and systematic literature review. Clim Dev 7:297–309

Arduini F, Amine A (2014) Biosensors based on enzyme inhibition. In: Gu MB, Kim H (eds), Biosensors based on aptamers and enzymes. Berlin, pp 299–326, ISSN: 0724-6145

Arlett JL, Myers EB, Roukes ML (2011) Comparative advantages of mechanical biosensors. Nat Nanotechnol 6:203–215

Arora N (2013) Recent advances in biosensors technology: a review. Octa J Biosci 1(2):147–150

Bagal-Kestwal D, Karve MS, Kakade B, Pillai VK (2008) Invertase inhibition based electrochemical sensor for the detection of heavy metal ions in aqueous system: application of ultra-microelectrode to enhance sucrose biosensor's sensitivity. Biosens Bioelectron 24(4):657–664

Bahadır EB, Sezgintürk MK (2015) Applications of commercial biosensors in clinical, food, environmental, and biothreat/biowarfare analyses. Anal Biochem 478:107–120

Bandodkar AJ, Wang J (2014) Non-invasive wearable electrochemical sensors: a review. Trends Biotechnol 32:363–371

Borgmann S, Schulte A, Neugebauer S, Schuhmann W (2011) Amperometric biosensors. In: Alkire RC, Kolb DM, Lipkowski J (eds), Advances in electrochemical science and engineering: Bioelectrochemistry, 13, Wiley-VCH Verlag GmbH & Co. KGaA

Citartan M, Gopinath SC, Tominaga J, Tang TH (2013) Label-free methods of reporting biomolecular interactions by optical biosensors. Analyst 138:3576–3592

Clark LC Jr, Lyons C (1962) Electrode systems for continuous monitoring in cardiovascular surgery. Ann N Y Acad Sci 102:29–45

Criswell JT, Campbell J, Luper C (2013) Toxicity of pesticides. Oklahoma Cooperative Extension Service, EPP-7457

Dias AD, Kingsley DM, Corr DT (2014) Recent advances in bioprinting and applications for biosensing. Biosensors (Basel) 4:111–136

Dasgupta N, Ranjan S, Mundekkad D, Ramalingam C, Shanker R, Kumar A (2015) Nanotechnology in agro-food: from field to plate. Food Res Int 69:381–400

Duhan JS, Kumar R, Kumar N, Kaur P, Nehra K, Duhan S (2017) Nanotechnology: the new perspective in precision agriculture. Biotechnol Rep 15:11–23

El-Said WA, Abdelshakour M, Choi JH, Choi JW (2020) Application of conducting polymer nanostructures to electrochemical biosensors. Molecules 25(307):1–11

Fracchiolla NS, Artuso S, Cortelezzi A (2013) Biosensors in clinical practice: focus on oncohematology. Sensors (Basel) 13:6423–6447

Fanen T, Olalekan A (2014) Assessing the role of climate-smart agriculture in combating climate change, desertification and improving rural livelihood in Northern Nigeria. Afr J Agric Res 9:1180–1191

Field CB, Barros VR, Dokken D, Mach K, Mastrandrea M, Bilir T, Chatterjee M, Ebi KL, Estrada YO, Genova RC et al. (2014) IPCC: climate change 2014: impacts, adaptation, and vulnerability. Part A: global and sectoral aspects. Contribution of working group II to the fifth assessment report of the intergovernmental panel on climate change. Cambridge University Press, Cambridge, UK and New York, NY, USA, pp 1–32

Food and Agriculture Organisation of the United Nations (FAO (2015) Climate smart agriculture sourcebook. Italy, Rome

Fraceto LF, Grillo R, Medeiros DAG, Scognamiglio V, Rea G, Bartolucci C (2016) Nanotechnology in agriculture: which innovation potential does it have? Front. Environ Sci, 4:20

Gammoudi I, Raimbault V, Tarbague H, Morote F, Grauby-Heywang C, Othmane A, Kalfat R, Moynet D, Rebiere D, Dejous C, Cohen-Bouhacina T (2014) Enhanced bio-inspired microsensor based on microfluidic/bacterial/love wave hybrid structure for continuous control of heavy metals toxicity in liquid medium. Sens Actuators B 198:278–284

Garnett T (2013) Food sustainability: problems, perspectives and solutions. Proc Nutr Soc 72:29–39

Ghica ME, Carvalho RC, Amine A, Brett CMA (2013) Glucose oxidase enzyme inhibition sensors for heavy metals at carbon film electrodes modified with cobalt and copper hexacyanoferrate. Sensd Actuators, B 178:270–278

Givaudan N, Binet F, Bot BL, Wiegand C (2014) Earthworm tolerance to residual agricultural pesticide contamination: field and experimental assessment of detoxification capabilities. Environ Pollut 192:9–18

Grabowska I, Malecka K, Jarocka U, Radecki J, Radecka H (2014) Electrochemical biosensors for detection of avian influenza virus—current status and future trends. Acta Biochim Pol 61:471–478

He HZ, Leung KH, Yang H, Chan DSH, Leung CH, Zhou J, Bourdoncle A, Mergny JL, Ma DL (2013) Label-free detection of sub-nanomolar lead (II) ions in aqueous solution using a metal-based luminescent switch-on probe. Biosens Bioelectron 41:871–874

Herndon JM, Whiteside M (2019) Further evidence that particulate pollution is the principal cause of global warming: humanitarian considerations. J Geogr Environ Earth Sci Int 21(1):1–11

Karimi V, Karami E, Keshavarz M (2018) Climate change and agriculture: impacts and adaptive responses in Iran. J Integr Agric 17(1):1–15

Kim J, Imani S, de Araujo WR, Warchall J, Valdes-Ramirez G, Paixao TR, Mercier PP, Wang J (2015a) Wearable salivary uric acid mouthguard biosensor with integrated wireless electronics. Biosens Bioelectron 74:1061–1068

Kim M, Lim JW, Kim HJ, Lee SK, Lee SJ, Kim T (2015b) Chemostat-like microfluidic platform for highly sensitive detection of heavy metal ions using microbial biosensors. Biosens Bioelectron 65:257–264

Kumar H, Rani R (2013) Development of biosensors for the detection of biological warfare agents: its issues and challenges. Sci Prog 96:294–308

Kunzelmann S, Solscheid C, Webb MR (2014) Fluorescent bio-sensors: design and application to motor proteins. Exp Suppl 105:25–47

Kwon SJ, Bard AJ (2012) DNA analysis by application of Pt nanoparticle electrochemical amplification with single label response. J Am Chem Soc 134:10777–10779.

Lawal AT, Adeloju SB (2012) Progress and recent advances in fabrication and utilization of hypoxanthine biosensors for meat and fish quality assessment: a review. Talanta 100:217–228

Li M, Zhou X, Ding W, Guo S, Wu N (2013) Fluorescent aptamer-functionalized graphene oxide biosensor for label-free detection of Mercury (II). Biosens Bioelectron 41:889–893

Long F, Zhu A, Shi H (2013a) Recent advances in optical biosensors for environmental monitoring and early warning. Sensors (Basel) 13:13928–13948

Long F, Zhu A, Shi H, Wang H, Liu J (2013b) Rapid on-site/in-situ detection of heavy metal ions in environmental water using a structure-switching DNA optical biosensor. Sci Rep 3:2308

Marrazza G (2014) Piezoelectric biosensors for organophosphate and carbamate pesticides: a review. Biosensors (Basel) 4:301–317

Mazzei F, Antiochia R, Botre F, Favero G, Tortolini C (2014) Affinity-based biosensors in sport medicine and doping control analysis. Bioanalysis 6:225–245

Neethirajan S, Ragavan V, Weng X, Rohit Chand R (2018) Biosensors for sustainable food engineering: challenges and perspectives. Biosensors 8:23

Nwankwo W, Olayinka SA, Ukhurebor KE (2020a) Green computing policies and regulations: a necessity? Int J Sci Technol Res 9(1):4378–4383

Nwankwo W, Olayinka AS, Ukhurebor KE (2020b) Nanoinformatics: why design of projects on nanomedicine development and clinical applications may fail? In: Proceeding of the 2020 international conference in Mathematics, Computer Engineering and Computer Science (ICMCECS), Lagos, Nigeria, IEEE Xplore, pp 1–7

Nwankwo W, Ukhurebor KE (2019) An x-ray of connectivity between climate change and particulate pollutions. J Adv Res Dyn Control Syst 11(8) Special Issue, 3002–3011

Oldach L, Zhang J (2014) Genetically encoded fluorescent biosensors for live-cell visualization of protein phosphorylation. Chem Biol 21:186–197

Peng F, Su Y, Zhong Y, Fan C, Lee ST, He Y (2014) Silicon nanomaterials platform for bioimaging, biosensing, and cancer therapy. Acc Chem Res 47:612–623

Prasad R, Bhattacharyya A, Nguyen QD (2017) Nanotechnology in sustainable agriculture: recent developments, challenges, and perspectives. Front Microbiol 8:1014

Pundir CS, Chauhan N (2012) Acetylcholinesterase inhibition-based biosensors for pesticide determination: a review. Anal Biochem 429:19–31

Randriamampita C, Lellouch AC (2014) Imaging early signalling events in T lymphocytes with fluorescent biosensors. Biotechnol J 9:203–212

Reyes De Corcuera JI, Cavalieri JR (2003) Prototype instruments for laboratory and on-line measurement of lipoxygenase activity. Food Sci Technol Int 9(1):5–9

Rodriguez-Mozaz S, Alda MJ, Marco MP (2005) Biosensors for environmental monitoring: a global perspective. Talanta 65(2):291–297

Salek-Maghsoud A, Vakhshiteh F, Torabia R, Hassani S, Ganjali MR, Norouzi P, Hosseini M, Abdollahi M (2018) Recent advances in biosensor technology in assessment of early diabetes biomarkers. Biosens Bioelectron 99:122–135

Sang S, Wang Y, Feng Q, Wei Y, Ji J, Zhang W (2015) Progress of new label-free techniques for biosensors: a review. Crit Rev Biotechnol 15:1–17

Scheller FW, Yarman A, Bachmann T, Hirsch T, Kubick S, Renneberg R, Schumacher S, Wollenberger U, Teller C, Bier FF (2014) Future of biosensors: a personal view. Adv Biochem Eng Biotechnol 140:1–28

Sbartai A, Namour P, Errachid A, Krejci J, Sejnohova R, Renaud L, Hamlaoui ML, Loir A-S, Garrelie F, Donnet C, Soder H, Audouard E, Granier J, Jaffrezic-Renault N (2012) Electrochemical boron-doped diamond film microcells micromachined with femtosecond laser: application to the determination of water framework directive metals. Anal Chem 84(11):4805–4811

Senthilkumaran B (2015) Pesticide- and sex steroid analogue-induced endocrine disruption differentially targets hypothalamo-hypophyseal-gonadal system during gametogenesis in teleosts—a review. Gen Comp Endocrinol 219:136–142

Shen MY, Li BR, Li YK (2014) Silicon nanowire field-effect-transistor based biosensors: from sensitive to ultra-sensitive. Biosens Bioelectron 60:101–111

Sun JZ, Peter KG, Si RW, Zhai DD, Liao ZH, Sun DZ, Zheng T, Yong YC (2015) Microbial fuel cell-based biosensors for environmental monitoring: a review. Water Sci Technol 71:801–809

Turner AP (2013) Biosensors: sense and sensibility. Chem Soc Rev 42:3184–3196

Turner APF, Karube I, Wilson GS (1987) Biosensors fundamentals and applications. Oxford University Press, Oxford

Vanderroost M, Ragaert P, Devlieghere F, De Meulenaer B (2014) Intelligent food packaging: the next generation. Trends Food Sci Technol 39:47–62

Ukhurebor KE (2020) The role of biosensor in climate smart organic agriculture towards agricultural and environmental sustainability. In: Meena RS (ed) Agrometeorology, IntechOpen, London, UK. https://doi.org/10.5772/intechopen.93150

Ukhurebor KE, Abiodun IC (2018) Variation in annual rainfall data of forty years (1978-2017) for South-South, Nigeria. J Appl Sci Environ Manage 22(4):511–518

Ukhurebor KE, Azi SO, Aigbe UO, Onyancha RB, Emegha JO (2020a) Analysing the uncertainties between reanalysis meteorological data and ground measured meteorological data. Measurement 165:108110

Ukhurebor KE, Aigbe UO, Olayinka AS, Nwankwo W, Emegha JO (2020b) Climatic change and pesticides usage: a brief review of their implicative relationship. Assumption Univ eJournal Interdisc Res 5(1):44–49

Ukhurebor KE, Azi SO (2019) Review of methodology to obtain parameters for radio wave propagation at low altitudes from meteorological data: new results for Auchi area in Edo State, Nigeria. J King Saud Univ—Sci 31(4): 1445–1451

Ukhurebor KE, Nwankwo W (2020) Estimation of the refractivity gradient from measured essential climate variables in Iyamho-Auchi, Edo State, South-South Region of Nigeria. Indonesian J Electr Eng Comput Sci 19(1):276–284

Ukhurebor KE, Olayinka SA, Nwankwo W, Alhasan C (2019) Evaluation of the effects of some weather variables on UHF and VHF receivers within Benin City, South-South region of Nigeria. J Phys IOP Conf Ser 1299:012052

Ukhurebor KE, Umukoro OJ (2018) Influence of meteorological variables on UHF radio signal: recent findings for EBS, Benin City, South-South, Nigeria. IOP Conf Ser Earth Environ Sci 173:012017

Verma N, Bhardwaj A (2015) Biosensor technology for pesticides—a review. Appl Biochem Biotechnol 175:3093–3119

Vigneshvar S, Sudhakumari CC, Senthilkumaran B, Prakash H (2016) Recent advances in biosensor technology for potential applications—an overview. Front Bioeng Biotechnol 4(11):1–9

Wang S, Poon GM, Wilson WD (2015) Quantitative investigation of protein-nucleic acid interactions by biosensor surface plasmon resonance. Methods Mol Biol 1334:313–332

Wang X, Liu M, Wang X, Wu Z, Yang L, Xia S, Chen L, Zhao J (2013) p-Benzoquinone-mediated amperometric biosensor developed with Psychrobacter sp. for toxicity testing of heavy metals. Biosens Bioelectron 41:557–562

Wong LS, Wong CS (2015) A new method for heavy and aluminium detection using biopolymer-based optical biosensor. IEEE Sens J 15(1):471–475

Yada RY (2015) Improving and tailoring enzymes for food quality and functionality, 1st edn. Elsevier Science, Burlington

Yan H, Tang N, Jairo GA, Chakravarty S, Blake DA, Chen RT (2016) High-sensitivity high-throughput chip-based biosensor array for multiplexed detection of heavy metals. Proc. In: Frontiers in biological detection: from Nanosensors to systems VIII, 972501–972508, United States, Proceedings of SPIE

Zhang H, Shi Y, Lan F, Pan Y, Lin Y, Lv J, Zhu Z, Jiang Q, Gqing C (2014) Detection of single-digit foodborne pathogens with the naked eye using carbon nanotubebased multiple cycle signal amplification. Chem Commun 50(15):1848–1850

Zhang P, Zhao X, Ji Y, Ouyang Z, Wen X, Li J, Su Z, Wei G (2015) Electrospinning graphene quantum dots into a nanofibrous membrane for dual-purpose fluorescent and electrochemical biosensors. J Mater Chem B 3:2487–2496

Chapter 8
New Trends in Biosensor Development for Pesticide Detection

Narlawar Sagar Shrikrishna, Subhasis Mahari, Naina Abbineni, S. A. Eremin, and Sonu Gandhi

Abstract Increased use of pesticides globally has led to harmful consequences on the environment, livestock, animals, as well as humans. Hence, its detection in animal feed, vegetables, agricultural crops, milk, eggs, and meat has become a pre-requisite. Conventional techniques such as High-Performance Liquid Chromatography (HPLC), Gas Chromatography (GC), Fourier transform Infrared Spectroscopy (FTIR), Mass Spectrometry (MS), Spectrophotometry, Polarography, etc., are generally being used in detection of pesticides. Alternatively, biosensors can be employed which are cost-effective and an easy way to detect pesticides. Biosensors can be differentiated based upon its biorecognition elements ranging from antibodies to cells. This chapter gives an overview of recent advances in biosensor development for the detection of pesticides with major emphasis on electrochemical biosensors.

Keywords Pesticides · Biosensors · Aptamers · Antibodies · Nanocomposites

8.1 Introduction

Pesticides, as the name suggest, are organic toxic compounds (chemicals) that are used to manage variety of insects, insect's larva, nematodes, rodents, fungi, weeds, and numerous other pests. Farmers use various chemical compounds in agriculture, such as pesticides, fumigants, plant growth regulators, and fertilizers, to protect their crops from hazardous pests (Sassolas et al. 2012). Residues of pesticides may come into the food chain by means of water, air, and soil (Kumar et al. 2019). It impacts the ecosystem and induces a variety of human and animal health issues (Özkara et al. 2016). They can cause diseases of the respiratory system, bone marrow, and nerves, infertility, immunological illness (Tsatsakis et al. 2019). Many pesticides

---

N. S. Shrikrishna · S. Mahari · N. Abbineni · S. Gandhi (✉)
DBT-National Institute of Animal Biotechnology, Hyderabad 500032, India
e-mail: gandhi@niab.org.in

S. A. Eremin
Faculty of Chemistry, M. V. Lomonosov,
Moscow State University, Leninsky Gory, 1, Moscow, 119991, Russia

R. N. Pudake et al. (eds.), *Biosensors in Agriculture: Recent Trends and Future Perspectives*, Concepts and Strategies in Plant Sciences,
https://doi.org/10.1007/978-3-030-66165-6_8

are persistent organic pollutants (POPs), primarily pesticides of the organochlorine family which are noted for their bioaccumulating effects. Pesticides that are used for vector management bioaccumulate in conserved areas. There is a substantial chance of human cancer correlated with the ingestion of such fish from the rivers (Gerber et al. 2016). POPs are lipophilic in nature, so they are found in the milk of animals (Hussain 2018) and their effective detection is the need of the hour.

Detection of pesticides at standards set by the Environmental Protection Agency (EPA) remains a challenge. Chromatographic approaches coupled with specific detectors have historically been used to evaluate pesticides because of their safety, durability, and performance. Nonetheless, they are time consuming and laborious, and require highly skilled technicians and costly machinery. Hence the development of biosensors for the sensitive detection of pesticides has been given significant attention in recent years as a promising tool. A biosensor is a self-contained system that incorporates an immobilized biological entity (e.g., whole cell, DNA fragment (Kasoju et al. 2020a, b), enzyme, biomembranes, antibody (Talan et al. 2018)) that detects the analyte (e.g., antigen, DNA, enzyme substrate, enzyme inhibitors such as some pesticides (Islam et al. 2019; Thakur et al. 2011; Mahendra et al. 2010; Tey et al. 2010; Raman et al. 2009)) and a transduction elements used to transform the biochemical signal arising from the analyte's contact with the bioreceptor into an electrical signal (Sassolas et al. 2012). Biosensors are categorized as electrochemical, electrical, piezoelectric, and mechanical biosensors, depending on the signal transduction technique (Raman et al. 2009). Because of their high sensitivity, electrochemical transductors have been commonly used in biosensors for pesticide detection. It is a low cost, basic design and compact size also make them excellent candidates for the construction of portable biosensors. Based on the biorecognition element biosensors are classified as enzymatic, immunochemical, whole cell, and DNA biosensors.

Classical methods such as chromatography, Mass spectroscopy, UV-Visible spectroscopy, Nuclear Magnetic Resonance (NMR), matrix-assisted laser desorption/ionization (MALDI-TOF) can be performed to detect those pesticides. However, these techniques are laborious, costly, time consuming, and require skilled personnel. In order to overcome these difficulties, certain new techniques have been developed that are based on bio recognition of molecules using biosensors.

8.2 Classical Methods of Pesticide Detection

Food and Drug Administration (FDA) and Food Safety Inspection Service (FSIS) monitor the presence of pesticides in different food commodities. There have been several advancements in technologies for the identification and quantification of pesticides in different products. FSIS provides suitable criteria for pesticide detection for example the detection should not take more than 24 hours, should require easily available equipment and reagents, and should be able to detect below the tolerance level (Pesticide Residues in Food: Technologies for Detection - U.S Congress 1988).

Different chromatogenic methods specifically thin layer chromatography (TLC), gas chromatography (GC), liquid chromatography, gel permeation chromatography, and high-performance chromatography helps in the separation of different products to check the presence of pesticides in it. These methods were developed in the early and mid-twentieth century and are hence classified under a classical method of pesticide detection.

8.2.1 Chromatographic Methods

8.2.1.1 Selective Gas Chromatography Detectors

Selective gas chromatography detectors depend upon different chemical actions on the sample before the actual detection. Selective detectors are usually liable to changes in the operating framework apart from the two main universal detectors, i.e., katharometer and the Flame ionization detector (FID). These detectors require calibration for the linearity of the data. To improve on this flaw, advancements have been made on the detectors in the last few years. Selective gas chromatography detectors include pulsed flamed photometric detector (PFPD), Halogen specific detector (XSD™, manufactured by OI Analytical), Tandem GC detectors, and Electrolyte conductivity detector (ELCD) (Adlard and Juvet 1975).

8.2.1.2 Pulsed Flame Photometric Detectors (PFPD)

PFPD is considered as a new tool in the hands of an analytical chemist. It was developed 30 years ago and uses a flame for high sensitivity and better selectivity for sulphur and phosphorus (Amirav and Jing 1995). In PFPD, the factors affecting sensitivity are analysed with a specific detector without interference of hydrocarbons. PFPD is coupled with a pulsed flame ionization detector (PFID) to produce combined pulse flamed photometric ionization detector (PFPID) (Fan 2005). PFPD also features light emission time having the ability to separate in time the emission of carbon species from sulphur and phosphorus which leads to enhancement in detection selectivity. It also has characteristics of reduced background noise that gets filtered in time, increased signal because of higher brightness of pulsed flame, less flow of combustible gas, and the using broad band filters. Again, air and hydrogen consumption is reduced to a great extent for increased selectivity and sensitivity in selective detection of nitrogen using Nitric acid flame chemiluminiscence (Qian and Burbank 2007). PFPD consists of a main body mounted on GC detector base, two hydrogen gas mixtures, a gas mixture combustor, a quartz light guide rod, a colored glass filter, and a photomultiplier for detection (Amirav and Jing 1995). A combustible gas mixture is eluted and at the same time, the combustor gets filled with that gas which is surrounded by a separate air-richer combustible gas mixture which bypasses the combustor but mixes with other gases from the previous pulse and helps in filling the complete igniter light shield element. A heated wire igniter ignites the pulsed flame and disseminates via an igniter light shield, passing through the combustor and terminating at the combustor holder. A hole of around 0.9 mm does not allow further movement (Qian and Burbank 2007).

Jing and Amirav (1997) analysed methyl parathion in eggplant extract where PFPD had superior sensitivity of detection that could be used in achieving lower detection limits and increased stability. Further, M. Salvador and his group verified the presence of organophosphorus in vegetables finding 24 pesticides with 0.01 mg/kg of limit of detection (Martínez Salvador et al. 2006).

8.2.1.3 Halogen Specific Detectors (XSD^{TM})

Halogen specific detector (XSD^{TM}) detects halogen-containing compounds from the elution of GC capillary column. It neither uses any organic solvent nor has any radioactive source. It does not require any catalyst tube, solvent, resin cartridges, pumps, or transfer lines. XSD^{TM} consists of a high detector that allows low-level selective analysis of halogens and a unique jet design that reduces the peak tailing and requires only air for operation. It has a very low range of detection of around 2 pg chlorine and has a high selectivity of 10^6 molecules of chlorinated fatty acids. Compounds with concentrations larger than 1 ng μL^{-1} has a specific peak broadening and peak trailing shows inappropriate results is a major drawback of XSD^{TM} (Marquardt et al. 2019; Zhuang et al. 2005). XSD^{TM} consists of an alkali glass-ceramic rod, a cathodic platinum bead attached to the bottom of the rod, an anodic platinum coil wound around the rod above the bead, and a reactor core that helps in maintaining a working temperature for the electrodes (Zhuang et al. 2005). The current that is emitted from XSD^{TM} consists of negative, positive, and free electrons. The background current is used for thermal electron emission of the cathode and when halogen atoms are produced from the combustion of halogenated analytes, they get adsorbed onto the surface of the cathode sensitized by alkali metal (Nilsson et al. 2001).

Pesticides such as chlorothalonil was detected by Brown and colleagues from fruit and vegetable samples including Tomato, Potato, and Cabbage using Halogen specific detectors with a limit of detection of 4 mg mL^{-1} (Brown et al. 2011).

8.2.1.4 Electrolyte Conductivity Detector (ELCD)

Electrolyte conductivity detector (ELCD) was first described by Piringer and Pascalau in 1962 for the detection of organic compounds (Piringer and Paşcalău 1962). Here, the compounds eluted from the column are collected in a quartz capillary producing carbon dioxide. The carrier gas in a long capillary tube is mixed with carbon dioxide and the conductivity of the aqueous solution is measured (De Vos et al. 2011). In 1962, Sternberg explained an ELCD coupled with a flame ionization detector (De Vos et al. 2011) where water flowing on the surface of reactants was positioned above the burner of the flame ionization detector. The products were entrapped in a liquid film and the change in conductivity was measured. With time, different models enhanced from the initials. In 1974, a micro detector (ELCD) was first described by Hall (Pape et al. 1977). The size was significantly small, and it showed increased sensitivity. It also carried space for liquid and gaseous phase distinction. Here, it contained a temperature regulating system and the solvent could be non-aqueous. With further improvement, an increased sensitivity was achieved using a bipolar pulse differential detector, where the conductivity of the solvent was measured in the first detector cell and solvent with products in the second cell (Dressler 1986). This enhanced the capability of the detector in solvent purity and temperature variations. Recently, a further improved design of ELCD was described as suitable for high resolution capillary gas chromatography (Verhaar et al. 1984). In this detector, no phase separator was present leading to liquid and gas mixture passing directly through the conductivity cell. The ELCD could be used for selective detection of halogens, sulphur, and nitrogen containing compounds. Dolan and Seiber (1977) used electrolyte conductivity detector for detection of different pesticides including Mirex, Heptachlor, Aldrin, Dieldrin, Lindane, and Endrin, with detection limits in the range of nanograms.

8.2.1.5 Tandem GC Detectors

Tandem GC detectors have always been tools for clinical assessment of steroids since the last 3 decades (Nyiredy 2004). Production of simultaneous chromatograms for aromatics and aliphatics is possible using the detectors that eliminate the need for separate analysis. Tandem GC detectors can be used as packed or capillary columns. The stream that elutes samples from GC column to the reaction chamber provides a continuous irradiation of high energy ultraviolet light, increasing the ionization potential of the samples. An electric field inside it collects the ions formed, leading to the production of ion current which can be amplified, and output could be checked by the GC's electrometer. Tandem GC detectors are non-destructive detectors that do not destroy the structural and elemental characteristics of the compounds. Next, the samples are passed inside a chamber where the analytes are combusted in a hydrogen-air flame, after which the ionized combustion products pass through a negatively biased annular electrode, producing an ion current proportional to the sample concentration in the flame. Tandem GC detectors can acquire two simultaneous chromatograms from a single injection leading to the elimination of transfer lines between detectors which improves the shape and performance of the peak.

8.2.1.6 Thin Layer Chromatography (TLC)

It is a separation technique used mainly for the separation of non-volatile mixtures and is used on a glass or plastic coated with an absorbent material usually silica gel, aluminium oxide, or cellulose. TLC has always been an important aspect in pesticide analysis like the determination of quantitative structure-activity relations (QSAR), molecular structure, and the biological activity of the compound. Analyses using TLC have been done on food samples, pesticides, fungicides, herbicides, and many more (Sherma and Fried 2005). It has also played an important role in pesticide separation, detection, identification, and quantification (Sherma and Fried 2005). Analysis of pesticides has always been one of the major applications of TLC for both qualitative and quantitative analysis. The idea behind TLC came from Izmailov and Shraiber (1938) when they used a powdered absorbent, like calcium chloride, magnesium oxide, or alumina to acquire a chromatographic zone from a single drop of analyte (Poole et al. 1986). Modifying the set-up, Williams (1946) used circular chromatograms to produce different layers of adsorbent in between two glass plates. Later, in 1952, researchers initiated a spreading technique to prepare alumina on glass plates (Grinberg 1990). Finally, the modern-day TLC was generated by Meinhard and Hall in 1949 following the research by Williams (1946) who used starch as the binding agent for mechanical stability (Abbott and Thomson 1965). The use of TLC in detection of pesticides started around the 1960s for pesticide residue analysis but its application reduced by a great extent due to the availability of gas-liquid chromatography (GLC) and High-performance liquid chromatography (HPLC) easily. In recent years, improvisation has been observed in the coating of materials, detection systems, and extraction of pesticides which has once again increased the use of TLC globally (Aktar et al. 2009; Institute of Medicine 1997). In addition to a separation technique, TLC is also used in the cleanup procedures prior to gas chromatography (GC) or HPLC (Sherma 1982).

Procedure of TLC: In the primary step, the sample is applied on the coated plate and then a solvent is drawn upward by capillary action. Commercially, pre-coated organic-bound, layers of silica are more favorable than laboratory made silica gel due to their stability and uniformity. Use of silica gel is usually preferred but there is an increasing shift toward high-performance silica gel and chemically bonded C_{18} reversed phase (Gabriëls and Plaizier-Vercammen 2004). Certificate of Analysis program recommends that silica gel, High-performance silica gel, and reversed phase plates with preadsorbent spotting areas are better suitable for pesticide analysis by TLC due to uniform spotting of samples and control (Lau and Dunn 2018; Kogan et al. 1989; Sherma and Fried 1987). Selection of the ideal mobile phase is quite necessary for the movement of solutes. Mobile phase is usually made up of pure organic or mixed aqueous—organic solvent systems differing in ratio of the solvent mixture (Sherma 1982). Mobile phase is usually carried out in rectangular chambers (usually 10×10 cm) but results vary when the size changes (Nyiredy 2004). The analytes in the sample move at different rates which helps in its separation (Tong et al. 1985). The properties of stationary phase are different from mobile phase, i.e., if the mobile phase is polar then the stationary phase is nonpolar, helping in proper movement of the molecules. This flow of molecules through the capillary action leads to separation of analytes. Components of pesticides are detected on the layers because of their natural color, fluorescence, or ultraviolet light by fluorescence quenching or fluorescence or color zone. A gaseous electrical discharge technique for fluorescence is used in some organophosphate insecticide zones on TLC plates can be produced (Hauck and Halpaap 1980). For identification of zones of pesticides on TLC plate, few drops of a reducing agent were added to the spots and the plate was heated at 100 °C for 10 min. The compound could be identified on the plate after cooling (Sherma 1982).

Applications of TLC in detection of pesticides: Use of TLC in forensic sciences is possible for estimation of the distribution of carbamate pesticides (Tewari 1980), organophosphate pesticides, urea-related pesticides, and triazine herbicides as well as different fungicides. Abbott and Thomson (1966) checked water samples were also analysed using TLC to check the presence of pesticides.

8.2.1.7 Gas Chromatography Coupled with Mass Spectrometry (GC-MS)

GC-MS is a technique that collaborates with the work of gas chromatography and mass spectrometry for sample analysis (Sparkman et al. 2011). There are several applications of GC-MS including drug detection, environmental analysis, fire and explosives investigation, identification of trace elements from materials, etc. GC–MS is made up of two blocks, i.e., gas chromatograph (GC) and mass spectrometer (MS). GC consists of a capillary column which is relative to the phase properties and column's dimensions. The diversity in chemical properties in molecules and their affinity toward the stationary phase enhances the separation of molecules during their travel in the column. Then the molecules are eluted at different time intervals from the column. This helps MS in capturing, analysing, and detecting the ionized molecules separately. MS breaks each ionized molecule and measures the fragments in mass to charge ratio. The two components connected together provide a greater degree of substance identification than when used individually. The only drawback is that MS cannot differentiate multiple samples having same travel time in the column as they elute at the same time and hence, combining the two methods ceases this chance of error as it is very rare that more than one molecule will show the same characteristics in both GC and MS.

In the year 1950, Fred McLafferty and Roland Gohlke, two researchers improved the analytical power of GC by coupling it with MS (https://www.acs.org/content/acs/en/education/whatischemistry/landmarks/gas-chromatography-mass-spectrometry.html). Coupling both the instruments lead to the determination of unambiguous identification of each component. Taking an unknown mixture of samples, the mass spectrum of each peak can reduce the possible identity of each component of the mixture. Having the known standard can help in understanding the retention time and a mass spectra matching with the. Components of GC–MS include capillary GC system, mass selective detector which constitutes an ionization source, mass separator, and ion detector. Two common mass analyzers or separators that are available for GC–MS is quadrapole and the ion trap.

Analysis of sample in GC–MS can be of two types, i.e., full scan or selective ion monitoring. The main goal of the instrument is to quantify the amount of substance that is possible by comparing different concentrations of the molecules in the analyte. Two kinds of analysis are usually done which are comparative and original. A comparative analysis compares the given spectrum with the spectrum library to correlate the characteristics while the original analysis is a simple analysis of the analytes. A full spectrum analysis checks all the peaks in the spectrum while selective ion monitoring only monitors specific ions linked with a specific substance. It is a fast mode of analysis based upon the previous data about the sample (Alves et al. 2018).

8.2.1.8 HPLC Coupled with Mass Spectrometry (HPLC–MS)

It is a technique that combines liquid chromatography with mass spectrometry to provide a proper analysis of a mixture of samples. Liquid chromatography helps in the separation of the molecules and mass spectrometry helps in the identification of the molecules. This technique helps in analysing biochemical, organic, and inorganic compounds commonly found in complex samples present in the environment (Jacob et al. 2014; Dass 2007). HPLC and MS are not compatible with each other as the mobile phase of liquid chromatography is a highly pressurized liquid while the MS works under high vacuum conditions. So, it is not possible to elute samples from HPLC system to MS source and hence an interface is required between the two. The interface could be done by transferring the maximum amount of elute from the mobile phase of the HPLC column while making sure that the interface should not interfere with the ionizing efficiency and vacuum conditions of the MS system (Dass 2007). Setting up HPLC–MS for analysis requires specific conditions and parameters. Each analyte requires specific optimization even when the standardized conditions are present. Analyte's nature determines the optical ionization and polarity, and the detection depends upon derivatisation (Gao et al. 2005; Zaikin and Halket 2006; Johnson 2007).

HPLC coupling with MS started around the 1950s. MS was originated in 1952 by A.J.P. Martin and A.T. James who tried to separate molecules based on analysis of mass (James and Martin 1952). Commercially available HPLC–MS was made available during the 1970s when capillaries were used to connect LC columns with MS ion sources (Niessen 2006). The preliminary interface between HPLC–MS was limited to volatile substances and nonpolar analytes with a less molecular mass. The interface between HPLC in the liquid phase and MS in a vacuum took a long time. The innovation toward electrospray ionization was a major step in this. Nowadays, almost all HPLC–MS utilizes electrospray ionization (ESI), atmospheric pressure photo-ionization (APPI), and atmospheric pressure chemical ionization (APCI) as the interface (Arpino 1992; Pitt 2009). Recently, various drying and deposition techniques have evolved like MALDI deposition (Arpino 1989; Murray 1997). An approach called direct—EI LC–MS interface coupled with nano HPLC and electron ionization attached to mass spectrometer is under development (Cappiello et al. 2008; James and Martin 1952). Barrek and colleagues analysed pesticide residues in olive oil using HPLC–MS. The group could separate 11 pesticides from the mixture in olive oil (Barrek et al. 2003).

8.2.1.9 Gel Permeation Chromatography (GPC)

This is a separation technique that helps in the separation of analytes based upon weight or size. It was first developed by Lathe and Ruthven (1956). In 1973, Jim Waters was the first person to construct the first commercial GPC based upon John Moore's design. It works on the basis of the hydrodynamic volume of the analytes. The analytes gets separated depending upon their physical and chemical interactions (Skoog et al. 1998). Porous beads packed in the column help in separation. If the size of the analytes is small, they enter into the beads in the column leading to an increase in retention time while larger-sized analytes do not get trapped in the column and are eluted first. But if the analytes are too large, they get eluted immediately and if the analytes are too small, they do not get eluted at all. This method of separation leads to differentiation of a broad size range of analytes. GPC consists of two phases-mobile phase and the stationary phase. The stationary phase constitutes of polymer matrix having pores completely filled with the solvent required for the mobile phase. Determination of the size of the pore is essential as the basis of separation of the molecules greater than a certain size do not enter the pores, and the interior of the pores are available, partly or completely to molecules having a smaller size. The theory of separation is based upon an equation:

$V_t = V_0 + V_i + V_m$

where,

V_t = The column's maximum capacity

V_0 = Liquid present outside the gel matrix

V_i = Liquid present inside the matrix

V_m = Volume of the gel matrix

GPC consists of a stationary phase composed of porous polymer gel beads, semi-permeable in nature having specific pore sizes. These gel beads are mechanically stable, chemically inert, and have an ideal homogenous pore structure and size. Mobile phase of GPC consists of a liquid that helps in dissolving the biomolecules allowing high and accurate detection. Pump used in GPC is either a syringe pump or a reciprocating pump with a constant rate of flow. The detectors can be of three types, i.e., refractive index detector, ultraviolet absorption detector, and evaporating light scattering detector. There are certain advantages of GPC including, it takes a short time for sample analysis, provides narrow bands showing high sensitivity, low or no loss of samples, requires less amount of mobile phase and the flow rate can be set. Along with all these GPC has some disadvantages like a limited number of peaks that can be resolved in a small GPC run. It requires filtration of samples before use which may otherwise lead to biased results. It also provides no conclusive data when the molecular masses of the analytes are very close. Apart from these disadvantages, GPC is used in protein fractionation, purification, and molecular weight determination. Along with that GPC is used in separation of sugars, proteins, peptides basing upon their size and even helpful in determination of the quaternary structure of the purified proteins. Jongenotter and his group analysed the concentration of organophosphates in olive oil. Pesticides with a limit of detection of 0.0002 mg kg^{-1} could be detected using gel permeation chromatography (Jongenotter et al. 1999).

8.2.2 Matrix-Assisted Laser Desorption/Ionization–Time of Flight (MALDI–TOF MS)

It is a tool for pesticide analysis with advantages like simplicity of obtained spectra, broad applicability, a better tolerance toward contaminants, and an ability to analyse complex mixtures without separating the analytes even before prior analysis (Madla et al. 2012). It was first termed by Franz Hillenkemp, Michael Karas, and their colleagues in 1985 (Karas et al. 1985). It must be noted that, no specific protocol has been designed for sample preparation in MALDI-TOF MS but the spectra depend upon the matrices. However, commercially available matrices are available which are analyte specific (Ayorinde et al. 1999). Mostly, organic matrices are used for MALDI-TOF. Matrix of MALDI consists of crystallized molecules like α-cyano-4-hydroxycinnamic acid, 3,5-dimethoxy-4-hydroxycinnamic acid, and 2,5-dihydroxybenzoic acid (DHB). Selection of matrix is usually based upon characteristics like low molecular weight and a strong optical absorption range either in the infrared (IR) or ultraviolet (UV) range for efficient absorption of laser light (Zenobi and Knochenmuss 1998). Sample preparation for MALDI undergoes the following procedure the sample needs to be mixed with the matrix solution and around 1 μL of the mixture is to be spotted on the plate and dried before taking the readings by MALDI–TOF MS. Formation of MALDI includes laser, time of flight, and atmospheric pressure. UV lasers are nitrogen lasers of 337 nm and frequency-tripled and quadrupled neodymium-doped yttrium aluminium garnet (Nd:YAG) with 355 nm and 266 nm (Dreisewerd 2014) and IR laser constitutes 2.94 μM Er:YAG laser, mid-IR optical parametric oscillator, and 10.6 μM carbon dioxide laser (Bahr et al. 1992). Madla and colleagues determined pesticides including Atrazine, Azoxystrobin, Flutolanil, Tricyclazole, etc., using MALDI–TOF. They reported that detection of small molecular weight pesticides can be analysed using MALDI–TOF that shows better sensitivity than other methods with a less amount of sample (Madla et al. 2012).

8.3 New Methods of Pesticide Detection and Analysis

Pesticide detection at low concentrations, decided by Environmental Protection Agency (EPA) has always been a challenge even after the development of certain chromatogenic methods since the early 1900s. There have been certain developments for pesticide detection which would be cost-effective, less laborious, require less expensive equipment, and be less time consuming. A biosensor can be a good alternative for pesticide detection which would contain an immobilized bio-element converting the biological signal into an electrical signal. Basing upon the transduction technique, they can be classified into electrochemical, piezoelectric, optical, and mechanical biosensors. Classifying on the basis of biorecognition elements, biosensors can be divided into enzymatic whole cell, immunochemical, and DNA sensors. Recent reports provide data on the identification of pesticides on the basis of molecular imprinted polymers (MIPs) and artificial neural networks (ANNs) coupled with sensor arrays. Along with that, nanotubes and nanorods help in improvising the efficiency of biosensors for pesticide detection. Further, different kinds of biosensors available commercially are described in Table 8.1.

8.3.1 Enzyme-Based Biosensors

Conventional methods of pesticide detections are laborious, time consuming, have a high operation cost, and require skilled personnel to operate machines, and hence some new methods that are based on biosensors are being developed. Enzyme-based biosensor is one such technique among the new methods.

Several pesticides have been engineered (designed) to inhibit the main enzymatic mechanism, and therefore biosensors that are based on the principle of enzyme inhibition are especially important as they accomplish quantitative detection of a pesticide. So, these enzyme-based biosensors take advantage of an enzyme's inhibition and molecular affinity toward its substrate (pesticide) in order to quantify the amount of pesticide present as shown in Fig. 8.1. Following are some of the enzymes used in pesticide detection.

8.3.1.1 Detection of Cholinesterase Enzyme Inhibiting Neurotoxic Insecticides

The application of insecticides on crops is stringently supervised and monitored by various authorities. For e.g., organophosphorus insecticides in the EU (European Union) are either allowed (dimethoate, malathion, chlorpyrifos) or prohibited (diazinon, coumaphos, parathion). A similar scenario exists for carbamate insecticide as both approved (formatanate, methomyl, pirimicarb) or prohibited (carbofuran, carbaryl, aldicarb) examples. (Commission, E. Eu Pesticides Database).

Because of its relevant toxicity, there is a need for rapid and flexible analytical methods for the analysis of a vast number of environmental samples and food tainted with small quantities of specific insecticides and their metabolites. The principle mechanism of toxicity of theses insecticides are based on the inhibition of the enzyme Acetylcholinesterase (King and Aaron 2015).

Table 8.1 Different types of biosensors developed for pesticide detection along with its limit of detection

Transduction system (Analyte)	Limit of detection (LOD)	References
Optical biosensor		
Isoproturon	0.046 ng mL^{-1}	Rodriguez-Mozaz et al. (2004)
Atrazine	0.155 ng mL^{-1}	Rodriguez-Mozaz et al. (2004)
Herbicide subclasses	3 nM	Giardi et al. (2005)
Diuron	0.01 ng mL^{-1}	Mallat et al. (2001)
Linuron	0.14 ng mL^{-1}	Mallat et al. (2001)
DDT	15 pg mL^{-1}	Mauriz et al. (2006)
Surface Plasmon resonance (SPR)		
2-4 D	4 ng/mL^{-1}	Kim et al. (2006)
Cyanazine	0.095 nM	Saylan et al. (2017)
Atrazine	1 μg mL^{-1}	Nakamura et al. (2003)
Simazine	0.031 nM	Saylan et al. (2017)
Atrazine	0.091 nM	Saylan et al. (2017)
Chloropyrifos	0.056 ng mL^{-1}	Li, Q. et al. (2019a)
Carbendazim	0.44 ng mL^{-1}	Li, Q. et al. (2019b)
Triazophos	0.096 ng mL^{-1}	Guo et al. (2018)
Fluorescence/chemiluminescence		
2,4-D, linuron and simazine	0.02 ng mL^{-1}	Herranz et al. (2008)
2,4-D	4 ng mL^{-1}	Marquette and Blum (2000)
Carbaryl	26.8×10^{-12} mol L^{-1}	Li et al. (2016)
Parathion-methyl	3.8×10^{-6}	Yan et al. (2015)
Glyphosate	12 ng mL^{-1}	Wang et al. (2016)
Dichlorvos	4.4 nM	Meng et al. (2013)
Paraquat	0.0117 nM	Durán et al. (2013)
Quartz crystal microbalance		
Carbaryl	1×10^{-7} M	Abad et al. (1998)
2,4-D	0.5 nM	Skládal et al. (1994)
Paraoxon	5×10^{-8} M	Abad et al. (1998)
Carbofuran	4.5×10^{-6} M	Jia et al. (2013)
Carbamate and Organophosphate	1 ng mL^{-1}	Karousos et al. (2002)
Endosulfan	5.59 μg L^{-1}	Liu et al. (2013)
Pirimicarb	5×10^{-7} M	Sun and Fung (2006)
Metolcarb	0.019 mg L^{-1}	Pan et al. (2013)
Parathion	4 μg L^{-1}	Funari et al. (2013)
Electrochemical Biosensors		

(continued)

Table 8.1 (continued)

Transduction system (Analyte)	Limit of detection (LOD)	References
2,4-D	50 ng mL^{-1}	Dzantiev et al. (1996)
Simazine	5-175 ng mL^{-1}	Starodub et al. (2000)
Monocrotophos	0.3 ng mL^{-1}	Du et al. (2008)
Dichlorvos	7×10^{-12}	Valdés-Ramírez et al. (2008)
Acephate	5.45×10^{-14}	Ishii et al. (2008)
Malathion	0.31 ppt (Parts per trillion)	Bao et al. (2019)
Chlorpyrifos	10 fM	Talan et al. (2018)
Micromechanical cantilever		
Atrazine	0.1 ng mL^{-1}	Raman et al. (2008)
Carbaryl	2×10^{-10} M	Chen et al. (2018)
Paraoxon	10^{-7} M	Karnati et al. (2007)
Rapid screening (dipstick and lateral flow assay biosensor)		
Atrazine	< 1.0 ng mL^{-1}	Kaur et al. (2007)
Sulfamethazine	50–100 ng mL^{-1}	Kandimalla et al. (2008)
Benzothiostrobin	25 μg L^{-1}	Wu et al. (2019)
Imidacloprid	0.02 ng mL^{-1}	Tan et al. (2020)
Profenophos	75 nM	Kim et al. (2018)
Chlorpyrifos	3.3 μg mL^{-1}	Fu et al. (2019)

AChE act as a biorecognition element for the analyte (Pesticide). When pesticide molecules come in close proximity to enzyme, chemical changes take place in reacting elements. These chemical changes lead to the formation of electroactive species, consumption of oxygen, reduced forms of by-products, and so forth (Wang 2008). These changes are detected and displayed on the controlling system (Fig. 8.1). There are reports of various types of biosensors developed for the detection of organophosphate and carbamate insecticides based on the mechanism of cholinesterase enzyme (Štěpánková et al. 2016).

8.3.1.2 Detection of Herbicides (Inhibiting the Process of Photosynthesis)

Herbicides as the name suggest kills a variety of grass (weed) species in agricultural practises, they are widely used in agriculture for the controlling of weeds. Though the lists of permitted herbicides and their permissible levels of residues accepted in food, feed, and water in different parts of the world are continuously being updated, certain herbicides such as atrazine, which have been prohibited in the European Union since 2000, are still permitted for use in many parts of the world, for

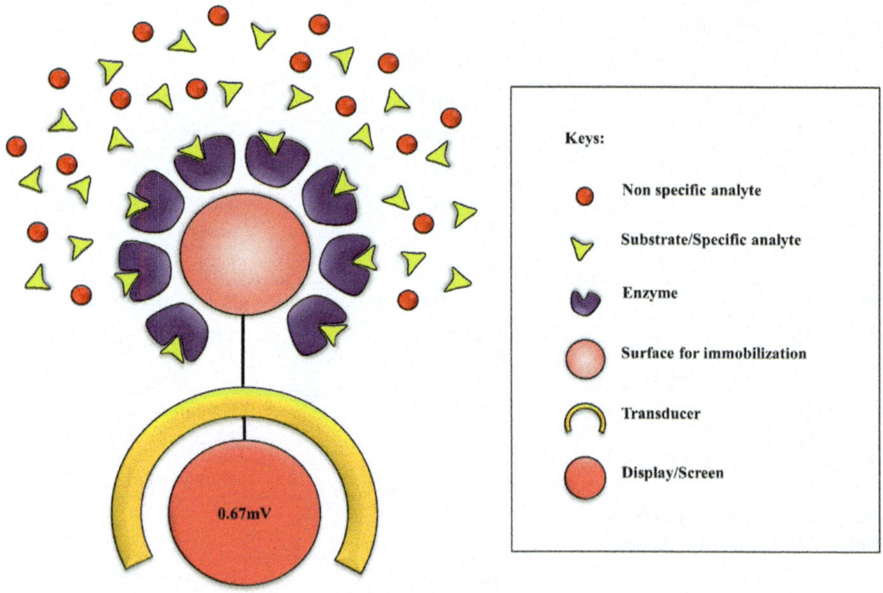

Fig. 8.1 Enzyme biosensor—A schematic illustration showing the enzyme-based detection of a specific analyte (Pesticide molecule), it illustrates enzyme coated on conductive surface recognizes the analyte by binding to it, transducer senses a subsequent change in electrochemical parameters upon binding and resultant value displays on the screen

example, in Asia and North America. Existing herbicide detection methods depend on Enzyme-linked immunosorbent assay (ELISA), Capillary electrophoresis liquid or gas chromatography linked with mass spectrometry (LC–MS or GC–MS) (Giardi et al. 2001). Herbicides pertaining to the class of phenylurea (e.g., chlorotoluron, chlorbromuron, diuron, fenuron, etc.), class of diazine (lenacil, bromacil, etc.), class of triazine (cyanazine, simazine, atrazine, etc.), and group of synthetic phenols (bromoxynil dinoseb, ioxynil, etc.) inhibit photosynthetic diatoms, algae, cyanobacteria, and plants. Several biosensors for the detection of these herbicides have been developed (Giardi et al. 2001). Table 8.1 shows the various biosensors which detect herbicides along with their Limit of detection (LOD). In the process of photosynthesis plants, diatoms, algae, and cyanobacteria transform light into chemical energy. The first step in this process is to break water (photolysis) to produce protons and oxygen as a byproduct. This takes place in Photosystem II complex where water, plastoquinone, oxidoreductase enzyme, and protein D1 present. Herbicide's molecules interact and bind to D1 in PSII, which blocks the electron transfer and therefore inhibits the process of photosynthesis.

8.3.1.3 Detection of Organophosphorus Hydrolase (OPH) Inhibiting Pesticides

Organophosphate hydrolase is an organophosphotriester hydrolyzing enzyme that uses organophosphate pesticides as a substrate and degrades it. The enzyme has a broad substrate specificity toward analyte and is able to hydrolyze monocrotophos, chlorpyrifos, parathion, dursban, diazinon, etc. OPH enzyme-based biosensors have low sensitivity but high detection values as compared to cholinesterase enzyme inhibition-based biosensors. Limit of detection of those biosensors is given in Table 8.1.

8.3.1.4 Glutathione-S Transferase (GST)

GST primarily metabolizes organophosphorus. It metabolizes endogenous and exogenous toxicants leading to the formation of Glutathione conjugates with the compounds or with their reactive intermediates (Fujioka and Casida 2007). It is used for the detection of atrazine by developing a fiber optic biosensor (Andreou and Clonis 2002).GST is immobilized on the glass using an intermediate binder sol-gel layer. Nucleophile attack of atrazine is catalyzed by GST helping in release of H^+ ions. This leads to a change in pH which changes the color measured by bromocresol green.

8.3.1.5 Electrochemical Biosensors

Biosensors based on amperometric detection of pesticides: In amperometric detection technic a constant potential to the working electrode is provided to the working electrode and resultant current is measured by the device as a function of time. Such amperometric biosensors have been constructed to detect biodegradable organic pollutants in aqueous samples by measuring biochemical oxygen demand (BOD). These amperometric biosensors used microbes as biosensing elements, examples of which includes *Serratia marcescens, Bacillus subtilis,* and *Arxula adeninivorans* (Liu and Mattiasson 2002; Chan et al. 2000; Kim and kwon 1999; Riedel et al. 1988; Jia et al. 2003; Hikuma et al. 1979; Liu et al. 2004).

Similar microbial amperometric biosensors were also developed for the detection of organophosphate pesticide residues (e.g., fenitrothion, parathion, methyl parathion, and ethyl p-nitrophenol thiobenzenephosphonate (EPN) (Su et al. 2011). Those biosensors were developed by co-immobilizing microorganisms and organophosphate hydrolase enzyme (OPH). OPH hydrolyses organophosphate pesticide molecules and due to which p-nitrophenol releases. Released p-nitrophenol is then oxidized by those microorganisms (e.g., *Pseudomonas putida* JS444). Oxidation of p-nitrophenol by microorganisms generates electroactive species, which is in turn detected by amperometric biosensors. (Lei et al. 2004). Some microbes consume oxygen, e.g., *Arthrobacter* sp. JS443, decrease in oxygen concentration in a system

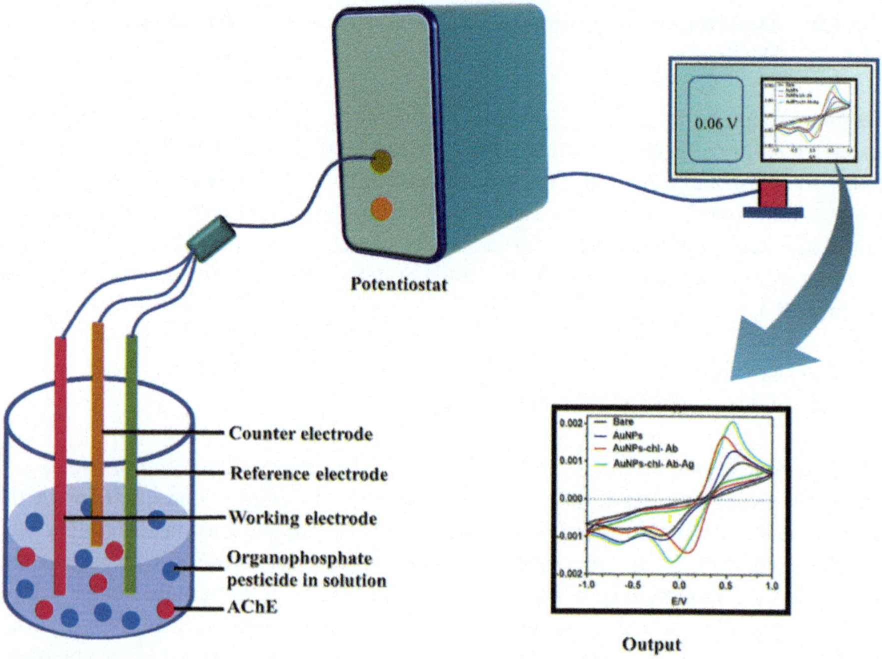

Fig. 8.2 Amperometric detection of pesticide—A schematic illustration showing the enzyme-based recognition of Organophosphate pesticideand amperometric detection, it illustrates enzyme Acetylcholinesterase (AChE) inhibited by organophosphate pesticide and resultant change in current is measured by potentiostat and displayed on monitor

is measure by Clark oxygen electrode as shown in Fig. 8.2 (working electrode) (Lei et al. 2005, 2006; Mulchandani et al. 2005).

Potentiometric detection: Ion-selective electrodes or gas selective electrodes have been developed with an immobilized microbial layer and the bio-analytes are converted into electrical signals and the change in current is measured. Potentiometric sensors measure the difference between two electrodes as well as one reference electrode having constant potential. The change in current is because of the change in sample dropped on the positive electrode (working electrode) and the counter electrode. Direct detection of paraoxon using potentiometric detection was done based upon immobilizing recombinant *E. coli* on a glass electrode. *Opd* gene containing OPH enzyme was inserted into *E. coli.* OPH enzyme activated bacteria which hydrolyzed organophosophorus pesticides leading to the production of two protons. Concentration of paraoxon was related to the quantity of hydrogen ions produced (Rainina et al. 1996). Again, a potentiometric biosensor was developed for the detection of terbuthylazine (TBA). Urea labeled antibody was fixed while free TBA and TBA conjugated BSA competed among each other for binding. Addition of urea (substrate) helped in the measurement of ammonia. Production of urea is inversely proportional to the amount of TBA present in a sample (Sassolas et al. 2012).

Conductometric detection: It is associated with the measurement of conductivity at different frequencies. Conductometric sensors depend upon the electrical conductivity of the material that affects the conductivity of the analytes. Advantages including low cost, simplicity, and no requirement of electrodes make conductometric sensors predominant (Stradiotto et al. 2003). Jardim et al. used conductometric sensors coupled with flow injection for detection of CO_2 in atmosphere (Almeida et al. 2001). Further, Valera et al. utilized conductometric sensors for detection of atrazine. Immobilization of Atrazine on interdigited μ-electrodes. Detection of free atrazine could be verified after a competitive reaction of immobilized atrazine with the antibody present in the solution. The method of detection used gold nanoparticles for labeling. The presence of gold nanoparticles produced red wine color which detected atrazine (Valera et al. 2010; Sassolas et al. 2012; Valera et al. 2008).

8.3.1.6 Piezoelectric Biosensors/QCM Biosensors

Piezoelectric biosensors are mass-based biosensors that uses quartz crystal as a transducer. Piezoelectric biosensors use quartz crystal as s principle sensing component and being highly sensitive, it is called Quartz Crystal Microbalance (QCM). QCM can sense or measure up to nanogram level of changes in mass per unit area. QCM's crystal is made of silicon dioxide (SiO_2) which can be tuned to oscillate with desired frequency by providing appropriate voltage to metal electrodes coupled to the crystal. Upon addition or deletion of mass (analyte) on to the surface of crystal coupled area, frequency of oscillation of quartz crystal alters. This alters in the frequency of the oscillation can be monitored in real time to acquire important information about interaction/reaction of analyte happening at the sensor surface, e.g., antigen-antibody interaction, etc. QCM has been reported to use under vacuum and gas phase (Sauerbrey et al. 1959) and this technique was shown to be pertinent in liquid media, as well (Nomura and Hattori 1980).

Molecular adsorption of the analyte onto the sensor surface can be done in gas or vacuum phase, the result of which is the formation of rigid films adhered to the sensor surface. The change in mass of such films is linearly related to the change in frequency of oscillation. The change in oscillation frequency of quartz crystal is given by infamous Sauerbrey Equation. Quartz crystal microbalance (QCM) provide accurate and useful information with respect to change in mass deposited or mass removal (it actually provides the rate of deposition or rate of removal of rigid films) (Sauerbrey et al. 1959).

When QCM is operated in a liquid medium, water molecules may additionally impart its mass on rigid films by hydration or by entrapment of water molecules. Because of this viscoelastic or soft films can generate. The mass of such resultant films cannot be determined accurately and the frequency of oscillation of such films are not accurate because of the dampening of frequency or energy loss. To overcome this problem new technic was developed. It called Quartz Crystal Microbalance with Dissipation (QCM-D) (Rodahl et al. 1996). Figure 8.3 shows the over-all design of the QCM biosensor.

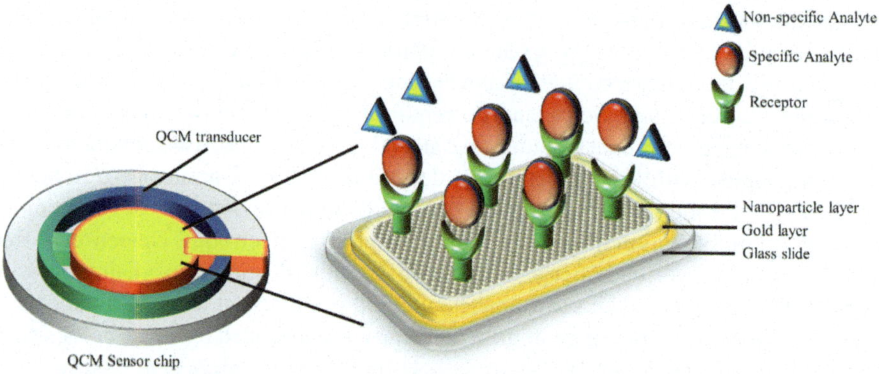

Fig. 8.3 QCM-based biosensor—A schematic illustration showing the QCM sensor chip-based detection of specific analyte (Pesticide). QCM sensor chip is composed of gold-coated glass slide on which analyte (Pesticide) specific antibody is adsorbed, this antibody recognizes the pesticide molecule and bind with it. The change in mass in response to pesticide binding with the selective antibody is detected as a change in frequency of QCM transducer

As the maximum mass sensitivity of QCM in the liquid is less than 1 ng/cm^2, it has been used to detect a number of different pesticides. QCM has been used to analyse pesticide contamination in water and traces of deltametrin, fenthion, methiocarb, and triadimol were found (Erbahar et al. 2012). QCM has been also used to detect organophosphate pesticides. The presence of paraoxone or carbaryl is detected by decreasing in the signal, i.e., resulting from its inhibitory effects. The standard curve gives the detection limit of pesticide by plotting the percent inhibition vs. pesticide logarithm concentration. Detection limits of 5.0×10^{-8} and 1.0×10^{-7} M were obtained for paraoxon and carbaryl, respectively (Abad et al. 1998).

8.3.1.7 Surface Plasmon Resonance (SPR)

SPR is a non-radiative electromagnetic surface wave that moves in parallel to the negative permittivity of the interface of the material. SPR involves reflection of free electrons and protons in a specific angle which is absorbed in the film, called as SPR angle (Sassolas et al. 2012). SPR-based biosensors can have real time monitoring values during the detection of pesticides and notably it does not require any labeling of molecules (Zhao et al. 2015; Homola et al. 1999). Illumination of polarized light in specific conditions of complete reflection, having a thin film in between two media having different refractive indices causes the production of SPR as shown in Fig. 8.4. For the first time, Farré and fellows used real samples for detection using SPR (Farré et al. 2007). Pesticides were primarily conjugated with BSA and was assembled

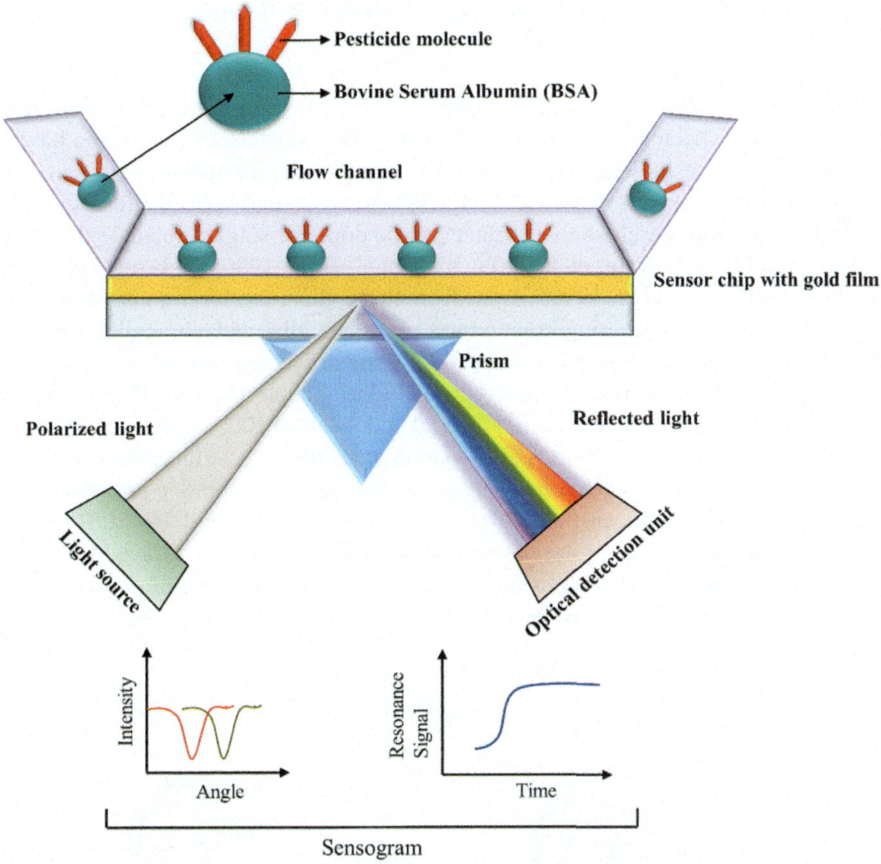

Fig. 8.4 SPR biosensor—A schematic illustration showing the SPR-based detection of a pesticide molecule conjugated to Bovine Serum Albumin (BSA) protein and adsorbed on sensor chip. The binding of BSA-Pesticide complex to the sensor surface increase the refractive index, which induces shift of the SPR angle. The shift is directly proportional to the mass increase

on a monolayer of gold electrode which is reusable. Change in concentration of samples leads to change of refractive indices (Sassolas et al. 2012). Detection of pesticides using SPR was a tremendous advancement as real samples could be detected without pre-treatment modifications (Cortina-Puig et al. 2010; Zhang et al. 2009). A low concentration of any molecule can also be detected using antigen-antibody recognition using SPR biosensor.

8.3.1.8 Fluorescence Polarization (Total Internal Reflection Fluorescence-TIRF)

TIRF uses properties of induced evanescent wave or field in a specific region associated with the specimen that is present close to the interface of two media having different refractive indices. TIRF is used to observe the composition, extent, and position of the substrate (Weis et al. 1982). spectroscopy and visualization of single-molecule fluorescence close to a surface can be done for single-molecule detection as shown in Fig. 8.5 (Kuter et al. 2008; Khan et al. 2000, 1996; Dickson et al. 1998; Sako et al. 2000; Ha et al. 1999). Apart from these, there are several applications of TIRF including (a) measuring kinetic rates of intracellular and extracellular protein binding to cell surface receptors and artificial membranes (Thompson et al. 1981; Burghardt and Axelrod 1981; Axelrod 2001; Fulbright and Axelrod 1993; Stout and Axelrod 1994; Mc Kiernan et al. 1997; Sund and Axelrod 2000; Nollert et al. 1995), (b) Structural dynamics of living cells (Mathur et al. 2000), (c) Fluorescence of cells during cultures (Wang and Axelrod 1994), (d) comparing membrane proximal ionic transients and transients in the cytoplasm (Omann and Axelrod 1996). Recently, europium-chelate-dyed polystyrene nano-composites are being produced for detection of atrazine while TIRF has been used for the development of immunosensors to estimate water pollution (Sassolas et al. 2012; Barzen et al. 2002).

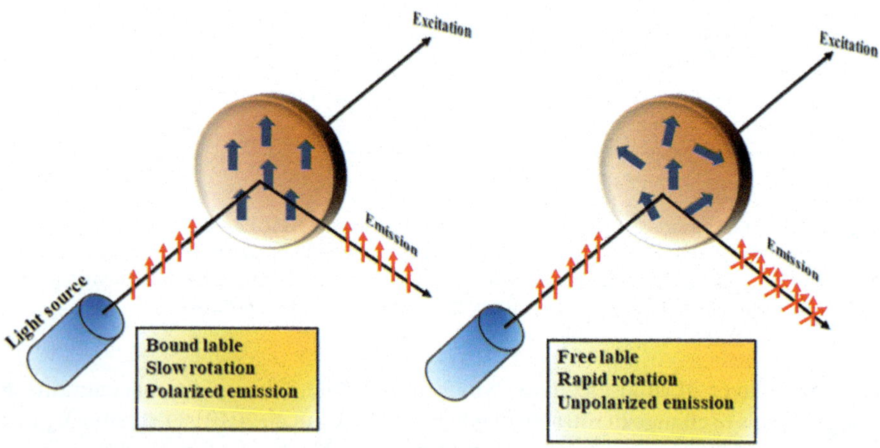

Fig. 8.5 Fluorescence Polarization: Fluorescence polarization—Excitation of fluorescent labels leads to selective absorption of polarized light by appropriately oriented molecules. Polarized emission of these oriented molecules occurs when their rate of rotation is low relative to the rate of fluorescence emission

8.3.1.9 Cantilever-Based Biosensors

Cantilever-based biosensors detect biomolecular interactions in a label-less procedure with high efficiency. Cheap and reliable standard silicon technologies are used in fabricating cantilevers. The working principle of cantilever-based biosensors is based upon translating biochemical reactions into mechanical movement. Biomolecular recognition shows a quantitative change in the cantilever beam, when there is an interaction between an analyte and its receptor (Ziegler 2004). Fabrication of cantilever can be possible in any shape like crystalline, polymer, silicon nitride or silicon oxide. After fabrication, the preliminary setting of beam was quite complicated, so newly developed cantilever biosensors are made up of single crystal silicon as shown in Fig. 8.6. Applications of cantilever-based biosensors are observed in environmental fields and in immunological detection. Biosensors are based upon antigen-antibody interaction directly, by specific immobilization of haptens on the surface and modulating the binding of antigen; or indirectly, by checking the inhibition proportions of the antibody with the antigen present on the immobilized surface (Alvarez et al. 2009).

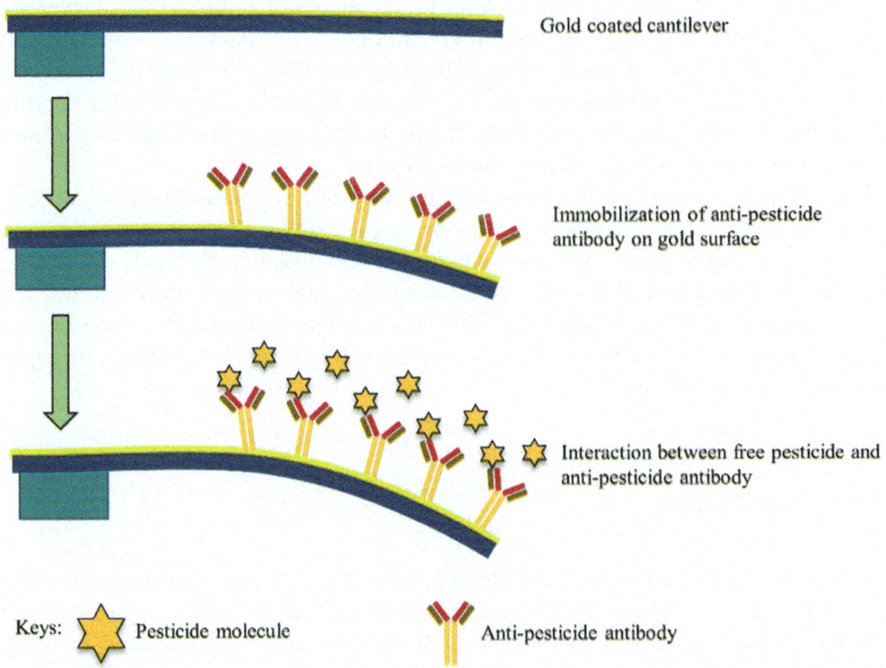

Fig. 8.6 Cantilever-based biosensor—A schematic illustration showing the nanomechanical detection of an anti-pesticide antibody-pesticide interactions on a cantilever. It illustrates the cantilevers coated with anti-pesticide antibody on a gold surface. Pesticide in solution forms a binding event with the antibody, causing the cantilever to bend downward due to compressive surface stress

Stimulation of Cantilever and Fabrication: Δz is called a cantilever deflection that is generated by change in surface stress, Δs, that depends upon microcantilever's dimensions, i.e., length l, thickness h, and properties of the material. According to Stroney's equation, a greater change in surface stress will be higher for longer and thinner cantilevers (Alvarez et al. 2009).

8.3.1.10 Lateral Flow Biosensors

Lateral flow biosensors are considered as improvisations in immunochromatogenic assay (Mishra et al. 2018; Gandhi et al. 2009; Parolo and Merkoçi 2013). Lateral flow biosensors have a low limit of detection, low sample required for analysis, high sensitivity, better selectivity, low cost, and quick assay with single-step detection (Quesada-González and Merkoçi 2015). Further, lateral flow biosensors can be used in the detection of biomarkers that includes nucleic acids, proteins, and even whole cells. Recent research shows evidence in the detection of metal ions and pesticides. Along with all these advantages, lateral flow biosensors have a few disadvantages. Results in lateral flow are qualitative but are not quantitative and solid or gaseous samples cannot be analysed in these sensors. It consists of a nitrocellulose layer and unspecific absorption on the pores can give biased results. However, commercially available lateral flow biosensors are used for pregnancy detection and HIV test (Pesce and Spitlnik 2007) as well as malaria (Cordray and Richards-Kortum 2012; Kersting et al. 2014). Advancement in the field of nanotechnology has provided opportunity in improvement of performance by these sensors.

Working of lateral flow biosensors: Manufacturing of lateral flow assays occurs in the form of strips. It consists of four parts, (a) cellulose pad, where the sample is placed, (b) a conjugate pad that helps in conjugation, usually made up of fiber, (c) detection pad, and (d) nitrocellulose sheet consisting both control and test line as shown in Fig. 8.7 (Ahmad et al. 2009). Furthermore, lateral flow biosensors that could be connected to electrical devices by integrating carbon nanotubes on top of them (Zhu et al. 2014).

8.4 Conclusion

Biosensors have always been a vital apparatus for pesticide detection. Development of different sensors with increased efficiency and accuracy has remarkably increased. Biological elements including antibodies, aptamers, enzymes, or whole cells are being used for detection. In recent times, development of genetically modified enzymes like AChEs have a better attachment and efficiency in comparison to normal enzymes and are being widely used in enzyme inhibition for pesticide detection. Further, genetically engineered microorganisms are being used for detection of pesticides. Nanotechnology has been playing a significant role in developing biosensors for pesticide detection. Nanomaterials such as nanorods and nanotubes

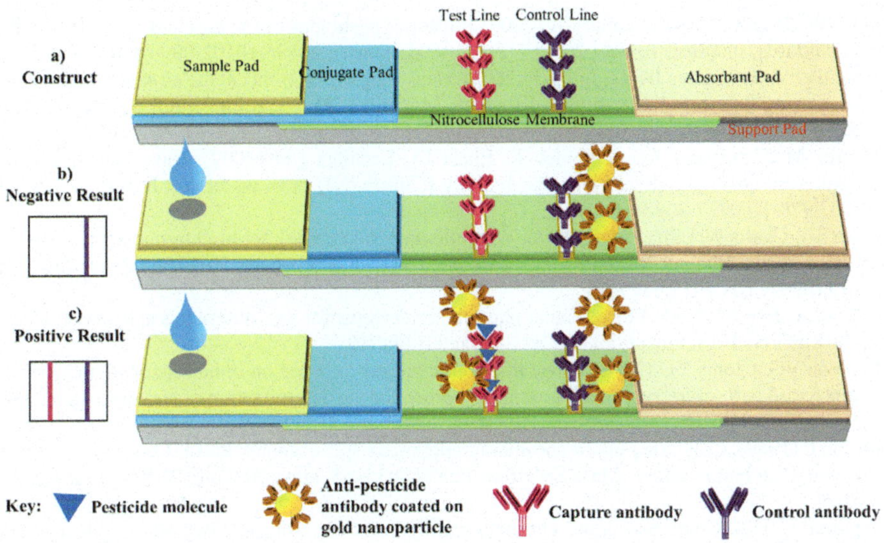

Fig. 8.7 Lateral flow assay (LFA)—**(a)** A schematic illustration showing design of LFA pad. **(b)** if pesticide is absent in the test sample then only one line (Control) appears **(c)** if pesticide is present in test sample then pesticide bind to capture antibody which is specific for that pesticide, after pesticide captured on capture antibody, it is recognized by anti-pesticide antibody which is coated on gold nanoparticle and test line appears along with control line

are being used for increasing the efficiency of the sensors. However, biosensors in the field of pesticide detection are not as enhanced as in medical applications. Hence, development of low detection limit, user-friendly, and cost-effective biosensors are required.

References

Abad JM, Pariente F, Hernández L, Abruna HD, Lorenzo E (1998) Determination of organophosphorus and carbamate pesticides using a piezoelectric biosensor. Anal Chem 70:2848–2855. https://doi.org/10.1021/ac971374m

Abbott DC, Thomson J (1966) Pesticide residue analysis by thin-layer chromatography. Pest Articles News Summ, Section A: Insect Control 12(1–2):21–34. https://doi.org/10.1080/04345546609415378

Abbott DC, Thomson J (1965). The application of thin-layer chromatographic techniques to the analysis of pesticide residues. In: *Residue reviews* (Vol. 11, pp 1–59). https://doi.org/10.1007/978-1-4615-8401-8_1

Adlard ER, Juvet RS (1975) A review of detectors for gas chromatography part II: selective detectors. C R C Crit Rev Anal Chem 5(1):13–37. https://doi.org/10.1080/10408347508542679

Ahmad AL, Low SC, Abd Shukor SR, Ismail A (2009) Morphological and thermal-mechanical stretching properties on polymeric lateral flow nitrocellulose membrane. Ind Eng Chem Res 48(7):3417–3424. https://doi.org/10.1021/ie801282t

Aktar W, Sengupta D, Chowdhury A (2009) Impact of pesticides use in agriculture: their benefits and hazards. Interdisc Toxicol 2(1):1–12. https://doi.org/10.2478/v10102-009-0001-7

Almeida FV, Guimarães JR, Jardim WF (2001) Measuring the CO_2 flux at the air/water interface in lakes using flow injection analysis. J Environ Monit 3(3):317–321. https://doi.org/10.1039/b010065j

Alvarez M, Carrascosa LG, Zinoviev K, Plaza JA, Lechuga LM (2009). Biosensors based on cantilevers. In: Methods in molecular biology (Clifton, NJ), 504, pp 51–71. https://doi.org/10.1007/978-1-60327-569-9_4

Alves SA, Caarls W, Lima PMV (2018). Weightless neural network for high frequency trading. In: Proceedings of the international joint conference on neural networks, 2018-July. https://doi.org/10.1109/IJCNN.2018.8489445

Amirav A, Jing H (1995) Pulsed flame photometer detector for gas chromatography. Anal Chem 67(18):3305–3318. https://doi.org/10.1021/ac00114a030

Andreou VG, Clonis YD (2002) Novel fiber-optic biosensor based on immobilized glutathione S-transferase and sol-gel entrapped bromcresol green for the determination of atrazine. Anal Chim Acta 460(2):151–161. https://doi.org/10.1016/S0003-2670(02)00250-7

Arpino P (1989). Combined liquid chromatography mass spectrometry. Part I. Coupling by means of a moving belt interface. Mass Spectrom Rev 8(1):35–55. https://doi.org/10.1002/mas.1280080103

Arpino P (1992). Combined liquid chromatography mass spectrometry. Part III. Applications of thermospray. Mass Spectrom Rev 11(1):3–40. https://doi.org/10.1002/mas.1280110103

Axelrod D (2001) Total internal reflection fluorescence microscopy in cell biology. Traffic 2(11):764–774. https://doi.org/10.1034/j.1600-0854.2001.21104.x

Ayorinde FO, Hambright P, Porter TN, Keith QL (1999) Use of meso-tetrakis(pentafluorophenyl)porphyrin as a matrix for low molecular weight alkylphenol ethoxylates in laser desorption/ionization time-of-flight mass spectrometry. Rapid Commun Mass Spectrom 13(24):2474–2479. https://doi.org/10.1002/(SICI)1097-0231(19991230)13:24%3c2474:AID-RCM814%3e3.0.CO;2-0

Bahr U, Deppe A, Karas M, Hillenkamp F, Giessmann U (1992) Mass spectrometry of synthetic polymers by UV-matrix-assisted laser desorption/ionization. Anal Chem 64(22):2866–2869. https://doi.org/10.1021/ac00046a036

Bao J, Huang T, Wang Z, et al (2019) 3D graphene/copper oxide nano-flowers based acetylcholinesterase biosensor for sensitive detection of organophosphate pesticides. Sensor Actuator B Chem 279:95–101. https://doi.org/10.1016/j.snb.2018.09.118

Barrek S, Paisse O, Grenier-Loustalot MF (2003) Determination of residual pesticides in olive oil by GC-MS and HPLC-MS after extraction by size-exclusion chromatography. Anal Bioanal Chem 376(3):355–359. https://doi.org/10.1007/s00216-003-1917-y

Barzen C, Brecht A, Gauglitz G (2002) Optical multiple-analyte immunosensor for water pollution control. Biosens Bioelectron 17(4):289–295. https://doi.org/10.1016/S0956-5663(01)00297-4

Brown AN, Cook JM, Hammack WT, Stepp JS, Pelt JV, Gerard G (2011) Analysis of pesticides residues in fresh produce using buffered acetonitrile extraction and aminopropyl cleanup with gas chromatography/triple quadrupole mass spectrometry, liquid chromatography/triple quadrupole mass spectrometry, gas chromatography/ion trap detector mass spectrometry, and GC with a halogen-specific detector. J AOAC Int 94(3):931–941

Burghardt TP, Axelrod D (1981) Total internal reflection/fluorescence photobleaching recovery study of serum albumin adsorption dynamics. Biophys J 33(3):455–467. https://doi.org/10.1016/S0006-3495(81)84906-5

Cappiello A, Famiglini G, Palma P, Pierini E, Termopoli V, Trufelli H (2008) Overcoming matrix effects in liquid chromatography – mass spectrometry. Anal Chem 80(23):9343–9348. https://doi.org/10.1021/ac8018312

Chan C, Lehmann M, Chan K, et al (2000) Designing an amperometric thick-film microbial BOD sensor. Biosens Bioelectron 15:343–353. https://doi.org/10.1016/S0956-5663(00)00090-7

Chen D, Yang L, Yu W, et al (2018) Micro-Electromechanical acoustic resonator coated with polyethyleneimine nanofibers for the detection of formaldehyde vapor. Micromachines 9:62. https://doi.org/10.3390/mi9020062

Cordray MS, Richards-Kortum RR (2012). Review: Emerging nucleic acid-based tests for point-of-care detection of malaria. Am J Trop Med Hyg 87(2): 223–230. https://doi.org/10.4269/ajtmh.2012.11-0685

Cortina-Puig M, Istamboulie G, Noguer T, Marty JL (2010). Analysis of pesticide mixtures using intelligent biosensors. Intell Biosens, 205–216. InTech. https://doi.org/10.5772/7154

Dass C (2007) Fundamentals of contemporary mass spectrometry. Wiley-Interscience. https://doi.org/10.1002/0470118490

De Vos J, Gorst-Allman P, Rohwer E (2011) Establishing an alternative method for the quantitative analysis of polychlorinated dibenzo-p-dioxins and polychlorinated dibenzofurans by comprehensive two dimensional gas chromatography-time-of-flight mass spectrometry for developing countries. J Chromatogr A 1218(21):3282–3290. https://doi.org/10.1016/j.chroma.2011.03.050

Dickson RM, Norris DJ, Moerner WE (1998) Simultaneous imaging of individual molecules aligned both parallel and perpendicular to the optic axis. Phys Rev Lett 81(24):5322–5325. https://doi.org/10.1103/PhysRevLett.81.5322

Dickson RM, Norris DJ, Tzeng YL, Moerner WE (1996) Three-dimensional imaging of single molecules solvated in pores of poly(acrylamide) gels. Sci 274(5289):966–969. https://doi.org/10.1126/science.274.5289.966

Dolan JW, Seiber JN (1977) Chlorine-selective detection for liquid chromatography with a Coulson electrolytic conductivity detector. Anal Chem 49(2):326–31

Dreisewerd K (2014) Recent methodological advances in MALDI mass spectrometry. Anal Bioanal Chem 406(9–10):2261–2278. https://doi.org/10.1007/s00216-014-7646-6

Dressler M (1986) J Chromatogr Librar, Elsevier (36). https://doi.org/10.1016/S0301-4770(08)60338-3

Du D, Chen S, Song D, Li H, Chen X (2008) Development of acetylcholinesterase biosensor based on CdTe quantum dots/gold nanoparticles modified chitosan microspheres interface. Biosens Bioelectron 24:475–479. https://doi.org/10.1016/j.bios.2008.05.005

Durán GM, Contento AM, Ríos Á (2013) Use of Cdse/ZnS quantum dots for sensitive detection and quantification of paraquat in water samples. Anal Chim Acta 801:84–90. https://doi.org/10.1016/j.aca.2013.09.003

Dzantiev BB, Zherdev AV, Yulaev MF, et al (1996) Electrochemical immunosensors for determination of the pesticides 2,4-dichlorophenoxyacetic and 2,4,5-trichlorophenoxyacetic acids. Biosens Bioelectron 11:179–185. https://doi.org/10.1016/0956-5663(96)83725-0

Erbahar DD, Gürol I, Gümüş G, et al (2012) Pesticide sensing in water with phthalocyanine based QCM sensors. Sensors Actuators B Chem 173:562–568. https://doi.org/10.1016/j.snb.2012.07.041

Fan X (2005) Nature, cause, and control of irradiation-induced off-odor in ready-to-eat meat products. Process React Flavors Chapter 17:208–221. https://doi.org/10.1021/bk-2005-0905.ch017

Farré M, Martínez E, Ramón J, Navarro A, Radjenovic J, Mauriz E, Lechuga L, Marco MP, Barceló D (2007) Part per trillion determination of atrazine in natural water samples by a surface plasmon resonance immunosensor. Anal Bioanal Chem 388(1):207–214. https://doi.org/10.1007/s00216-007-1214-2

Fu Q, Zhang C, Xie J, et al (2019) Ambient light sensor based colorimetric dipstick reader for rapid monitoring organophosphate pesticides on a smart phone. Anal Chim Acta 1092:126–131. https://doi.org/10.1016/j.aca.2019.09.059

Fujioka K, Casida JE (2007) Glutathione S-transferase conjugation of organophosphorus pesticides yields S-phospho-, S-aryl-, and S-alkylglutathione derivatives. Chem Res Toxicol 20(8):1211–1217. https://doi.org/10.1021/tx700133c

Fulbright RM, Axelrod D (1993) Dynamics of nonspecific adsorption of insulin to erythrocyte membranes. J Fluoresc 3:1–16

Funari R, Della Ventura B, Schiavo L, et al (2013) Detection of parathion pesticide by quartz crystal microbalance functionalized with UV-activated antibodies. Anal Chem 85:6392–6397. https://doi.org/10.1021/ac400852c

Gabriëls M, Plaizier-Vercammen J (2004) Development of a reversed-phase thin-layer chromatographic method for artemisinin and its derivatives. J Chromatogr Sci 42(7):341–347. https://doi.org/10.1093/chromsci/42.7.341

Gandhi S, Caplash N, Sharma P, Suri CR (2009) Strip-based immunochromatographic assay using specific egg yolk antibodies for rapid detection of morphine in urine samples. Biosens Bioelectron 25:502–505. https://doi.org/10.1016/j.bios.2009.07.018

Gao S, Zhang ZP, Karnes HT (2005) Sensitivity enhancement in liquid chromatography/atmospheric pressure ionization mass spectrometry using derivatization and mobile phase additives. J Chromatogr B: Anal Technol in the Biomed Life Sci 825(2):98–110. https://doi.org/10.1016/j.jchromb.2005.04.021

Gerber R, Smit NJ, Van Vuren JHJ, et al (2016) Bioaccumulation and human health risk assessment of DDT and other organochlorine pesticides in an apex aquatic predator from a premier conservation area. Sci Total Environ 550:522–533. https://doi.org/10.1016/j.scitotenv.2016.01.129

Giardi MT, Koblízek M, Masojídek J (2001) Photosystem II-based biosensors for the detection of pollutants. In: Biosensors and bioelectronics. Elsevier, pp 1027–1033

Giardi MT, Guzzella L, Euzet P, Rouillon R, Esposito D (2005) Detection of herbicide subclasses by an optical multibiosensor based on an array of photosystem II mutants. Environ Sci Technol 39:5378–5384. https://doi.org/10.1021/es040511b

Grinberg N (1990). Modern thin-layer chromatography. CRC Press, Boca Raton

Guo Y, Liu R, Liu Y, et al (2018) A non-competitive surface plasmon resonance immunosensor for rapid detection of triazophos residue in environmental and agricultural samples. Sci Total Environ 613–614:783–791. https://doi.org/10.1016/j.scitotenv.2017.09.157

Ha T, Ting AY, Liang J, Caldwell WB, Deniz AA, Chemla DS, Schultz PG, Weiss S (1999) Single-molecule fluorescence spectroscopy of enzyme conformational dynamics and cleavage mechanism. Proc Nat Acad Sci USA 96(3):893–898. https://doi.org/10.1073/pnas.96.3.893

Hauck HE, Halpaap H (1980) HPTLC on inert pre-coated layers of cellulose and of silica 50000. Chromatogr 13(9):538–548. https://doi.org/10.1007/BF02311857

Herranz S, Ramón-Azcón J, Benito-Peña E, et al (2008) Preparation of antibodies and development of a sensitive immunoassay with fluorescence detection for triazine herbicides. Anal Bioanal Chem 391:1801–1812. https://doi.org/10.1007/s00216-008-1895-1

Hikuma M, Suzuki H, Yasuda Y, Karube I, Suzuki S (1979) Amperometric estimation of BOD by using living immobilized yeasts. Eur J Appl Microbiol Biotechnol 8:289–297. https://doi.org/10.1007/BF00508793

Homola J, Yee SS, Gauglitz G (1999) Surface plasmon resonance sensors: review. Sens Actuators, B: Chem 54(1):3–15. https://doi.org/10.1016/S0925-4005(98)00321-9

Hussain A (2018) A perspective on organochlorine pesticide residues in milk produced in pakistan. EC Nutrition 13:402–410

Institute of Medicine (1997) Toxicology and environmental health information resources: the role of the national library of medicine. The National Academies Press, Washington, DC. https://doi.org/10.17226/5496

Ishii A, Takeda S, Hattori S, Sueoka K, Mukasa K (2008) Ultrasensitive detection of organophosphate insecticides by carbon nanotube field-effect transistor. Colloids Surfaces A Physicochem Eng Asp 313–314:456–460. https://doi.org/10.1016/j.colsurfa.2007.05.071

Islam S, Shukla S, Bajpai VK, et al (2019) Microfluidic-based graphene field effect transistor for femtomolar detection of chlorpyrifos. Sci Rep 9:1–7. https://doi.org/10.1038/s41598-018-36746-w

Izmailov NA, Shraiber MS (1938) The application of analysis by drop chromatography to pharmacy. Farmezia 3:1

Jacob C, Burkholz T, Kirsch G, Slusarenko A, Winyard PG (2014) Introduction. In: Recent advances in redox active plant and microbial products (pp 1–11) Springer Netherlands. https://doi.org/10.1007/978-94-017-8953-0_1

James AT, Martin AJ (1952) Gas-liquid partition chromatography; the separation and micro-estimation of volatile fatty acids from formic acid to dodecanoic acid. Biochem J 50(5):679–690. https://doi.org/10.1042/bj0500679

Jia K, Adam P, Ionescu RE (2013) Sequential acoustic detection of atrazine herbicide and carbofuran insecticide using a single micro-structured gold quartz crystal microbalance. Sensor Actuator B: Chem 188:400–404. https://doi.org/10.1016/j.snb.2013.07.033

Jia J, Tang M, Chen X, Qi L, Dong S (2003) Co-immobilized microbial biosensor for BOD estimation based on sol-gel derived composite material. Biosens Bioelectron 18:1023–1029. https://doi.org/10.1016/S0956-5663(02)00225-7

Jing H, Amirav A (1997) Pesticide analysis with the pulsed-flame photometer detector and a direct sample introduction device. Anal Chem 69(7):1426–1435. https://doi.org/10.1021/ac961110k

Johnson DW (2007) A modified Girard derivatizing reagent for universal profiling and trace analysis of aldehydes and ketones by electrospray ionization tandem mass spectrometry. Rapid Commun Mass Spectrom 21(18):2926–2932. https://doi.org/10.1002/rcm.3175

Jongenotter GA, Kerkhoff MAT, van der Knaap HCM, Vandeginste BGM (1999) Automated on-line GPC-GC-FPD involving co-solvent trapping and the on-column interface for the determination of organophosphorus pesticides in olive oils. J High Resolut Chromatogr 22(1):17–23. https://doi.org/10.1002/(SICI)1521-4168(19990101)22:1%3c17:AID-JHRC17%3e3.0.CO;2-7

Kandimalla VB, Kandimalla K, Hruska K, Franek M (2008) Detection of sulfamethazine in water, milk and pig manure by dipstick immunoassay. Veterinární Medicína 52 (No. 10):445–450

Karas M, Bachmann D, Hillenkamp F (1985) Influence of the wavelength in high-irradiance ultraviolet laser desorption mass spectrometry of organic molecules. Anal Chem 57(14):2935–2939. https://doi.org/10.1021/ac00291a042

Karnati C, Du H, Ji HF, et al (2007) Organophosphorus hydrolase multilayer modified microcantilevers for organophosphorus detection. Biosens Bioelectron 22:2636–2642. https://doi.org/10.1016/j.bios.2006.10.027

Karousos NG, Aouabdi S, Way AS, Reddy SM (2002) Quartz crystal microbalance determination of organophosphorus and carbamate pesticides. Anal Chim Acta 469:189–196. https://doi.org/10.1016/S0003-2670(02)00668-2

Kasoju A, Shahdeo D, Khan AA, et al (2020a) Fabrication of microfluidic device for Aflatoxin M1 detection in milk samples with specific aptamers. Sci Rep 10. https://doi.org/10.1038/s41598-020-60926-2

Kasoju A, Shrikrishna NS, Shahdeo D, et al (2020b) Microfluidic paper device for rapid detection of aflatoxin B1 using an aptamer based colorimetric assay. RSC Adv 10:11843–11850. https://doi.org/10.1039/d0ra00062k

Kaur J, Singh KV, Boro R, et al (2007) Immunochromatographic dipstick assay format using gold nanoparticles labeled protein-hapten conjugate for the detection of atrazine. Environ Sci Technol 41:5028–5036. https://doi.org/10.1021/es070194j

Kersting S, Rausch V, Bier FF, Von Nickisch-Rosenegk M (2014) Rapid detection of Plasmodium falciparum with isothermal recombinase polymerase amplification and lateral flow analysis. Malaria J 13(1):99. https://doi.org/10.1186/1475-2875-13-99

Khan S, Pierce D, Vale RD (2000) Interactions of the chemotaxis signal protein CheY with bacterial flagellar motors visualized by evanescent wave microscopy. Curr Biol: CB 10(15):927–930. https://doi.org/10.1016/s0960-9822(00)00629-1

Kim MN, Kwon HS (1999) Biochemical oxygen demand sensor using serratia marcescens LSY 4. Biosens Bioelectron 14:1–7. https://doi.org/10.1016/S0956-5663(98)00107-9

Kim SJ, Gobi KV, Tanaka H, Shoyama Y, Miura N (2006) Enhanced sensitivity of a surface-plasmon-resonance (SPR) sensor for 2,4-D by controlled functionalization of self-assembled monolayer-based immunosensor chip. Chem Lett 35:1132–1133. https://doi.org/10.1246/cl.2006.1132

Kim HJ, Kim Y, Park SJ, Kwon C, Noh H (2018) Development of Colorimetric Paper Sensor for Pesticide Detection Using Competitive-inhibiting Reaction. Biochip J 12:326–331. https://doi.org/10.1007/s13206-018-2404-z

King AM, Aaron CK (2015) Organophosphate and Carbamate Poisoning. Emerg Med Clin North Am 33:133–151

Kogan MJ, Pierson DJ, Durkin MM, Willson NJ (1989). Thin-layer chromatography of benzoylecgonine: a rapid qualitative method for confirming the EMIT Urine Cocaine Metabolite Assays. J Chromatogr B: Biomed Sci Appl 490(C): 236–242. https://doi.org/10.1016/S0378-4347(00)82781-4

Kumar A, Thakur A,VS-J of AF (2019) Undefined pesticide residues in animal feed: status. Safety Scope 103118697

Kuter DJ, Bussel JB, Lyons RM, Pullarkat V, Gernsheimer TB, Senecal FM, Aledort LM, George JN, Kessler CM, Sanz MA, Liebman HA, Slovick FT, de Wolf JTM, Bourgeois E, Guthrie TH, Newland A, Wasser JS, Hamburg SI, Grande C, … Nichol JL (2008) Efficacy of romiplostim in patients with chronic immune thrombocytopenic purpura: a double-blind randomised controlled trial. The Lancet 371: 395–403 https://doi.org/10.1016/S0140-6736(08)60203-2

Lathe GH, Ruthven CR (1956) The separation of substances and estimation of their relative molecular sizes by the use of colums of starch in water. Biochem J 62(4):665–674. https://doi.org/10.1042/bj0620665

Lau JL, Dunn MK (2018) Therapeutic peptides: Historical perspectives, current development trends, and future directions. Bioorg Med Chem 26(10):2700–2707. https://doi.org/10.1016/j.bmc.2017.06.052

Lei Y, Mulchandani P, Chen W, Wang J, Mulchandani A (2004) Whole Cell-Enzyme Hybrid Amperometric Biosensor for Direct Determination of Organophosphorous Nerve Agents with p-Nitrophenyl Substituent. Biotechnol Bioeng 85:706–713. https://doi.org/10.1002/bit.20022

Lei Y, Mulchandani P, Chen W, Mulchandani A (2005) Direct determination of p-nitrophenyl substituent organophosphorus nerve agents using a recombinant Pseudomonas putida js444-modified clark oxygen electrode. J Agric Food Chem 53:524–527. https://doi.org/10.1021/jf048943t

Lei Y, Mulchandani P, Chen W, Mulchandani A (2006) Biosensor for direct determination of fenitrothion and EPN using recombinant pseudomonas putida js444 with surface expressed organophosphorus hydrolase. 1. Modified clark oxygen electrode. In: Sensors. MDPI AG, pp 466–472

Liu J, Mattiasson B (2002) Microbial BOD sensors for wastewater analysis. Water Res 36:3786–3802

Liu J, Olsson G, Mattiasson B (2004) Short-term BOD (BODst) as a parameter for on-line monitoring of biological treatment process. Biosens Bioelectron 20(3):562–570

Li H, Sun C, Vijayaraghavan R, et al (2016) Long lifetime photoluminescence in N, S co-doped carbon quantum dots from an ionic liquid and their applications in ultrasensitive detection of pesticides. Carbon N Y 104:33–39. https://doi.org/10.1016/j.carbon.2016.03.040

Liu N, Han J, Liu Z, Qu L, Gao Z (2013) Rapid detection of endosulfan by a molecularly imprinted polymer microsphere modified quartz crystal microbalance. Anal Methods 5:4442–4447. https://doi.org/10.1039/c3ay40697k

Li Q, Dou X, Zhang L, et al (2019a) Oriented assembly of surface plasmon resonance biosensor through staphylococcal protein A for the chlorpyrifos detection. Anal Bioanal Chem 411:6057–6066. https://doi.org/10.1007/s00216-019-01990-0

Li Q, Dou X, Zhao X, et al (2019b) A gold/Fe 3 O 4 nanocomposite for use in a surface plasmon resonance immunosensor for carbendazim. Microchim Acta 186:1–8. https://doi.org/10.1007/s00604-019-3402-0

Madla S, Miura D, Wariishi H (2012) Potential applicability of MALDI-MS for low-molecular-weight pesticide determination. Anal Sci 28(3):301–304. https://doi.org/10.2116/analsci.28.301

Mahendra Wijaya IP, Nie TJ, Gandhi S, et al (2010) Femtomolar detection of 2,4-dichlorophenoxyacetic acid herbicides via competitive immunoassays using microfluidic based carbon nanotube liquid gated transistor. Lab Chip 10:634–638. https://doi.org/10.1039/b918566f

Mallat E, Barzen C, Abuknesha R, Gauglitz G, Barcelo D (2001) Part per trillion level determination of isoproturon in certified and estuarine water samples with a direct optical immunosensor. In: Analytica Chimica Acta. Elsevier, pp 209–216

Martínez Salvador I, Garrido Frenich A, Egea González FJ, Martínez Vidal JL (2006) Determination of organophosphorus pesticides in vegetables by GC with pulsed flame-photometric detection, and confirmation by MS. Chromatogr 64(11–12):667–672. https://doi.org/10.1365/s10337-006-0104-7

Marquardt RM, Kim TH, Shin JH, Jeong JW (2019) Progesterone and estrogen signaling in the endometrium: What goes wrong in endometriosis? Int J Mol Sci 20(15):3822. https://doi.org/10.3390/ijms20153822

Marquette CA, Blum LJ (2000) Regenerable immunobiosensor for the chemiluminescent flow injection analysis of the herbicide 2,4-D. Talanta 51:395–401. https://doi.org/10.1016/S0039-9140(99)00298-2

Mathur AB, Truskey GA, Reichert WM (2000) Atomic force and total internal reflection fluorescence microscopy for the study of force transmission in endothelial cells. Biophys J 78(4):1725–1735. https://doi.org/10.1016/S0006-3495(00)76724-5

Mauriz E, Calle A, Abad A, et al (2006) Determination of carbaryl in natural water samples by a surface plasmon resonance flow-through immunosensor. Biosens Bioelectron 21:2129–2136. https://doi.org/10.1016/j.bios.2005.10.013

Mc Kiernan AE, MacDonald RI, MacDonald RC, Axelrod D (1997) Cytoskeletal protein binding kinetics at planar phospholipid membranes. Biophys J 73(4):1987–1998. https://doi.org/10.1016/S0006-3495(97)78229-8

Meng X, Wei J, Ren X, Ren J, Tang F (2013) A simple and sensitive fluorescence biosensor for detection of organophosphorus pesticides using H2O2-sensitive quantum dots/bi-enzyme. Biosens Bioelectron 47:402–407. https://doi.org/10.1016/j.bios.2013.03.053

Mishra P, Banga I, Tyagi R, et al (2018) An immunochromatographic dipstick as an alternate for monitoring of heroin metabolites in urine samples. RSC Adv 8:23163–23170. https://doi.org/10.1039/C8RA02018C

Mulchandani P, Hangarter CM, Lei Y, Chen W, Mulchandani A (2005) Amperometric microbial biosensor for p-nitrophenol using Moraxella sp.-modified carbon paste electrode. Biosens Bioelectron 21:523–527. https://doi.org/10.1016/j.bios.2004.11.011

Murray KK (1997) Coupling matrix-assisted laser desorption/ionization to liquid separations. Mass Spectrom Rev 16(5):283–99. https://doi.org/10.1002/(SICI)1098-2787(1997)16:5%3c283:AID-MAS3%3e3.0.CO;2-D

Nakamura C, Hasegawa M, Nakamura N, Miyake J (2003) Rapid and specific detection of herbicides using a self-assembled photosynthetic reaction center from purple bacterium on an SPR chip. In: Biosensors and Bioelectronics. Elsevier Ltd, pp 599–603

Niessen WMA (2006) Liquid chromatography–mass spectrometry. CRC/Taylor and Francis. https://doi.org/10.1201/9781420014549

Nilsson GA, Nilsson O, Odenbrand I, Wesén C (2001) New halogen-specific detector applied to the analysis of chlorinated fatty acids. J Chromatogr A 912(1):99–106. https://doi.org/10.1016/s0021-9673(01)00519-2

Nollert P, Kiefer H, Jähnig F (1995) Lipid vesicle adsorption versus formation of planar bilayers on solid surfaces. Biophys J 69(4):1447–1455. https://doi.org/10.1016/S0006-3495(95)80014-7

Nomura T, Hattori O (1980) Determination of micromolar concentrations of cyanide in solution with a piezoelectric detector. Anal Chim Acta 115:323–6

Nyiredy S (2004) Chapter 6 planar chromatography. J Chromatogr Libr 69(C):253–296. https://doi.org/10.1016/S0301-4770(04)80012-5

Omann GM, Axelrod D (1996) Membrane-proximal calcium transients in stimulated neutrophils detected by total internal reflection fluorescence. Biophys J 71(5):2885–2891. https://doi.org/10.1016/S0006-3495(96)79485-7
Özkara A, Akyil D, Konuk M (2016) Pesticides, environmental pollution, and health. In: Environmental health risk - hazardous factors to living species. InTech
Pan M, Kong L, Liu B, et al (2013) Production of multi-walled carbon nanotube/poly(aminoamide) dendrimer hybrid and its application to piezoelectric immunosensing for metolcarb. Sensor Actuator B Chem 188:949–956. https://doi.org/10.1016/j.snb.2013.07.116
Pape BE, Rodgers DH, Flynn TC (1977) The electrolytic conductivity detector as an element-selective gas chromatographic detector. J Chromatogr A 134(1):1–24. https://doi.org/10.1016/S0021-9673(00)82565-0
Parolo C, Merkoçi A (2013) Paper-based nanobiosensors for diagnostics. Chem Soc Rev 42:450–457. https://doi.org/10.1039/c2cs35255a
Pesce MA, Spitalnik SL (2007) Saliva and the clinical pathology laboratory. Ann NY Acad Sci 1098(1):192–199. https://doi.org/10.1196/annals.1384.032
Pesticide Residues in Food: Technologies for Detection - U.S Congress 1988
Piringer O, Paşcalău M (1962) A new detector for gas chromatography. J Chromatogr A, 8(C):410–412. https://doi.org/10.1016/s0021-9673(01)99277-5
Pitt JJ (2009) Principles and applications of liquid chromatography-mass spectrometry in clinical biochemistry. Clini Biochem Rev 30(1):19–34. http://www.ncbi.nlm.nih.gov/pubmed/19224008
Poole CF, Butler HT, Coddens ME, Schuette SA (1986) Instrumental Evaluation of Thin-Layer Chromatograms. In: Analytical and chromatographic techniques in radiopharmaceutical chemistry (pp. 3–37). Springer New York. https://doi.org/10.1007/978-1-4612-4854-5_1
Qian MC, Burbank HM (2007). Volatile sulfur compounds in cheddar cheese determined by headspace solid-phase microextraction—gas chromatography—PFPD. Flavor of Dairy Products, Chapter 8: 135–154. https://doi.org/10.1021/bk-2007-0971.ch008
Quesada-González D, Merkoçi A (2015) Nanoparticle-based lateral flow biosensors. Biosens Bioelectron 73:47–63. https://doi.org/10.1016/j.bios.2015.05.050
Rainina EI, Efremenco EN, Varfolomeyev SD, Simonian AL, Wild JR (1996) The development of a new biosensor based on recombinant E. coli for the direct detection of organophosphorus neurotoxins. Biosens Bioelectron 11(10):991–1000. https://doi.org/10.1016/0956-5663(96)87658-5
Raman SC, Boro R, Nangia Y, et al (2009) Immunoanalytical techniques for analyzing pesticides in the environment. TrAC - Trends Anal Chem 28:29–39. https://doi.org/10.1016/j.trac.2008.09.017
Raman SC, Kaur J, Gandhi S, Shekhawat GS (2008) Label-free ultra-sensitive detection of atrazine based on nanomechanics. Nanotechnology 19. https://doi.org/10.1088/0957-4484/19/23/235502
Riedel K, Renneberg R, Kühn M, Scheller F (1988) A fast estimation of biochemical oxygen demand using microbial sensors. Appl Microbiol Biotechnol 28:316–318. https://doi.org/10.1007/BF00250463
Rodriguez-Mozaz S, Reder S, Lopez de Alda M, Gauglitz G, Barceló D (2004) Simultaneous multi-analyte determination of estrone, isoproturon and atrazine in natural waters by the RIver ANAlyser (RIANA), an optical immunosensor. Biosensors and Bioelectronics 19(7):633–640. https://doi.org/10.1016/s0956-5663(03)00255-0
Sako Y, Minoghchi S, Yanagida T (2000) Single-molecule imaging of EGFR signalling on the surface of living cells. Nat Cell Biol 2:168–172. https://doi.org/10.1038/35004044
Sassolas A, Prieto-Simón B, Marty JL (2012) Biosensors for pesticide detection: new trends. Am J Anal Chem 03(03):210–232. https://doi.org/10.4236/ajac.2012.33030
Sauerbrey G (1959) Verwendung von Schwingquarzen zur Wägung dünner Schichten und zur Mikrowägung. Zeitschrift für Phys 155:206–222. https://doi.org/10.1007/BF01337937
Saylan Y, Akgönüllü S, Çimen D, et al (2017) Development of surface plasmon resonance sensors based on molecularly imprinted nanofilms for sensitive and selective detection of pesticides. Sensor Actuator B Chem 241:446–454. https://doi.org/10.1016/j.snb.2016.10.017

Sherma J (1982) Thin layer chromatography of pesticides - A review of recent techniques and applications. J Liq Chromatogr 5(6):1013–1032. https://doi.org/10.1080/01483918208067565

Sherma J, Fried B (1987) Preparative thin layer chromatography. J Chromatogr Libr 38(C):105–127. https://doi.org/10.1016/S0301-4770(08)60365-6

Sherma J, Fried B (2005) Thin layer chromatographic analysis of biological samples. Rev J Liq Chromatogr Relat Technol 28:2297–2314. https://doi.org/10.1080/10826070500187491

Skládal P, Minunni M, Mascini M, Kolar V, Franek M (1994) Characterization of monoclonal antibodies to 2,4-dichlorophenoxyacetic acid using a piezoelectric quartz crystal microbalance in solution. J Immunol Methods 176:117–125. https://doi.org/10.1016/0022-1759(94)90356-5

Skoog DA, Holler FJ, Nieman TA (1998) Principles of instrumental analysis, Brooks/Cole

Sparkman OD, Penton Z, Kitson FG (2011) Gas chromatography and mass spectrometry: a practical guide. Academic Press. https://doi.org/10.1016/C2009-0-17039-3

Starodub NF, Dzantiev BB, Starodub VM, Zherdev AV (2000) Immunosensor for the determination of the herbicide simazine based on an ion-selective field-effect transistor. Anal Chim Acta 424:37–43. https://doi.org/10.1016/S0003-2670(00)01143-0

Štěpánková Š, Vorčáková K, S, těpánkovárka S (2016) Journal of Enzyme Inhibition and Medicinal Chemistry Cholinesterase-based biosensors Cholinesterase-based biosensors. J Enzym Inhib Med Chem 31:180–193. https://doi.org/10.1080/14756366.2016.1204609

Stout AL, Axelrod D (1994) Reversible binding kinetics of a cytoskeletal protein at the erythrocyte submembrane. Biophys J 67(3):1324–1334. https://doi.org/10.1016/S0006-3495(94)80604-6

Stradiotto NR, Yamanaka H, Zanoni MVB (2003). Electrochemical sensors: A powerful tool in analytical chemistry. J Braz Chem Soc 14(2):159–173. Sociedade Brasileira de Quimica. https://doi.org/10.1590/S0103-50532003000200003

Su L, Jia W, Hou C, Lei Y (2011) Microbial biosensors: a review. Biosens Bioelectron 26:1788–1799

Sun H, Fung Y (2006) Piezoelectric quartz crystal sensor for rapid analysis of pirimicarb residues using molecularly imprinted polymers as recognition elements. Anal Chim Acta 576:67–76. https://doi.org/10.1016/j.aca.2006.04.058

Sund SE, Axelrod D (2000) Actin dynamics at the living cell submembrane imaged by total internal reflection fluorescence photobleaching. Biophys J 79(3):1655–1669. https://doi.org/10.1016/S0006-3495(00)76415-0

Talan A, Mishra A, Eremin SA, et al (2018) Ultrasensitive electrochemical immuno-sensing platform based on gold nanoparticles triggering chlorpyrifos detection in fruits and vegetables. Biosens Bioelectron 105:14–21. https://doi.org/10.1016/j.bios.2018.01.013

Tan G, Zhao Y, Wang M, et al (2020) Ultrasensitive quantitation of imidacloprid in vegetables by colloidal gold and time-resolved fluorescent nanobead traced lateral flow immunoassays. Food Chem 311:126055. https://doi.org/10.1016/j.foodchem.2019.126055

Tey JN, Wijaya IPM, Suri CR, et al (2010) Liquid gated carbon nanotubes field effect transistors (LGCNTFET) platform for herbicide sensing. In: ASME international mechanical engineering congress and exposition, proceedings. American Society of Mechanical Engineers (ASME), pp 249–254

Tewari SN (1980) Distribution of pesticides in human organs as determined by quantitative thin-layer chromatography. Pesticide Anal Methodol Chapter 14:259–276. https://doi.org/10.1021/bk-1980-0136.ch014

Thakur S, Gandhi S, Paul AK, Suri CR (2011) A flow injection immunosensor for the detection of atrazine in water samples. Sensor Transducers J 131:91–100

Thompson NL, Burghardt TP, Axelrod D (1981) Measuring surface dynamics of biomolecules by total internal reflection fluorescence with photobleaching recovery or correlation spectroscopy. Biophys J 33(3):435–454. https://doi.org/10.1016/S0006-3495(81)84905-3

Tong HY, Centen JD, Karasek FW, Jellum E, Helland P (1985) Identification of trace organic compounds in dimethyl sulphoxide solution using high-performance liquid chromatography and gas chromatography-mass spectrometry. J Chromatogr A, 324(C):373–383. https://doi.org/10.1016/S0021-9673(01)81336-4

Tsatsakis A, Docea AO, Constantin C, et al (2019) Genotoxic, cytotoxic, and cytopathological effects in rats exposed for 18 months to a mixture of 13 chemicals in doses below NOAEL levels. Toxicol Lett 316:154–170. https://doi.org/10.1016/j.toxlet.2019.09.004

Valdés-Ramírez G, Fournier D, Ramírez-Silva MT, Marty JL (2008) Sensitive amperometric biosensor for dichlorovos quantification: application to detection of residues on apple skin. Talanta 74:741–746. https://doi.org/10.1016/j.talanta.2007.07.004

Valera E, Muñiz D, Rodríguez Á (2010) Fabrication of flexible interdigitated μ-electrodes (FIDμEs) for the development of a conductimetric immunosensor for atrazine detection based on antibodies labelled with gold nanoparticles. Microelectron Eng 87(2):167–173. https://doi.org/10.1016/j.mee.2009.07.001

Valera E, Ramón-Azcón J, Sanchez FJ, Marco MP, Rodríguez Á (2008) Conductimetric immunosensor for atrazine detection based on antibodies labelled with gold nanoparticles. Sens Actuators, B: Chem 134(1):95–103. https://doi.org/10.1016/j.snb.2008.04.023

Verhaar LAT, Kuster BFM, Claessens HA (1984) Retention behaviour of carbohydrate oligomers in reversed-phase chromatography. J Chromatogr A, 284(C): 1–11. https://doi.org/10.1016/S0021-9673(01)87797-9

Wang J (2008) Electrochemical glucose biosensors. Chem Rev 108:814–825. https://doi.org/10.1021/cr068123a

Wang MD, Axelrod D (1994) Time-lapse total internal reflection fluorescence video of acetylcholine receptor cluster formation on myotubes. Dev Dyn 201(1):29–40. https://doi.org/10.1002/aja.1002010104

Wang L, Bi Y, Hou J, et al (2016) Facile, green and clean one-step synthesis of carbon dots from wool: application as a sensor for glyphosate detection based on the inner filter effect. Talanta 160:268–275. https://doi.org/10.1016/j.talanta.2016.07.020

Weis RM, Balakrishnan K, Smith BA, McConnell HM (1982) Stimulation of fluorescence in a small contact region between rat basophil leukemia cells and planar lipid membrane targets by coherent evanescent radiation. J Biol Chem 257(11):6440–5

Williams TI (1946) An introduction to chromatography. London and Glasgow, Blackie and Son

Wu JK, Ma JW, Wang H, et al (2019) Rapid and visual detection of benzothiostrobin residue in strawberry using quantum dot-based lateral flow test strip. Sensor Actuator B Chem 283:222–229. https://doi.org/10.1016/j.snb.2018.11.137

Yan X, Li H, Wang X, Su X (2015) A novel fluorescence probing strategy for the determination of parathion-methyl. Talanta 131:88–94. https://doi.org/10.1016/j.talanta.2014.07.032

Zaikin VG, Halket JM (2006). Derivatization in mass spectrometry - 8. Soft ionization mass spectrometry of small molecules. European J Mass Spectrom 12(2):79–115. https://doi.org/10.1255/ejms.798

Zenobi R, Knochenmuss R (1998) Ion formation in MALDI mass spectrometry. Mass Spectrom Rev 17(5):337–366

Zhang X, Guo Q, Cui D (2009) Recent advances in nanotechnology applied to biosensors. Sens 9(2):1033–1053. https://doi.org/10.3390/s90201033

Zhao B, Gandhi S, Yuan C, et al (2015) Mapping the topographic epitope landscape on the urokinase plasminogen activator receptor (uPAR) by surface plasmon resonance and X-ray crystallography. Data Br 5:107–113. https://doi.org/10.1016/j.dib.2015.08.027

Zhu X, Shah P, Stoff S, Liu H, Li CZ (2014) A paper electrode integrated lateral flow immunosensor for quantitative analysis of oxidative stress induced DNA damage. Analyst 139(11):2850–2857. https://doi.org/10.1039/c4an00313f

Zhuang W, McKague AB, Reeve DW, Carey JH (2005) Evaluation of halogen-specific detector (XSD^{TM}) for trace analysis of chlorinated fatty acids in fish. Instrum Sci Technol 33(4):481–507. https://doi.org/10.1081/CI-200066205

Ziegler C (2004) Cantilever-based biosensors. Anal Bioanal Chem 379(7–8):946–59. https://doi.org/10.1007/s00216-004-2694-y

Chapter 9
Application of Biosensor for the Identification of Various Pathogens and Pests Mitigating Against the Agricultural Production: Recent Advances

Charles Oluwaseun Adetunji, Wilson Nwankwo, Kingsley Eghonghon Ukhurebor, Akinola Samson Olayinka, and Ayodeji Samuel Makinde

Abstract The growing global population which is estimated to hit 9 billion by 2050 is considered a time bomb considering the scarce resources at the disposal of the nations. Hence, there is a need to identify a sustainable tool that could lead to an increase in the production of food production globally. Smart monitoring using biosensors ensure that chemical and other forms of contaminants are kept at bay from compromising the quality and safety of the stored food as well as the pest and pathogens that could affect fresh agricultural produce. Biosensors are also deployed for the purpose of measuring carbohydrates, alcohol, acids. Though present techniques in chromatography like high-performance liquid chromatography, capillary electrophoresis, flow immunoanalysis, gas chromatography, and spectroscopic techniques (Ultraviolet-visible spectroscopy, nuclear magnetic resonance spectroscopy, fourier-transform infrared spectroscopy) are advantageous because they are highly responsive, reliable, and efficient for pesticide residue analysis as well as the application of immunoanalysis in plant pathogen detection, and utilization of electrophoresis for plant nucleotide evaluation. However, these techniques are time-consuming,

C. O. Adetunji (✉)
Applied Microbiology, Biotechnology and Nanotechnology Laboratory, Department of Microbiology, Edo University Iyamho, P.M.B. 04, Auchi, Edo State, Nigeria
e-mail: adetunji.charles@edouniversity.edu.ng

W. Nwankwo · A. S. Makinde
Climatic/Environmental/Telecommunication Physics Unit, Department of Physics, Edo University Iyamho, P.M.B. 04, Auchi, Edo State, Nigeria

K. E. Ukhurebor
Informatics and CyberPhysical Systems Laboratory, Department of Computer Science, Edo University Iyamho, P.M.B. 04, Auchi, Edo State, Nigeria

A. S. Olayinka
Department of Physics, Edo University Iyamho, P.M.B. 04, Auchi, Edo State, Nigeria

R. N. Pudake et al. (eds.), *Biosensors in Agriculture: Recent Trends and Future Perspectives*, Concepts and Strategies in Plant Sciences,
https://doi.org/10.1007/978-3-030-66165-6_9

arduous, costly, complicated, and voluminous and involve the operation of qualified technicians. Hence, the application of biosensors has been identified as a reliable technology for effective detection of pathogen and pest affecting the increase in the agricultural production as well as utilized to reduce the impact of pesticides on farmlands and on the health of the farmers and residents. Therefore, this chapter intends to provide comprehensive detail about the application of biosensors for the identification of pest and pathogens affecting the increase in agricultural production. Numerous types of biosensors with high relevance in the detection of pest and pathogen were also highlighted.

Keywords Pathogen · Pest · Agriculture · Food · Biosensor · Global population

9.1 Introduction

Food security and safety have been a major concern of researchers as the world population is estimated to peak at 9.7 billion by the year 2050 from the current estimate of 7.7 billion. This is due to the fact that an increase in population will naturally translate to a corresponding increasing in food demand. It has been stated that adequate insight on the nutrition and the varying nature of malnutrition have enhanced curiosity in diet quality. Most countries around the globe do not have an adequate sub-national level and national data about diet and diet quality. Nevertheless, there is minimal insight about dietary outlines, as well as little knowledge on health policy, food industry, trade, and agriculture could be utilized for enhancement of diet quality (Vandevijvere et al. 2013). At present, it has been validated that the implementation of up-to-date and forthcoming technologies together with intelligent policy execution will be the operative and workable path to deciphering food security trials (Barretto et al. 2013; Charlebois 2015).

Also, pests and diseases have been identified as one of the biotic factors responsible for numerous agricultural losses. So, there is a need for an all-encompassing approach to address this devasting effect on food security and sustenance. Concerted effort has been geared towards curtailing the spread of disease and pests across borders by International Plant Protection Convention (IPPC), but the *in situ* pest and disease remains a menace in agricultural production system. Since the invention of biosensors, which is dated back to 1950, much attention has been drawn to its applications in various strata of human endeavours, medicine, drug discovery, ecological pollution control, and agriculture. It has been successfully used in the food industry as a quantifier for biological signals in quality control developments.

As discussed in other chapters, biosensors are bioelectronics devices used for signal detection and processing of biological materials which may be classified based on different sensing techniques. Piezoelectric biosensors, which are also branded as acoustic biosensors work on the principle of sound vibrations. Electrical signals are produced when vibrational force is sensed by a piezoelectric biosensor. In electrochemical biosensors, the probing surface is coated with biological (sensing) molecules and held in place with non-interfering membrane. Optical biosensors allow signal sensing using various properties of light like scattering, absorption, and diffraction. Smart biosensors, these types of biosensors belong to recent advances in biosensing technology. These emerging biosensors have been termed smart because of its inherent intelligence derived from machine learning and other computational techniques. Smart biosensors can be optical, piezoelectric, or electromechanical based (Adetunji et al. 2018).

Hence, this chapter intends to provide comprehensive detail about the application of biosensors for the identification of pests and pathogens affecting the increase in Agricultural production. Numerous types of biosensors with high relevance in the detection of pests and pathogens were also highlighted.

9.2 Principles of Biosensor Operation

The first biosensor was in the early 1950s by Leland Clark, a biochemist who wanted a device that could effectively detect the amount of oxygen circulating in the blood (Nastuk 2013; Clark et al. 1950). His biosensor was then called the Clark electrode. The Clark electrode was purely an oxygen detector and nothing more largely because only oxygen-permeable membranes could be utilized (Miniaev et al. 2013; Holt 2007). Following the creation of Clark electrode, there have been several developments thereafter. The first improvement on the oxygen detector was the overlaying of a glucose-sensitive enzyme film over the existing electrode and this further enhanced the functionality of the Clarke electrode towards acting as a blood sugar sensor. The blood sugar sensor is used for detecting and measuring the level of blood glucose (Ghosh et al. 2020; Zhang and Hoshino 2014). In furtherance to this development was the use of a layer of urease enzyme which was able to detect urea in blood, urine, etc. (Morales-Cruz et al. 2019; Tricoli and Neri 2018).

The concept of biosensor is derived from the two keywords that constitute the word i.e. *bio* and *sense*. *Bio* connotes life or living whereas *sense* connotes detection or recognition. In other words, biosensing implies the ability to recognize something within or around a living environment or living organism. Thus, a biosensor is any device designed to detect, recognize, measure, and relay information on a biophysical or biochemical substance e.g. oxygen, glucose, urea, protein, neurotransmitter, etc., around or within a living organism or environment, through electrical signals. Invariably, a biosensor can work inside or outside a living organism however, the target characteristic parameter to which the device is applied to detect/measure often emanate from the living organism or ecosystem (Cinti et al. 2017; Ali et al. 2017; Kaur et al. 2018).

Biosensors may be categorized according to the following characteristics:

a. **Targeted analyte**: is the target substance to be sensed e.g. glucose biosensor, oxygen biosensor, nitrogen biosensor, DNA biosensor. Note that the analyte must have an affinity for the sensing surface to which it may react chemically, and electrochemically.

b. **Bioreceptor**: This is the deactivated medium (an immobilized hard or soft biocomponent) often selectivity specific, that is to be exposed to the targeted substance or analyte (Kahn and Plaxco 2010). The common bioreceptor categories are enzymes (oxidoreductases, peroxidases, polyphenol oxidases aminooxidases, etc.) (Mehrotra 2016), tissues, antibodies, DNA, microbes (Kaur et al. 2018), biomimetics (Bharat 2009; Chauhan et al. 2020), nanoparticles. The bioreceptor is often interfaced to the transducer.
c. **Generation** (the period during when it was launched; biosensors manufactured and introduced into the market over a given period with similar characteristics belong to a generation). There are first generation, second generation, third generation, etc. Like computers and other electronic devices, the later generation often shows more sophistication than the earlier generation. The sophistication may be expressly defined in terms of precision, energy consumption, multifunctionality, accuracy, size, the scale of integration, and refinement.
d. **Sensing or transduction strategy**: The sensing strategy depicts the principle of detection adopted and embedded in the biosensor for detection. Ordinarily, the strategy is based on the behaviour of the transducer. Modern strategies include electrical, optical, piezoelectric, photoelectric, thermometric, photoelectrochemical, nano, and electrochemical sensing strategies (Monosik et al. 2012; Niu et al. 2018; Farzin et al. 2020; Long et al. 2020; Zhou and Tang 2020).

Biosensors are not just detectors; they also serve analytical functions depending on the domain of application. For example, as an analytical tool, a biosensor may be deployed to detect the specific existence of specific biomolecules, microorganisms, biological structures, or any other biological analytes. Biosensors operate on the principle of signal transduction (a process involving the relay of physical or biochemical signals that reflects molecular events from a cell or substance in such a manner that is measurable (Mehrotra 2016). Biosensors are more advanced regarding selectivity and sensitivity if compared with any other currently existing diagnostic tool. Some of the significant benefits of biosensors include fast and continuous measurement, reduced use of reagents, elimination of calibration and re-calibration, high precision, quick response time, ability to measure non-polar molecules that other traditional devices cannot estimate.

9.3 Electronic Principles of Biosensor

As stated earlier, biosensors operate on signal transduction principle. The main components of a typical biosensor include an element of biorecognition, a biotransducer, and an electronic component that may integrate, microprocessor, amplifier, and display/monitor, respectively (Gupta and Kakkar 2020). The biorecognition or sensing component is essentially a bioreceptor designed to detect or act on a particular analyte—the target substance whose relevant characteristics is to be detected or measured. The transducer receives input from the bioreceptor and depending on the transducer employed, a signal is relayed to the signal processor. The signal output amplitude is proportional to an analyte's concentration, and then amplified by the electronic device and processed (Sode et al. 2016).

For instance, in an amperometric sensor, the bioreceptor comprises a deactivated enzyme or specific biomaterial, and said deactivated biomaterial is in close contact with the transducer. The analyte reacts with the biomaterial. In this way, a new analyte may be formed which in effect gives the calculable electronic reaction. Sometimes, the analyte is transferred to a system that can be attached to the gas, heat, electron ions, or hydrogen ions discharge. In this, the transducer can adjust the device linked, then converts to electrical signals that could be altered and computed.

Biosensors could be distinguished by way of the transduction sensing strategy employed in its construction, based on the principle of translating biological signal into an electronic signal. Common families are electrochemical, calorimetric, optical. In some cases, there one family may dovetail into the other in which case there may not be very clear boundaries of distinction. One remarkable feature of any biosensor is specificity and selectivity to the analyte. A particular bioreceptor needs to be activated towards its attachment to the transducer/electrode to ensure that only the specific analyte is detected. Bioreceptors may be proteins (antibodies, enzymes) (Song et al. 2018), cellular material (cells, tissue culture, etc.) (Soleymani et al. 2018, 2020), nucleic acids (DNA, RNA) (Mittal et al. 2017; Saadati et al. 2019), nanomaterial (Zamora-Galvez et al. 2017), or biomimetic material (Moro et al. 2019).

Seven major types of biosensors based on sensing strategy utilized in the construction of the biosensor have been identified (Akgönüllü et al. 2020) There is also an evolving category, nanobiosensors (Sharifi et al. 2020; Sung et al. 2019), which are derived from nanomaterials. A nanobiosensor may be a hybrid device, that may utilize different sensing strategies available in the other major types highlighted above.

9.4 Mode of Operation of Biosensors

The components of a biosensor are instrumental to its operation. Together with physicochemical transducers, biosensor harness the exquisite sensitivity and specificity of the biology to deliver complex bioanalytical measurements in simple, and easy-to-use formats (Turner 2013). First, every biosensor is designed to perform specific functions. The way and manner a biosensor operates is dependent on the bio receptor (enzyme, DNA, phage, antibody, and many more) and the sensing strategy i.e. the workings of the transducer. The transducer's signalling is always small and overlays on a fairly high baseline. Usually, the processing of the signal requires the deduction of a baseline location signal, collected from a similar transducer without covering any biocatalyst. The direct output is an analogue signal at this point (Malhotra et al. 2017). However, the signal may be digitized and relayed to the microprocessor unit where the information is processed, guided by preferred units, and exported to a data store.

9.5 Biosensors in Agriculture

Agriculture is a major sector that pervades other sectors. It contributes to the development of all other economic sectors (Veldhuizen et al. 2020). Like education, agricultural innovations are desired for economic sustainability (Nwankwo and Njoku 2020). Agriculture contributes to socioeconomic development through products and raw materials, foreign exchange, market, poverty alleviation, employment (Nwankwo et al. 2019).

9.5.1 Applications of Biosensors in Agriculture

Summary of areas to which biosensors are deployed in agriculture is given in Table 9.1. Some other applications are discussed herein.

9.5.2 Biotoxicity Detection

Annually, a significant percentage of the citizenry in developing societies suffers from pesticide contamination. Various pesticides are commonly utilized in the fields by farmers around the world to prevent damages due to pests and diseases during the various phases of plant growth. However, the application or use of these chemicals has some significant effect on the farmer, as the drift of sprayers contributes to air pollution and the worker's safety is severely affected by the inhalation of the spray particles. The floating particles and spray droplets often contaminate the water bodies.

Biosensors are now used to reduce the impact of pesticides on farmlands and on the health of the farmers and residents (Kundu et al. 2019). A good number of available pesticides has carbamates and organophosphates as their main components. Biosensors have been applied to enhance the monitoring of toxicity and risks that may ensue from the use of pesticides on farmlands. Three things are achieved through pesticide toxicity biosensors (Songa and Okonkwo 2016; Garg and Mehrotra 2017; Uniyal and Sharma 2018), first, to prevent the excessive absorption of harmful pesticides by crops, fruits, and stored products, second, to prevent soil toxicity and subsequent destruction of useful soil flora; and third, to prevent local and remote health hazards arising from toxic substances and environmental pollution.

Several methods have been reported recently in the development of biotoxicity detection biosensor. Gao et al. (2017) fabricated a double-mediator based whole-cell electrochemical biosensor for acute biotoxicity detection of wastewater. Biosensor sensitivity was improved using ferricyanide-menadione double-mediator. Also, a chitosan hydrogel polymer film with boron-doped nanocrystalline diamond particle

Table 9.1 Some biosensor applications in agriculture

Transduction type	Electrode	Analyte detected	Benefits	References
Electrochemical magneto-immunosensing	Magnetic graphite-epoxy composite (m-GEC)	*Salmonella* in milk	No requirement of reagents and is rapid detection	Liébana et al. (2009), Brandão et al. (2015)
Electrochemical magnet immunosensing	Magnetic graphite-epoxy composite (m-GEC)	b-lactamase resistance in *Staphylococcus aureus*	GEC products have weak non-specific adsorption for either DNA samples or enzymes labels. They do not require blocking steps on the transducer's free sites to mitigate non-specific adsorption	Pividori and Alegret (2005), Adetunji et al. (2018)
Gold nanoparticle-based	graphite-epoxy composite	*Salmonella* IS200	Gold nanocomposite is a good substrate for improved and directed immobilisation of biomolecules with excellent transductive properties for the construction of a wide variety of electrochemical biosensors, such as immunosensors, genosensors, and enzyme sensors	Brasil de Oliveira Marques et al. (2009), Wang et al. (2016)
Amperometric electrochemical immunoassay	Working and counter electrode of Platinum (Pt), and Ag/AgCl reference electrode	*Staphylococcus aureus* in milk, cheese, and meat	The electrochemical assay technique has been shown to be fast, effective, and reproducible, and can be used to detect certain pathogenic microorganisms (*Salmonella spp*, *Lysteria monocytohenes*, or *Staphylococcus aureus*) by using antibodies against antigens	Majumdar et al. (2012), Silva et al. (2018)

(continued)

Table 9.1 (continued)

Transduction type	Electrode	Analyte detected	Benefits	References
Multiplexing optical (luminescence)	NA	*E. coli* O157: H7, *S. typhimurium* and *Legionella pneumophila*	The total measurement and calibration assay time is 18 min, enabling very rapid analysis	Karsunke et al. (2009), Maas et al. (2018)
High-density microelectrode array	NA	*E. coli* O157: H7 bacteria in food materials	Field ready device that is easy to use, compact, and reagent-less. Can provide rapid result within minutes	Radke and Alocilja (2005)
Flow-type antibody sensor	quartz crystal microbalance chip	*E. coli* in drinking water, beef, pork, and dumpling	The sensor measures changes in frequency due to mass deposits that are formed by antigen-antibody interaction	Kim and Park (2003)
Acoustic-based biosensor (the Quartz Crystal Microbalance)	NA	DNA detection	Device significantly reduces the processing of time by avoiding gel electrophoresis and can be combined in a diagnostic laboratory or an automated lab-on-a-chip device for plant pathogen diagnostics	Papadakis et al. (2015)

NA Not available

was electrodeposited onto a glassy carbon (GC) electrode to immobilize *Saccharomyces cerevisiae* cells and the mediators. Zhang et al. (2019) developed a microfluidic paper-based analytical biosensor device with benzoquinone using *E. coli* as a direct biotoxicity indicator through its respiratory inhibition induced by toxicants to measure the biotoxicities of pollutants. Detection samples in the form of a solution, vegetable juice, and soil were also applicable to the proposed microfluidic paper-based analytical biosensor device.

Qiu et al. (2018) demonstrate the feasibility of using an aptamer-invertase biosensor integrated with a personal glucose meter (PGM) to quantify quinine detection in reclaimed wastewater. The limit of detection for quinine detection in pure water and recycled sewer are 0.13 μM and 0.32 μM, respectively. Chen et al. (2020) developed an Escherichia coli-based electrochemical biosensor for detecting mycotoxin (aflatoxin B_1 and zearalenone) toxicity. *E. coli* acts as a signal recognition

Table 9.2 Biotoxicity detection sensors

Biosensor	Biotoxicity detected	Mediator	References
Double-mediator based whole-cell electrochemical	Acute toxicity of real wastewater	Menadione	Gao et al. (2017)
Microfluidic paper-based analytical	Biotoxicities of pollutants	Benzoquinone	Zhang et al. (2019)
E. coli-based electrochemical	Mycotoxin	p-benzoquinone	Chen et al. (2020)
Aptamer-invertase biosensor integrated with a personal glucose meter (PGM)	Quinine	NA	Qiu et al. (2018)
Mediated microbial electrochemical	Toxicity detection in water	1, 4-benzoquinone	Yu et al. (2020)

NA Not available

element, while p-benzoquinone acts as the mediator. A two-step reaction procedure has been developed to separate the mediator from the mycotoxins. The various types of a biosensor used in the detection of biotoxicity are listed in Table 9.2.

9.5.3 *Sustainable Food Safety and Quality*

Food quality and safety are important during the harvest and postharvest stages. The farm produces such as grains, tubers, etc. should be protected from contamination by chemicals. Smart monitoring using biosensors ensures that chemical and other forms of contaminants are kept at bay from compromising the quality and safety of the stored food. Biosensors are also deployed for the purpose of measuring carbohydrates, alcohol, acids, etc. The use of these facilities helps food technologists and producers to conduct quality assurance in that the food content or mixture is certified based on its measured quality parameters in respect of the various nutrients e.g. carbohydrates, proteins, alcohols, vitamins, gases, etc.

9.5.3.1 Immunosensors

Immunosensors are deployed in food processing and packaging for quality assurance and successful contamination detection purposes. One of its major applications is to enable adequate interaction between antibodies and antibodies (Zhu et al. 2019). This immunoassay functions similarly to the enzyme-linked immunosorbent enzyme-linked assay (ELISA). Detection of antigens is achieved in food items through observation of the antibody-antigen binding that is connected to the substrate-bearing

Table 9.3 Immunosensors development for food quality monitoring

Immunosensor	Test samples	Detected	Limit of detection	References
Label-free immunosensor	Drinking water and orange juice	*Escherichia coli*	94 CFU mL^{-1}	Kaushik et al. (2019)
Laser-Induced Graphene Electrochemical Immunosensors	Chicken Broth	*Salmonella enterica*	13 ± 7 CFU mL^{-1}	Soares et al. (2020)
Electrochemical immunosensor	Food	*Escherichia coli* O157: H7	~30 CFU mL^{-1}	Güner et al. (2017)
electrochemical gold nanoparticle-coated screen-printed immunosensor	Cookies and chocolate	Ara h 1	3.8 ng mL^{-1}	Alves et al. (2015)
electrochemical label-free immunosensor	Paste, rice flour, wheat flour, cornflour, quaker oats	Gliadin	1.2 ng mL^{-1}	Chekin et al. (2016)
magnetoimmunosensing platform	Peanut based derivatives, Hazelnuts	Ara h 1	6.3 ng mL^{-1}	Montiel et al. (2015)

enzyme. The antigen quantity or volume is known to exhibit a proportional relationship with the fluorescence produced by the interaction of antigen and antibodies (Sakamoto et al. 2018).

Several studies have been conducted recently on immunosensors development for food quality monitoring. Seo et al. (2016) fabricated an immunosensor system using CMOS image sensor (CIS) to pocket-size dimensions for detecting food-borne pathogen (*Vibrio parahaemolyticus*). The CIS helps in resolving the detection issue hindering monitoring via the internet of things (IoT). Information collected from the sensor is uploaded as data to the server to be shared with the public. Kaushik et al. (2019) proposed a fibre optic surface plasmon resonance immunosensor based on biofunctionalized molybdenum disulphide (MoS_2) nanosheets for rapid detection of *E. coli*. The immunosensor uses wavelength interrogation method with a strong linear relationship of $R^2 = 0.994$ was observed between the spectral response of immunosensor and different concentrations of *E. coli*. It was recorded that the method shows promising applications in regular water and food quality monitoring for different pathogenic microorganisms.

Soares et al. (2020) reported the fabrication of highly sensitive label-free laser-induced graphene electrochemical immunosensors to quantify the food-borne pathogen *Salmonella ecterica* in chicken broth. The list of immunosensors utilized for food quality monitoring is given in Table 9.3.

9.5.3.2 e-nose

This is commonly used for detecting, recognizing, and classifying volatile compounds. They may be utilized to categorize odoriferous agricultural products. The e-nose is a smart tool. It mimics the human olfactory organs. e-nose consists of the sensing element, signal collection unit, and an effective algorithm for pattern recognition. An e-nose was successfully implemented in agricultural applications for the determination of fruit maturity, identification of soil-borne pathogens, inspection of fish, and many more. The e-nose detection principle is that the sensor array identifies an odour composed of many different volatiles in the headspace of a sample. It then produces an output that reflects a fingerprint of all the components for the sample. The fingerprint identified by the e-nose sensors is used for detection of possible sample information based on suitable algorithms provided (Qiu and Wang 2017).

Across many fields, e-nose has become an increasingly common detection device with significant advantages such as high sensitivity, simple construction, high response, and recovery time, operate at room temperature, and cost-effectiveness (Deshmukh et al. 2015). Huang et al. (2017) forecast the days until peach fruit decay using near-infrared spectroscopy and an e-nose. Their study shows that the integration of near-infrared spectroscope and e-nose gives reliable and rapid forecasting of the day before decay of peach fruit with 82.26% accuracy. Tan and Kerr (2019) employed the use of e-nose to continuously monitor the volatile compounds of cocoa samples going through refining and conching that differs in degree of roasting and mass. The e-nose uses kernel distribution model to characterize volatiles profiles. Dong et al. (2019) measured the quality properties effect of roasted coffee beans under different drying process conditions (solar drying, hot-air drying, room temperature drying, heat pump drying, and freeze-drying) using e-nose. It was recorded that the drying process affected the total titratable acidity, pH, total soluble solids, and total solids. Wen et al. (2019) developed a sweeping electronic nose system (SENS) for detecting early infestation in citrus fruits. The e-nose based system utilized eigenvalue extraction method for sensor signal, INV, and RSVA. The infestation citrus fruits are classified using linear discriminate analysis (LDA) and principal component analysis (PCA). Their research demonstrated the possible feasibility of the e-nose technology under market conditions for in-filed detection of postharvest pest infestation in citrus fruit. The numerous types of e-nose development for food quality monitoring is given in Table 9.4.

9.5.3.3 Surface Plasmon Resonance (SPR) Sensor

SPR biosensors have been widely applied to calculate kinetics and bimolecular binding affinity in real-time and label-free mode with low reagent consumption. Furthermore, the SPR biosensor acts as an essential tool in multichannel or array-imaging instrumental set-ups for the high-throughput study of food allergens (Zhou et al. 2019a, b). However, conventional SPR technologies suffer from many

Table 9.4 e-nose development for food quality monitoring

e-nose sensor	Test sample	Application	References
Metal oxide sensor	Peach	Fruit decay	Huang et al. (2017)
Metal oxide gas sensor	Potatoes	Soft-rot infection detection	Rutolo et al. (2018)
Gas sensor	Spinach	Odour Classification	Shen and Tao (2018)
Gas sensor	Cocoa	Monitoring volatile compounds during the refining.	Tan and Kerr (2019)
MOS sensor	Coffee	Profiled volatile and Taste properties.	Dong et al. (2019)
MOS sensor	Citrus	Early detection of infestation in citrus fruits.	Wen et al. (2019)
MOS gas sensor	Broccoli	Freshness Analysis	Chen et al. (2019)
MOS sensor	Galangal	Odour classification	

performance limitations: (1) several allergens are challenging to examine in food samples at the same time; (2) SPR equipment is too large and costly and thus cannot be used to detect food allergen in a consumer-friendly way (Zhou et al. 2019a, b).

Farka et al. (2016) developed an SPR immunosensor based biocatalyzed precipitation for the analysis of Salmonella in dairy products. The specific capture antibody on the SPR chip was immobilized, which allowed a precipitation-enhanced detection of Salmonella with a limit of detection ranging from 100 to 10^6 CFU mL^{-1}. The total analytical time, including the binding and enhancement stage for the bacteria, was less than 60 min. Zhou et al. (2019a, b) reviewed four types of surface plasmon resonance (SPR) biosensors for allergens detection in milk, egg, peanut, and seafood. The four studied SPR include localized surface plasmon resonance (LSPR), fiber-optic surface plasmon resonance, transmission surface plasmon resonance (TSPR), and surface plasmon resonance imaging (SPRI). The SPRI biosensor can be joined with LSPR and TSPR with the advancement of technologies to provide an accurate, secure, economical, and highly sensitive method for controlling hundreds of food allergens in one or more food matrices. The typical examples of SPR sensor utilised for food quality monitoring is listed in Table 9.5.

9.5.3.4 Quartz Crystalline Microbalance Biosensor

This is an acoustic biosensor used to detect bacterial phytopathogens of tomato, pepper, etc. such as *Pseudomonas syringae, Xanthomonas vesicatoria, Xanthomonas campestris,* and *Ralstonia solanacearum* (Papadakis et al. 2015). Compagnone et al. (2015) carried out quality control test on two series of chocolate samples having different formulated products over two sets of sensors including metalloporphyrins and nanoparticles peptide coated quartz crystal microbalances. The

Table 9.5 SPR development for food quality monitoring

SPR sensor	Test sample	Capturing ligand	Limit of detection	References
SPR based biocatalyzed precipitation	Powder Milk	Antibody	100 - 10^6 CFU mL^{-1}	Farka et al. (2016)
SPR	Milk	Antibody	57.8 ng mL^{-1}	Ashley et al. (2017)
Gold-coated U-shaped plastic optical fiber (POF)	Water	*E. coli*	1.5×10^3 CFU mL^{-1}	Arcas et al. (2018)
SPR	Egg	Antibody	0.66 nM	Saylan et al. (2017)
SPR	Milk	NanoMIPs	127 ± 97.6 ng mL^{-1}	Ashley et al. (2018)
SPR	Shellfish	Antibody	1.0 μg mL^{-1}	Zhou et al. (2019a, b)

Table 9.6 QCM sensor development for food quality monitoring

QCM sensor	Test sample	Detection method	Limit of detection	References
QCM	mango	Beta-caryophyllene	0.32 ppm	Ali et al. (2018)
Piezoelectric	Banana, ethylene pear, orange	ethylene	0.42 ppm	Tolentino et al. (2018)
QCM	tomato	trans-2-hexenal	NA	Debabhuti et al. (2016)
QCM	melon	Ethyl butyrate or limonene	0.2 mmol L^{-1}	Sasaki et al. (2018)

NA Not available

first series of chocolate was obtained under standard process conditions while the second comprising sample with some volatile compounds associated with low-quality raw materials. It was recorded that the best performance was obtained using the gold nanoparticle peptide sensors. Cervera-Chiner et al. (2018) proposed a high fundamental frequency quartz crystal microbalance (HFF-QCM) immunosensor for detecting pesticide in honey. The HFF-QCM is known to have high sensitivity, low cost, and fast detection. The HFF-QCM is based on piezoelectric sensors with frequencies of 100 MHz. Liu and Zhang (2020) have reviewed the state-of-the-art volatile organic compounds gas sensor based on quartz crystal microbalance for detecting fruit freshness. The QCM-based gas sensor has been widely applied to judge the freshness, shelf life, and maturity of fruits because of its high integration level, low cost, and ease of access to changing the sensitive membrane material. The QCM sensor development for food quality monitoring is listed in Table 9.6.

9.5.4 *The Utilization of e-nose for Plant Diagnosis*

The plant volatile organic compounds have been identified to play a crucial role in their reaction against pest detection. The composition of volatile organic compounds liberated by plants could be linked to types of damage. It has been stated that volatile organic compounds are biologically generated due to numerous physiological processes that occur in different types of plant tissue. Whenever agricultural pathogen invades the plant defence mechanism after infection there is a tendency that the composition of the volatile organic compounds might change. Plant needs numerous ranges of defence mechanisms to fight against numerous mechanical and herbivorous insects (Choudhary et al. 2008).

It has been discovered that the liberation of volatile organic compounds depends on the response of the potential pathogen and pests whenever they attach plant which area triggers in response to their attack. These constitute the changes of volatile organic compounds which give reliable principles of pest discovery using e-nose most especially from bacterial and fungi infections as well as insect infestations in plants (Pare and Tumlinson 1999). Several scientists have validated the merits and demerits on the profile of the volatile organic compounds when utilized as a biosensor which includes their non-destructive, reliability and their application in the real-time plant disease and insect management and monitoring (Rajendran et al. 2016; Cellini et al. 2016; Fang et al. 2014; Fang and Ramasamy 2015).

9.5.4.1 Bacterial and Fungal Disease Infection

The detection of numerous agricultural diseases and plant infection before the onset of all the symptom associated are very paramount management techniques and pest control strategies for effective mitigation of the diseases (Sankaran et al. 2010; Shaikh et al. 2017; Ray et al. 2017). e-noses have been identified as crucial in the identification of bacterial and fungal infections. This technique involves the identification of volatile organic compounds liberated from isolated microorganisms utilizing e-nose; while the indirect techniques entail evaluating changes in volatile organic compounds liberated from the infected plants that have been inoculated with particular bacteria or fungus. The techniques using e-noses have been utilized for the quick identification of blossom blight which is caused by *Pseudomonas syringae* PV. Syringae and fire blight caused by *Erwinia amylovor* most especially on apple trees when carried out in a field (Spinelli et al. 2010) and controlled laboratory environment (Fuentes et al. 2018).

MOS sensors could be utilized for the detection of numerous types of compounds which entails some reducing compounds and oxidizing compounds. Also, the MOS sensors possess a very high sensitivity which could detect sub-part per billion ppb levels for some gases to the oxidizing and some reducing compounds. Some other merits include rapid rejoinder time to the analytes, the capability to operates at high temperature and pressure and they could easily be fabricated (Mielle 1996). The application of e-nose based on MOS sensors has been highlighted to detect fire-blight-infected pear trees when they are infected at an early stage. The inoculation of tumorigenic strains of *Agrobacterium vitis* has been shown to variation when subjected to the uninoculated ones which display 83.3% effectiveness through the application of portable e-nose system (Spinelli et al. 2010, 2012). Tomato plant has been identified as one of the most important crops which has been utilized in the greenhouse crop globally because it has drawn the attention of scientist for validating the VOC fingerprints for evaluating the status of plants health. Zhang and co-workers utilized the influence of powdery mildew on the volatile organic compounds of the tomato plant that has been affected most especially under greenhouse condition. The outcome indicated that these diseases had a significant influence on the volatile organic compounds profile and the application of the e-nose could be able to detect between healthy and infected ones most especially with the accuracy of 94% (Ghaffari et al. 2011; Zhang et al. 2011). While the rate of infection induced by *Ceratocystis fagacearum* have been detected with 78.65% accuracy using e-nose recognition (Wilson et al. 2004).

9.5.4.2 Insect Damage

The application of e-noses has been demonstrated for their practical application for identification of infested plants as well as the insect population dynamics. The utilization of e-nose technology has been affirmed in the effective differentiation of *Nilaparva talugens* (brown planthopper) and *Chilo suppressalis* (striped rice stem borer) that are responsible for the damage of rice plant. Interestingly, the number of volatiles liberated from insects such as stink bugs and brown planthoppers could be used in the numerous aspects of insect-infested plants. Xu et al. (2014) demonstrated the possibility of estimating the age and number of brown planthoppers using an e-nose with classification accuracies of 100% and 48.93%, respectively. Also, the e-nose could be utilized for quick identification of the gender of an insect in a quick and practical way, that is very paramount importance in the mating disruption (Lan et al. 2008; Zhou and Wang 2011).

9.6 Conclusion and Future Recommendation

This chapter has provided a detailed information on the application of biosensor for mitigating the pest and pathogens problems affecting the food and crop production. The various types of biosensor that are being developed in this regard have been discussed in detail. The application of e-nose as a typical sensor for the detection of pest and pathogen were also highlighted. However, there is a need to perform numerous experiments most especially on the field and compare the laboratory results to validate its field application for the detection of pest and pathogen. This will go a long way towards the development of sustainable technology, that is sensitive, portable, eco-friendly, and economical. Moreover, there is a need to adopt and invest more in futuristic approaches such as smartphone-based sensing or wireless sensing.

References

Adetunji CO, Phazang P, Sarin NB (2018) Biosensors: a fast-growing technology for pathogen detection in agriculture and food sector. In: Biosensing technologies for the detection of pathogens: a prospective way for rapid analysis, p 37. https://doi.org/10.5772/intechopen.74668

Ali SB, Ghatak B, Debabhuti N, Sharma P, Ghosh A, Tudu B, Bhattacharya N, Bandyopadhyay R (2018) Detection of β-caryophyllene in mango using a quartz crystal microbalance sensor. Sens Actuators B, Chem 255:3064–3073

Ashley J, Piekarska M, Segers C, Trinh L, Rodgers T, Willey R, Tothill IE (2017) An SPR based sensor for allergens detection. Biosens Bioelectron 88:109–113

Ashley J, Shukor Y, D'Aurelio R, Trinh L, Rodgers TL, Temblay J, Pleasants M, Tothill IE (2018) Synthesis of molecularly imprinted polymer nanoparticles for α-casein detection using surface plasmon resonance as a milk allergen sensor. ACS Sens 3:418–424

Arcas ADS, Dutra FDS, Allil RC, Werneck MM (2018) Surface plasmon resonance and bending loss-based U-shaped plastic optical fiber biosensors. Sensors 18(2):648

Alves RC, Pimentel FB, Nouws HP, Marques RC, González-García MB, Oliveira MBP, Delerue-Matos C (2015) Detection of Ara h 1 (a major peanut allergen) in food using an electrochemical gold nanoparticle-coated screen-printed immunosensor. Biosens Bioelectron 64:19–24

Akgönüllü S, Duygu Ç, Monireh B, Nilay B, Handan Y, Adil D (2020) Commercial sensors for pathogen detection. In Sezgintürk MK (ed) Commercial biosensors and their applications. Elsevier, Amsterdam, pp 89–106

Ali J, Najeeb J, Ali MA, Aslam MF, Raza A (2017) Biosensors: their fundamentals, designs, types and most recent impactful applications: a review. J Biosens Bioelectron 8:235. https://doi.org/10.4172/2155-6210.1000235

Barretto AGOP, Berndes G, Sparovek G, Wirsenius S (2013) Agricultural intensification in Brazil and its effects on land-use patterns: an analysis of the 1975–2006 period. Glob Change Biol 19:1804–1815. https://doi.org/10.1111/gcb.12174

Brandão D, Liébana S, Campoy S, Alegret S, Pividori MI (2015) Immunomagnetic separation of Salmonella with tailored magnetic micro and nanocarriers. A Comp Study. Talanta 143:198–204

Bharat B (2009) Biomimetics: lessons from nature—an overview. Phil Trans R Soc A 367:1445–1486. http://doi.org/10.1098/rsta.2009.0011

Brasil de Oliveira Marques PR, Lermo A, Campoy S, Yamanaka H, Barbe J, Alegret S, Pividori MI (2009) Double-tagging polymerase chain reaction with a thiolated primer and electrochemical genosensing based on gold nanocomposite sensor for food safety. Anal Chem 81(4):1332–1339

Cellini A, Biondi E, Blasioli S, Rocchi L, Farneti B, Braschi I, Savioli S, Rodriguez-Estrada MT, Biasioli F, Spinelli F (2016) Early detection of bacterial diseases in apple plants by analysis of volatile organic compounds profiles and use of electronic nose. Ann Appl Biol 168:409–420. https://doi.org/10.1111/aab.12272

Choudhary DK, Johri BN, Prakash A (2008) Volatiles as priming agents that initiate plant growth and defence responses. Curr Sci 94:595–604

Charlebois S (2015) Is 3-D printing the future of global food [Internet]. The Globe and Mail; [cited 2016 Jan 26]. Available from http://www.theglobeandmail.com/report-on-business/rob-commentary/is-3-d-printing-the-future-of-global-food/article24981139/

Cinti S, Volpe G, Piermarini S, Delibato E, Palleschi G (2017) Electrochemical biosensors for rapid detection of foodborne Salmonella: a critical overview. Sensors 2017(8):1910. https://doi.org/10.3390/s17081910

Chen Y, Yang Y, Wang Y, Peng Y, Nie J, Gao G, Zhi J (2020) Development of an Escherichia coli-based electrochemical biosensor for mycotoxin toxicity detection. Bioelectrochemistry 133:107453

Chen HZ, Zhang M, Guo Z (2019) Discrimination of fresh-cut broccoli freshness by volatiles using electronic nose and gas chromatography-mass spectrometry. Postharvest Biol Technol 148:168–175

Compagnone D, Faieta M, Pizzoni D, Di Natale C, Paolesse R, Van Caelenberg T, Beheydt B, Pittia P (2015) Quartz crystal microbalance gas sensor arrays for the quality control of chocolate. Sens Actuators B: Chemical 207:1114–1120

Cervera-Chiner L, Juan-Borrás M, March C, Arnau A, Escriche I, Montoya Á, Jiménez Y (2018) High fundamental frequency quartz crystal microbalance (HFF-QCM) immunosensor for pesticide detection in honey. Food Control 92:1–6

Chauhan N, Soni S, Agrawal P, Balhara YPS, Jain U (2020) Recent advancement in nanosensors for neurotransmitters detection: present and future perspective. Process Biochem 91:241–259. https://doi.org/10.1016/j.procbio.2019.12.016

Clark LC, Gollan F, Gupta VB (1950) The oxygenation of blood by gas dispersion. Science 111(2874):85–87

Chekin F, Singh SK, Vasilescu A, Dhavale VM, Kurungot S, Boukherroub R, Szunerits S (2016) Reduced graphene oxide modified electrodes for sensitive sensing of gliadin in food samples. ACS Sensors 1(12):1462–1470

Dong W, Hu R, Long Y, Li H, Zhang Y, Zhu K, Chu Z (2019) Comparative evaluation of the volatile profiles and taste properties of roasted coffee beans as affected by drying method and detected by electronic nose, electronic tongue, and HS-SPME-GC-MS. Food Chem 272:723–731

Deshmukh S, Bandyopadhyay R, Bhattacharyya N, Pandey RA, Jana A (2015) Application of electronic nose for industrial odors and gaseous emissions measurement and monitoring—an overview. Talanta 144:329–340

Debabhuti N, Ghatak B, Parasrampuria V, Rafiqul SM, Hassan PA, Gupta SD, Sharma P, Tudu B, Bandyopadhyay R, Ali SB, Bhattacharyya N (2016) Development of a QCM sensor for detection of trans-2-hexenal in tomatoes. In: 2016 International Conference on Intelligent Control Power and Instrumentation (ICICPI), IEEE, pp 93–97

Farzin L, Shamsipur M, Samandari L, Sadjadi S, Sheibani S (2020) Biosensing strategies based on organic-scaffolded metal nanoclusters for ultrasensitive detection of tumor markers. Talanta 214, 120886. https://doi.org/10.1016/j.talanta.2020.120886

Farka Z, Juřík T, Pastucha M, Skládal P (2016) Enzymatic precipitation enhanced surface plasmon resonance immunosensor for the detection of salmonella in powdered milk. Anal Chem 88(23):11830–11836

Fang Y, Umasankar Y, Ramasamy RP (2014) Electrochemical detection of p-ethylguaiacol, a fungi infected fruit volatile using metal oxide nanoparticles. Analyst 139:3804–3810. https://doi.org/10.1039/C4AN00384E

Fang Y, Ramasamy RP (2015) Current and prospective methods for plant disease detection. Biosensors 5:537–561. https://doi.org/10.3390/bios5030537

Fuentes MT, Lenardis A, de la Fuente EB (2018) Insect assemblies related to volatile signals emitted by different soybean – weeds – herbivory combinations. Agric Ecosyst Environ 255:2026. https://doi.org/10.1016/j.agee.2017.12.007

Ghaffari R, Zhang F, Iliescu DD, Hines EL, Leeson MS, Napier R (2011) Intelligent systems for machine olfaction: tools and methodologies. In: Detection of diseases and volatile discrimination of plants: an electronic nose and self-organizing maps approach. IGI Global, Hershey, PA, USA, pp 214–230

Ghosh P, Bhattacharjee D, Nasipuri M (2020) Intelligent toilet system for non-invasive estimation of blood-sugar level from urine. IRBM 41(2):94–105. https://doi.org/10.1016/j.irbm.2019.10.005

Garg M, Mehrotra S (2017) Biosensors. In: Principles and applications of environmental biotechnology for a sustainable future. Springer, Singapore, pp 341–363

Gupta S, Kakkar V (2020) Development of environmental biosensors for detection, monitoring, and assessment. In: Nanomaterials and environmental biotechnology. Springer, Cham, pp 107–125
Güner A, Çevik E, Şenel M, Alpsoy L (2017) An electrochemical immunosensor for sensitive detection of Escherichia coli O157: H7 by using chitosan, MWCNT, polypyrrole with gold nanoparticles hybrid sensing platform. Food Chem 229:358–365
Gao G, Fang D, Yu Y, Wu L, Wang Y, Zhi J (2017) A double-mediator based whole cell electrochemical biosensor for acute biotoxicity assessment of wastewater. Talanta 167:208–216
Huang L, Meng L, Zhu N, Wu D (2017) A primary study on forecasting the days before decay of peach fruit using near-infrared spectroscopy and electronic nose techniques. Postharvest Biol Technol 133:104–112
Holt KB (2007). Clarke oxygen microelectrode. In: Zoski CG (ed) Handbook of electrochemistry. Elsevier, Amsterdam, pp 243–249
Kahn K, Plaxco KW (2010) Principles of biomolecular recognition. In: Zourob M (ed) 9 Recognition receptors in biosensors. Springer, New York, pp 3–46
Kim N, Park IS (2003) Application of a flow-type antibody sensor to the detection of Escherichia coli in various foods. Biosens Bioelectron 18(9):1101–1107
Karsunke XY, Niessner R, Seidel M (2009) Development of a multichannel flow-through chemiluminescence microarray chip for parallel calibration and detection of pathogenic bacteria. Anal Bioanal Chem 395(6):1623
Kundu M, Krishnan P, Kotnala RK, Sumana G (2019) Recent developments in biosensors to combat agricultural challenges and their future prospects. Trends Food Sci Technol 88:157–178
Kaushik S, Tiwari UK, Pal SS, Sinha RK (2019) Rapid detection of Escherichia coli using fiber optic surface plasmon resonance immunosensor based on biofunctionalized Molybdenum disulfide (MoS2) nanosheets. Biosens Bioelectron 126:501–509
Kaur H, Bhosale A, Shrivastav S (2018) Biosensors: classification, fundamental characterization and new trends: a review. Int J Health Sci Res 8(6):315–333. https://doi.org/10.1007/978-3-319-97852-9_23
Lan YB, Zheng XZ, Westbrook JK, Lopez J, Lacey R, Hoffmann WC (2008) Identification of stink bugs using an electronic nose. J Bionic Eng 2008(5):172–180. https://doi.org/10.1016/S1672-6529(08)60090-6
Long R, Guo Y, Xie L, Shi S, Xu J, Tong C, Lin Q, Li T (2020) White pepper-derived ratiometric carbon dots for highly selective detection and imaging of coenzyme A. Food Chemistry 315. https://doi.org/10.1016/j.foodchem.2020.126171
Liébana S, Lermo A, Campoy S, Cortés MP, Alegret S, Pividori MI (2009) Rapid detection of Salmonella in milk by electrochemical magneto-immunosensing. Biosens Bioelectron 25(2):510–513
Liu K, Zhang C (2020) Volatile organic compounds gas sensor based on quartz crystal microbalance for fruit freshness detection: a review. Food Chem 334, 127615
Majumdar T, Agarwal S, Chakraborty R, Raychaudhuri U (2012) Study on electrochemical immunoassay methodology for protein a: a modified approach towards detection and quantification of Staphylococcus Aureus in food samples. Res J Chem Sci 2:62–68
Monosik R, Streansky M, Surdik E (2012) Biosensors-classification, characterization and new trends. Acta Chimica Slovaca 5:109–120
Mehrotra P(2016) Biosensors and their applications—a review. J Oral Biol Craniofac Res 6(2):153–159
Mielle P (1996) Electronic noses: towards the objective instrumental characterization of food aroma. Trends Food Sci Technol 7:432–438
Malhotra S, Verma A, Tyagi N, Kumar V (2017). Biosensors: principle, types and applications. Int J Adv Res Innov Ideas Educ 3(2):3639–3644
Maas MB, Maybery GH, Perold WJ, Neveling DP, Dicks LM (2018) Borosilicate glass fiber-optic biosensor for the detection of Escherichia coli. Curr Microbiol 75(2):150–155
Mittal S, Kaur H, Gautam N, Mantha AK (2017) Biosensors for breast cancer diagnosis: a review of bioreceptors, biotransducers and signal amplification strategies. Biosens Bioelectron 88:217–231

Montiel VRV, Campuzano S, Pellicanò A, Torrente-Rodríguez RM, Reviejo AJ, Cosio MS, Pingarrón JM (2015) Sensitive and selective magnetoimmunosensing platform for determination of the food allergen Ara h 1. Anal Chim Acta 880:52–59

Morales-Cruz M, Solis-Marcano NE, Binder C, Priest C, Cabrera CR (2019) Electrochemical Proteus vulgaris whole cell urea sensor in synthetic urine. Curr Res Biotechnol 1:22–27. https://doi.org/10.1016/j.crbiot.2019.07.002

Moro G, De Wael K, Moretto LM (2019) Challenges in the electrochemical (bio) sensing of nonelectroactive food and environmental contaminants. Curr Opin Electrochem 16:57–65

Miniaev MV, Belyakova MB, Kostiuk NV, Leshchenko DV, Fedotova TA (2013) Non-obvious problems in Clark electrode application at elevated temperature and ways of their elimination. J Anal Methods Chem 249752. https://doi.org/10.1155/2013/249752

Niu X, Zhong Y, Chen R, Wang F, Liu Y, Luo D (2018) A "turn-on" fluorescence sensor for Pb2+ detection based on graphene quantum dots and gold nanoparticles. Sens Actuators B: Chemical 255:1577–1581

Nastuk WL (2013) Special methods: physical techniques in biological research. Elsevier Science, Amsterdam

Nwankwo W, Njoku CC (2020) Sustainable development in developing societies: the place of ICT-driven computer education. Int J Emerg Technol Learn 15(12): 290–297

Nwankwo W, Olayinka AS, Umezuruike C (2019) Boosting self-sufficiency in maize crop production in osisioma ngwa local government with IOT-climate messaging: a model. Afr J Agric Res (AJAR) 14(7):406–416. https://doi.org/10.4172/2155-6210.1000235

Pare PW, Tumlinson JH (1999) Update on plant-insect interactions plant volatiles as a defense against insect herbivores. Plant Physiol 1999(121):325–331. https://doi.org/10.1104/pp.121.2.325

Pividori MI, Alegret S (2005) Electrochemical genosensing based on rigid carbon composites: a review. Anal Lett 38(15):2541–2565

Papadakis G, Skandalis N, Dimopoulou A, Glynos P, Gizeli E (2015) Bacteria murmur: application of an acoustic biosensor for plant pathogen detection. PLoS ONE 10(7):e0132773

Qiu S, Wang J (2017) The prediction of food additives in the fruit juice based on electronic nose with chemometrics. Food Chem 230:208–214

Qiu Y, Gu C, Li B, Shi H (2018) Aptameric detection of quinine in reclaimed wastewater using a personal glucose meter. Anal Methods 10(24):2931–2938

Ray M, Ray A, Dash S, Mishra A, Achary KG, Nayak S, Singh S (2017) Fungal disease detection in plants: traditional assays, novel diagnostic techniques and biosensors. Biosens Bioelectron 87:708–723. https://doi.org/10.1016/j.bios.2016.09.032

Radke SM, Alocilja EC (2005) A high-density microelectrode array biosensor for detection of E. coli O157: H7. Biosens Bioelectron 20(8):1662–1667

Rajendran DK, Park E, Nagendran R, Hung NB, Cho BK, Kim KH, Lee YH (2016) Visual analysis for detection and quantification of Pseudomonas cichorii disease severity in tomato plants. Plant Pathol J 2016(32):300–310. https://doi.org/10.5423/PPJ.OA.01.2016.0032

Rutolo MF, Clarkson JP, Covington JA (2018) The use of an electronic nose to detect early signs of soft-rot infection in potatoes. Biosys Eng 167:137–143

Silva NF, Magalhães JM, Freire C, Delerue-Matos C (2018) Electrochemical biosensors for Salmonella: State of the art and challenges in food safety assessment. Biosens Bioelectron 99:667–682

Seo SM, Kim SW, Jeon JW, Kim JH, Kim HS, Cho JH, Lee WH, Paek SH (2016) Food contamination monitoring via internet of things, exemplified by using pocket-sized immunosensor as terminal unit. Sens Actuators B, Chem 233: 148–156

Soares RR, Hjort RG, Pola CC, Parate K, Reis EL, Soares NF, McLamore ES, Claussen JC, Gomes CL (2020). Laser-induced graphene electrochemical immunosensors for rapid and label-free monitoring of salmonella enterica in chicken broth. ACS Sens 5(7): 1900–1911

Saylan Y, Yılmaz F, Derazshamshir A, Yılmaz E, Denizli A (2017) Synthesis of hydrophobic nanoparticles for real-time lysozyme detection using surface plasmon resonance sensor. J Mol Recognit 30(9):e2631
Sasaki K, Furusawa H, Nagamine K, Tokito S (2018) Detection of odorant molecules in the gaseous phase using α-, β-, and γ-cyclodextrin films on a quartz crystal microbalance. Technologies 6(3):63
Songa EA, Okonkwo JO (2016) Recent approaches to improving selectivity and sensitivity of enzyme-based biosensors for organophosphorus pesticides: a review. Talanta 155:289–304
Sakamoto S, Putalun W, Vimolmangkang S, Phoolcharoen W, Shoyama Y, Tanaka H, Morimoto S (2018) Enzyme-linked immunosorbent assay for the quantitative/qualitative analysis of plant secondary metabolites. J Nat Med 72(1):32–42
Sharifi M, Hasan A, Attar F, Taghizadeh A, Falahati M (2020) Development of point-of-care nanobiosensors for breast cancers diagnosis. Talanta 217:121091
Sung ES, Fatemehsadat T, Seon JP, Behnam A, Kyung HK, et al (2019) Smartphone with optical, physical, and electrochemical nanobiosensors. J Ind Eng Chem 77:1–11
Shen H, Tao J (2018) Applying electronic nose based on odour classification and identification technology in detecting the shelf life of fresh fruits. Chem Eng Trans 68:217–222
Sankaran S, Mishra A, Ehsani R, Davis C (2010) A review of advanced techniques for detecting plant diseases. Comput Electron Agric 72:1–13. https://doi.org/10.1016/j.compag.2010.02.007
Spinelli F, Noferini M, Vanneste JL, Costa G (2010) Potential of the electronic-nose for the diagnosis of bacterial and fungal diseases in fruit trees. EPPO Bull 2010(40):59–67. https://doi.org/10.1111/j.1365-2338.2009.02355
Spinelli F, Cellini A, Vanneste JL, Rodriguez-Estrada MT, Costa G, Savioli S, Harren FJM, Cristescu SM (2012) Emission of volatile compounds by Erwinia amylovora: biological activity in vitro and possible exploitation for bacterial identification. Trees Struct Funct 26:141–152. https://doi.org/10.1007/s00468-011-0667-2
Sode K, Yamazaki T, Lee I, Hanashi T, Tsugawa W (2016) BioCapacitor: a novel principle for biosensors. Biosens Bioelectron 76:20–28
Shaikh PMM, Butala N, Sharma S, Sukumaran S (2017) Review on diagnosis the plant disease at early stage. Int J Adv Res Comput Commun Eng 2017(6):321–324. https://doi.org/10.17148/IJARCCE.2017.6372
Song J, Dailey J, Li H, Jang HJ, Russell L, Zhang P, Searson PC, Wang JTH, Everett AD, Katz HE (2018). Influence of bioreceptor layer structure on myelin basic protein detection using organic field effect transistor-based biosensors. Adv Funct Mater 28(37):1802605
Soleymani J, Hasanzadeh M, Somi MH, Ozkan SA, Jouyban A (2018) Targeting and sensing of some cancer cells using folate bioreceptor functionalized nitrogen-doped graphene quantum dots. Int J Biol Macromol 118:1021–1034
Soleymani J, Hasanzadeh M, Somi MH, Jouyban A (2020) Differentiation and targeting of HT 29 cancer cells based on folate bioreceptor using cysteamine functionalized gold nano-leaf. Mater Sci Eng, C 107:110320
Saadati A, Hassanpour S, de la Guardia M, Mosafer J, Hashemzaei M, Mokhtarzadeh A, Baradaran B (2019) Recent advances on application of peptide nucleic acids as a bioreceptor in biosensors development. TrAC Trends Anal Chem 114:56–68
Tan J, Kerr WL (2019) Characterizing cocoa refining by electronic nose using a Kernel distribution model. LWT 104:1–7
Tolentino MAKP, Albano DRB, Sevilla FB III (2018) Piezoelectric sensor for ethylene based on silver (I)/polymer composite. Sens Actuators, B Chem 254:299–306
Turner AP (2013) Biosensors: sense and sensibility. Chem Soc Rev 42(8):3184–3196
Tricoli A, Neri G (2018) Miniaturized bio-and chemical-sensors for point-of-care monitoring of chronic kidney diseases. Sensors (Basel, Switzerland) 18(4):942. https://doi.org/10.3390/s18040942

Uniyal S, Sharma RK (2018) Technological advancement in electrochemical biosensor based detection of Organophosphate pesticide chlorpyrifos in the environment: a review of status and prospects. Biosens Bioelectron 116:37–50

Veldhuizen LJL, Giller KE, Oosterveer P, Brouwer ID, Janssen S, van Zanten HHE, Slingerland MA (2020) The missing middle: connected action on agriculture and nutrition across global, national and local levels to achieve sustainable development goal 2. Global Food Secur 24:100336

Vandevijvere S, Monteiro C, Krebs-Smith SM et al (2013) Monitoring and benchmarking population diet quality globally: a step-wise approach. Obes Rev 14(Suppl 1):135–149. https://doi.org/10.1111/obr.12082

Wilson AD, Lester DG, Oberle CS (2004) Development of conductive polymer analysis for the rapid detection and identification of phytopathogenic microbes. Phytopathology 94:419–431. https://doi.org/10.1094/PHYTO.2004.94.5.419

Wang W, Liu L, Song S, Xu L, Kuang H, Zhu J, Xu C (2016) Gold nanoparticle-based strip sensor for multiple detection of twelve Salmonella strains with a genus-specific lipopolysaccharide antibody. Sci China Mater 59(8):665–674

Wen T, Zheng L, Dong S, Gong Z, Sang M, Long X, Luo M, Peng H (2019) Rapid detection and classification of citrus fruits infestation by Bactrocera dorsalis (Hendel) based on electronic nose. Postharvest Biol Technol 147:156–165

Xu S, Zhou Z, Lu H, Luo X, Lan Y, Zhang Y, Li Y (2014) Estimation of the age and amount of brown rice plant hoppers based on bionic electronic nose use. Sensors 14:18114–18130. https://doi.org/10.3390/s141018114

Yu D, Li J, Kang Z, Liu L, He J, Fang Y, Dong S (2020) An unexpected discovery of 1, 4-benzoquinone as a lipophilic mediator for toxicity detection in water. Analyst 145:5266–5272. https://doi.org/10.1039/D0AN00991

Zhang J, Yang Z, Liu Q, Liang H (2019) Electrochemical biotoxicity detection on a microfluidic paper-based analytical device via cellular respiratory inhibition. Talanta 202:384–391

Zhou J, Wang Y, Qian Y, Zhang T, Zheng L, Fu L (2019a) Quantification of shellfish major allergen tropomyosin by SPR biosensor with gold patterned Biochips. Food Control 107:106547

Zamora-Galvez A, Morales-Narváez E, Mayorga-Martinez CC, Merkoçi A (2017) Nanomaterials connected to antibodies and molecularly imprinted polymers as bio/receptors for bio/sensor applications. Appl Mater Today 9:387–401

Zhou J, Qi Q, Wang C, Qian Y, Liu G, Wang Y, Fu L (2019b) Surface plasmon resonance (SPR) biosensors for food allergen detection in food matrices. Biosens Bioelectron 142:111449

Zhu G, Yin X, Jin D, Zhang B, Gu Y, An Y (2019) Paper based immunosensors: current trends in the types and applied detection techniques. TrAC Trends Anal Chem 111:100–117

Zhang F, Iliescu DD, Hines EL, Leeson MS (2011) Intelligent systems for machine olfaction: tools and methodologies. In: Tomato plant health monitoring: an electronic nose approach. IGI Global, Hershey, PA, USA, pp 231–248

Zhou B, Wang J (2011) Discrimination of different types damage of rice plants by electronic nose. Biosyst Eng 109:250–257. https://doi.org/10.1016/j.biosystemseng.2011.03.003

Zhou Q, Tang D (2020) Recent advances in photoelectrochemical biosensors for analysis of mycotoxins in food. TrAC—Trends Anal Chem 124. https://doi.org/10.1016/j.trac.2020.115814

Zhang JXJ, Hoshino K (2014) Molecular sensors and nanodevices principles, designs and applications in biomedical engineering, Elsevier, pp 169–232. https://doi.org/10.1016/C2012-0-07668-5

Chapter 10
Gold Nanoparticles-Based Point-of-Care Colorimetric Diagnostic for Plant Diseases

Ravi Mani Tripathi and Prashant Sharma

Abstract Plant diseases caused by pathogens such as fungi, viruses, and bacteria create a substantial loss worldwide that leads to a severe threat to food security and human health. Therefore, disease management is a priority in agricultural-based countries. The detection of plant-pathogens is the primary step in disease management, which is crucial for plant health monitoring. Presently, few immunological methods are utilized to detect plant pathogens such as direct tissue blot immunoassay, enzyme-linked immunosorbent assay, and fluorometric immunoassay. DNA-based detection methods are also keenly applied for plant pathogen identification and detection, such as polymerase chain reaction, real-time PCR, and dot blot hybridization. However, these detection methods are inherited with some disadvantages like costly antibodies, complicated procedures, challenge to produce specific antibodies, expensive equipment, and susceptibility to contamination. This chapter reviews an instrument-free, instant, highly selective, and sensitive colorimetric detection of plant pathogens. The colorimetric detection is working on localized surface plasmon resonance properties of AuNPs, which causes the colour change of assay.

Keywords Plant pathogens · Colorimetric detection · Gold nanoparticles · Properties · Functionalization · Mechanism

R. M. Tripathi (✉)
Amity Institute of Nanotechnology, Amity University Uttar Pradesh, Sector 125, Noida, UP 201303, India
e-mail: rmtripathi@amity.edu

P. Sharma
Department of Pediatrics, Division of Pediatric Hematology, Oncology & Bone Marrow Transplant, University of Wisconsin–Madison, Madison, WI 53706, USA

R. N. Pudake et al. (eds.), *Biosensors in Agriculture: Recent Trends and Future Perspectives*, Concepts and Strategies in Plant Sciences,
https://doi.org/10.1007/978-3-030-66165-6_10

10.1 Introduction

Plant pathogens like fungi, viruses, and bacteria cause significant agricultural product loss in developing countries (Agrios 2005). The fungi, bacteria, and viruses infect various plant parts like leaves, stems, tubers, fruits, with various agronomic impacts and finally lead to severe symptoms such as etiolation, necrosis, and blight (Fox and Mumford 2017; Khater et al. 2017; Martinelli et al. 2015). These plant diseases can create a significant loss in crop production and lead to severe threats to food security and human health. The control of diseases is challenging due to the lack of efficient chemical treatment products under field conditions. Therefore, this problem is making a priority for disease management in agriculture-based countries. The detection of plant diseases is a crucial technique to prevent the loss of crop production.

Several techniques have been established to identify plant pathogens (Punja et al. 2007). Plant pathogens can be detected by skilled persons/farmers based on symptoms or by laboratory experts by microscopic, molecular, or culturing techniques (Horsfall 2012). The physiological, biochemical, and pathological parameters are also examined on the plant diseases by culturing pathogens on selective media. The traditional techniques are time-consuming and need skilled plant pathologists to analyse and confirm the particular disease symptoms and responsible pathogens. Therefore, these shortcomings compelled us to develop detection methods with cost-effective, higher sensitivity, specificity, and rapid process to identify plant pathogen.

Extensive assays such as direct tissue blot immunoassay (DTBIA), enzyme-linked immunosorbent assay (ELISA), and fluorometric immunoassay used to detect plant pathogens (Fang and Ramasamy 2015; Martín et al. 2002; Shojaei et al. 2016). These methods are inherited with some drawbacks like costly antibodies, complicated operation, and challenge to generate specific antibodies that restrict immunoassay application for plant disease management (Toh et al. 2015). DNA-based approaches have been applied to detect plant pathogens after the invention of polymerase chain reaction (PCR) in the 1980s. Several PCR-based detection methods have reported for the identification of plant pathogens (Vincelli and Tisserat 2008; Ward et al. 2004). Issues related to specificity and sensitivity to the amplification of pathogen-specific sequences were overcome by the coupling of PCR with other techniques (Brasileiro et al. 2004; Carrasco-Ballesteros et al. 2007; Merighi et al. 2000). Another strategy, immunocapture-PCR (IC-PCR), has been adopted for the detection of viruses by fusing conventional PCR amplification with antibody capturing virus. This strategy enhanced the sensitivity by 250-fold compared to PCR amplification and developed a platform to detect the bacterial blight disease in Anthurium propagation material (Khoodoo et al. 2005). However, PCR-based plant pathogen detection is associated with some limitations, such as expensive equipment and susceptibility to cross contamination (Yang et al. 2016).

Nowadays, the nanomaterials-based approach for the detection of plant pathogens is getting more popular among researchers. Researchers have used gold nanoparticles (AuNPs) for various applications, such as catalysts, antibacterial, colorimetric sensors, and also in drug delivery systems, because of their unique chemical and physical properties (Mehrotra and Tripathi 2015; Tripathi and Chung 2019; Tripathi et al. 2014b, 2015a, 2018b, 2019a). The colorimetric detections, non-biological, and biological analytes using AuNPs by surface modification with ligands and biological receptors (Tripathi et al. 2019a; Verma et al. 2015). The principle behind the colorimetric detection is the change in the localized surface plasmon resonance properties of AuNPs, which causes the change in colour of the assay (Tripathi et al. 2019a). AuNPs based colorimetric detection is getting extensive exploration due to its simplicity, rapid detection, and cost-effectiveness.

In this chapter, we discussed the relevant physical and chemical properties of AuNPs for the detection of plant pathogens such as fungi, bacteria, and viruses. This chapter reviews the colorimetric detection of various plant pathogens. The principle behind the colorimetric detection is localized surface plasmon resonance properties of AuNPs. As per the requirement application, AuNPs are surface functionalized with different ligands, such as polymers and biomolecules. Nanoparticle-based colorimetric detection techniques are instrument-free, instant, highly selective, and sensitive for detecting plant pathogens.

10.2 AuNPs and Their Properties

AuNPs are most exploited as compared to other metallic nanoparticles for various purposes, such as catalysis fields (Tripathi et al. 2018b), antibacterial agents (Tripathi et al. 2018b), drug carriers (Tripathi et al. 2015a), colorimetric sensor (Tripathi et al. 2014b, 2019a), etc. AuNPs show several excellent characteristics like large surface-to-volume ratio, shape-dependent optoelectronic properties, exceptional biocompatibility, and low toxicity. The shape and physico-chemical properties of AuNPs are associated with optical features, and their size affects the colour of the colloidal solution. The most applicable optical properties of AuNPs are the surface plasmon resonance (SPR) (Tripathi et al. 2014b) and fluorescence quenching (Xia et al. 2016). SPR is a unique property of AuNPs, and it depends on the morphology and size of nanoparticles. SPR phenomenon produces a strong characteristic absorption for AuNPs in the UV-visible region. The appropriate term for the SPR in this condition is localized surface plasmon resonance (LSPR), which was recognized as Mie-scattering in 1904 (Tripathi et al. 2019a). LSPR has theoretically anticipated for the spherical metal particles with dimensions similar to the light wavelength. LSPR is well-defined by the quantization of surface plasmons in different types of nanometal. The most exciting property of AuNPs is their LSPR, which is the critical principle of colorimetric detection of a biological and chemical analytes. Generally, the AuNPs colloidal solution is red or pink, but agglomeration caused the colour change to purple blue (Fig. 10.1). These properties of AuNPs make them a suitable candidate for applications in sensing various analytes.

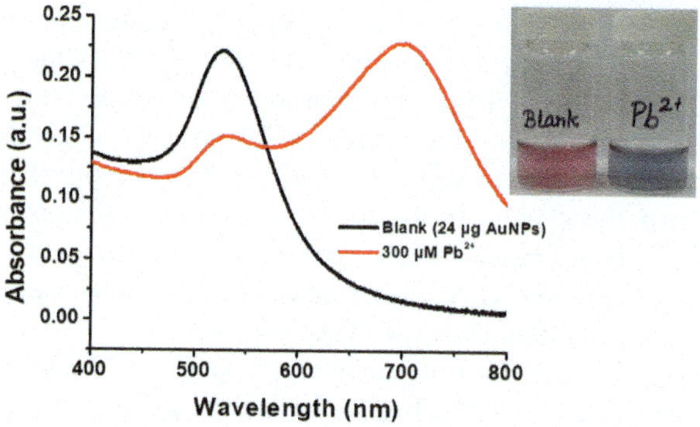

Fig. 10.1 A representative example of calorimetric detection of an analyte by AuNPs. The graph shows the UV–Vis spectra of the blank sample and the assay containing targets. The inset shows the colour change of the assay in the presence of target (the figure was reproduced from Tripathi et al. 2019 with the permission of Elsevier Science)

10.3 Synthesis of AuNPs

The nanoparticles are synthesized by two leading approaches: top-down and bottom-up. The creation of particles at the nanoscale by crushing or cutting bulk materials is a top-down approach, whereas the creation of nanoscale materials by assembling them atom by atom, molecule by molecule, or cluster by cluster is bottom-up. Further, the type of nanomaterials synthesis categorized into three methods: physical, chemical, and biological. The physical methods required high energy and pressure, whereas chemical methods are hazardous and create serious environmental problems.

We have synthesized AuNPs with a narrow size distribution ranging from 5 to 25 nm, and the average nanoparticle size was 10 nm using the Turkevich method (Fig. 10.2). The synthesized nanoparticles have been used for instrument-free, instant, highly selective, and sensitive colorimetric detection of lead (Tripathi et al. 2019a).

Nowadays, biosynthesis of the nanomaterial is getting more popular because of the issues related to physical and chemical methods (Tripathi et al. 2013, 2015a, b, 2018a). Biologically originated nanomaterials have been shown excellent properties for colorimetric detection and do not need special surface modification (Tripathi et al. 2014a, b). Typically, bacteria, fungi, and plants are used for the intra and extracellular synthesis of various metallic and semiconductor nanomaterials (Agarwal et al. 2016; Mahajan et al. 2016; Tripathi et al. 2014a, 2015a; Tripathi and Chung 2020).

Gold exists in three unique forms, gold (0), gold(I), and gold (III) in an aqueous solution (El-Brolossy et al. 2008; Tripathi and Chung 2019). AuNPs show exceptional chemical and physical properties as compared to their bulk due to quantum effects. AuNPs can be synthesized by using the biomass of fungi and bacteria. The various morphologies of AuNPs have been synthesized aqueous extract of *Rhizopus*

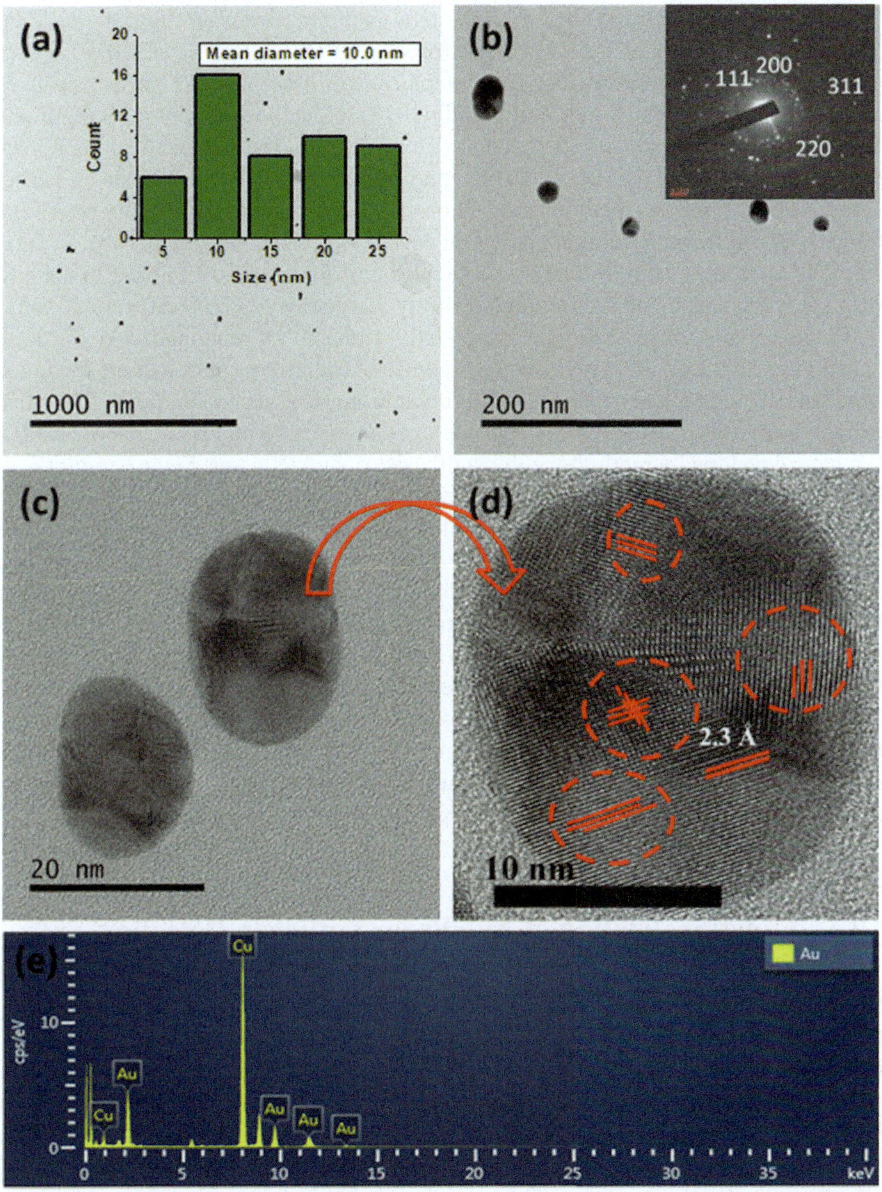

Fig. 10.2 TEM images of the AuNPs at different magnifications: **a** sample large view and the inset shows the TEM histogram; **b** focused on a few particles and the inset shows the SAED pattern; **c** HR-TEM image; **d** polycrystalline nature and d-spacing; **e** EDX spectrum (figures were reproduced from Tripathi et al. 2019 with the permission of Elsevier Science)

oryzae by changing the synthesis parameters, including reaction time, the concentration of gold salt, and the pH of the reaction mixture (Das et al. 2010). *Pseudomonas aeruginosa* and *Rhodopseudomonas capsulate* biomass have been used to synthesis of AuNPs (Singh and Kundu 2014). *Coriolus versicolor* (Sanghi and Verma 2010) and *Fusarium oxysporum* (Thakker et al. 2012) have also been used for the synthesis AuNPs, but these are pathogenic fungi. Therefore, in an earlier study, AuNPs synthesis method using a non-pathogenic and agriculturally beneficial fungus; *Trichoderma harzianum* has been developed (Tripathi et al. 2014b). Synthesized nanoparticles using *T. harzianum* showed narrow size distributions (26–34 nm) with spherical shape and having high stability were noticed (Tripathi et al. 2014b).

However, the microorganism mediated synthesis of nanoparticles is time-consuming and needs to grow enough biomass. Therefore, plant extract mediated synthesis gained advancement over microbial biomass assisted synthesis. The plant extract is used as a reducing and capping agents for the synthesis of nanoparticles. We have developed a method for the synthesis of AuNPs using *Ficus benghalensis* leaf extract, and synthesized nanoparticles had a spherical shape with a size ranging from 17 to 50 nm (Tripathi et al. 2012). The shape of nanoparticles can also be changed according to their application. Recently we have synthesized hexagonal and triangular AuNPs using *Erigeron annuus* leaf extract (Tripathi et al. 2019b). For this, fresh leaves of *E. annuus* were taken and cut into small pieces. Thereafter, 8.5 g leaves were dispersed in 100 mL DI water and boiled at continuous stirring. Finally, the obtained solution was filtered and stored at 4 °C for the synthesis of AuNPs. The synthesized nanoparticles were characterized by transmission electron microscopy and micrographs show triangular AuNPs synthesized (Fig. 10.3). These studies make the biosynthesized nanoparticles a putative and robust candidate for different calorimetric detection.

10.4 Functionalization of AuNPs for Pathogen Detection

Surface modification of AuNPs with a functional group is a crucial step in detecting pathogens, which provides high selectivity to the method. Chemically anchor nanoparticles conjugate functionalizing molecules, and another end remains free, which binds with the target molecule of interest (Wang and Yu 2013). AuNPs can be functionalized with a variety of ligands, such as polymers, biomolecules, etc. AuNPs can easily be conjugated with DNA, peptides, and antibodies due to its strong binding affinity with the thiols group. The surface can be modified by covalent bonding (Bhamore et al. 2015; Zhang et al. 2016), electrostatic interaction (Liu et al. 2016), or attachment of a molecule for precise identification of the interested target (Liu et al. 2018). The most widely adopted surface modification of nanoparticles is the covalent form. The AuNPs are functionalized with the molecule comprising sulfhydryl groups. The sulfhydryl group interaction is found between Au and S groups. Thiol (Shahrivari et al. 2018), cysteamine (Gukowsky et al. 2018), glutathione (Zhang et al. 2012), thioglycolic acid (Tolessa et al. 2018), cysteine (Raj et al. 2015), and protein

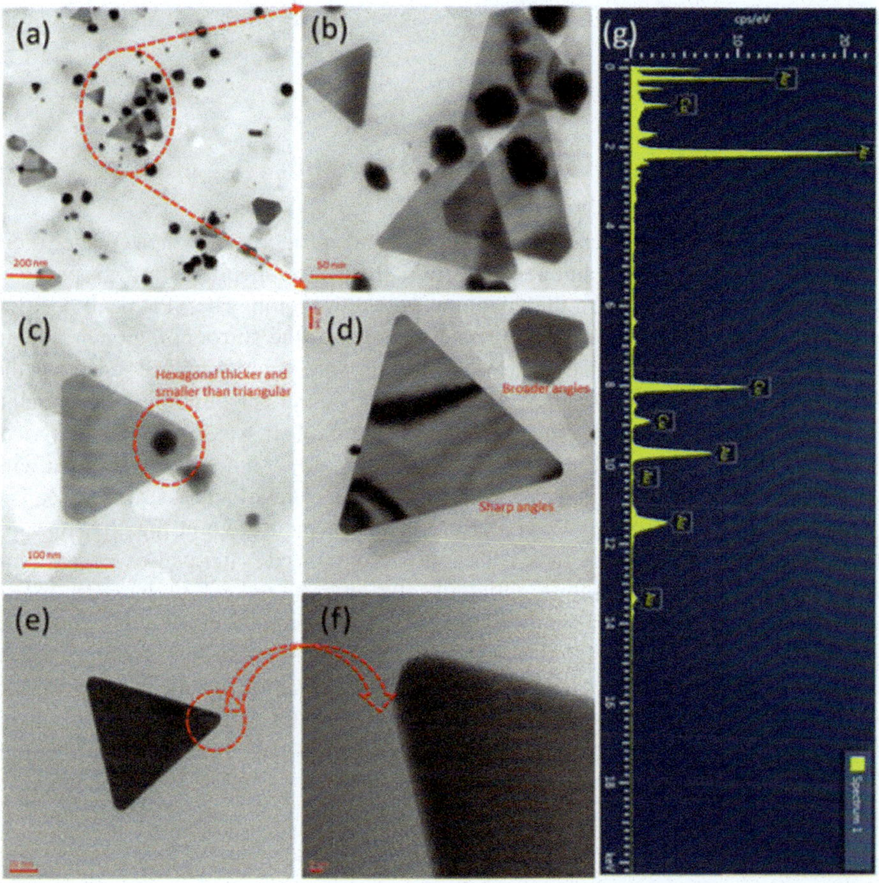

Fig. 10.3 TEM micrographs of triangular AuNPs. **a** Large view of sample; **b** magnified view; **c** thickness difference between hexagonal and triangular nanoparticles; **d** broad and sharp angles; **e** high magnification of single triangular; **f** angle feature of triangle; **g** EDX of triangular (Tripathi et al. 2019)

(Hu et al. 2017) are widely used sulfhydryl ligands. AuNPs prepared by $NaBH_4$ reduction and stabilized by thiol surfactant can be easily substituted by other thiol-terminated groups (Gupta et al. 2012). The reaction for ligand exchange is performed in a toluene solvent under ambient temperature for a long time (Stobiecka et al. 2010). AuNPs surface modification by ionic modifiers or ionic capping agents is providing stability by electrostatic repulsion. The surface charge magnitude (SCM) influenced repulsion strength and stability. SCM depends on the ionic strength and pH value solution and the pKa value of immobilized ligands (Liu and Lämmerhofer 2019). The electrostatic interactions are weak compared to the covalent approach, and electrostatic modified nanoparticles have poor stability. The ligands are easily detached under the high salt circumstances, which caused aggregation of AuNPs (Xu et al. 2017).

10.5 Mechanism of Colorimetric Detection

AuNPs possess unique chemical and physical properties, which make them promising materials for pathogen detection. LSPR is an interesting property of AuNPs show LSPR peaks in the visible wavelength region. The SPR peak position of AuNPs shows a red-shift as the refractive index of the surrounding environment increases (Nguyen et al. 2016). The working principle of colorimetric sensors is the sensitivity of their LSPR to the refractive index of the surrounding environment. The LSPR depends on the size and shape of AuNPs. It was found the spherical nanoparticles having an average size ranging from 9 to 99 nm, which observed with absorbance peaks from 517 to 575 nm, respectively (Daniel and Astruc 2004). The surface modification of AuNPs is a critical factor which brought together through the formation of linkages between the individual particles and causes aggregation AuNPs. The synthesized AuNPs colloidal solution has a red colour, but after aggregation, the colour turned into purple-blue which give impact on the absorbance and size of nanoparticles (Fig. 10.4) (Assah et al. 2018). The intensity of colour change can be quantified by UV-visible spectroscopy. In our previous study, the synthesized AuNPs absorbance occurs at 528 nm, but after aggregation peak shifts ranging from 630 to 700 nm depending on the degree of aggregation was observed (Tripathi et al. 2019a).

10.6 Colorimetric Detection of Plant Pathogens

10.6.1 Fungi

Fungi are the most responsible disease-causing agents in crop plants worldwide. They can survive for a longer time in diseased and dead plant tissues in a dormant state and proliferate in favourable conditions. The spores of fungi can easily spread in several ways, such as water, wind, insects, soil, and other invertebrates (Jain et al. 2019). The crops and plant diseases are caused by more than 19,000 species of fungi (Jain et al. 2019). A biosensor has been developed for the detection of a plant pathogen fungus *Phakopsora pachyrhizi*, which causes Soybean rust (Mendes et al. 2009). Potato late blight caused by *Phytophthora infestans* which is a most destructive plant pathogen. Therefore, the detection of *P. infestans* is a critical concern for crops and plant disease management. Researchers are attracted to colorimetric detection because of rapid, instrument-free, visual detection, and fieldable estimation of the plant pathogens. Chang and coworkers have reported colorimetric detection

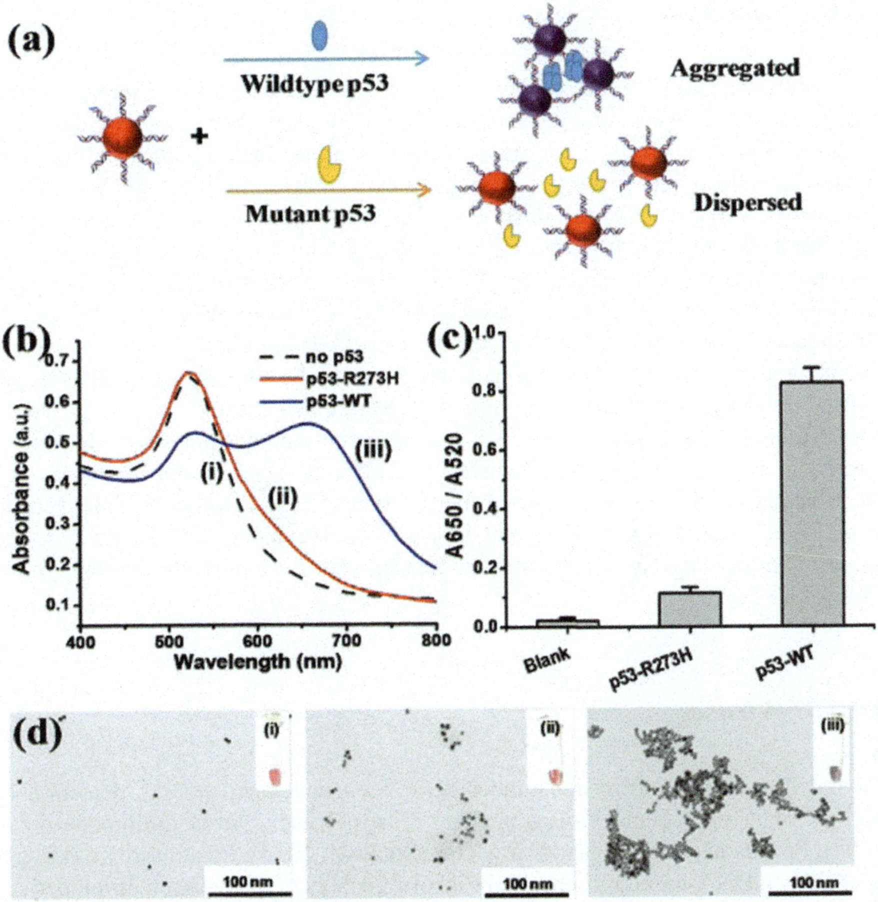

Fig. 10.4 Representative example of colorimetric detection by using Au NPs. **a** Schematic diagram showing working principle of the colorimetric assay. **b** UV-vis absorption spectra of RE-AuNP probes (i) absence of p53 protein (black dashed line), (ii) wildtype p53 protein (p53-WT) (blue line), and (iii) mutant p53 protein (p53-R273H) (red line), **c** aggregation ratio, **d** TEM micrographs of samples (i) to (iii) and the insets show corresponding assay colour (figures were reproduced from Assah et al. 2018 with the permission of Elsevier Science)

for the specific nucleotide sequences of *P. infestans* (Chang et al. 2019). They have used AuNPs as optical probes and CRISPR/Cas9-triggered isothermal amplification. Cas9/sgRNA complex has broken the recognized target DNA and used for isothermal amplification. The oligonucleotide functionalized AuNPs were hybridized with a specific amplified product, which caused aggregation of AuNPs and produces colour change from red to purple.

10.6.2 Bacteria

Plant diseases caused by bacteria can be broadly classified into four major groups, based on plant tissue damage and the symptoms that they cause, which may comprise vascular wilt, soft rot, necrosis, and tumour. Bacteria are invading plant tissues by wounds produced by adverse weather, human, insects, and nematodes. The natural ways are also responsible for bacterial infection such as natural openings such as stomata, hydathodes, lenticels, and nectar-producing glands. The bacterial plant diseases caused by *Pseudomonas, Agrobacterium, Xanthomonas, Erwinia, Proteobacteria,* etc. are responsible for a big loss in agricultural yields. Nowadays, colorimetric, or visible detection methods are popular due to advancements over conventional detection methods. Besides the AuNPs aggregation mechanism, bridging flocculation is a very well-recognized strategy in colloid chemistry (Ruehrwein and Ward 1952). Several detection methods have been developed based on aggregation mechanism using an antibody, DNA probe modified- and electrostatic-mediated aggregation of nanoparticles (Li and Rothberg 2004; Mirkin et al. 1996; Xu et al. 2009). The rapid detection of plant pathogen has been reported based on reversible adsorption to differentiate between long and short DNA polymers (Wee et al. 2015).

10.6.3 Viruses

Lateral flow immunoassay techniques have been developed for the detection of various plant pathogenic viruses such as Citrus Tristeza virus (Salomone et al. 2004), Prunus necrotic ringspot virus (Jarocka et al. 2013), Potato virus x (Drygin et al. 2012). DNA-based biosensors for label-free detection have been developed for rapid and sensitive detection of the pathogen, which offered qualitative and quantitative investigation of pathogens. The colorimetric nucleic acid detection has been developed using the AuNPs-labelled DNA probe, which is simple, rapid, and high sensitivity (Khater et al. 2017). Unmodified AuNPs based probes have been used to detect RT-PCR amplified Cucumber green mottle mosaic virus (Wang et al. 2017). The limit of detection (LOD) 30 pg/μL of CGMMV RNA was reported in this study. Dharanivasan and coworkers reported AuNPs conjugated bifunctional oligo probe for the colorimetric detection of PCR amplified DNA of Tomato leaf curl New Delhi virus (Dharanivasan et al. 2016). The LOD of this assay was about ~7.2 ng dsDNA target, has potential for the development of ultra-sensitive nanoassay method. The colorimetric detection method has been developed, based on LSPR activity of AuNPs for analysis of unamplified *Tomato yellow leaf curl virus* (TYLCV) genome in infected plants (Razmi et al. 2019). In this method, the AuNP-DNA probe hybridized with specific complementary targets and caused the colour change of assay due to LSPR, which can measure using UV-visible spectroscopy. This technique provides advancement over the conventional detection method by eliminating the used of

PCR amplification and detection equipment. Similarly, the Maize chlorotic mottle virus is a serious pathogen that can be transmitted via infected maize seeds. So, to prevent its further spread a rapid and sensitive assay was needed, and to achieve this Liu and colleagues explored the use of unmodified gold nanoparticles (AuNPs) to visual detection (Liu et al. 2015). In their study, the interaction between the species-specific probes and RT-PCR target products changed the colour of dispersed AuNPs from red to grey blue. The LOD of their assay was as low as 30 pg μL^{-1} of viral RNA.

10.7 Conclusion and Prospects

In this chapter, the colorimetric detection of plant pathogens using AuNPs have been described. A variety of surface characteristics of AuNPs have been exploited for the colorimetric detection of pathogens either by sensing nucleic acids, surface proteins, or whole pathogens. The surface modification of AuNPs is essential in respect of selective detection of the target. The most adopted nanoparticle surface modification is the covalent approach. The AuNPs are functionalized with the molecule comprising sulfhydryl groups, having interaction between Au and S groups. The mechanism behind colorimetric detection is the sensitivity of AuNPs LSPR towards the refractive index of the surrounding environment. The LSPR depends on the size and shape of nanoparticles, and agglomeration cause sufficient signal for the detection of targets. This technique is simple, cost-effective, and rapid for the detection of plant pathogens such as fungi, bacteria, and viruses. In the future, this colorimetric method might be used in field detection of plant pathogens and a promising technique for clinical diagnosis of human and animal pathogens.

References

Agarwal M et al (2016) Catalytic degradation of methylene blue by biosynthesised copper nanoflowers using F. benghalensis leaf extract. IET Nanobiotechnol 10:321–325

Agrios GN (2005) Plant pathology. Elsevier Academic Press, Amsterdam, The Netherlands

Assah E et al (2018) Rapid colorimetric detection of p 53 protein function using DNA-gold nanoconjugates with applications for drug discovery and cancer diagnostics. Colloids Surf, B 169:214–221

Bhamore J, Rawat KA, Basu H, Singhal RK, Kailasa SK (2015) Influence of molecular assembly and NaCl concentration on gold nanoparticles for colorimetric detection of cysteine and glutathione. Sens Actuators B: Chem 212:526–535

Brasileiro BTRV, Coimbra MRM, Morais Jr MAd, Oliveira NTd (2004) Genetic variability within Fusarium solani specie as revealed by PCR-fingerprinting based on PCR markers. Braz J Microbiol 35:205–210

Carrasco-Ballesteros S, Castillo P, Adams BJ, Pérez-Artés E (2007) Identification of Pratylenchus thornei, the cereal and legume root-lesion nematode, based on SCAR-PCR and satellite DNA. Eur J Plant Pathol 118:115–125

Chang W, Liu W, Liu Y, Zhan F, Chen H, Lei H, Liu Y (2019) Colorimetric detection of nucleic acid sequences in plant pathogens based on CRISPR/Cas9 triggered signal amplification. Microchim Acta 186:243

Daniel M-C, Astruc D (2004) Gold nanoparticles: assembly, supramolecular chemistry, quantum-size-related properties, and applications toward biology, catalysis, and nanotechnology. Chem Rev 104:293–346

Das SK, Das AR, Guha AK (2010) Microbial synthesis of multishaped gold nanostructures. Small 6:1012–1021

Dharanivasan G, Riyaz SM, Jesse DMI, Muthuramalingam TR, Rajendran G, Kathiravan K (2016) DNA templated self-assembly of gold nanoparticle clusters in the colorimetric detection of plant viral DNA using a gold nanoparticle conjugated bifunctional oligonucleotide probe. RSC Adv 6:11773–11785

Drygin YF et al (2012) Highly sensitive field test lateral flow immunodiagnostics of PVX infection. Appl Microbiol Biotechnol 93:179–189

El-Brolossy T, Abdallah T, Mohamed MB, Abdallah S, Easawi K, Negm S, Talaat H (2008) Shape and size dependence of the surface plasmon resonance of gold nanoparticles studied by Photoacoustic technique. Eur Phys J Special Topics 153:361–364

Fang Y, Ramasamy RP (2015) Current and prospective methods for plant disease detection. Biosensors 5:537–561

Fox A, Mumford R (2017) Plant viruses and viroids in the United Kingdom: an analysis of first detections and novel discoveries from 1980 to 2014. Virus Res 241:10–18

Gukowsky JC, Tan C, Han Z, He L (2018) Cysteamine-Modified Gold Nanoparticles as a Colorimetric Sensor for the Rapid Detection of Gentamicin. J Food Sci 83:1631–1638

Gupta RK, Srinivasan M, Dharmarajan R (2012) Synthesis of 16-Mercaptohexadecanoic acid capped gold nanoparticles and their immobilization on a substrate. Mater Lett 67:315–319

Horsfall JG (2012) Plant disease: an advanced treatise: how pathogens induce disease. Elsevier

Hu T, Lu S, Chen C, Sun J, Yang X (2017) Colorimetric sandwich immunosensor for Aβ (1–42) based on dual antibody-modified gold nanoparticles. Sens Actuators B: Chem 243:792–799

Jain A, Sarsaiya S, Wu Q, Lu Y, Shi J (2019) A review of plant leaf fungal diseases and its environment speciation. Bioengineered 10:409–424

Jarocka U, Radecka H, Malinowski T, Michalczuk L, Radecki J (2013) Detection of Prunus Necrotic Ringspot Virus in plant extracts with impedimetric immunosensor based on glassy carbon electrode. Electroanalysis 25:433–438

Khater M, de la Escosura-Muñiz A, Merkoçi A (2017) Biosensors for plant pathogen detection. Biosens Bioelectron 93:72–86

Khoodoo M, Sahin F, Jaufeerally-Fakim Y (2005) Sensitive detection of Xanthomonas axonopodis pv. dieffenbachiae on Anthurium andreanum by immunocapture-PCR (IC-PCR) using primers designed from sequence characterized amplified regions (SCAR) of the blight pathogen. Eur J Plant Pathol 112:379–390

Li H, Rothberg L (2004) Colorimetric detection of DNA sequences based on electrostatic interactions with unmodified gold nanoparticles. Proc Natl Acad Sci 101:14036–14039

Liu S, Lämmerhofer M (2019) Functionalized gold nanoparticles for sample preparation: a review. Electrophoresis 40:2438–2461

Liu Z, Xia X, Yang C, Wang L (2015) Visual detection of Maize chlorotic mottle virus using unmodified gold nanoparticles. RSC Adv 5:100891–100897

Liu G et al (2016) Colorimetric sensing of atrazine in rice samples using cysteamine functionalized gold nanoparticles after solid phase extraction. Anal Methods 8:52–56

Liu C, Zhang J, Zhang X, Zhao L, Li S (2018) Enantiomeric separation of adrenaline, noradrenaline, and isoprenaline by capillary electrophoresis using streptomycin-modified gold nanoparticles. Microchim Acta 185:227

Mahajan R, Bhadwal AS, Kumar N, Madhusudanan M, Pudake RN, Tripathi RM (2016) Green synthesis of highly stable carbon nanodots and their photocatalytic performance. IET Nanobiotechnol 11:360–364

Martín S, Alioto D, Milne R, Guerri J, Moreno P (2002) Detection of Citrus psorosis virus in field trees by direct tissue blot immunoassay in comparison with ELISA, symptomatology, biological indexing and cross-protection tests. Plant Pathol 51:134–141

Martinelli F et al (2015) Advanced methods of plant disease detection: a review. Agron Sustain Dev 35:1–25

Mehrotra N, Tripathi RM (2015) Short interfering RNA therapeutics: nanocarriers, prospects and limitations. IET Nanobiotechnol 9:386–395

Mendes R, Carvalhal R, Stach-Machado D, Kubota L (2009) Surface plasmon resonance immunosensor for early diagnosis of Asian rust on soybean leaves. Biosens Bioelectron 24:2483–2487

Merighi M, Sandrini A, Landini S, Ghini S, Girotti S, Malaguti S, Bazzi C (2000) Chemiluminescent and colorimetric detection of Erwinia amylovora by immunoenzymatic determination of PCR amplicons from plasmid pEA29. Plant Dis 84:49–54

Mirkin CA, Letsinger RL, Mucic RC, Storhoff JJ (1996) A DNA-based method for rationally assembling nanoparticles into macroscopic materials. Nature 382:607–609

Nguyen NLT, Kim EJ, Chang S-K, Park TJ (2016) Sensitive detection of lead ions using sodium thiosulfate and surfactant-capped gold nanoparticles. BioChip Journal 10:65–73

Punja ZK, De Boer S, Sanfaçon H (2007) Biotechnology and plant disease management. Cabi

Raj V, Vijayan AN, Joseph K (2015) Cysteine capped gold nanoparticles for naked eye detection of E. coli bacteria in UTI patients. Sens Bio-sensing Res 5:33–36

Razmi A, Golestanipour A, Nikkhah M, Bagheri A, Shamsbakhsh M, Malekzadeh-Shafaroudi S (2019) Localized surface plasmon resonance biosensing of tomato yellow leaf curl virus. J Virol Methods 267:1–7

Ruehrwein R, Ward D (1952) Mechanism of clay aggregation by polyelectrolytes. Soil Sci 73:485–492

Salomone A, Mongelli M, Roggero P, Boscia D (2004) Reliability of detection of Citrus tristeza virus by an immunochromatographic lateral flow assay in comparison with ELISA. J Plant Pathol 86:43–48

Sanghi R, Verma P (2010) pH dependant fungal proteins in the 'green'synthesis of gold nanoparticles. Adv Mater Lett 1:193–199

Shahrivari S, Faridbod F, Ganjali MR (2018) Highly selective and sensitive colorimetric determination of Cr3+ ion by 4-amino-5-methyl-4H-1, 2, 4-triazole-3-thiol functionalized Au nanoparticles. Spectrochim Acta Part A Mol Biomol Spectrosc 191:189–194

Shojaei TR, Salleh MAM, Sijam K, Rahim RA, Mohsenifar A, Safarnejad R, Tabatabaei M (2016) Fluorometric immunoassay for detecting the plant virus Citrus tristeza using carbon nanoparticles acting as quenchers and antibodies labeled with CdTe quantum dots. Microchim Acta 183:2277–2287

Singh PK, Kundu S (2014) Biosynthesis of gold nanoparticles using bacteria. Proc Nat Acad Sci, India B Biol Sci 84:331–336

Stobiecka M, Coopersmith K, Hepel M (2010) Resonance elastic light scattering (RELS) spectroscopy of fast non-Langmuirian ligand-exchange in glutathione-induced gold nanoparticle assembly. J Colloid Interface Sci 350:168–177

Thakker JN, Dalwadi P, Dhandhukia PC (2012) Biosynthesis of gold nanoparticles using *Fusarium oxysporum* f. sp. *cubense* JT1, a plant pathogenic fungus. ISRN biotechnology. 2013, 515091

Toh SY, Citartan M, Gopinath SC, Tang T-H (2015) Aptamers as a replacement for antibodies in enzyme-linked immunosorbent assay. Biosens Bioelectron 64:392–403

Tolessa T, Tan Z-Q, Liu J-F (2018) Hydride generation coupled with thioglycolic acid coated gold nanoparticles as simple and sensitive headspace colorimetric assay for visual detection of Sb (III). Anal Chimica Acta 1004:67–73

Tripathi R, Chung SJ (2019) Biogenic nanomaterials: Synthesis, characterization, growth mechanism, and biomedical applications. J Microbiol Methods 157:65–80

Tripathi R, Chung SJ (2020) Reclamation of hexavalent chromium using catalytic activity of highly recyclable biogenic Pd (0) nanoparticles. Scientific reports 10:1–14

Tripathi R, Archana S, Shrivastavc B (2012) Biofabrication of gold nanoparticles using leaf extract of Ficus benghalensis and their characterization. Int J Pharma Bio Sci 3:551–558
Tripathi RM, Ranac D, Shrivastav A, Singh RP, Shrivastav BR (2013) Biogenic synthesis of silver nanoparticles using saraca indica leaf extract and evaluation of their antibacterial activity. Nano Biomed Eng 5(1):50–56
Tripathi R, Bhadwal AS, Gupta RK, Singh P, Shrivastav A, Shrivastav B (2014a) ZnO nanoflowers: novel biogenic synthesis and enhanced photocatalytic activity. J Photochem Photobiol, B 141:288–295
Tripathi R, Gupta RK, Singh P, Bhadwal AS, Shrivastav A, Kumar N, Shrivastav B (2014b) Ultra-sensitive detection of mercury (II) ions in water sample using gold nanoparticles synthesized by Trichoderma harzianum and their mechanistic approach. Sens Actuators B Chem 204:637–646
Tripathi R, Kumar N, Bhadwal AS, Gupta RK, Shrivastav B, Shrivastav A (2015a) Facile and rapid biomimetic approach for synthesis of HAp nanofibers and evaluation of their photocatalytic activity. Mater Lett 140:64–67
Tripathi RM, Gupta RK, Bhadwal AS, Singh P, Shrivastav A, Shrivastav B (2015b) Fungal biomolecules assisted biosynthesis of Au–Ag alloy nanoparticles and evaluation of their catalytic property. IET Nanobiotechnol 9:178–183
Tripathi R, Rao RP, Tsuzuki T (2018a) Green synthesis of sulfur nanoparticles and evaluation of their catalytic detoxification of hexavalent chromium in water. RSC advances 8:36345–36352
Tripathi RM, Shrivastav BR, Shrivastav A (2018b) Antibacterial and catalytic activity of biogenic gold nanoparticles synthesised by *Trichoderma harzianum*. IET Nanobiotechnol 12:509–513
Tripathi R et al (2019a) Metal-induced redshift of optical spectra of gold nanoparticles: An instant, sensitive, and selective visual detection of lead ions. Int Biodeterior Biodegradation 144:104740
Tripathi R, Yoon S-Y, Ahn D, Chung SJ (2019b) Facile synthesis of triangular and hexagonal anionic gold nanoparticles and evaluation of their cytotoxicity. Nanomaterials 9:1774
Verma MS, Rogowski JL, Jones L, Gu FX (2015) Colorimetric biosensing of pathogens using gold nanoparticles. Biotechnol Adv 33:666–680
Vincelli P, Tisserat N (2008) Nucleic acid–based pathogen detection in applied plant pathology. Plant Dis 92:660–669
Wang C, Yu C (2013) Detection of chemical pollutants in water using gold nanoparticles as sensors: a review. Rev Anal Chem 32:1–14
Wang L, Liu Z, Xia X, Yang C, Huang J, Wan S (2017) Colorimetric detection of Cucumber green mottle mosaic virus using unmodified gold nanoparticles as colorimetric probes. J Virol Methods 243:113–119
Ward E, Foster SJ, Fraaije BA, Mccartney HA (2004) Plant pathogen diagnostics: immunological and nucleic acid-based approaches. Ann Appl Biol 145:1–16
Wee E, Lau H, Botella J, Trau M (2015) Re-purposing bridging flocculation for on-site, rapid, qualitative DNA detection in resource-poor settings. Chem Commun 51:5828–5831
Xia N, Zhou B, Huang N, Jiang M, Zhang J, Liu L (2016) Visual and fluorescent assays for selective detection of beta-amyloid oligomers based on the inner filter effect of gold nanoparticles on the fluorescence of CdTe quantum dots. Biosens Bioelectron 85:625–632
Xu X, Wang J, Yang F, Jiao K, Yang X (2009) Label-free colorimetric detection of small molecules utilizing DNA oligonucleotides and silver nanoparticles. Small 5:2669–2672
Xu L, Dong S, Hao J, Cui J, Hoffmann H (2017) Surfactant-modified ultrafine gold nanoparticles with magnetic responsiveness for reversible convergence and release of biomacromolecules. Langmuir 33:3047–3055
Yang L et al (2016) Highly selective and sensitive electrochemiluminescence biosensor for p 53 DNA sequence based on nicking endonuclease assisted target recycling and hyperbranched rolling circle amplification. Anal Chem 88:5097–5103
Zhang J, Xu X, Yang X (2012) Highly specific colorimetric recognition and sensing of sulfide with glutathione-modified gold nanoparticle probe based on an anion-for-molecule ligand exchange reaction. Analyst 137:1556–1558
Zhang Y et al (2016) Colorimetric sensor for cysteine in human urine based on novel gold nanoparticles. Talanta 161:520–527

Chapter 11
Advancements in Biosensors for Fungal Pathogen Detection in Plants

Utkarsh Jain, Ramesh Namdeo Pudake, Nidhi Chauhan, and Sakshi Pareek

Abstract Globally, the pathogenic diseases caused by fungi in plants are the major constraints for healthy harvesting and plantation. Therefore, to lessen these constraints, it is required to monitor these fungal pathogens at an early phase. Currently, ample methods to identify, detect, and quantify various fungal pathogens in the infected plant incorporate ELISA, PCR, RT-PCR are available. But these have their own limitations, for instance, lack of sensitivity, time-consuming, expensive, labor-intensive, requires highly equipped instruments, and skilled personnel for their handling. Thus, there is a strong demand for fabricating biosensing methods at early stage detection with great sensitivity and specificity for different plant diseases. In context, this chapter provides an outline of biosensors based on various bio-recognition elements, like DNA, antibodies, and enzymes. They are emerging as a powerful tool for the initial stage monitoring of fungal disease because of their beni-ficial properties like specificity, sensitivity, rapidness, cost-effectiveness, easy to use when compared to the available traditional methods. Thus, biosensors are considered to be a reliable and promising tool for the diagnosis of these pathogens and increasing the quantity as well as the quality of crop yield.

Keywords Fungi · Biosensors · Plant diseases · DNA-based biosensor · Antibody-based biosensor

11.1 Introduction

Diseases in plants occur due to the multiplex interaction of plant host, pathogen, and the environmental conditions, causing the privation in yields as well as harming the health of humans. Fungi, viruses, parasite plants, bacteria, and nematodes are the causes of various plant diseases. The diversity in the variety of pathogenic interactions between fungi and plants considered by the formation of a parasitic relation of about 400 million years back (Kumar et al. 2016). There is a vital loss of production

U. Jain (✉) · R. N. Pudake · N. Chauhan · S. Pareek
Amity Institute of Nanotechnology, Amity University Uttar Pradesh, 201303 Noida, UP, India
e-mail: ujain@amity.edu

R. N. Pudake et al. (eds.), *Biosensors in Agriculture: Recent Trends and Future Perspectives*, Concepts and Strategies in Plant Sciences,
https://doi.org/10.1007/978-3-030-66165-6_11

and revenue in the agricultural and forestry fields due to the diseases. For instance, soybean rust, a fungal infection causes a huge economical loss and by preventing the infection only by 20%, there may be a profit of 11 million dollars (Sankaran et al. 2010). Fungal plant pathogens are a great economical threat and it is evaluated that agriculture production globally bears an average loss of 15% annually (Neeraja et al. 2010). With the developing results of environmental change, the loss is anticipated to increase (Lo Presti et al. 2015). Many fungal diseases have come up from small to large importance and were discovered recently. More than 40 fungi are causing major disease problems globally (Grau et al. 2004).

Plants create a significant environment for microorganisms as they provide an extensive range of domains for growth, including the aerial plant part, i.e., phyllosphere, the root system i.e., rhizosphere, and the inner transport system i.e., endosphere (Ferreira et al. 2006). Plant–fungus correlation described as vulnerable if the fungus process characteristic effects and regenerates; or as resistant if a plant can counter the disease to limit side effects of pathogen proliferation. Understanding the mechanism of pathogenicity and the qualities that control them may permit the utilization of recombinant DNA innovation to adjust the plant's defense system to encourage resistance to the attack. The characterization of fungal pathogenic genes is a demanding area in plant pathology at present (Martínez-Medina et al. 2019).

The life cycle of the fungal pathogen is complex, together with sexual and asexual reproduction and development of various vegetative and reproductive forms during infection stages. The essential stages of infection development are colonization, growth and propagation, dissemination, and endurance of the pathogen without the involvement of the host. Although, reaching every stage varies from pathogen to pathogen (González-Fernández et al. 2010).

Fungal plant pathogen harms the plants in the vegetative phase as well as in the post-harvest infection form (Fig. 11.1). The main plant pathogen for economic aspects are fungi which include *Botrytis cinerea*, and species of *Fusarium, Magnaporthe, and Rhizoctonia* (Hua et al. 2018; Sharon and Shlezinger 2013). Ascomycete genus, Cercospora members, Mycosphaerellaceae, Dothideomycetes, Capnodiales appear universally causing leaf spots on dicot and plant families of monocot, ferns, and gymnosperm ranking in the main cause of plant pathogens destruction (Cai et al. 2011). Wheat harvests in middle age were usually ruined when the grains got infected with spores of fungus known as bunt or stinking smut which was a dark and dusty powder. During the 1870s, a pandemic of downy mildew, which occurred by the organism *Plasmopara viticola*, struck the grape vineyards of central Europe, effecting grape cultivators and wine producers. In the United States itself, a huge number of bushels of wheat have been lost in epidemic for a very long time to stem rust (Ellis et al. 2008).

Hence, the accessibility of rapid, sensitive, and precise techniques for the identification and detection of fungi is increasingly important to improve the decision-making process for infection control. The traditional methods used for the identification of plant pathogens depend on culture-based structural methodologies. But these methods are time utilizing, difficult, and require broad information of traditional groupings. Even these methods have other disadvantages including problems

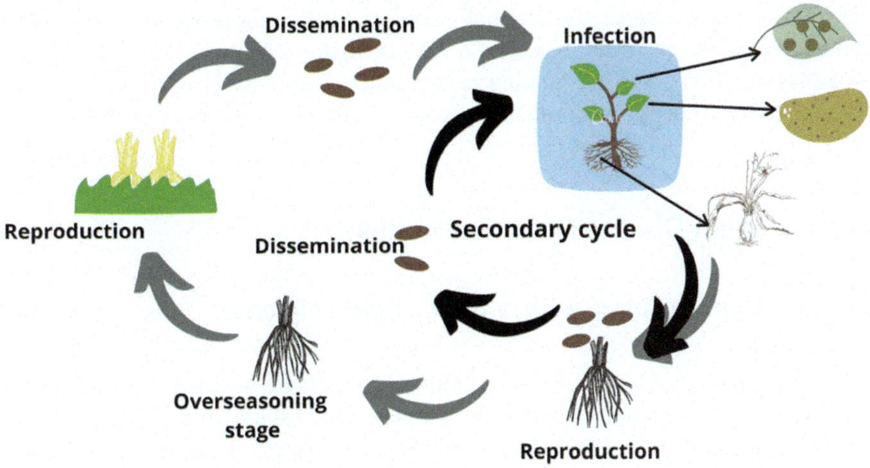

Fig. 11.1 Fungi infection cycle in plants

to in vitro culture certain species and the failure to precisely measure the pathogen (Capote et al. 2012) (Tsui et al. 2011). New innovations, for example, nanoscale platforms, biosensors, miniaturization of detection devices, and nanoscale play a major role in pathogen detection and management (Abd-Elsalam 2013). The requirement for rapid, on-line, and precise detecting opens up doors for biosensors in a wide range of agricultural fields in situ investigation of contaminations in yields and soils, recognition and distinguishing proof of irresistible illnesses in harvests and domesticated animals, on-line estimations of significant food handling boundaries, checking creature ripeness and screening restorative medications in veterinary testing (Velasco-Garcia and Mottram 2003).

By definition, a biosensor is a self-contained combined system, which recognizes an analyte of biological element and quantitatively by utilizing a bioreceptor component (called a biorecognition component) and a sign transduction component. The analytes of benefits can be biological, such as DNA, protein, cell. When a bioreceptor component perceives a particular analyte and binds, a transducer component gives quantitative or semi-quantitative data about the objective by changing over the physicochemical signal produced during the biorecognition into a quantifiable electrical signal (Ali et al. 2017; Chen and Shamsi 2017; Turner 2013). The term biosensor has been connected to numerous innovations. They can be both subjective or quantitative, encrypted, or solid physical devices, giving concurrent time, or time-fractioned information, in vivo or ex vivo, and of numerous different structures (Sadanandom and Napier 2010). Fundamentally, a biosensor couples to a recognition element or a chemical response by a biological element with a transducer, which then converts this response into a readable form (Singh and Verma 2019). Biosensors are getting significant in a wide scope of analysis. Diverse biosensor designs have been created for single target analytes and wide range monitoring (Tothill 2001). The materials utilized in biosensors are distinguished into three categories depending on their mechanisms: biocatalytic consisting of enzymes, bio affinity group, with antibodies and nucleic acids, and microbe-based microorganisms (Mehrotra 2016; Rodriguez-Mozaz et al. 2006). The outstanding performance factors for biosensors enhancing are sensitivity, limit of detection, and specificity (Bhatt and Bhattacharya 2019).

11.2 Fungal Disease Detection in Plants

11.2.1 Conventional Assays for Fungal Detection

Few fungal pathogens have various symptoms which are noticeable to the normal eye, or at less amplification. These symptoms include stain, withering, and fissure on the fungal pathogens. For example, soybean seeds decay is caused by *Phomopsis* spp. as well as *Cercospora kikuchii* and the spots on peanut seeds that are infected by *Cylindrocladium parasiticum* (Mancini et al. 2016).

Also, incubation methods are often utilized for pathogens, particularly for high-frequency fungi that can happen on seed tests at a level >1%, additionally giving data about the practicality of the inoculum on the seeds (Mancini et al. 2016). Despite being inexpensive and convenient in use, the precision and reliability of these techniques rely upon the experience and skill of the individual doing it (McCartney et al. 2003). Also, these conventional methods require much time as it takes days and weeks for the verification, making them not suitable for instance and fast detection (Ray et al. 2017). These traditional methods for the identification of plant pathogens are dependent on culture-based methods. The main drawbacks of these approaches were the ability of the organism to be cultured, contamination prone, requiring more time and arduous nature, and the necessity for broad taxonomical information (Atkins and Clark 2004; Lievens and Thomma 2005) (Table 11.1). Thus, there is an urge in the search for some other diagnostic approaches (Atkins and Clark 2004).

Table 11.1 Some examples of fungal disease identification using different conventional techniques

Plant/Tree	Pathogen	Technique	Reference
Soybean	*Phhakopsora pachayrhizi*	IF	(Baysal-Gurel et al. 2008)
Timber	*Serpula lacrymans and Coniophora puteana*	GC–MS	(Ewen et al. 2004)
Tomato	*Lycopersicon esculentum*	RT-PCR	(Lievens et al. 2006)
Apple	*Ventura ineaequalis*	Hyperspectral	(Delalieux et al. 2007)
Onion	*Sclerotium cepivorum*	PCR	(Haq et al. 2003)

11.2.2 Molecular Assays for Fungal Detection

Molecular Assay is classified into two categories—direct and indirect detection methods. Various direct traditional assays for detection post-harvest fungi involve polymerase chain reaction (PCR), real-time PCR (RT-PCR), ELISA, and magnetic capture hybridization (MCH-PCR, which is accurate and sensitive. Having complex protocols with more cost, these methods are less used. Moreover, they detect the fungal infection in crops, but cannot evaluate their pathogenicity state (Kumar et al. 2016; Zamir et al. 2020). Also, a plant can be infected by different other pathogens, therefore PCR and DNA arrays are used for multiplexing. Although the diagnostic methods comprising molecular tools are reduced by removing cultured cultivation, still this process of molecular validation of target genes is time utilizing and expensive (Kumar et al. 2016). Although PCR is considered to be a sensitive important method that helps with problems related to control, identify, and containment in plants (Schena et al. 2004) but recurrent use of PCR inhibitors in the soil or plant tissues, lessening the sensitivity of reactions resulting in false results. This false result even occurs when DNA or PCR products toxin are present (Narayanasamy 2011; Schena et al. 2004). Lately, quantitative PCR (qPCR) identification techniques have been broadly used to identify phytopathogenic parasites and have incredibly added to the progression of information on plant pathology. However, significant errors and limitations, are still there. Errors at the whole procedure for the advancement of qPCR strategies, at the degree of choice of proper DNA extraction and purification method, distinguishing suitable target regions, decision of chemistry, design, and approval of particular primers and probes, choice of an absolute quantification method, and analysis of the risk of identifying target DNA from dead sources (Schena et al. 2013). Fast detection assays emerged as an easy and effective qualitative test. In spite of their great specificity, the need for reproducibility, reliability, and sensitivity are regular constraints, also frequently false-positive outcomes have appeared and are related to cross-responses among antibody and fungal metabolites (Oliveira et al. 2019). Indirect strategies are basically imaging methods, they are sensitive and very much utilized for huge information data, hyperspectral imaging is applied the most for ailment location in plants (Madufor et al. 2017).

11.3 Biosensors for Fungi Detection in Plants

11.3.1 DNA-Based Biosensors

The DNA-based biosensors consists of hybridization, amplification, and recombination on the double helix DNA structure. Nucleic acid hybridization is the basic principle of DNA biosensors. Arrangement hybridization is faster than hybridization on strong support, yet, except if the measure is a homogeneous test, a separation step is required before the final identification (Zhai et al. 1997) (Fig. 11.2). DNA

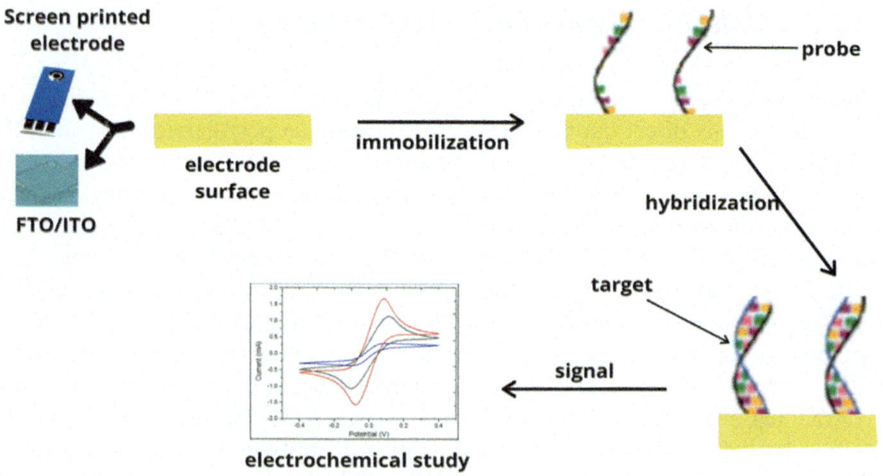

Fig. 11.2 Scheme for DNA based biosensors

biosensors are promising in rapid, simple, and cost-effective detection of the target molecules (Chao et al. 2016).

Basal Stem Rot and Upper Stem Rot infections are caused due to pathogenic fungus *Ganoderma boninense and are* a major threat to the palm oil industry. Therefore, an automatic high DNA output extraction method using a microfluidics device for disposable and rapid, without label detection was done within 2 hrs. It was examined using current and potential measurements, UV visible, FTIR. With the help of PCR, results were compared between extracted and commercial ssDNA of *G. boninense* and appeared differently in relation to unknown fungus, demonstrating an effective DNA extraction convention through microfluidics (Ayoib et al. 2020).

Alternaria panax Whetz is one of the most destructive diseases in ginseng and may lead to major economic loss in ginseng production. Therefore, a fast and precise atomic symptomatic technique, a single-tube nested PCR-lateral flow biosensor assay (STNPCR-LFBA) was fabricated to detect the pathogen. The sensitivity of STNPCR-LFBA was compared with conventional PCR-LFBA, and it was found to be more effective. Other than that, the PCR sample was checked by lateral flow-based biosensor assay, *which gave* the migration of the detection technology to a point-of-care test (POCT) format. STNPCR-LFBA was specific to *A. panax* Whetz, and no cross-response was seen in any other non-targeted samples with the limit of detection 0.01 pg. STNPCR-LFBA could also be used for particular identification in real samples (Wei et al. 2018).

Phytophthora infestans is the causal agent for late blight in potatoes and tomatoes, with a global impact on agriculture. A visual detection method for *P. infestans* has been designed by combining unique primer mediated asymmetric PCR with gold nanoparticle modified lateral flow biosensor. Quantitative analysis reported the optical intensity of the red band with a detection limit of 0.1 pg μL^{-1}. The specificity of this biosensor was measured by three different *Phytophthora* species as well as two pathogenic fungi. It was assumed that this method was possible for early stage prediction of potato late blight disease and prompting of the administrator's activities to lessen the danger of scourge improvement (Zhan et al. 2018).

11.3.2 Antibody-Based Biosensors

The ways to deal with direct-acting immunosensors are the utilization of their antibodies simply to give specific binding of analyte to a surface and property of that surface detected by the transducing component (North 1985) (Fig. 11.3). Few studies to detect the plant pathogens using the antibody-based sensors are reviewed below.

An artificial cellulose membrane, particularly deciding if a phytopathogenic pythium is present in an aqueous mixture of Pythium (phytopathogenic and non-pathogenic) was proposed in a recent study (Yamaguchi et al. 2017). Dark Sigatoka is an infection that occurs in banana cultivation globally and is occurred by the hemibiotrophic fungus *Pseudocercospora fijiensis*. An exceptionally sensitive SPR immunosensor was fabricated to recognize *P. fijiensis* in real samples of leaf separates in the beginning phase of the infection. A polyclonal antibody produced against HF1 was covalently immobilized on a gold-covered chip by mixing self-assembled monolayer (SAM) of alkanethiols with EDC/NHS method. The immunosensor obtained a detection limit of 11.7 μg mL^{-1}, with sensitivity of 0.0021 units of reflectance per ng mL^{-1} and linear range from 39.1 to 122 μg mL^{-1} (Luna-Moreno et al. 2019). *Pectobacterium atrosepticum* (Pba) is an isolated and terrifying phytopathogen known agent of blackleg and soft rot infection in potatoes in many areas. Current

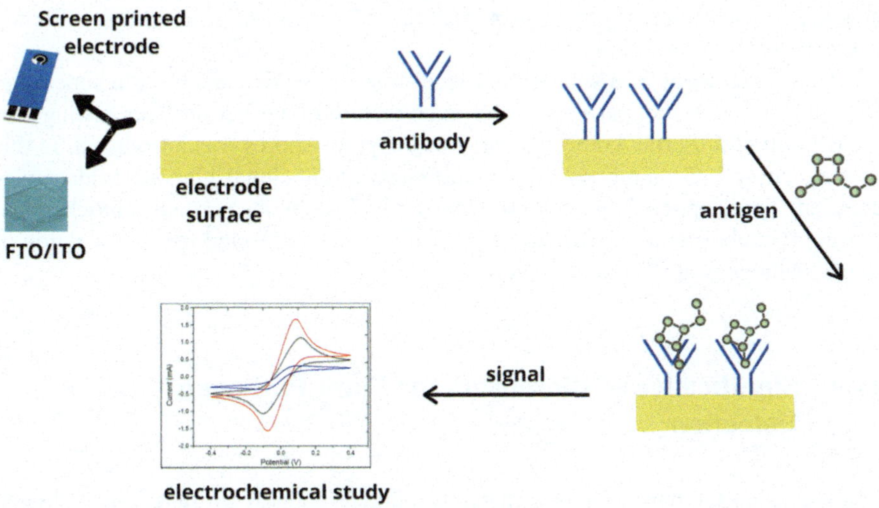

Fig. 11.3 Scheme for antibody-based biosensors

Table 11.2 Various biosensors for the detection of fungal pathogens

Biosensor	Pathogen	Detection limit	Reference
Electrochemcial	*Ganoderma boninense*	NA	(Ayoib et al. 2020)
Electrochemcial	*Alternaria panax*	0.01 pg	(Wei et al. 2018)
Electrochemcial	*Phytophthora* species	0.1 pg μL^{-1}	(Zhan et al. 2018)
Electrochemcial	Phytopathogenicity	NA*	(Yamaguchi et al. 2017)
Immunosensor	*Pseudocercospora fijiens*	1.7 $\mu g\ mL^{-1}$	(Luna-Moreno et al. 2019)
Immunosensor	*Pectobacterium atrosepticum*	10^4 CFU mL^{-1}	(Hashemi Tameh et al. 2020)
Impedimetric	*Sclerotinia sclerotiorum*	7.8×10^4 ascospores mL^{-1}	(Shoute et al. 2018)
Optical	*Aspergillus niger*	NA*	(Gambhir et al. 2018)

**NA* Not available

conventional techniques, for example, ELISA and PCR are utilized for Pba identification; however, they are costly and time-consuming. Therefore, an electrochemical impedance immunosensor was reported with a microfluidic module and microelectrodes array with the various advantages of cost-efficiency, usability, and portability (Hashemi Tameh et al. 2020). Sclerotinia stem decay, occurred by parasitic pathogen *Sclerotinia sclerotiorum*, is a dangerous disease of canola and numerous other broadleaf yields. Hence, advancement in impedimetric non-faradaic biosensor dependent on hostile to *S. sclerotiorum* polyclonal antibodies as tests to specifically catch the ascospores and sense their attachments by an impedance-based interdigitated electrodes which were found in straightforward and unambiguously relate the quantity of ascospores on the sensor surface with the impedance reaction (Shoute et al. 2018).

The advancement of a smaller and label-free sensor dependent on a superficial level adjustment of copper vapor laser manufactured long-period fiber gratings for recognition of airborne *Aspergillus niger* contagious spores was introduced. In this optical sensor IgG1 antibody with immobilized monoclonal glucose oxidase was detected. Thus, optical sensors have solved many problems related to biochemical sensing (Gambhir et al. 2018). Some of the biosensors reported for the detection of fungal pathogens are listed in Table 11.2.

11.4 Significance of Biosensors for Plant Pathogen Detection

In this chapter we have reviewed the conventional methods for detection of fungal diseases caused, for instance, PCR, ELISA, immunoassays with their limitations to monitor various fungal diseases. In addition, these methods do not provide real-time detection making them less suitable for use. Thus, biosensors usage for these fungal pathogens has gained importance as they help in increasing crop quality and economic improvement. Early stage detection has gained individual importance in the detection of fungi. With these biosensors potential to detect various fungal pathogens enhancing the use of antibodies and DNA has been comprehensively reviewed with the present approaches' comparison.

11.5 Conclusion

The chapter outlines the early detection of various diseases caused by fungal pathogens. The effect of these fungi harms the yields and economic largely. Therefore, early monitoring plays an important role in disease management, leading to lessen the harm that is caused by these pathogens worldwide. To estimate and control the solemnity of fungal plant infections, it is required to design and fabricate fast, precise, futuristic, and progressive detection approaches for these diseases. The current strategies are not potentially abundant for detecting fungal pathogens as these conventional detection techniques are time-consuming, related to special equipment, and require skilled professionals with some experience. The main aspect of developing advanced methods is high sensitivity. To overcome all these difficulties, recent increase and enhancement in micro and nanotechnology have opened paths for the development of biosensors for determination of fungi infections. The use of biosensor with additional efforts has gained significant importance in the development of highly sensitive and selective tools for real-time monitoring of plant diseases in different field conditions and infections. With these more than one fungal disease in infected plant can be diagnosed reducing risk of false results. DNA-based biosensors and antibody-based biosensors discussed here have gained much importance due to their speed, sensitivity of even detecting samples at small amounts, and help in securing the quality of plants. Moreover, biosensors for enormous fungus disease screening assessment has paid attention to several aspects such as cost, accessibility of on-site detection and pre- and post-test of risk caused by infection. Lastly, we conclude that biosensing platform for these fungus pathogens provides an essential role in overcoming the limitations of the present methods and enhancing crop harvesting.

References

Abd-Elsalam KA (2013) Nano platforms for plant pathogenic fungi management. App Phys A 100:829–834

Ali J, Najeeb J, Ali MA, Aslam MF, Raza A (2017) Biosensors: their fundamentals, designs, types and most recent impactful applications: a review. J Biosens Bioelectron 8:1–9

Atkins SD, Clark IM (2004) Fungal molecular diagnostics: a mini review. J Appl Genet 45:3–15

Ayoib A, Hashim U, Gopinath SC (2020) Automated, high-throughput DNA extraction protocol for disposable label free, microfluidics integrating DNA biosensor for oil palm pathogen, *Ganoderma boninense*. Process Biochem 92:447–456

Baysal-Gurel F et al (2008) An immunofluorescence assay to detect urediniospores of *Phakopsora pachyrhizi*. Plant Dis 92:1387–1393

Bhatt G, Bhattacharya S (2019) Biosensors on chip: a critical review from an aspect of micro/nanoscales. Journal of Micromanufacturing 2:198–219

Cai L, Giraud T, Zhang N, Begerow D, Cai G, Shivas RG (2011) The evolution of species concepts and species recognition criteria in plant pathogenic fungi. Fungal Divers 50:121

Capote N, Pastrana AM, Aguado A, Sánchez-Torres P (2012) Molecular tools for detection of plant pathogenic fungi and fungicide resistance. Plant pathology: 151–202

Chao J, Zhu D, Zhang Y, Wang L, Fan C (2016) DNA nanotechnology-enabled biosensors. Biosens Bioelectron 76:68–79

Chen S, Shamsi MH (2017) Biosensors-on-chip: a topical review. J Micromech Microeng 27:083001

Delalieux S, Van Aardt J, Keulemans W, Schrevens E, Coppin P (2007) Detection of biotic stress (Venturia inaequalis) in apple trees using hyperspectral data: non-parametric statistical approaches and physiological implications. Eur J Agron 27:130–143

Ellis SD, Boehm MJ, Mitchell TK (2008) Fungal and fungal-like diseases of plants. Fact Sheet (PP401 07) Agriculture and Natural Resources, The Ohio State University

Ewen RJ, Jones PR, Ratcliffe NM, Spencer-Phillips PT (2004) Identification by gas chromatography-mass spectrometry of the volatile organic compounds emitted from the wood-rotting fungi *Serpula lacrymans* and *Coniophora puteana*, and from *Pinus sylvestris* timber. Mycol Res 108:806–814

Ferreira RB et al (2006) Fungal pathogens: the battle for plant infection. Crit Rev Plant Sci 25:505–524

Gambhir M, Gupta S, John P, Mahakud R, Kumar J, Prakash O (2018) Surface modified long period fiber grating sensor for rapid detection of aspergillus niger fungal spores. Fiber Integr Opt 37:79–91

González-Fernández R, Prats E, Jorrín-Novo JV (2010) Proteomics of plant pathogenic fungi. BioMed Research International 2010, Article ID 932527

Grau CR, Dorrance AE, Bond J, Russin JS (2004) Fungal diseases. Pages 732–734. In: Soybeans: improvement, production, and uses, 3rd Ed. H. R. Boerma and J. E. Specht, eds. Monogr. 16. Am. Soc. of Agron., Madison, WI

Haq MA, Collin HA, Tomsett AB, Jones MG (2003) Detection of *Sclerotium cepivorum* within onion plants using PCR primers. Physiol Mol Plant Pathol 62:185–189

Hashemi Tameh M, Primiceri E, Chiriacò MS, Poltronieri P, Bahar M, Maruccio G (2020) *Pectobacterium atrosepticum* biosensor for monitoring blackleg and soft rot disease of potato. Biosensors 10:64

Hua C, Zhao J-H, Guo H-S (2018) Trans-kingdom RNA silencing in plant–fungal pathogen interactions. Mol Plant 11:235–244

Kumar P, Akhtar J, Kandan A, Kumar S, Batra R, Dubey S (2016) Advance detection techniques of phytopathogenic fungi: current trends and future perspectives. In: Current trends in plant disease diagnostics and management practices. Springer, pp 265–298

Lievens B, Brouwer M, Vanachter AC, Cammue BP, Thomma BP (2006) Real-time PCR for detection and quantification of fungal and oomycete tomato pathogens in plant and soil samples. Plant Sci 171:155–165

Lievens B, Thomma BP (2005) Recent developments in pathogen detection arrays: implications for fungal plant pathogens and use in practice. Phytopathology 95:1374–1380

Lo Presti L et al (2015) Fungal effectors and plant susceptibility. Annu Rev Plant Biol 66:513–545

Luna-Moreno D, Sánchez-Álvarez A, Islas-Flores I, Canto-Canche B, Carrillo-Pech M, Villarreal-Chiu JF, Rodríguez-Delgado M (2019) Early detection of the fungal banana black sigatoka pathogen Pseudocercospora fijiensis by an SPR immunosensor method. Sensors 19:465

Madufor N, Perold W, Opara U (2017) Detection of plant diseases using biosensors: a review. In: VII International Conference on Managing Quality in Chains (MQUIC2017) and II International Symposium on Ornamentals in 1201, pp 83–90
Mancini V, Murolo S, Romanazzi G (2016) Diagnostic methods for detecting fungal pathogens on vegetable seeds. Plant Pathol 65:691–703
Martínez-Medina A, Pescador L, Terrón-Camero LC, Pozo MJ, Romero-Puertas MC (2019) Nitric oxide in plant–fungal interactions. J Exp Bot 70:4489–4503
McCartney HA, Foster SJ, Fraaije BA, Ward E (2003) Molecular diagnostics for fungal plant pathogens. Pest Management Science: Formerly Pesticide Science 59:129–142
Mehrotra P (2016) Biosensors and their applications–a review. Journal of Oral Biology and Craniofacial Research 6:153–159
Narayanasamy P (2011) Detection of fungal pathogens in plants. In: Microbial plant pathogens-detection and disease diagnosis. Springer, pp 5–199
Neeraja C, Anil K, Purushotham P, Suma K, Sarma P, Moerschbacher BM, Podile AR (2010) Biotechnological approaches to develop bacterial chitinases as a bioshield against fungal diseases of plants. Crit Rev Biotechnol 30:231–241
North JR (1985) Immunosensors: antibody-based biosensors. Trends Biotechnol 3:180–186
Oliveira IS, da Silva Junior AG, de Andrade CAS, Oliveira MDL (2019) Biosensors for early detection of fungi spoilage and toxigenic and mycotoxins in food. Current Opinion in Food Science 29:64–79
Ray M, Ray A, Dash S, Mishra A, Achary KG, Nayak S, Singh S (2017) Fungal disease detection in plants: traditional assays, novel diagnostic techniques and biosensors. Biosens Bioelectron 87:708–723
Rodriguez-Mozaz S, de Alda MJL, Barceló D (2006) Biosensors as useful tools for environmental analysis and monitoring. Anal Bioanal Chem 386:1025–1041
Sadanandom A, Napier RM (2010) Biosensors in plants. Curr Opin Plant Biol 13:736–743
Sankaran S, Mishra A, Ehsani R, Davis C (2010) A review of advanced techniques for detecting plant diseases. Comput Electron Agric 72:1–13
Schena L, Li Destri Nicosia M, Sanzani S, Faedda R, Ippolito A, Cacciola S (2013) Development of quantitative PCR detection methods for phytopathogenic fungi and oomycetes. Journal of Plant Pathology: 7–24
Schena L, Nigro F, Ippolito A, Gallitelli D (2004) Real-time quantitative PCR: a new technology to detect and study phytopathogenic and antagonistic fungi. Eur J Plant Pathol 110:893–908
Sharon A, Shlezinger N (2013) Fungi infecting plants and animals: killers, non-killers, and cell death. PLoS Pathog 9:e1003517
Shoute LC et al (2018) Immuno-impedimetric biosensor for onsite monitoring of ascospores and forecasting of sclerotinia stem rot of canola. Scientific Reports 8:1–9
Singh AK, Verma N (2019) Plants and plant-derived materials used for biosensor development. In: Book: industrial biotechnology plant systems, resources and products, Publisher: De Gruyter, Germany, pp 250. https://doi.org/10.1515/9783110563337-004
Tothill IE (2001) Biosensors developments and potential applications in the agricultural diagnosis sector. Comput Electron Agric 30:205–218
Tsui CK et al (2011) Molecular techniques for pathogen identification and fungus detection in the environment. IMA Fungus 2:177–189
Turner AP (2013) Biosensors: sense and sensibility. Chem Soc Rev 42:3184–3196
Velasco-Garcia MN, Mottram T (2003) Biosensor technology addressing agricultural problems. Biosys Eng 84:1–12
Wei S, Sun Y, Xi G, Zhang H, Xiao M, Yin R (2018) Development of a single-tube nested PCR-lateral flow biosensor assay for rapid and accurate detection of *Alternaria panax* Whetz. PLoS ONE 13:e0206462
Yamaguchi K, Uriu Y, Kageyama K, Shimizu M (2017) Development of a biosensor for selective detection of phytopathogenic pythiums. In: 2017 IEEE 7th International Conference Nanomaterials: Application & Properties (NAP), IEEE, pp 04NB03-01-04NB03-03

Zamir D, Galsarker O, Alkan N, Eltzov E (2020) Detection of quiescent fungi in harvested fruit using CMOS biosensor: a proof of concept study. Talanta: 120994

Zhai J, Cui H, Yang R (1997) DNA based biosensors. Biotechnol Adv 15:43–58

Zhan F, Wang T, Iradukunda L, Zhan J (2018) A gold nanoparticle-based lateral flow biosensor for sensitive visual detection of the potato late blight pathogen, *Phytophthora infestans*. Anal Chim Acta 1036:153–161

Chapter 12
Journey of Agricultural Sensors—From Conventional to Ultra-Modern

Ashish Mathur, Shikha Wadhwa, Shalini Nagabooshanam, and Souradeep Roy

Abstract Agriculture has a major contribution to the evolution of human civilization. Owing to the rise of the human population, the demand for food increases simultaneously which requires the development of advanced techniques to augment food production. Sensors play a vital role in crop production and are used very successfully in almost all main sectors of agriculture. From the production to the consumers, sensors are used in every process. Monitoring the agronomic environment for several factors such as humidity, temperature, moisture, pH, salinity, and other factors will aid sustainable crop production. As a conventional approach to monitor these factors during farming, individuals are employed physically to check them at regular time intervals. Currently, the conscious employment of advanced technologies for significant improvement in the agronomic environment has increased. Collecting manual data for the desired environmental factors leads to irregular and incorrect measurements which will deviate from the actual values. Thus, leading to difficulties in controlling vital environmental factors. The best solution would be to use wireless distinctive sensor nodules which can minimize the time utilization and physical effort for monitoring the agronomical environment effectively. The data logging ensures the safety and security of the sorted data. Also, a possibility of engagement in dangerous situations without the need to put personnel would reduce the hazardous health risk. Hence, monitoring systems ensure faster response times to environmental factors, improved quality control of the produce, and a lower labor cost. Various factors affecting food production include temperature, moisture, atmospheric pressure, and level of ground water. Advanced technology would definitely be a great support for remote measurement of these vital factors. The aim of this

A. Mathur (✉) · S. Wadhwa
Department of Physics, School of Engineering, University of Petroleum and Energy Studies (UPES), Bidholi Campus, Dehradun, India
e-mail: nanoashish@gmail.com

S. Nagabooshanam · S. Roy
Amity Institute of Nanotechnology, Amity University Uttar Pradesh, Noida, UP, India

S. Wadhwa
Department of Chemistry, School of Engineering, University of Petroleum and Energy Studies (UPES), Bidholi Campus, Dehradun, India

R. N. Pudake et al. (eds.), *Biosensors in Agriculture: Recent Trends and Future Perspectives*, Concepts and Strategies in Plant Sciences,
https://doi.org/10.1007/978-3-030-66165-6_12

chapter is to cover the most advanced and commonly used sensors in agriculture in all processes involved.

Keywords Agriculture sensors · Smart technologies · Agriculture · Precision farming · MEMS · NEMS

12.1 Introduction

Agriculture plays a vital role in our economy and in the survival of the nation. People in rural areas rely on agriculture for their nutrition, employment, income generation, and trade. With rising population, food demand has risen abundantly. For increasing population, economic, and well-managed agriculture is desirable. The global population will rise to 9.55 billion by 2050 and 11.2 billion in 2100 as anticipated by the United Nations. Therefore, agricultural methods will need to change dramatically to meet associated food demand without triggering irreversible environmental damage. At present agriculture face a number of important challenges which include a dearth of labors, climatic variability, poor land fertility due to drought, erosion, urbanization, and the ever-increasing communal anticipations for sustainable practices like marginal use of water and agrochemicals.

Agricultural growth is necessary for country's development and conversion from an old-fashioned to a contemporary economy. The technical advancement in methods incorporated by farmers not only raises the value and scope of agri-production but also enhances the standard of the economy and the livelihood of farmers. Sensor technology is one such method used in varied fields. Sensors have wide scope in agriculture. They can maneuver in a wide range of environments such as bare fields, vineyards, orchards, and from flat to complex topography and varied weather conditions. Though the maintenance of sensors and their usage is complex but once installed, this functionality is always beneficial.

Agriculture sensors include sensors which operate in smart farming providing monitoring of the produce, soil and moisture conditions, plant diseases, and other important environmental factors. The data provided by sensors guide farmers to effectively monitor and grow crops by regulating according to the environmental variations. Sensors can be mounted over the weather stations, drones, and robots which are employed in the agricultural industry. Specifically developed mobile apps are used to control these sensors using wifi connections or cellular towers. Some of the basic usages of sensors in agriculture include, the installation of them at weather stations aids to provide vital data such as temperature of soil/air, moisture/humidity of soil, chlorophyll content, climate/wind predictions, atmospheric pressure, etc. They can also be integrated to several agricultural equipment's like dendrometer which is utilized to measure the diameter of tree or plant trunk, leaf wetness, etc., agricultural drones to spray insecticides and pesticides and so on. Utilization of solar pumps which are mobile operated is quite prevalent in agriculture for the sake of electricity cost minimization. Also, e-fences can be beneficial for rural Indian farmers which can saves crops from wild animals such as elephants (Fig. 12.1 and Table 12.1).

Fig. 12.1 Schematic of interconnected sensors in the field and collection of data through cloud computing (RF Wireless World 2012)

Table 12.1 List of agricultural sensors utilized for various purposes in agricultural industry

Agricultural Sensors	Functional depiction
Locality Sensors	To find out the latitude, longitude, and the altitude of any position in a specific region. GPS satellites are used for assistance
Optical Sensors	To monitor various soil properties and they can be easily installed on satellites, drones, or robots to regulate the properties of the soil
Electrochemical Sensors	To gather chemical data of the soil by detecting specific ions in the soil. The information is obtained in the form of pH and soil nutrient levels
Mechanical Sensors	To monitor soil compaction
Dielectric Soil Moisture Sensors	To monitor moisture levels
Air Flow Sensors	To check air permeability

Fig. 12.2 Schematic of integrated wireless sensors for water content in agri-field (Damas et al. 2001)

12.2 Few Applications of Sensors in Agricultural Sector

12.2.1 Irrigation

Irrigation refers to a man-made application of water in agricultural land and forms a key resource in agriculture. Due to the global scarcity of water, the use of it in agriculture must be monitored properly for its judicious use. Several means of irrigation are followed such as drip irrigation, sprinkler irrigation, etc, which aids in solving water-related problems during flood and drought. Spanish researchers worked on developing a device controlled, automatic irrigation method for a specific region (Damas et al. 2001). The region was further divided equally into seven small regions. Wireless Local Access Network (WLAN) (Fig. 12.2) was connected to a controller in which all the control sectors are interlinked to one another. The results demonstrated good water conservation up to 30–60%.

A number of sensor projects have been formulated, keeping in view the requirement of controlling irrigated water usage for improved yield production. Another researcher (Basu et al. 2006) developed an advanced computer-controlled drip irrigation system which has the ability to acquire data remotely. The saved sensor data is utilized for the statistical analysis which aids in determining the irrigation method required for a wide variety of crops. Automated irrigation integrated to Internet-of-Things (IoT) provides the best solution to conserve water in agriculture and also cost-effective.

12.2.2 Fertilizers

Soil fertility is an important factor for produce as it varies the rate of plant growth and essence of food. Soil fertility can be affected to a desirable level by the application of fertilizers. Broadcasting, manual spreading, and spraying are some of the various ways of the use of fertilizers. The supply of fertilizer to the necessary place requires sensing ability to make it more effective. Many researchers have demonstrated a variety of solutions for optimal fertilization.

A mechanical sensor developed for an area specific fertilization (Pendulum meter) by a German research team (Ehlert et al. 2004). This pendulum meter was used to analyze the density of plantations and was mounted in the forepart of the tractor. A fertilizer spreader and onboard computer were used along with the pendulum meter to determine the level of N applied in the region. Another agri-researcher (He et al. 2011) developed an integrated support system to make decisions related to the use of fertilizers using wireless sensors LAN with IEEE entente and GPS server analysis. Real time soil moisture, temperature, conductivity, air moisture, pH value, radiance, etc., can be monitored using this sensor system.

12.2.3 Pest Control

All over the globe the average crop production loss accounted due to pests and insects during pre-harvest is about 35% (Oerke 2006) and an additional loss of 35% is accounted for loss during transportation, food processing, storage, packaging, and marketing (Popp 2011). The global demand for the increase in food production, crop yield, and quality raw agricultural commodities intended the utilization of pesticides in agricultural sector. The judicious usage of pesticides in agricultural practice helps in protecting crops from pests and insects with unambiguous multi-fold increase in the crop yield (Hassani et al. 2017). Controversially, the rampant usage of pesticides has created havoc for the environment and human beings with serious health hazards. The prolonged utilization of pesticides in agricultural fields affects soil fertility and soil organisms over a longer period.

The main area of interest in "pest control" methods and procedures remains on improving the products which help in better yield and advanced production. With the advancement in more compatible and ecological newer technology, the pest control techniques have become most optimal and best suited.

The overuse of pesticides contaminates the environment which causes various issues related to health. In addition to it, the pesticides reach non-targeted species and get accumulated and cause deleterious effects (Damalas and Koutroubas 2016). Pesticides get drifted from the sprayed field to the environment through air, water, soil, crops (fruits, vegetables, grains, pulses, and animal products), and human or living beings as depicted in Fig. 12.1. When pesticides are sprayed in agricultural fields or crops, the excessive amounts leach out into soil. They also get transported into soil by the plat residues (Fenik et al. 2011). The consumption of pesticides by human or living beings also occurs through residues present in fruits and vegetables. In addition, the toxic pesticides enter humans when it is inhaled, orally in taken and dermal exposure. By means of wind, dust, rain, volatilization, and evaporation the harmful pesticides enter the water bodies and also contaminates air. Despite the regulations formulated for the proper usage of pesticides in agricultural lands, the

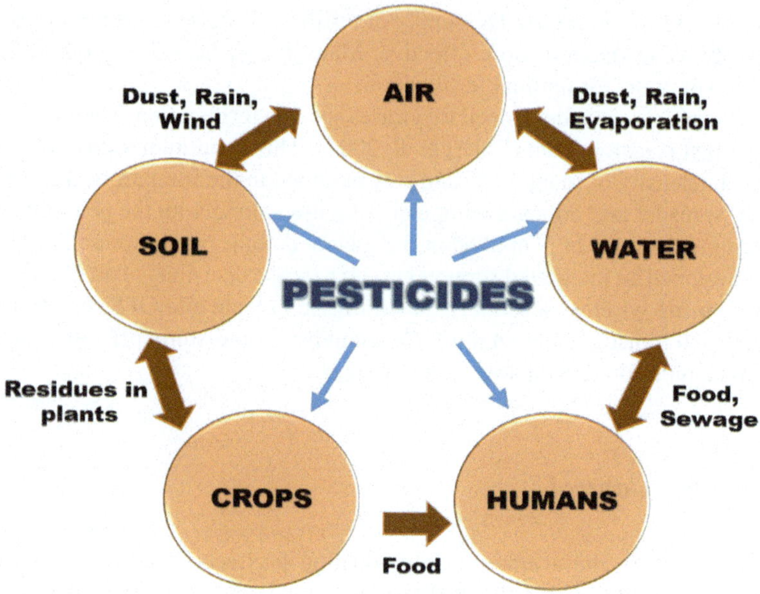

Fig. 12.3 Life cycle of pesticides in the environment

overuse or misuse of pesticides still remains a menace. The exposure or persistence of pesticides in the environment poses a high threat and therefore, proper management on the usage of pesticides is more important for sustainable agriculture (Fig. 12.3).

From the literature, it is seen that the generally used techniques for detection of pesticides were chromatographic techniques and other techniques include spectrophotometric sensors and immunochemical sensors (Stoytcheva and Zlatev 2011). Chromatography-based techniques such as Gas Chromatography (GC), Liquid Chromatography (LC), Mass Spectrometry (MS), Capillary Electrophoresis (CE), Ion Mobility Spectrometry (IMS), Gas Chromatography-Mass Spectrometry (GC-MS), Liquid Chromatography-Mass Spectrometry (LC-MS), and High Performance Liquid Chromatography (HPLC) are the extensively utilized among the traditional techniques for the determination and detection of pesticides (Stoytcheva and Zlatev 2011). The analytical techniques such as GC, LC, MS, CE, IMS are two-dimensional sensors and based on separation methods. Due to the complex chemical nature and their matrices of pesticides, the analytical separation methods which are integrated with an analytical detector provides a better specificity and broad analysis range (Hill and Martin 2002). Despite better specificity and broad analysis range, they fail to determine the unknown samples. In order to overcome these three-dimensional sensors which are upgraded versions and can detect unknown analytes are GC-MS, LC-MS, and HPLC. For vapor phase samples GC is utilized which has the separation time of about ~ 30 minutes and this time can be minimized by wisely choosing the selective detectors. However, the high-pressure requirement for the detection of complex compounds and the small particle size of samples make GC less interesting.

The LC can be utilized for compounds which are non-volatile and thermally stable. In LC, the slow diffusion rates lead to a longer detection time of about > 20 minutes. For ionic and ionizable substances CE is preferred which has faster analysis time < 10 minutes with the use of specific detector named flame photometric detector. HPLC and LC are similar in the process, but the difference is LC uses the gravitational force for solvent movement in the column, whereas HPLC uses a high-pressure pump for solvent movement in the column.

For field analysis, IMS are widely used which uses gas phase ions for sensing pesticides and nerve agents within 1-minute exposure time of radioactive ionization source. However, the use of radioactive ionization sources and low resolving power of these devices make them less effective. Compared to IMS, MS uses the same gas phase ions, but they measure the mass difference to the charge ratio traveling through the vacuum column. Despite of having better resolving power, MS needs more crucial separation process which makes it very tedious during sample analysis. The coupling of GC and LC with MS obviously improves the performance with more reliable data analysis. LC-MS can be employed for detecting unknown samples in the range of ppb and yet it has the disadvantages such as the tedious sample preparation process with slow diffusion rates.

There are few reports which utilize spectrophotometric techniques and immunochemical assays. The general colorimetric test involves the detection of pesticides, simply based on color change. In brief, the pre-treated spots are developed and when the analyte gets absorbed on the pre-treated spots there will be a noticeable change in color, representing the presence of OPs. This approach is less sensitive and not a reliable way for the practical utilization of spectrophotometers. It is based on luminance and fluorescence, in which a change in signal is detected upon the absorption of OPs. The advantage of spectrophotometric techniques is good sensitivity, relatively simple, and low detection limits in the range of ppb. The disadvantage of this technique is the detection time of about ~ 10–20 minutes, less selectivity, and unsatisfactory quantitative detection. The immunological assays, for example, the Enzyme-Linked Immuno-Sorbent Assay (ELISA), overcome the limitations of these techniques. It is based on the principle of immune-assays which utilizes antigen-antibody specific interaction. With the help of ELISA, we can determine the measure of antigen or antibody concentration. The advantages of ELISA are that they are highly selective, sensitive, accurate, and do not use any radioactive agents or sophisticated devices. The recent interest is on developing a bio-sensing platform which provides several benefits such as extraordinary selectivity, sensitivity, lower detection limits, a wider detection range, cost-effectiveness, user-friendly, and non-requirement of a sophisticated analytical laboratory. Nagabooshanam and colleagues fabricated an electrochemical sensing platform to detect organophosphates utilizing functional materials such as carbon nanocomposites and metal organic frameworks (Nagabooshanam et al. 2019, 2020). Other types of pesticide sensors are based on an economical paper-based 3D origami electrochemical sensor developed in recent studies (Arjmand et al. 2017; Arduini et al. 2019).

12.2.4 Use of Sensors in Horticulture

The practice of cultivation, production, distribution, and use of flower, fruit, greenhouse, ornamental crops are considered as horticulture. Horticulture represents low scale or low levered farming. Greenhouse is a key area under horticulture as the research community has shown core concern toward this. A wireless network developed by Zhang et al. (2004) tracks various key parameters such as temperature, moisture/humidity of the soil, ambient light, etc., which helps in crop production and yield.

12.2.5 Soil and Crop Sensor

Essential soil properties can be determined using agriculture sensors. Such sensors are utilized to create field maps to investigate the properties of soil which can be done by using control equipment in a real-time basis or along with Global Positioning System (GPS). Also, determines the factors such as the frequency of sampling or measurement, the space in between the passes, and the travel speed, and the total measurement of points per acre. This provides a cost-effective route for field mapping. The efforts had been made by scientists and equipment manufacturers in order to modify the existing laboratory and other indirect monitoring techniques for on-the-go soil mapping. But, as of now, very few types of sensors have been developed and investigated which includes electrochemical, electromagnetic, optical, acoustic, mechanical, and airflow sensors to name a few.

Electrochemical sensors detect the soil pH and soil nutrients. The soil samples testing at the laboratory involves various standard laboratory procedures, sample preparation methods, and measurement techniques. Some of the techniques utilize ion-selective electrodes which measure the activity of specific ions. Determination of soil pH, nitrogen, and potassium can be done using an ion-selective electrode. Several traditional soil preparation and measurement procedures are adapted to conduct a laboratory test on-the-go which eventually increases the overall accuracy of soil nutrient or pH testing.

Electromagnetic sensors utilize electromagnetic circuits to monitor the conductivity or accumulated electrical charge of the soil particles. In these types of sensors, the soil becomes part of an electromagnetic circuit itself and the small variations in the local soil conditions can be readily measured. Several commercial sensors that available in the market. One such is the Soil EC 3100 which is an on-the-go soil mapping sensor developed by Veris Technologies, Salina, Kansas (https://www.veristech.com/the-sensors/v3100). Another commercial device is EM-38 developed by Geonics Ltd, Ontario, Canada which maps the electromagnetic response to investigate soil properties. The electromagnetic properties of soil alter according to its texture, moisture/humidity, salinity, and organic matter constituents. Electromagnetic sensors can also be utilized to monitor soil pH and nutrients.

Optical sensors utilize light, near/mid-infrared light, and polarized light reflectance to measure soil properties such as moisture, nutrients, and organic constituents. The light reflectance varies in between the healthy or deficient soil and once the data is procured it is analyzed by mapping method to measure various soil properties such as moisture, nutrient level, etc. Few companies offer a remote sensing facility which aids to determine soil reflectance by engaging a satellite or airplane platform. Despite, the advantages of these platforms there are several factors such as the cost, timing, clouds, and ground covered plant residues limit the usage of rare soil imagery. Nearby-range, subsurface, and on-the-go vehicle-based optical sensors have the potential to measure the data points or information certainly at more than one portion of the spectrum at a time.

Acoustic sensors measure the change in acoustic pulse transmission which happens due to a change in the soil properties. It helps to monitor soil texture, moisture, nutrient constituents, etc. The major limitation of acoustic sensors is that the emitted acoustic signal is very weak and noisy. Mechanical sensors measure the mechanical resistance or compaction of the soil. Due to the force exerted by strain gauges or load cells at the time of measurement a cut/penetration which propagates through the soil is measured. Several device prototypes have been developed which clearly demonstrates the applicability of uninterrupted mapping of soil resistance. Still, these devices are not yet available commercially.

Airflow sensors measure the soil air permeability. In these types of sensors, the pressure required to squeeze a certain volume of air into the soil at a fixed depth is associated with various soil properties. Airflow sensors can be utilized for analyzing varied soil types, moisture levels, and soil structure/compaction.

12.3 Gas Sensors Used in Agriculture

Gas sensors provide an important platform for measurements within the agricultural industry for various applications. Temperature and Carbon Dioxide (CO_2) concentration can be monitored using commercial greenhouse control systems for better crop growth. Low range CO_2 Guardian NG DC developed a device for continuous monitoring of CO_2 which provides an economical solution to increase crop growth (https://edinburghsensors.com/). In pig and poultry farming the monitoring CO_2 is of utmost importance during the process of gas stunning. For this, the Gascard NG is utilized which can monitor the high concentrations of CO_2 during poultry food processing. CO_2 measurement is also vital to detect the spoilage of stored grains wherein gas sensors are employed. The example of commercial gas sensors for the measurement of CO_2 is shown in Fig. 12.4.

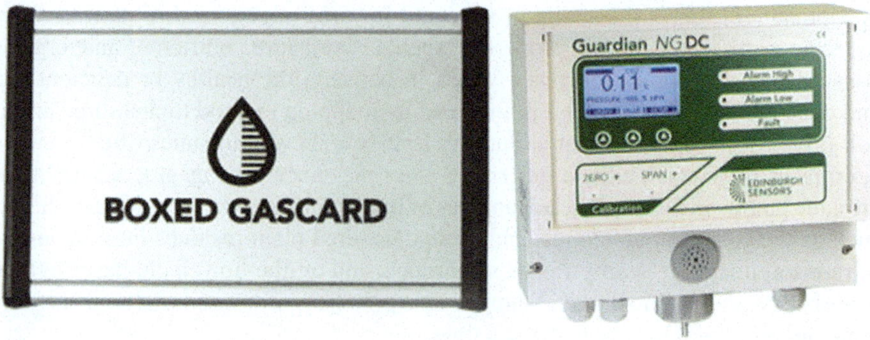

Fig. 12.4 Commercial gas sensors available for monitoring CO_2 (https://edinburghsensors.com/)

12.4 Sensors in Food Packaging, Transportation, and Inspection

Economical RFID biosensor tags were generally utilized to maintain inventory and examine food products. Few reports demonstrated a prospective of RFID tags for sophisticated appliances, recycling, packaging, automatic checkout, and marketing/furthering opportunities (Kumar et al. 2009; Smits et al. 2012). The RFID technology aids to provide safe, productive, and traceable produce which leads to operational savings and increased capital. Geers et al. (1998) developed a monitoring system that functions on the road for animals during transportation and handling. This technology involved the installation of sensors in the animal section in order to detect the animal variety, to check on the air-quality and animal behavior. The location of the vehicle during the transportation is provided by GPS. Regular merits are sent to service centers through GSM network by a data transfer unit regularly. Najjar et al. (1997) developed a portable PC for the quality check (QC) inspectors working in food-processing plants. This type of system helps the QC inspectors to select a suitable for filling the data and send it to the plant manager through full duplex audio and wireless data communication. Recently, wireless sensors are being used to monitor and for quality control purposes in the food processing industry. The data of the sensor can later be transferred to the central data controller wirelessly. Marra and Romano (2003) designed a systemic mathematical model in order to investigate the effects of various methods which interposes wireless temperature sensors into conductive tinned food for inspecting thermal sterilization.

12.5 Smartphone-Based Sensors in Agriculture

Farming activities are more inclined toward the growth of plants, killing weeds/pests, identifying and correcting plant diseases, applying fertilizers, and estimating growth/yield of crops. Recent progress in smartphone application development and increasing availability of smartphones have been very helpful in lifting some of these agricultural burdens and provide necessary guidance. For instance, smartphone applications can be used to examine color of crop leaves and farmers can calculate the amounts of fertilizers to be applied to the crop fields. Many tasks in the field such as seeding, weeding, fertilizing, and watering, appear repetitive, mundane, and laborious. However, these tasks usually require precision and involve decision-making steps prior to the actual activities so that the farming cycle becomes effective (Yang and Li 2008). Prior knowledge on identifying crop diseases, the location of the disease in crops, and their prevention and cure is necessary to save the time of farmers and costs in the practice of farming. This can be made possible by smartphone applications in farming. The subcategories of smartphone applications in farming are discussed below:

Disease Detection and Diagnosis: Smartphones enabled with sensors can be utilized for disease detection/diagnosis in farms. Prasad et al. (2014) developed a mobile vision system for plant disease identification. With this system, images of diseased plant leaves are captured and preprocessed. Then the images are transmitted and sent to remote laboratories. The image preprocessing step saves the transmission cost of sending diseased leaf images to plant pathologists in remote laboratories. Leaf images are further segmented into three areas: background, non-diseased and diseased portions of the leaf. Suggestions on cure and prevention methods for plant diseases are then provided to farmers. A mobile application *Magri* was suggested by Wu and Chang (2013), which uses pest/disease information previously self-reported by farmers. *Magri*'s report module facilitated the generation of advice for the cure of pests and pest alerts in many nearby areas using GPS.

Fertilizer Calculator: Application of fertilizers on crops is an important farming activity which has a great potential to affect farm productivity. Farmers make a key decision about the chemicals to apply and their crop-specific suitable amounts needed. Sumriddetchkajorn (2013a) reviewed a few mobile device-based optical applications for agriculture. It has mentioned a smartphone-based color estimator named *BaiKhao*, dedicated for rice leaves' chlorophyll evaluation. The newer released version *BaiKhaoNK* could evaluate the color level of rice leaves and suggested the amounts of nitrogen fertilizer needed onto the rice field (Intaravanne and Sumriddetchkajorn 2012; Sumriddetchkajorn 2013a, b).

Soil Study: Soil forms a key component in farming that has a huge effect on the agricultural output. Soil data supplied to farmers gives them an advantage in farming, especially in precision agriculture. A large variety of smartphone applications are exemplified in the literature that uses smartphone sensors to study soil properties and collects soil data for agricultural purpose. Mobile phones were used as soil color sensors (Gómez-Robledo et al. 2013) which could read soil color information based on the images captured by built-in digital cameras on smartphones and processed using image-processing techniques. One more smartphone application is 'Soil Indicators for Scottish Soils', *SIFSS*, has demonstrated the capability to study soil (Aitkenhead et al. 2013). Farmers were able to receive soil's detailed information, such as pH, soil carbon, N, P, and K, based on their locations in Scotland. *SOCiT*, another application is aimed at users in Scotland, provided information about soil's carbon content based on users' geographical positions (Aitkenhead et al. 2013).

Water Study: The quality of water is an important parameter affecting farming and agriculture in various regions. A smartphone application has been developed under a project in Scotland, *iDee*, which is devoted to inspire users to provide information of water conditions, such as water level, clarity, obstruction in river, algae cover, temperature, and nonnative plants in water, and accompanying photographs of the River Dee (Aitkenhead et al. 2013).

Crop Water Needs Estimation: Crop water requirements of an agricultural land rely on numerous conditions including crop types, season, climate, and growth stages of crops (Allen et al. 1998). Crops lose water through transpiration and nearby soil, water, and canopy lose water through evaporation. Collectively, the loss of water through both these processes is termed as evapotranspiration (Allen et al. 1998). Crop water requirements are verified to supplement water loss so that crops grow according to farmers' needs. *PocketLAI*, a smartphone application is developed to help farmers to determine Leaf Area Index (LAI), one of the key factors to evaluate crop water needs (Confalonieri et al. 2013). The application *PocketLAI* evaluated LAI by an indirect method based on sensors equipped on modern smartphones. Image sensors (cameras) captures the pictures of leaf canopy to evaluate Leaf Area Index while accelerometers obtains the angle of smartphones as the devices are rotated to evaluate LAI and the gap fraction was measured at 57.5° (Confalonieri et al. 2013). Farmers can regulate their watering/irrigation tools accordingly after water requirements were estimated from Leaf Area Index. Sumriddetchkajorn et al. (2013a, b) devised an application to evaluate brightness smartphone cameras. Another mobile application *RaGPS* was developed by Molina-Martínez et al. (2011), for devices with Windows Mobile operating system to determine the extraterrestrial solar radiation and similar other parameters in real time. Such radiation parameters offer a useful tool to estimate crop water requirements and validating weather data provided by the agricultural weather stations. The experiments demonstrated that *RaGPS* was able to estimate radiation parameters quite similar to those obtained from the weather station in clear-sky days.

Crop Produce Readiness Analysis: Another novel application of smartphone application is to determine the ripeness of fruits (Sumriddetchkajorn 2013a, b). Various images of fruits under white and UV-A light sources were captured to evaluate ripeness levels of green fruits. The system could be integrated by farmers into their farms to determine different ripeness levels of various fruits before sending them to markets. The computer vision techniques helped farmers to complete the process at a bulk scale rather than manual inspection of each fruit by farmers.

Smartphone applications are useful in sensing the environment using a number of sensors under various farming categories and processes of planting crop. Observed data may be in the form of snapped pictures, specific locations in the farm, colors of soil, water, and plant leaves, and light. Farmers are the key target users of smartphone applications as they are involved in mundane and laborious farming activities. Such applications aim at enhancing farm productivity, by evaluating agricultural specimens, supporting agricultural decisions, and solving task-specific problems. (Intaravanne and Sumriddetchkajorn 2012; Confalonieri et al. 2013; Sumriddetchkajorn 2013a, b; Gómez-Robledo et al. 2013; Prasad et al. 2014).

12.6 Conventional Sensors in Agriculture

12.6.1 Soil Quality Sensors

Soil properties play a key role in promoting crop growth and yield. Hence the properties need to be continuously monitored for precision agricultural practice. For a higher yield and quality production, NPK as fertilizers are utilized (Prativa and Bhattarai 2012). Despite the availability of NPK naturally in the soil to further enhance the growth of crops man-made formulations are incorporated (Mengistu et al. 2017). These formulations help to improve the soil quality in the nutrient scarcity areas. The blend of organic manures and the man-made NPK formulations are known to increase the yield but, as a controversy the injudicious use of NPK is a major threat to soil quality and fertility (Bhati et al. 2018). Therefore, soil nutrient management is of utmost importance and there are some traditional techniques to monitor the NPK values in the soil for sustainable agriculture one such is the random grid sampling technique (Wollenhaupt and Wolkowski 1994). In the random grid technique, the soil samples were collected in the random grid area and the samples were sent to analytical laboratory. Here, the samples go through the extraction process and extensive chemical analysis for the determination of NPK values. The other conventional technics utilized for the NPK sensing are chromatographic techniques, spectral imaging analysis and ion-selective field effect transistors (Adamchuk et al. 2005; Chen et al. 2014; de Castro Vasconcellos et al. 2015). The aforementioned techniques are selective, and sensitive but they are time consuming, expensive, and require sophisticated laboratory equipment's. In order to overcome above limitations, various sensor techniques such as optical, electrochemical, spectroscopic, etc., are explored which are economical, portable, and user-friendly. Lanfranco and colleagues utilized Raman spectroscopic technique to detect the mixed-metal phosphates in soils at low concentrations (Lanfranco et al. 2003). Raman spectroscopy involves a simple detection technique depends on Raman scattering upon the interaction with various samples. Other than spectroscopic detection electrochemical sensors are used for NPK detection. Chauhan and Urooj (2019) developed an ion-selective field effect transistor (ISFET)-based electrochemical sensor for the analysis of NPK values in real time.

Apart from the soil nutrients as mentioned above other parameters of soil such as pH, moisture, salinity, etc., are also of equal importance for the overall crop production and sustainable agriculture. The pH values of the soil influence the crop growth, i.e., the growth is good at higher values of pH and the growth is declined when the pH values are lower (Kumar et al. 2015). The optimum pH value is between

6–7.5 and it varies for different types of crops. Rani et al. (2008) reported a sensitive pH sensor based on multi-finger ISFET sensor. In this ISFET the change in the H^+ ion concentration is detected between the electrode and electrolyte which influences the change in potential at the gate.

12.6.2 *Monitoring Moisture Content and Temperature of Soil*

Soil moisture is also important along with the soil nutrients, pH, salinity, etc., and hence it is necessary to monitor the soil moisture levels regularly (Chadha et al. 2019). The conventional laboratory method to test the soil moisture is gravimetric water content and volumetric water content analysis (Almaw et al. 2018). In former, it measures the mass of water per mass of soil sample and the later measures the volume. The exchange of thermal and hydro energy between the surface of land and atmosphere is controlled by soil temperature and moisture content by means of evaporation and water/nutrient transportation in plants. Consequently, moisture content and soil temperature play a vital role in the regulation of weather conditions, the latter being a key player in crop production and growth. Micro-machined cantilever beams consisting of a water responsive nano-polymer integrated with piezo-resistive temperature sensor form the basic components of MEMS-based temperature and moisture sensing devices.

Soil moisture levels and corresponding electrical conductivity are the two most important parameters employed in agriculture. However, recently soil pH is often employed as a criterion for assessing crop productivity as well as being used as a tool to scrutinize areas where adequate fertilization has occurred. Few other parameters, namely, water flux and thermal properties of soil (thermal diffusivity, thermal conductivity, and specific heat capacity) are also employed for ensuring enhanced crop productivity (Heitman et al. 2007; Bristow et al. 1994; Ren et al. 1999; Ham and Benson 2004). Electrical conductivity readings can also be monitored using the same conducting needle-like probe (Mortensen et al. 2005; Valente et al. 2006). Ion-Sensitive Field Effect Transistors (ISFETs) are normally used for measuring the soil pH as shown in Fig. 12.5.

12.7 MEMS Devices Used in Agriculture

In 1986, the concept of Micro Electro Mechanical System (MEMS) was first presented. Since then, numerous devices were commercialized throughout the globe. Nowadays, multiple devices incorporate MEMS technology, namely, inkjet printing heads, accelerometers, gyroscopes as well as modern surgical tools and Lab-on-Chips (LoCs) for healthcare applications. Some other applications of MEMS-based devices include modern car airbags, touch screen-based smart phones, drones, or UAVs (Unmanned Arial Vehicles), gaming consoles, digital cameras to name a few.

Fig. 12.5 A typical MEMS-based ISFET employed for monitoring soil temperature, nutrient content, moisture levels, and pH toward smart agro-applications (https://www.davisinstruments.com/)

It can therefore be concluded that consumer goods, electronics, and healthcare are the prime areas of research for MEMS-based device applications. In agriculture sector, MEMS is used in drone technology for monitoring crop quality and growth. However, when it comes to Precision Agriculture (PA) which states that farmlands should be categorized into zones rather than whole fields in order to increase production efficiency, Internet-of-Things (IoT) technology has the capability to be integrated with MEMS devices for ensuring enhanced crop quality monitoring. MEMS market is expected to rise drastically from $11 billion in 2014 to $21 billion in 2020 (Bariáin et al. 2000). This is further supplemented due to the availability of highly efficient biomedical, biochemical sensors, and energy harvesters which can impart a huge impact on the agriculture sector. Owing to high-market value of MEMS-based devices facilitated by multitude of applications and advantages, such technologies find huge potential to be implemented in agriculture sector for enhanced outcomes.

12.7.1 MEMS for Monitoring Crop Root Growth

In order to quantitatively assess physical interaction between crop root and the soil, a silicon microchannel device integrated with motion sensors has been recently developed which helps in the analysis of certain barriers which may impede the uptake of minerals and nutrients from soil to crop roots.

12.7.2 MEMS Microchip-Based Capillary Electrophoresis Sensor

Plants generate a defense mechanism against numerous pathogens or biotic stress through different strategies. Plants respond to fungal attack by activating defensive responses associated with the accumulation of pathogenesis-related proteins (PRs), which prevent the occurrence of pathogen infection. Capillary electrophoresis is an important technique which aids in the detection as well as the separation of variety of species—from heavy metals to phenolic acid in plant extracts. Analysis of PRs by means of MEMS-based capillary electrophoresis tool in *Eruca sativa* cultivation to detect pathogen presence was investigated in 15 and 25 days old plants (Sharma et al. 2018, 2019).

12.8 3-D Imaging Systems for Agricultural Applications

Currently, 3-D sensors are becoming smaller, smarter, and highly affordable. Therefore, technological breakthroughs are viable provided sufficient research leads to commercialization since agricultural environments are substantially complex. In such scenario, 3-D vision can serve an important role in improved perception which can positively impact a huge number of applications in every agricultural aspect. The ultimate efficacy of 3-D data depends on the excellent sensing capabilities compared to 2-D strategies. Several market predictions have recently surfaced around topics such as 3D sensors, LED lighting, UAVs, and agricultural automation. Almost all of them predict a profitable market value for these interconnected topics by the end of the decade. Few 3-D techniques have either not been tested (i.e., shape polarization, photometric stereo, shape-from-zooming) or more research is yet to be performed with the rich variety of 3-D imaging techniques for agricultural applications (i.e., interferometry, light field, CTCs). 2-D and 3-D fusion are highly promising since it exploits the advantages of both the former and latter and has proved to be useful for either obtaining detailed information about the object's surface or facilitating the image segmentation process. All 3-D sensors are sensitive to sunlight; however, further research must be performed to reduce the latter's effects and reduce dependence on shading devices. A positive outcome of this disadvantage is enabling of a thorough investigation of night farming since 3-D sensors behave properly in this environment.

12.9 Use of Smart Cameras in Agriculture

While smart camera technology happens to be far from a typical analog sensor, it has been increasingly implemented in a variety of smart agro-applications. Industrial firms such as Blue River Technology, USA have incorporated smart camera technologies to monitor the presence of weeds and plants to release herbicides and fertilizer automatically and precisely. This process increases the overall productivity by decreasing chemical usage, the latter at high quantities might often damage crops and plants. Moreover, pest control has always been one of the most prolific challenges in agriculture. Nowadays, farmers employ smart cameras for real-time pest monitoring and to effectively minimize/eliminate pest presence without harming the crops. Smart cameras also possess the potential to replace sensing devices, namely, ambient light monitoring, which enables system simplification and a reduction in the component count.

12.10 Some Challenges

For a practical application, a sensor must be positioned in direct contact with, or in close proximity to the ground and coupled to a "black box" in order to analyze the sensor response with further processing of the data. A sensor must be able to provide real-time data to improve the total economic or agronomic effect of the production input. Nevertheless, the following difficulties observed in the "real world" are not considered while considering this approach:

- Many agronomic sensors and applicator controllers need a definite time for measurement, integration, and analysis. This lessens the operational speed or measurement density.
- Additional information is required by variable rate fertilizer and pesticide (such as yield potential) to design prescription algorithms.
- Presently, no site-specific management prescription algorithm is available that demonstrates compatibility for all variables involved in crop production.

The most appropriate approach, rather than utilizing real-time sensors, would be to use on-the-go sensors equipped with controllers and a map-based method. This will allow the possibility of data collection and analysis, creating the prescription, and carry out the variable rate application in minimal steps.

12.11 Limitations of Agriculture Sensors

Some of the drawbacks of agriculture sensors comprise the following:

- Continuous internet connectivity is required for smart farming and IoT technology which is more challenging in developing countries and other parts of the world.
- There is some resistance from the consumer's side to acknowledge and accept the latest IoT-based devices equipped with smart agricultural sensors.
- Some basic infrastructure necessities including smart grids, traffic systems, and cellular towers are unavailable at vast places. Therefore, further growth is hampered.

12.12 Future of Agriculture Sensor Technology

The smart agriculture industry is continuously growing, along with almost daily arrival of new innovative smart solutions to the market. Devices which accumulates sensor data, relay critical information to farmers and ranchers, which in-turn helps to optimize vast agricultural processes are continuously increasing in capability and importance. Smart sensors are used for precision agriculture to import critical data that helps farmers to monitor and optimize crop yield, and also to keep up with changing environmental conditions. By employing smart sensors, farmers can assess their crops at a micro scale, sustain resources, and reduce environmental impacts. Smart agriculture dates back to the 1980s when GPS became accessible for civic use. At that point, farmers were able to accurately examine their fields, as well as monitor and apply fertilizers accurately at desired areas.

Attaching a sensor and tracking device to a cow will facilitate the ability to track the cow's health and other critical behaviors (Fig. 12.6). A smart sensor will thus help with easy animal examination such as body temperature detection and health monitoring to separate and treat sick cows within the herd. A smart sensing device will further help with early mastitis detection to lower risk of milk loss.

Due to shortages in labor and the urgent need to feed an increasing global population, agricultural robots are now commonly implemented by farmers. In the USA,

Fig. 12.6 Representative pics showing applications of sensor with IoT technology for monitoring cow health and activity for enhanced production (Calderone 2019)

crop production has decreased by an estimated $3 billion a year due to labor shortages. At the same time, the world's population is expected to grow by 2.3 billion within the next 3 decades. Vision and machine learning now allows robots to watch and train using their surroundings, and due to the cost-effective nature of smart sensors, they are being deployed in an increased rate. Yet, we are in the early stages of an agriculture robotics revolution with most of the smart devices in the early phases of trial process and R&D mode.

Emerging smart sensing technologies allow farmers to remotely assess their fields' pest population in real-time and take immediate action to protect their crops, facilitated by online cloud services and a dashboard acting as the central hub (Fig. 12.7).

A trap employs a kind of lure (pheromone) to attract and imprison a particular pest. Information regarding pests is wirelessly transmitted to the central hub and eventually to farmer's smartphone or computer. The farmer will then be able to view a satellite image of their field with a quantitative measure of number of pests captured

Fig. 12.7 Representation of remote-controlled access of pest population in specific spots in farmlands in real-time facilitated by state-of-art cloud computing techniques (Calderone 2019)

in each trap as well as details on historic trends and pesticide use. This facilitates the process of quantitatively checking insect traps and figuring out where and how much pesticide to be used in a particular spot in the field, thereby eliminating billion dollars' worth of pest damage.

Optical Sensors employ light to analyze soil properties, by recording various frequencies of light in near-infrared and mid-infrared spectrums. These sensors can be placed on vehicles or drones which allows soil reflectivity and plant color data to be recorded and processed. Nature of clay, organic matter, and soil moisture can be determined by the implementation of optical sensors.

Electrochemical Sensors mounted on specially designed spots help collect, process, and map the chemical nature of soil. Electroanalytical sensors provide the information required for precision agriculture, that is facilitating the monitoring of soil pH and nutrient levels. Soil samples are sent to a soil-analyzing laboratory and an array of standard protocols are carried out. Certain measurements, especially the identification of pH, are performed by utilizing an ion-selective electrode. These electrodes monitor the activity of certain ionic species such as potassium, nitrate, or hydrogen.

Mechanical sensors are used to assess soil compactness which is another key parameter which affects the overall crop yield. These sensors employ a mechanism that penetrates the soil and quantifies the force monitored by load cells or strain gauges. When such a sensor probe penetrates the soil, it records the resistive forces resulting from the cutting, breaking, and displacing of soil. The soil mechanical resistance is then measured in pressure units and signifies the ratio of the force required for penetrating the soil medium, to the front surface area of the tool engaged with the soil.

Dielectric soil moisture detectors monitor humidity levels by measuring the dielectric constant, which varies depending on the soil moisture content. The moisture sensors are used in conjunction with rain gauge stations throughout the farm. This allows for the observation of soil moisture conditions when vegetation levels are low.

Agricultural weather stations primarily are self-contained sensors that are situated at different locations in particular farmland. Such stations consist of a variety of sensors that are suitable for examination of local climatic conditions, thereby leading to improved assessment of crop quality. Parameters such as soil and air temperature at various depths, leaf wetness, rainfall, chlorophyll content, dew point temperature, wind speed, wind direction, solar radiative power, relative humidity, and atmospheric pressure are determined and recorded at programmed intervals. Due to the miniaturized nature of such devices, the acquired data are then transmitted wirelessly to a central hub followed by transfer to nearby control stations for facilitating faster necessary actions.

There could be a loss of time and production should a farm vehicle break down. Farmers now have the ability to remotely gather and manage information from their field equipment. Equipment telemetry lets mechanical devices such as tractors to

Fig. 12.8 Representation for emerging smart sensing technologies employed to transmit soil-nature data to far away field supervisors by means of satellite communication (Calderone 2019)

notify mechanics that a failure is likely to happen soon. Most of the agricultural equipment companies are building telematics systems.

Electronic sensors are often mounted on tractors and other field equipment in order to monitor the former's operating conditions. Meanwhile, cellular and satellite communication systems are also used to transmit the data immediately to computers or e-mail it to certain individuals as represented in Fig. 12.8. The field supervisor can then retrieve the information on their office computers or cell phone and take necessary action.

Airflow sensors measure the permeability of air in the soil. Measurements can be made at one particular location or at multiple spots. The desired output is the pressure required to push a predetermined amount of air into the ground at a pre-programmed depth. Various types of soil properties, including compactness, chemical nature, and moisture level produce unique signatures.

The use of smart sensors, software, and eventual integration with IoT as well as artificial intelligence-based platforms possess huge potential to transform the agricultural market. Combinational sensors along with sensor fusion hardware possess tremendous potential to generate the required plant data in order to enhance crop production rate and implement disease-free high yield crops to maintain the synchronization with the increasing demand for food proportionally with the global population.

References

Adamchuk VI, Lund ED, Sethuramasamyraja B et al (2005) Direct measurement of soil chemical properties on-the-go using ion-selective electrodes. Comput Electron Agric 48:272–294. https://doi.org/10.1016/j.compag.2005.05.001

Aitkenhead M, Donnelly D, Coull M, Black H (2013) E-SMART: Environmental Sensing for Monitoring and Advising in Real-Time. IFIP advances in information and communication technology. Springer, Berlin, Heidelberg, pp 129–142

Allen RG, Pereira LS, Raes D, Smith M (1998) Crop evapotranspiration-guidelines for computing crop water requirements-FAO irrigation and drainage paper 56

Almaw AA, Naveen Kumar SK, Akshaya Kumar A (2018) Soil moisture sensors in agriculture and the possible application of nanomaterials in soil moisture sensors. IJAERT 6:134–142

Arduini F, Cinti S, Caratelli V et al (2019) Origami multiple paper-based electrochemical biosensors for pesticide detection. Biosens Bioelectron 126:346–354. https://doi.org/10.1016/j.bios.2018.10.014

Arjmand M, Saghafifar H, Alijanianzadeh M, Soltanolkotabi M (2017) A sensitive tapered-fiber optic biosensor for the label-free detection of organophosphate pesticides. Sensors Actuators, B Chem 249:523–532. https://doi.org/10.1016/j.snb.2017.04.121

Bariáin C, Matías IR, Arregui FJ, López-Amo M (2000) Optical fiber humidity sensor based on a tapered fiber coated with agarose gel. Sensors Actuators, B Chem 69:127–131. https://doi.org/10.1016/S0925-4005(00)00524-4

Basu T, Thool VR, Thool RC, Birajdar AC (2006) Computer based drip irrigation control system with remote data acquisition system. In: Computers in agriculture and natural resources—proceedings of the 4th world congress, pp 104–109

Bhati V, Yadav PK, Kumar R (2018) Effect of levels of inorganic fertilizers, organic manure and bio-fertilizers on plant growth attributes of nion (Allium cepa L.) cv. N-53 under Hot Arid Region of Western Rajasthan, India. Int J Curr Microbiol Appl Sci 7:3593–3601. https://doi.org/10.20546/ijcmas.2018.703.413

Bristow KL, Kluitenberg GJ, Horton R (1994) Measurement of soil thermal properties with a dual-probe heat-pulse technique. Soil Sci Soc Am J 58:1288–1294. https://doi.org/10.2136/sssaj1994.03615995005800050002x

Calderone L (2019) Top article for 2019—smart sensors in farming | agriTechTomorrow. In: AgritechTomorrow. https://www.agritechtomorrow.com/article/2019/02/top-article-for-2019-smart-sensors-in-farming/11247. Accessed 25 Aug 2020

Chadha A, Florentine SK, Chauhan BS et al (2019) Influence of soil moisture regimes on growth, photosynthetic capacity, leaf biochemistry and reproductive capabilities of the invasive agronomic weed; lactuca serriola. PLoS One 14:e0218191. https://doi.org/10.1371/journal.pone.0218191

Chauhan N, Urooj S (2019) The effect of Azadirachta Indica extract on the soil nutrients and the NPK value determination by electrochemical sensor. Int J Biosens Bioelectron 5. https://doi.org/10.15406/ijbsbe.2019.05.00158

Chen L, Lin L, Cai G et al (2014) Identification of nitrogen, phosphorus, and potassium deficiencies in rice based on static scanning technology and hierarchical identification method. PLoS ONE 9:113200. https://doi.org/10.1371/journal.pone.0113200

Confalonieri R, Foi M, Casa R et al (2013) Development of an app for estimating leaf area index using a smartphone: trueness and precision determination and comparison with other indirect methods. Comput Electron Agric 96:67–74. https://doi.org/10.1016/j.compag.2013.04.019

Damalas CA, Koutroubas SD (2016) Farmers' exposure to pesticides: toxicity types and ways of prevention. Toxics 4:1–10. https://doi.org/10.3390/toxics4010001

Damas M, Prados AM, Gómez F, Olivares G (2001) HidroBus® system: fieldbus for integrated management of extensive areas of irrigated land. Microprocess Microsyst 25:177–184. https://doi.org/10.1016/S0141-9331(01)00110-7

de Castro Vasconcellos P, da Rocha GO, Caramão EB et al (2015) Chromatographic techniques for organic analytes. Compr Anal Chem 70:267–309. https://doi.org/10.1016/bs.coac.2015.09.009

Ehlert D, Schmerler J, Voelker U (2004) Variable rate nitrogen fertilisation of winter wheat based on a crop density sensor. Precis Agric 5:263–273. https://doi.org/10.1023/B:PRAG.0000032765.29172.ec

Fenik J, Tankiewicz M, Biziuk M (2011) Properties and determination of pesticides in fruits and vegetables. TrAC—Trends Anal Chem 30:814–826. https://doi.org/10.1016/j.trac.2011.02.008

Geers R, Saatkamp HW, Goossens K et al (1998) TETRAD: an on-line telematic surveillance system for animal transports. Comput Electron Agric 21:107–116. https://doi.org/10.1016/S0168-1699(98)00029-5

Gómez-Robledo L, López-Ruiz N, Melgosa M et al (2013) Using the mobile phone as munsell soil-colour sensor: an experiment under controlled illumination conditions. Comput Electron Agric 99:200–208. https://doi.org/10.1016/j.compag.2013.10.002

Ham JM, Benson EJ (2004) On the construction and calibration of dual-probe heat capacity sensors. Soil Sci Soc Am J 68:1185–1190. https://doi.org/10.2136/sssaj2004.1185

Hassani S, Momtaz S, Vakhshiteh F et al (2017) Biosensors and their applications in detection of organophosphorus pesticides in the environment. Arch Toxicol 91:109–130. https://doi.org/10.1007/s00204-016-1875-8

He J, Wang J, He D et al (2011) The design and implementation of an integrated optimal fertilization decision support system. Math Comput Model 54:1167–1174. https://doi.org/10.1016/j.mcm.2010.11.050

Heitman JL, Horton R, Ren T, Ochsner TE (2007) An improved approach for measurement of coupled heat and water transfer in soil cells. Soil Sci Soc Am J 71:872–880. https://doi.org/10.2136/sssaj2006.0327

Hill HH, Martin SJ (2002) Conventional analytical methods for chemical warfare agents. Pure Appl Chem 74:2281–2291. https://doi.org/10.1351/pac200274122281

Intaravanne Y, Sumriddetchkajorn S (2012) BaiKhao (rice leaf) app: a mobile device-based application in analyzing the color level of the rice leaf for nitrogen estimation. In: Optoelectronic imaging and multimedia technology II. SPIE, p 85580F

Kumar P, Reinitz HW, Simunovic J et al (2009) Overview of RFID technology and its applications in the food industry. J Food Sci 74:R101–R106. https://doi.org/10.1111/j.1750-3841.2009.01323.x

Kumar S, Babankumar, Ritula T, Kumar M (2015) Soil pH sensing techniques and technologies—a review. Soil pH Sens Tech Technol A Rev 4:4452–4456. https://doi.org/10.15662/ijareeie.2015.0405077

Lanfranco AM, Schofield PF, Murphy PJ et al (2003) Characterization and identification of mixed-metal phosphates in soils: the application of Raman spectroscopy. Mineral Mag 67:1299–1316. https://doi.org/10.1180/0026461036760166

Marra F, Romano VA (2003) Mathematical model to study the influence of wireless temperature sensor during assessment of canned food sterilization. J Food Eng 59:245–252

Mengistu T, Gebrekidan H, Kibret K et al (2017) The integrated use of excreta-based vermicompost and inorganic NP fertilizer on tomato (Solanum lycopersicum L.) fruit yield, quality and soil fertility. Int J Recycl Org Waste Agric 6:63–77. https://doi.org/10.1007/s40093-017-0153-y

Molina-Martínez JM, Jiménez M, Ruiz-Canales A, Fernández-Pacheco DG (2011) RaGPS: a software application for determining extraterrestrial radiation in mobile devices with GPS. Comput Electron Agric 78:116–121. https://doi.org/10.1016/j.compag.2011.06.009

Mortensen AP, Hopmans JW, Mori Y, ˇ imů nek JS (2005) Multi-functional heat pulse probe measurements of coupled vadose zone flow and transport. https://doi.org/10.1016/j.advwatres.2005.03.017

Nagabooshanam S, John AT, Wadhwa S et al (2020) Electro-deposited nano-webbed structures based on polyaniline/multi walled carbon nanotubes for enzymatic detection of organophosphates. Food Chem 323:126784. https://doi.org/10.1016/j.foodchem.2020.126784

Nagabooshanam S, Roy S, Mathur A et al (2019) Electrochemical micro analytical device interfaced with portable potentiostat for rapid detection of chlorpyrifos using acetylcholinesterase conjugated metal organic framework using Internet of things. Sci Rep 9:1–9. https://doi.org/10.1038/s41598-019-56510-y

Najjar LJ, Thompson JC, Ockerman JJ (1997) Wearable computer for quality assurance inspectors in a food processing plant. In: International symposium on wearable computers, digest of papers, pp 163–164

Oerke EC (2006) Crop losses to pests. J Agric Sci 144:31–43. https://doi.org/10.1017/S0021859605005708

Popp J (2011) Cost-benefit analysis of crop protection measures. J fur Verbraucherschutz und Leb 6:105–112. https://doi.org/10.1007/s00003-011-0677-4

Prasad S, Peddoju SK, Ghosh D (2014) Energy efficient mobile vision system for plant leaf disease identification. In: IEEE wireless communications and networking conference, WCNC. Institute of Electrical and Electronics Engineers Inc., pp 3314–3319

Prativa KC, Bhattarai B (2012) Effect of integrated nutrient management on the growth, yield and soil nutrient status in tomato. Nepal J Sci Technol 12:23–28. https://doi.org/10.3126/njst.v12i0.6474

Rani RA, Syono MI, Ramli AS (2008) Multi-finger gate ISFET (Mf-ISFET) for pH sensor application. In: IEEE International Conference on Semiconductor Electronics, Proceedings, ICSE, pp 350–353

Ren T, Noborio K, Horton R (1999) Measuring soil water content, electrical conductivity, and thermal properties with a thermo-time domain reflectometry probe. Soil Sci Soc Am J 63:450–457. https://doi.org/10.2136/sssaj1999.03615995006300030005x

Sharma N, Pant BD, Mathur J (2018) MEMS or microchip based capillary electrophoresis tool for analysis of Pathogenesis-related (PRs) proteins in Eruca sativa (Miller) cultivars against Fusarium oxysporum. Microsyst Technol 24:4933–4944. https://doi.org/10.1007/s00542-018-3909-z

Sharma N, Pant BD, Mathur J (2019) MEMS devices used in agriculture—a review. Journal of Biosensors and Bioelectronics 10:1. https://doi.org/10.4172/2155-6210.1000267

Smits E, Schram J, Nagelkerke M, et al (2012) Development of printed RFID sensor tags for smart food packaging. 14th Int Meet Chem Sensors 403–406. https://doi.org/10.5162/IMCS2012/4.5.2

Stoytcheva M, Zlatev R (2011) Organophosphorus pesticides analysis. Pestic Mod World—Trends Pestic Anal. https://doi.org/10.5772/20892

Sumriddetchkajorn S (2013a) Mobile device-based optical instruments for agriculture. In: Sensing technologies for biomaterial, food, and agriculture 2013. SPIE, p 88810D

Sumriddetchkajorn S (2013b) How optics and photonics is simply applied in agriculture? In: ICPS 2013: International Conference on Photonics Solutions. SPIE, p 888311

Valente A, Morais R, Tuli A et al (2006) Multi-functional probe for small-scale simultaneous measurements of soil thermal properties, water content, and electrical conductivity. Sensors Actuators, A Phys 132:70–77. https://doi.org/10.1016/j.sna.2006.05.010

Wollenhaupt NC, Wolkowski RP (1994) Grid soil sampling. Better Crop 78:6–8

Wu Y, Chang K (2013) An empirical study of designing simplicity for mobile application interaction. Proceedings of the Nineteenth Americas Conference on Information Systems, Chicago, Illinois, August 15–17, 2013

Yang F, Li S (2008) Development of information support system for the application of new maize variety based on smartphone. In: IFIP International Federation for Information Processing, pp 817–824

Zhang W, Kantor G, Singh S (2004) Integrated wireless sensor/actuator networks in an agricultural application. In: SenSys'04—Proceedings of the Second International Conference on Embedded Networked Sensor Systems. Association for Computing Machinery, New York, New York, USA, p 317

Part II
Biosensors in Food Science

Chapter 13
Advances in Biosensors Based on Electrospun Micro/Nanomaterials for Food Quality Control and Safety

Aylin Altan and Meryem Yılmaz

Abstract The food industry needs fast, sensitive, simple, and low-cost techniques to ensure the safety and quality of food products throughout all the stages from production to dispatch. The application of different types of biosensors to monitor safety and quality of food has an increasing importance to meet these requirements. Electrospun nanomaterials produced by electrospinning technology have high potential for use in the development of biosensors due to their characteristic features such as a high surface area-to-volume ratio, controllable porosity, and interconnectivity. The large surface area of electrospun nanofibers provides the abundant active sites for immobilization of biomolecules and active species. Furthermore, electrospun nanofibers can be further functionalized by incorporation of nanomaterials within polymer solution or after electrospinning process to improve sensitivity, stability, and selectivity of biosensors. Recent studies show that electrospun fibers are promising for the design of the biosensor platform in many areas such as clinical, biomedical, and environmental fields, but there are limited number of studies for their use in food applications. This chapter aims to present the recent progress regarding the development of electrospun nanofibers-based electrochemical biosensors in monitoring food safety and quality, such as detection of pesticides, pathogens, and toxins, both detection and quantification of food components, and process control.

Keywords Biosensors · Electrospinning · Nanofiber · Food safety · Quality control

13.1 Introduction

Quality assurance and safety of food during manufacturing process are the essential requirements to preserve the quality of foods while preventing contamination and spread of foodborne illness. Food safety and quality relates to the microbiological, toxicological, chemical, physical, and organoleptic characteristics of foods, which

A. Altan (✉) · M. Yılmaz
Department of Food Engineering, Mersin University, Mersin, Turkey
e-mail: aaltan@mersin.edu.tr

R. N. Pudake et al. (eds.), *Biosensors in Agriculture: Recent Trends and Future Perspectives*, Concepts and Strategies in Plant Sciences,
https://doi.org/10.1007/978-3-030-66165-6_13

have to be ensured to guarantee in an acceptable level for adequate protection to consumers. Conventional methods have an important role in evaluating the safety and quality of foods, but they require highly trained staff and long procedures that limit their use for routine analysis. The demand for rapid, sensitive, and accurate measurements has been continuously rising in evaluating of food quality and safety during and after food processing. Nanotechnology has great potential to meet these demands of food industry. Recent studies have focused on the potential applications of nanotechnology in food and agriculture as encapsulation of active compounds, controlled release of active ingredients, smart packaging, and sensors. Sensors are devices used for the detection or measurement of the physical or chemical properties of the target analyte to which the device responds by giving a signal (Kress-Rogers 1998). These devices contain two main components which are a receptor and a transducer. The physical and chemical information is transformed into a form of energy by receptors, while this energy is converted into electrical signal by transducers (Caon et al. 2017). In biosensors, bioreceptors recognize a target analyte, while transducers convert biochemical signals into a measurable signal. The bioreceptor or biorecognition element in biosensors can be an organic or biological material (e.g., enzyme, antigen, microbe, hormone, or nucleic acid) (Yam et al. 2005). Biosensors developed based on nanomaterials are also called as nanobiosensors. Low detection limits can be achieved by nanobiosensors, leading to significant improvements in food quality control and safety (Pérez-López and Merkoçi 2011). Nanotechnology allows development of nanomaterials such as nanocomposites, carbon nanotubes, metal nanoparticles, and nanowires. These materials play a major role in the design of biosensing systems because they exhibit unique surface chemistry, thermal and electrical properties, and high surface area per unit mass (Asefa et al. 2009; Pérez-López and Merkoçi 2011; D'Souza et al. 2017). The unique properties of nanomaterials can help improve response times and the sensitivity of biosensors.

The versatility of electrospinning technology allows spinning of various polymers for the production of micro/nanofibers with large surface area, high porosity, and high molecular orientation (Zhao et al. 2020). Such submicron and nanofiber electrospun materials have many structural and functional advantages such as tunable fiber diameters and tailored morphology, efficient encapsulation, sustained and controlled release of active compound (Wen et al. 2017). These materials have been applied in a wide range of fields such as biomaterials, drug delivery, energy storage and conversion, environmental protection but the application of nanofibers in agriculture and foods is relatively new. Recent studies have demonstrated the great potential of electrospun fibers in encapsulation of bioactive compounds with enhanced stability and high encapsulation efficiency and active and intelligent food packaging applications (Fig. 13.1) (Echegoyen et al. 2017; Wen et al. 2017; Topuz and Uyar 2020). However, many structural and functional advantages of electrospun fibers make them potential candidates for biosensor applications (Yılmaz and Altan 2017). The large surface area of nanofibers can provide an increase in the availability of sites for the interaction or loading of catalysts or other reactive materials and this can increase the reactivity of the material and fast adsorption or release mechanism in the development of biosensors (Mercante et al. 2017). This chapter presents the electrospinning

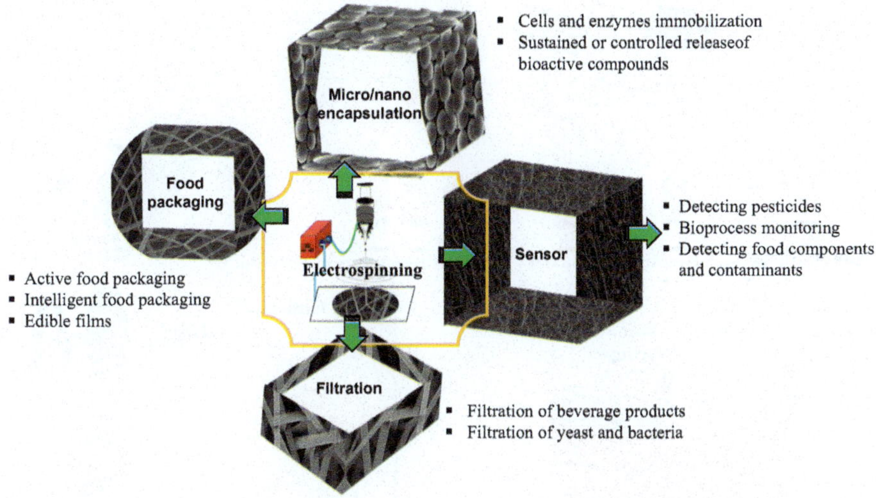

Fig. 13.1 The potential use of electrospun nanofibers in food applications

process and the recent applications of electrospun nanofibers in the development of electrochemical biosensors for food quality control and safety.

13.2 Electrospinning Process

Electrospinning is an effective and versatile process that produces fibers from a variety of polymers in submicron to nanoscale range. A power supply with high voltage, spinneret, syringe pump, and collector are basic parts of the electrospinning system (Fig. 13.2). During processing, the polymer solution is first pumped at a constant rate, polymer droplet is formed at the tip of the needle or nozzle. The droplet at the nozzle tip is subjected to two forces that are the Coulombic force of the external electric field and the electrostatic repulsion of similar charges, and hence the droplet shape changes from hemispherical to conical shape, known as the Taylor cone. After reaching a critical electric field value where the electrostatic force exceeds the surface tension, the charged liquid jet of polymer solution is ejected from the tip of the cone. The jet is elongated by whipping and bending motion as it travels to the grounded collector (Bhushani and Anandharamakrishnan 2014). As the jet moves toward the collector, the charged jet becomes thinner and solvent is evaporated rapidly. Consequently, the solidified ultrafine fibers are collected on the collector. The forces that are electrostatic, drag, gravitational, Coulombic repulsion, viscoelastic and surface tension act on the charged droplet (Mit-uppatham et al. 2004). The electrostatic force is responsible for carrying the charged jet from tip to the grounded collector, while the repulsive Coulombic force repels the adjacent charged species present in the jet and therefore stretches the charged jet. Both the surface tension and viscoelastic forces work against the stretching of charged jet.

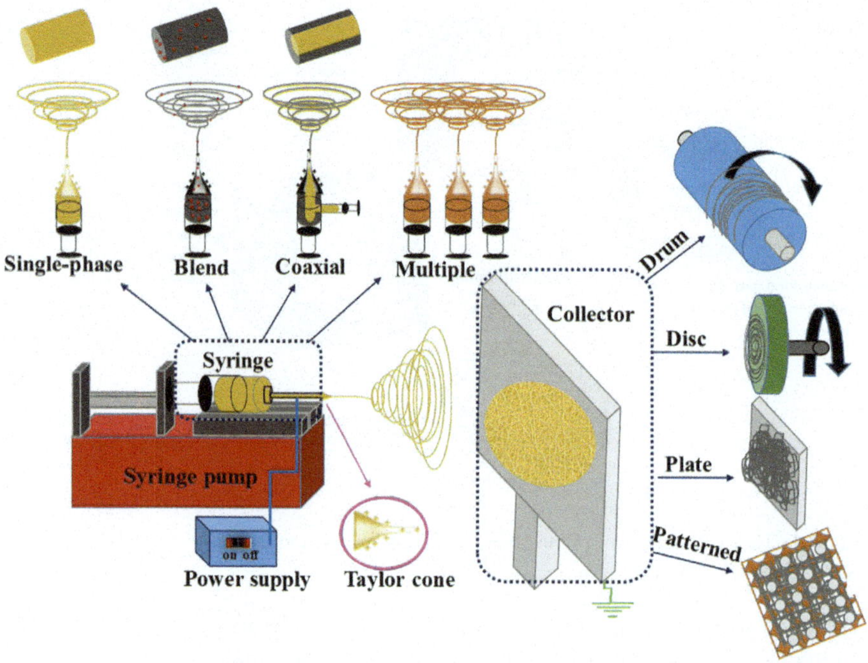

Fig. 13.2 Schematic illustration of the electrospinning process with different collectors and spinnerets

The jet is also subject to drag force due to friction between the charged jet and the surrounding air (Mit-uppatham et al. 2004).

The solution electrospinning and melt electrospinning are two major categories of the electrospinning systems (Topuz and Uyar 2020). The solution electrospinning has been commonly employed in the development of electrospun based micro/nanostructured materials for food applications due to feasibility, versatility, and absence of heat during process. The morphology and structure of electrospun fibers can be modified using different electrospinning designs (Fig. 13.2). These are blending electrospinning, coaxial electrospinning, emulsion electrospinning, edge electrospinning, and magnetic electrospinning (Zhang et al. 2020). In blending electrospinning, all ingredients are co-dissolved in one solution and processed using a single nozzle. Solution and processing parameters determine the morphology of the electrospun fibers, and the resulting fiber can be beaded, porous, or smooth depending on these properties (Fig. 13.3). There are two needles arranged concentrically in coaxial electrospinning to obtain the core-shell nanofibers. Emulsion electrospinning using a single nozzle is also another way to produce core-shell structured fibers. During the electrospinning of the emulsion, the continuous phase of emulsion produces the shell structure of fibers, while the core part of fibers is obtained from the droplet phase.

A typical electrospinning system has a lower throughput because a single needle can only produce one polymer jet. Mass production of nanofibers can be achieved

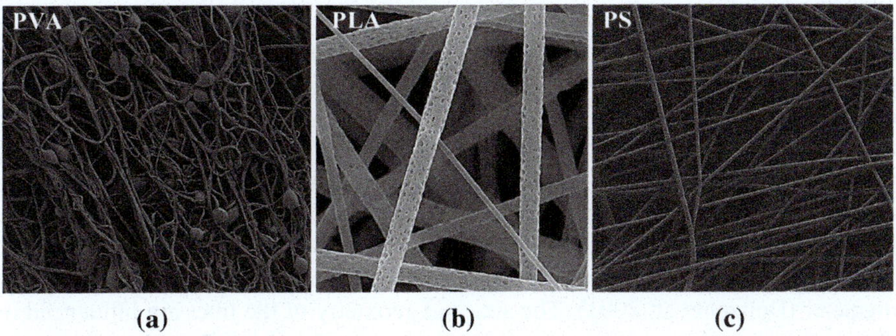

Fig. 13.3 Scanning electron microscopy (SEM) images of electrospun fibers with different morphologies (a) beaded; (b) porous, and (c) smooth (PVA: polyvinyl alcohol; PLA: polylactic acid; PS: polystyrene)

by the formation of multiple jet system with multiplying needles or multiple jets accomplished with various needless electrospinning methods (Wei et al. 2019). In edge electrospinning set up, the source of electrode is either the edge of flat plate or the edge of a bowl filled with the solution (Echegoyen et al. 2017). The polymer flow is gravity assisted in edge-plate configuration and dependent on the angle of plates and the volume of the solution in the reservoir and the size of pipette aperture (Thoppey et al. 2010). For bowl configuration, spinning takes place directly from the polymer solution surface within the bowl rather than utilizing falling or elongated droplets as in the edge-plate geometry (Thoppey et al. 2011). Highly aligned fibers can be obtained using rotating mandrel, drum and disk collectors that allow scalability and improved conformability (Quirós et al. 2016; Zhang et al. 2020). The electrospinning technique has many advantages over other fiber production methods such as cost effectiveness, easy operation and function, a wide range of polymer materials, mild processing conditions and easy incorporation of bioactive compounds into nanofibers (Wen et al. 2017; Zhao et al. 2020). The process parameters can be adjusted to prepare electrospun nanofibers with custom shapes and various orientations for requirements in various research fields (Zhang et al. 2020). The main parameters that affect the spinnability, fiber morphology, and diameter distribution are polymer solution properties, processing parameters, and ambient conditions.

13.2.1 Polymer Solution Properties

In the electrospinning process, solution parameters influencing the electrospinnability of polymer solution are viscosity, polymer concentration, polymer molecular weight, electrical conductivity, and surface tension. Solution viscosity which depends on the polymer concentration and the molecular weight of the polymer is an important parameter in electrospinning (Neo et al. 2012). The higher the molecular weight of the polymer, the more entangled polymer chains in the solution and hence the viscosity of polymer solution increases (Tan et al. 2005). Higher levels of chain entanglements as a result of both higher molecular weights of the polymer and increased polymer concentrations favor fiber formation. Therefore, there is a minimum concentration for a particular polymer solution that provides critical chain entanglements to obtain fiber during electrospinning (Shenoy et al. 2005). Beaded fibers are obtained at low concentration with low solution viscosity, but further increase in concentration or solution viscosity results in uniform fibers with increased diameter (Deitzel et al. 2001). The size and geometry of the fiber are influenced by the physical instabilities acting on the polymer solution jet. When the viscosity of the solution is not high enough or when the electric field strength is low, the axisymmetric instability, known as the Rayleigh instability, occurs due to the surface tension of solution. The charged polymer jet breaks up into droplets and leads to the beaded fibers morphology (Baji et al. 2010). Adding a surfactant to the solution can reduce the surface tension, and this facilitates the formation of fibers at a lower electric field (Shenoy et al. 2005).

The solution conductivity is another important parameter affecting the process and morphology of the resulting electrospun fibers. The polymer type, solvent and the availability of ionisable salts affect the solution conductivity (Bhardwaj and Kundu 2010). It determines the elongation level of a jet by an electrical force because it reflects a charge density on a jet (Tan et al. 2005). Solutions having higher conductivity results in higher elongation of jet due to high surface charge density. The self-repulsion between the excess charges present in the jet causes the bending and whipping instabilities. These instabilities promote the thinning and elongation of the jet and thus lead to smaller diameter fibers (Baji et al. 2010).

13.2.2 Electrospinning Process Parameters

The main processing parameters that affect electrospinning are solution feeding rate, applied voltage, distance from needle tip to collector. The critical voltage is required to induce necessary charge on the droplet and the electrospinning process begins when the electrostatic force exceeds the surface tension (Mazoochi and Jabbari 2011). The higher stretching of the jet occurs due to the greater Coulombic forces in the jet at high voltage and the resulting electric field, leading to a decrease in the diameter of fiber as well as rapid evaporation of the solvent from the fibers (Bhardwaj and Kundu 2010; Mazoochi and Jabbari 2011). Shenoy et al. (2005) suggested that the morphology varies from beads to beaded fibers to only fibers as the voltage is increased. However, any further increase in voltage may result in relatively large diameter fibers due to rapid removal of the solution from the needle tip (Haghi and Akbari 2007). Meanwhile, the tendency to bead formation increases at higher voltages, possibly due to jet instability in the high voltage electrical field (Deitzel et al. 2001; Shenoy et al. 2005).

The flow rate controls polymer solution available for electrospinning. A minimum solution flow rate is required to keep a stable Taylor cone (Zong et al. 2002). Megelski et al. (2002) reported that an increasing flow rate with a decrease in spinning voltage resulted in increased fiber size. Yuan et al. (2004) suggested that the lower flow rate gives the solvent much more time to evaporate so that the formation of beads in the electrospun fibers can be avoided.

The polymer jet behavior and the morphology of the resultant fibers are affected by changing the distance from the nozzle tip to the collector. For instance, if the distance is short, the electric field strength increases and speeds up the electrospinning process (Zaikov 2017). However, the time required for evaporation of the solvent cannot be provided. Consequently, wet electrospun fibers can be obtained due to insufficient evaporation of solvent with the short distance between the nozzle tip and the collector (Ghorani and Tucker 2015). Zhao et al. (2004) reported that increasing the distance causes a decrease in the flight speed of the jet and the jet flight takes longer time from the nozzle tip to the collector. As a result, the charged jet elongates more and thin fibers with a narrow diameter range are formed.

Other processing parameters such as needle or nozzle inner diameter can also influence the electrospinning process (Ghorani and Tucker 2015). Mo et al. (2004) observed clogging at the tip of nozzle when the nozzle diameter was increased. They found that using smaller diameter needle led to bead free fibers without clogging the needle tip. Zhao et al. (2004) suggested that decreasing the diameter of nozzle increased surface tension of the droplet. The flight time for the jet from the nozzle to the collector is increased due to the increased surface tension and small diameter can be obtained.

13.2.3 Environmental Conditions

Electrospinning process and fiber properties are affected by environmental conditions such as ambient temperature and relative humidity since these parameters may influence the properties of electrified jets (Quirós et al. 2016). De Vrieze et al. (2008) found that increasing ambient temperature causes a decrease in fiber diameter due to the effect of temperature on solvent evaporation rate and viscosity of the polymer solution. Chen and Yu (2010) investigated the effect of the temperature of the polymer solution on the fiber diameter. It has been reported that the electrospinning ability of the polymer solutions is improved by increasing the temperature. This is because the viscosity and surface tension decrease as the temperature increases (Chen and Yu 2010). Casper et al. (2004) showed that raising the relative humidity causes an increased number of pores on the surface of fibers. The volatile solvent evaporation can be fast at very low humidity. This can create a problem such as clogging of needle tip during electrospinning process. The discharge of the electrospun fibers may increase in a more humid environment (Bhardwaj and Kundu 2010).

13.3 Biosensors Based on Electrospun Micro/Nanomaterials

Biosensors can be defined as devices that can detect the presence of a particular biochemical analyte and convert a biological response into a readable electrical/optical signal. Biosensors basically consist of three components, a bioreceptor or biorecognition element, a transducer, and a detector. The function of bioreceptors is to recognize the target analyte, while transducers are used to convert the recognized event into a measurable electrical signal (Maftoonazad and Ramaswamy 2018). The steps required for the fabrication of a biosensor have been described by De Cesare and Macagnano (2015). These steps are as follows: (i) choosing a suitable bioreceptor that targets biological identification; (ii) immobilizing the receptor on the transducer; (iii) selecting the transduction mechanism; and (iv) using a suitable detector to obtain reproducible responses (De Cesare and Macagnano 2015). Biosensors can be categorized either on the basis of their transduction mechanisms or their bioreceptors used. The types of transduction mechanism include electrochemical, optoelectronic, piezoelectric, bioluminescent, resonant mirror, and thermistor. A variety of biomolecules can be used as a biorecognition element depending on the sensing mechanism. They include antibodies, aptamers, carbohydrates, and antimicrobial peptides (Gunasekaran 2014). Biosensors designed using nanomaterials are called nanobiosensors. Nanobiosensors can have very low detection limits and high stability due to the incorporation of nanomaterials because these nanomaterials can provide improved electrical conductivity and chemical reactivity with increased interacting surface area. Electrospinning technology can be one of the most effective routes to significantly increase the sensitivity of biosensors by taking advantage of the extremely large surface area provided by nanofibers (De Cesare and Macagnano 2015). It is possible to produce nanofibers with diameter ranging from micro to nanoscale, which leads to large surface area per unit mass. Electrospinning also allows to produce different structures such as hollow, porous, and core-sheath by adjusting process parameters and solution properties. Therefore, electrospun nanofiber membranes with such outstanding properties make these materials highly attractive to incorporate them within sensors designed for different applications (Ding et al. 2010). The use of nanofibers can remarkably increase the availability of sites for immobilization within biosensors due to their large surface area and porosity. The number of biorecognition sites also increases for biosensing through the immobilization of bioreceptors into or onto electrospun nanofibers (Matlock-Colangelo and Baeumner 2012; De Cesare and Macagnano 2015). Figure 13.4 shows a schematic representation of the electrochemical biosensor based on the electrospun fiber. Electrospinning technology allows to the use of various polymer materials to obtain fibers with controlled morphologies, sizes, and compositions.

Electrospun nanofibers can further be functionalized through the incorporation of nanoscale materials (e.g., magnetic nanoparticles, quantum dots) within spinning solutions or on the surface of nanofibers after spinning (Zhang et al. 2017). Therefore, the incorporation of these material in the biosensing platform can enable to

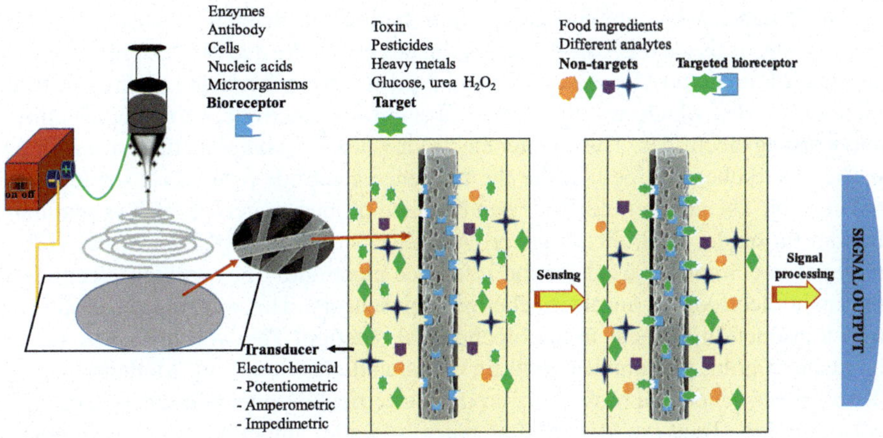

Fig. 13.4 Schematic representation of the electrochemical biosensor assembly

achieve low detection limits with high sensitivity of the resulting biosensors. Applications related to the use of nanobiosensors are mostly in diagnostic, biomedical, and environmental but they can also be powerful tools in food safety, quality control, and traceability by providing fast, high sensitivity and specificity in measurements (De Cesare and Macagnano 2015). Electrospun nanofibers have many applications in various fields such as material science, tissue engineering, biomedicine, environmental science, and sensing (Zhang et al. 2019). However, the use of nanofibers in food applications is very recent, especially in biosensors. All these unique features of electrospun nanomaterials make electrospinning technology highly attractive to nanotechnological applications in food field. This review provides the main recent electrospun nanobiosensors studies and applications related to the food safety and quality control.

13.3.1 Electrochemical Biosensors

By definition of the International Union of Pure and Applied Chemistry (IUPAC), an electrochemical biosensor is a self-contained device that can selectively provide quantitative or semi-quantitative analytic information using a bioreceptor being in direct spatial contact with an electrochemical transduction element (Thévenot et al. 1999). Electrochemical biosensors offer great promise for a wide variety of applications in clinical diagnosis, food and environmental analysis because they have many advantages such as high selectivity, wide detection range, low cost, and easy miniaturization (Viswanathan et al. 2009; Pilolli et al. 2013). Depending on the biological recognition process, affinity sensors and biocatalytic devices are the two main categories of electrochemical biosensors. In an affinity biosensor, a binding reaction occurs among biomolecules such as an antibody and targeted analyte. On the other hand, a biocatalytic device uses biological elements such as enzymes, whole cells, or tissue that can recognize the target analyte and produce electroactive species (Mondal and Sharma 2016). In a typical electrochemical sensing, there are three electrodes which are the counter or auxiliary electrode, reference electrode, and working electrodes. The counter electrode is used to complete the current circuit in the electrochemical cell, while the reference electrode provides a stable potential control. The counter electrode can be made of inert materials such as platinum, indium tin oxide, gold, etc. A silver (Ag)/silver chloride (AgCl) or glassy carbon material is commonly used for the reference electrode. The working electrode is the main electrode of an electrochemical biosensors. The working electrode acts as a transduction element in the biochemical reaction. The working electrode can be made of gold, platinum, indium tin oxide, and glassy carbon (Malhotra and Ali 2018). Electrochemical biosensors produce electrical outputs as a function of target concentration. They are classified as potentiometric, amperometric, voltammetric, and impedimetric (Thévenot et al. 1999; Pilolli et al. 2013).

13.3.1.1 Potentiometric Biosensors

Potentiometric sensors measure potential difference between two reference electrodes at near zero current, separated by a membrane (Pilolli et al. 2013). For biosensing, a biological element (enzyme, antibody, aptamers) within the sensor catalyzes a reaction, an ion formed can be detected using specific electrode. An ion-selective electrode is used to determine the change in ionic concentrations in these biosensors because the release and uptake of hydrogen ions occurs in many enzymatic reactions. In potentiometric biosensors, the potential difference between the potentiometric electrode and reference electrode is measured because this measured difference is proportional to the concentration of the substrate. There are three types of ion-selective electrodes (ISE) which are based on glass electrodes for cations, glass pH electrode layered with a gas-permeable membrane and solid-state electrode (Sahota et al. 2017). The ion-selective field effect transistors (ISFET) and light addressable potentiometric sensor (LAPS) are two other important types of potentiometric biosensor. They have basically adopted a similar operational principle and both systems consist of electrolyte/insulator/semiconductor structure. The potential is generated by the effect of surface ion concentration from a solution in the ISFET but, in the case of LAPS, the area of interest is illuminated by the light that generates photocurrent, which is correlated with surface ion concentration (Zaid et al. 2019).

13.3.1.2 Voltammetric/Amperometric Biosensors

Both voltammetric and amperometric techniques measure the electric current as a result of the electrochemical oxidation or reduction of an electroactive species

(Thévenot et al. 1999). Amperometric biosensors are a subclass of the voltammetric sensors (Viswanathan et al. 2009). The amperometric biosensor is based on monitoring the current as a function of time while keeping the working electrode at constant potential, but in voltammetric technique, the current is measured at varying potential (Pilolli et al. 2013). The continuous alteration of the potential is applied to the solution and the resulting current is measured. Cyclic voltammetry (CV) is the most popular technique, which is used to determine electrochemical behavior of analyte. The most commonly used voltammetric techniques are differential pulse (DPV), normal pulse (NPV), and square wave (SWV). In these techniques, the applied potential changes through pulsing from one potential to another rather than sweeping through varying potentials (Chen and Shah 2013). Amperometric measurements are usually performed on a platinum, gold, titanium, indium tin oxide, or carbon-based working electrode or an array of electrodes with respect to a reference electrode (Thévenot et al. 1999). The detection mechanism of amperometric biosensors is the movement of electrons due to enzyme-catalyzed redox reactions and the current flow resulting from the transferred electron by the enzymatic reaction substrate is measured (Sahota et al. 2017). The improvement of sensitivity in enzymatic biosensors can be provided by increasing the electrode surface area. The electrode surface area can be increased by using nanomaterials and thus this allows a higher amount of enzyme loading in the recognition layer (Rathee et al. 2016).

13.3.1.3 Impedimetric Biosensors

In impedimetric biosensors, the electrical impedance of the electrode-electrolyte interface is measured by applying a sinusoidal voltage to a sample and by simultaneously measuring the current that flows through it (Pilolli et al. 2013). The electrical impedance is equal to the ratio between the voltage and the corresponding current in a given frequency domain (Malhotra and Ali 2018). The impedimetric technique is generally referred to as electrochemical impedance spectroscopy (EIS), which is a very powerful technique for analysis of both resistive and capacitive properties arising from the biorecognition events at the interface between electrode and analyte (Tran et al. 2016). Biomolecular interactions such as DNA hybridization, antigen-antibody, enzymatic and protein-protein interactions can be quantified by using impedimetric biosensors (Malhotra and Ali 2018). The impedimetric biosensing is based on monitoring changes in interfacial properties such as electric current flow, charge transfer resistance, or double layer capacitance when the target analyte reacts with a specific bioreceptor on the surface of the sensor. The impedance value varies in proportion to both biochemical reaction rate and the related analyte concentration (Tran et al. 2016).

13.3.2 Electrochemical Biosensors Through Electrospinning

Electrochemical nanobiosensors with high efficiency (e.g., rapid response, low detection limit, high sensitivity) are nowadays being developed by the incorporation of nanomaterials (Hammond et al. 2016). The unique features of nanofibers make them an ideal candidate for incorporation within sensors designed for food analysis and food processing control as well as the tracing of contaminants. There is an increasing demand in the food industry for fast, real-time, simple, and low-cost techniques to ensure safety and quality of food products. Some of these advantages and potentials of electrospun materials open doors to the development of different biosensing tools for possible applications in the food industry. Table 13.1 summarizes some of the recent electrospun based electrochemical biosensor studies and applications for food safety and quality control.

Contamination of food products with physical, chemical, or biological hazards can happen at any point in the food production and processing chain. The source of chemical hazards can be ingredient-related (e.g., pesticides, mycotoxins and other natural toxins, animal drug residues, heavy metals, environmental contaminants), process-related, and facility-related (FDA 2018). These chemical contaminants may show adverse effects on human and animal health. Pesticides are generally used for pest control, especially in plant protection. Among pesticides, pirimiphos-methyl is an organophosphorus insecticide with both contact and fumigant action. It can be applied to a variety of plants such as citrus, cereals, olives, sugar canes, and vegetables (El-Moghazy et al. 2016). El-Moghazy et al. (2016) fabricated an enzyme-based electrochemical biosensor using electrospun nanofibers for detection of pirimiphos-methyl in olive oil. Electrospun fibers were prepared from a polymer blend of chitosan and poly (vinyl alcohol) and collected on the surface of screen-printed electrodes to immobilize genetically modified acetylcholinesterase (AChE). The electrospun based biosensor prototype was tested to determine pirimiphos-methyl concentration in olive oil. This study showed that the response of developed biosensor could be improved by twofolds compared to casted electrodes due to the use of nanofibers. The detection limit achieved for primiphos-methyl was 0.2 nM, which is well below the limit permitted by international regulations (164 nM).

In another study, a novel biosensor has been developed for detecting organophosphate pesticides (OPs) in aqueous system (Bao et al. 2016). The proposed biosensor was constructed based on nanocomposites of titanium dioxide nanofibers (TiO_2NFs) and carboxylic acid functionalized multi-walled carbon nanotubes (c-MWCNTs). The suspension of both TiO_2NFs and c-MWCNTs was casted onto the glassy carbon electrode surface. The mixture of elastin-like-polypeptide-organophosphate hydrolase (ELP-OPH) and bovine serum albumin (BSA) was coated on the surface of the prepared electrode to immobilize OPH enzyme. The ELP-OPH, acted as biocatalyst for OPs, was obtained by purification from genetically engineered *Escherichia coli* based on the unique thermal triggered phase transition of ELP. TiO_2NFs were produced by electrospinning technique and subsequent calcination process. They were preferred because of their affinity to the phosphoric group in organophosphates.

Table 13.1 Electrospun nanofiber-based electrodes in electrochemical biosensing and potential applications for food safety and quality control

Working electrode material	Transducer type	Analyte	Application	Detection limit	Linearity	Reference
AChE/CS-PVA NFM/SPE	Amperometric	Pirimisphos-methyl	Pesticide detection in olive oil	0.2 nM	1×10^{-10}–8×10^{-8} M	El-Moghazy et al. (2016)
ELP-OPH/BSA/TiO_2NFs/c-MWCNTs	Amperometric	Methyl parathion	Pesticide detection in lake water	12 nM	Up to 36.4 μM	Bao et al. (2016)
BCL@MOF nanofibers/chitosan/GCE	Voltammetric	Methyl parathion	Pesticide detection in vegetables	0.067 μM	0.1–38 μM	Wang et al. (2019a)
MWCNT-ZnO nanofibers/GCE	Impedimetric	Atrazine	Pesticide detection	5.368 zM	10 zM–1 μM	Supraja et al. (2020)
Aptamer modified CMCNFs FET	Potentiometric	Bisphenol A	Bisphenol A detection in bottled water	1 fM	10^0–10^4 fM	Kim et al. (2016)
PVA/GQD/GCE	Amperometric	H_2O_2	–	0.53 μM	0.1–200 mM	Zhang et al. (2015)
Gelatin/chitosan nanofibers/HRP	Voltammetric	H_2O_2	Disinfectant	0.05 mM	0.1–1.7 mM	Teepoo et al. (2017)
CuO/PANI nanofibers/GCE	Amperometric	H_2O_2	–	0.11×10^{-6} M	5 μM–9.25 mM	Liu et al. (2019)
CNFs/$MnCo_2O_{4.5}$-TMB	Colorimetric	Sulfite, Ascorbic acid	–	19 nM 0 M	0–1 μM 0–40 μM	Gao et al. (2017)

(continued)

Table 13.1 (continued)

Working electrode material	Transducer type	Analyte	Application	Detection limit	Linearity	Reference
AChE: acetylcholinesterase, CS-PVA NFM: chitosan-poly (vinyl alcohol) nanofibrous membranes, SPE: screen-printed electrodes, ELP-OPH: elastin-like-polypeptide-organophosphate hydrolase, BSA: bovine serum albumin, TiO2NFs: titanium dioxide nanofibers, c-MWCNTs: carboxylic acid functionalized multi-walled carbon nanotubes, BCL@MOF: Burkholderia cepacia lipase@Metal organic framework, GCE: glassy carbon electrode, MWCNT-ZnO: multi-walled carbon nanotube-Zinc oxide, CMCNFs: carboxyl-functionalized modified multichannel carbon nanofibers, FET: field-effect transistor, GQD: graphene quantum dots, HRP: Horseradish peroxidase, CuO/PANI: cupric oxide/polyaniline, CNFs: carbon nanofibers, TMB: 3,3',5,5' Tetramethylbenzi-dine						
FTO/PA6/PPy/ZnO/urease	Impedimetric	Urea	Urea detection in milk	0.011 mg/dL	0.1–250 mg/dL	Migliorini et al. (2018)
Laccase/A-CZUF/GCE	Voltammetric	Catechol	–	0.166 μM	0.166–7 μM	Chen et al. (2015)
BSA/anti-AFB1/μchannel/carbon-Pt	Impedimetric	Aflatoxin B1	–	1 pg/mL	1 pg/mL-μg/L	Mondal et al. (2016)
PAN nanofiber modified electrode	Voltammetric	Zearalenone	Mycotoxin detection in food simulant	5 nM	5–30 nM	Moradi et al. (2020)
PANi/PEO composite nanofiber-based interdigitated electrode	Chemiresistive	DENVCP, *Staphylococcus aureus* specific genes: nuc, mecA, vanA, protein A	DNA hybridization detection	1.9 fM 0.31 pM, 0.42 pM, 0.67 pM 0.17 pM	10 fM–1 μM 1 pM–1 μM	Tripathy et al. (2019)
PVA/PAA hydrogel NF-LAPS	Potentiometric	*Escherichia coli*	*Escherichia coli* detection in orange juice	10^2 CFU/mL	10^2–10^6 CFU/mL	Shaibani et al. (2018)
GN/CN/GelMA/RBL mast cell/SPE	Voltammetric	Milk allergen casein	Casein detection in milk	3.2×10^{-8} g/mL	1×10^{-7}–1×10^{-6} g/mL	Jiang et al. (2019)

(continued)

Table 13.1 (continued)

FTO: fluorine doped tin oxide, PA6: polyamide 6, PPy: polypyrrole, A-CZUF:gold-crosslinked zein ultrafine fibers, AFB1: Aflatoxin B1, Pt: platinum, PAN: Polyacrylonitrile, PANi/PEO: polyaniline/polyethylene oxide, DENVCP: dengue virus specific consensus primer, PAA: poly(acrylic acid), NF-LAPS: nanofiber-light addressable potentiometric sensor, GN/CN/GelMA/RBL: graphene/carbon nanofiber/gelatin methacryloyl/rat basophilic leukemia, SPE: Screen Printed Electrode

XOD@Cu-MOF/SA/GCE	Amperometric	Hypoxanthine Xanthine	Hypoxanthine and xanthine detection in seafood	0.0023 µM 0.0064 µM	0.01–10 µM 0.01–10 µM	Wang et al. (2019b)
GOX/CNT/NFM/GCE	Amperometric	Glucose	Glucose detection in commercial beverage and monitoring brewing process	20 µM	1–3 mM	Mason et al. (2016)
Ni-Co3O4/CPE	Amperometric	Maltose	Maltose detection in malt syrup	32.5 nM	0.2–100 µM	Xu et al. (2016)
GOD/TCNFs/GCE	Amperometric	Glucose	Glucose detection in human blood serum	3.7 µM	0.013–10.5 mM	Guo et al. (2018)
Ni-CoO/CNF/GCE	Voltammetric	Glucose		0.03 µM	0.25–600 µM	Mei et al. (2019)
Cu-nanoflower @AuNPs-GO NFs	Voltammetric	Glucose	–	0.018 µM	0.001–0.1 mM	Baek et al. (2020)
BSA/Anti-VD/Fe3O4-PANnFs/ITO	DPV	Vitamin-D_3	–	12 ng/mL	10–100 ng/mL	Chauhan et al. (2018)
PtNps/GCNF/LOx-SPCEs	Amperometric	Lactate	Lactate detection in wines and ciders	6 µM	10–2000 µM	Loaiza et al. (2015)

(continued)

Table 13.1 (continued)

XOD@Cu-MOF/SA: Xanthine oxidase@copper-based metal organic framework nanofibers/sodium alginate, GOX/CNT/NFM: Glucose oxidase/carbon nanotubes/nylon nanofibrous membrane, Ni-Co_3O_4/CPE: Nickel doped cobalt (II, III)/carbon paste electrode, AuNPs: Gold nanoparticles, GO: graphene oxide, Anti-VD: antibody specific to Vitamin-D_3, Fe_3O_4-PANnFs/ITO: Magnetite nanoparticles-polyacrylonitrile nanofibers/indium tin oxide, PtNps/GCNF/LOx-SPCEs: platinum nanoparticles/graphitized carbon nanofibers/lactate oxidase-screen-printed carbon electrodes

The functions of BSA and c-MWCNTs on the biosensor platform were stabilization of OPH activity, immobilization of ELP-OPH, and enhancement of the electron transfer in amperometric detection. The developed sensor exhibited excellent performance for methyl parathion and parathion detection in terms of response time (5 s), detection limits (12 nM and 10 nM), and reliability (94.67 ± 2.82%–106.73 ± 2.61%). The high performance of sensor has been attributed to the strong adsorption between organophosphates and nanofibers, and the improvement of electron transfer achieved by using both nanofibers and carbon nanotubes in the biosensor.

Wang et al. (2019a) fabricated enzyme-based biosensor for rapid determination of methyl parathion residues in vegetable samples. Metal-organic framework nanofibers with lipase from *Burkholderia cepacia* and chitosan were used for the fabrication of biosensor. The mixture of nanofibers, lipase, and chitosan was coated on the glassy carbon electrode surface. The prepared electrode was used as working electrode. The developed biosensor was tested with vegetable samples (lettuce and cabbage) spiked with methyl parathion solutions at different concentrations. The samples were immersed in methanol to dissolve residual methyl parathion and the methanolic extracts were used in differential pulse voltammetry measurement. The prepared biosensor responded within concentration range of 0.1–38 μm and a detection limit of 0.067 μm. In addition, the developed biosensor maintained at least 80% of its initial response after 3 weeks of storage at 4 °C.

Recently, Supraja et al. (2020) developed an electrochemical biosensor based on the multi-walled carbon nanotube (MWCNT)-ZnO composite nanofibers for atrazine (ATZ) detection. The composite of MWCNT-ZnO nanofibers were produced by electrospinning technique. The polymer solution containing polyacrylonitrile (PAN), MWCNTs, and zinc acetate dehydrate was prepared and electrospun via electrospinning. The resulting fibers were calcinated to remove PAN and also to obtain ZnO from zinc acetate dehydrate without decomposition of MWCNTs. The electrode was prepared by immobilizing the anti-ATZ antibody onto the glassy carbon working electrode (GCE). Before immobilization of antibody, MWCNT-ZnO nanofibers were used to modify the electrode and then carboxylic functionalization was applied by dipping in BSA. The concentration range of MWCNT-ZnO nanofiber-based electrochemical biosensor was 100 zM–1 μM, while the detection limit was found to be 5.368 zM.

Chemical contaminants must be properly controlled during processing and storage, because if they are above limits, they can be extremely harmful to the human body. Chemicals can become part of food products without intentionally added such as the inevitable migration of chemical compounds from packaging material into food and beverages. For example, there are major concerns about bisphenol A (BPA) because the migration of BPA into food can occur at high temperatures (Kim et al. 2016). Therefore, it is essential to develop rapid, reliable, and sensitive methods for detection of BPA. Kim et al. (2016) developed a field-effect transistor (FET) sensor for BPA detection. The sensor was designed using multichannel carbon nanofibers (MCNFs) and BPA-binding aptamers. The MCNFs were produced from polymer solutions of poly(acrylonitrile) and poly(methyl methacrylate) by a single-nozzle electrospinning technique and carboxyl-functionalized by oxidative treatment

(CMCNFs). They were served as a conductive signal transducer. The performance of sensor was tested on the real samples which are thermal printing paper and polycarbonate bottle filled with distilled water. Hot water and ethanol extracts of BPA from polycarbonate bottle and paper were used. Figure 13.5 shows the schematic diagram and FE-SEM and TEM images of functionalized multichannel carbon nanofibers (CMCNFs). The MCNFs increased the affinity of aptamer to the BPA molecules since they were covalently bonded between the electrode substrate and the aptamer. A large specific surface area offered by the multichannel structure increased the quantity of conjugated aptamers. Therefore, improved performance of the proposed sensor in terms of sensitivity and specificity has been achieved with a detection limit of 1 fM for BPA detection.

Hydrogen peroxide (H_2O_2) is one of the most widely used chemical sterilization agents in food industry for food packaging. After applying H_2O_2, a rinsing step with sterile water and/or blowing with sterile air is required to remove residual H_2O_2. The residual levels permitted in products are legally regulated. Additionally, some milk processors intentionally add H_2O_2 to milk to reduce microbial load. Therefore, the determination of H_2O_2 residues is of practical importance in the food industry and other applications such as clinical, biological, and chemical (Hsu et al. 2008). Teepoo et al. (2017) developed hydrogen peroxide biosensor using composite nanofibers of chitosan and gelatin biopolymers. The composite nanofibers were produced using electrospinning and deposited onto graphite electrode surface. The horseradish

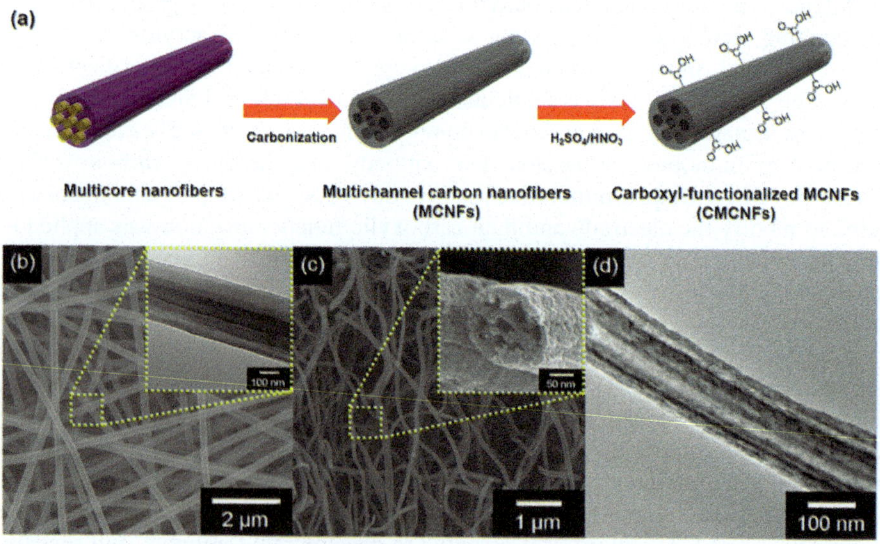

Fig. 13.5 **a** Schematic diagram of carboxyl-functionalized multichannel carbon nanofibers (CMCNFs). **b** FE-SEM and TEM images of PAN-PMMA nanofibers. **c** FE-SEM image of the multichannel carbon nanofibers (MCNFs) and a high magnification FE-SEM image of MCNF cross-section. **d** TEM image of CMCNF (Reprinted from Kim et al. 2016. Copyright with permission from American Chemical Society)

peroxidase enzyme was immobilized on the nanofibers to prepare the biosensor. The prepared electrodes were characterized by cyclic voltammetry. The hydrogen peroxide biosensor had a detection limit of 0.05 mM with a linear calibration range of 0.1–1.7 mM. The immobilized enzyme on the composite nanofibers possessed good bioactivity after two months of use. Recently, the CuO/polyaniline (CuO/PANI) nanocomposite-based enzyme-free electrochemical biosensor has been fabricated for H_2O_2 detection (Liu et al. 2019). To fabricate sensor template, polyamide acid (PAA) nanofibers were produced by electrospinning technique. The hollow PANI nanofibers were prepared by in situ polymerization of aniline monomers followed by removing of the PAA template via dissolving in N, N-dimethylacetamide solution. CuO was coated on the surface of hollow PANI nanofibers to improve the electrochemical properties of PANI fibers. The developed sensor detected H_2O_2 in the range of 5 μM–9.255 mM with a low detection limit of 0.110×10^{-6} M. It has been shown that the prepared sensor maintains its stability when stored at 4 °C for 10 days. These results suggested that the developed sensor can be used for food safety and environmental applications.

The detection of urea is of great importance in many areas, including clinical diagnosis, environmental analysis, and food. Urea is added to milk as an adulterant to increase solids-not-fat value and nitrogen content. Serious health problems such as indigestion, urinary tract obstruction, gastrointestinal bleeding, and cancer may arise when the urea concentration in milk exceeds 70 mg dL^{-1} (Nikoleli et al. 2010). Therefore, rapid and sensitive detection of urea in milk and other food products is essential to ensure food safety. Recently, an impedimetric biosensor based on the electrospun nanofibers has been developed for urea detection (Migliorini et al. 2018). Polymer solutions of polyamide 6 (PA6) and polypyrrole (PPy) were used to obtain electrospun nanofibers by electrospinning and the resulting fibers were collected onto fluorine doped tin oxide (FTO) electrodes. The prepared electrode was modified by the adsorption of zinc oxide nanoparticles (ZnO) on the surface of fibers. The urease immobilized FTO/PA6/PPy/ZnO electrode was used as the working electrode. Milk sample was used to test the fabricated biosensor. The results of this study showed that the urea biosensor based on electrospun nanofibers having a low detection limit (0.011 mg dL^{-1}) with a wide concentration range (0.1–250 mg dL^{-1}) can be used for monitoring urea in dairy products.

Phenolic compounds are important because of their environmental significance and biological roles. The concentration levels of phenolic compounds in community water need to be strictly monitored due to their toxicity (Sarika et al. 2017). So, an effective detection method is needed to monitor phenolic compounds. A novel laccase biosensor for catechol detection has been proposed by Chen et al. (2015). The ultrafine zein fibers cross-linked with glyoxal (CZUF) were produced by electrospinning and coated with gold nanoparticles (Au NPs) by in situ polymerization using poly(ethyleneimine) (PEI). The electrochemical performance of the proposed sensor was measured by square wave voltammetry and cyclic voltammetry. The selectivity of proposed sensor was determined in the presence of other phenolic compounds such as guaiacol, epicatechin, aminophenol, and phenol. The small interference of these phenolic compounds ($\leq 3\%$) toward the biosensor response demonstrated favorable selectivity of the biosensor in this study. The detection limit of catechol was 0.166 μM and a linear range of 0.166–7 μM. Catechol is a natural polyphenolic compound found in some plant products such as vegetables, fruits, and teas. The quantification of catechol is important because of biological roles such as antioxidation and antiviral properties (Wahab and Darain 2017). Therefore, the rapid estimation of polyphenol levels with a catechol biosensor can be used to evaluate the freshness, antioxidant content, and quality control of foods (Sarika et al. 2017; Wahab and Darain 2017).

Foods and feeds can be contaminated with mycotoxins at any stage of their handling or storage. Aflatoxins are mycotoxins that are produced by strains from the *Aspergillus* family. Aflatoxin B1 is one of the highly toxic food contaminants and known to be carcinogenic in human (Xu et al. 2016). The fast, sensitive, specific, and reliable methods for their detection and quantification are important to ensure food safety. Mondal et al. (2016) fabricated an electrochemical immuno-biosensor for aflatoxin B1 (AFB1) detection. The biosensor was fabricated based on the micro/nanochannels created by poly(methyl methacrylate) (PMMA) fibers that embedded in porous carbon electrodes. To fabricate microchannels, the PMMA fibers deposited on the Si wafer were spin coated with polyacrylonitrile (PAN) films. Carbonization was applied to the composite film to remove the PMMA fibers and produce amorphous macroporous carbon films. The channels created were doped with Pt nanoparticles (Pt NPs) to improve the performance of the carbon electrode. The widths of channels were in the range of 200 nm to tens of micrometers. Oxygen plasma treatment was applied to fabricated electrodes with and without microchannels to form –COOH groups on their surfaces and activated using 1-ethyl-3-(3-dimethylaminopropyl) carbodiimide (EDC) and *N*-hydroxysuccinimide (NHS). The –COOH groups created on the surface covalently interacted with the $-NH_2$ groups of the anti-AFB1 to form strong amide bond on the transducer surface. The AFB1 concentration was determined by an electrochemical impedance spectroscopy. The concentration of aflatoxin B1 was detected in the range of 10^{-12}–10^{-7} g mL^{-1} with a low detection limit of 1 pg mL^{-1}. The improved performance of biosensor due to the aligned channels in the porous carbon film was shown in this study. It has been reported that the channels provided by electrospun fibers enhanced antigen-antibody interactions and electron transport between the electrode and the electrolyte. Moradi et al. (2020) fabricated electrospun nanofibers-based sensor for zearalenone (ZEN) mycotoxin detection in the food simulant. Polyacrylonitrile (PAN) was electrospun to produce a layer of nanofiber using electrospinning and deposited on the graphite electrode surface. This modified electrode was used as a working electrode. The limit of quantification and limit of detection were found to be 5 and 1.66 nM, respectively.

Recently, electrospun nanofibers have been used to develop a chemiresistive DNA hybridization sensor for applications in healthcare and food analysis (Tripathy et al. 2019). In this chemiresistive biosensing, interdigitated metal electrodes were prepared using Si wafers and glass substrates. They were then subjected to a metal deposition followed by an optical photolithography process and metal etching. The composite conductive nanofibers were prepared from polymer blend of polyaniline (PANi) and polyethylene oxide (PEO) and used to modify metal electrodes. Graphene doped Mn_2O_3 nanofibers were drop casted onto surface of the electrodes as a second approach in this study. The surface modified nanofibers incubated with probe DNA were used as the capturing agent for the corresponding target DNA. Multiple target DNAs include dengue virus specific consensus primer (DENVCP) and specific genes of *Staphylococcus aureus* (*nuc*, *mecA*, *vanA*, and *protein A*). The concentration range of DENVCP detection was 10 fM–1 μM with a detection limit of 1.9 fM. The detection of four DNA targets was ranged between 1 pM and 1 μM.

Shaibani et al. (2018) developed a potentiometric sensor for the detection of *Escherichia coli* in orange juice. The light addressable sensor was fabricated based on a combination of hydrogel nanofibers as a pH sensitive layer and illumination with a red LED. The solution of poly(vinyl alcohol) and poly(acrylic acid) polymers was electrospun and collected on the Si substrates with SiO_2 layer. Sensor chips were annealed and functionalized with D-mannose. The working electrode was connected to the potentiostat, displaying the rectified values of the current at any given applied voltage. The developed sensor was tested with orange juice spiked with *Escherichia coli* at concentrations of 10^2–10^6 CFU mL^{-1}. The pH change was adjusted by adding glucose at a concentration of 80 μM. The selectivity of the sensor toward *Escherichia coli* was tested using *Salmonella typhimurium* as a sugar fermenting bacterium. The presence of *E. coli* was detected in less than 1 h with the developed sensor. The detection limit was 10^2 CFU mL^{-1} with a high selectivity. This study proposed that the sensor can be used for monitoring to ensure the safety of orange juice at various stages from production to consumption.

Certain food components can cause allergic reactions when foods are consumed. The severity of allergic reactions can range from minor to life threatening. The major potential allergenic foods are peanuts, wheat protein gluten, milk, egg, nuts, shellfish, and soybeans. Food allergies affect the quality of life of a significant part of the population (Neethirajan et al. 2018). Therefore, it is necessary to develop rapid and simple analysis tools for monitoring and detection of food allergens (Neethirajan et al. 2018; Jiang et al. 2019). An electrochemical paper biosensor has been developed to detect milk allergen casein (Jiang et al. 2019). The sensor platform was fabricated on wax-patterned cellulose paper. The composite hydrogel of carbon nanofibers, gelatin methacryloyl (GelMA), and graphene were prepared and coated on the screen-printed electrode. This modified electrode provided an effective layer to immobilize mast cells of rat basophilic leukemia (RBL-2H3) due to the biological affinity of the GelMA hydrogel. The performance of the sensor was tested by exposing RBL mast cells to casein. In the study, an inverse relationship was found between the casein concentration and the current signal. The paper-based sensor showed a casein detection limit of 3.2×10^{-8} g mL^{-1} with a linear concentration range of 1×10^{-7}–1×10^{-6} g mL^{-1}. The results of this study proposed that various food allergens can be monitored on-site and determined by the developed paper-based sensor.

In a recent study, an electrochemical biosensor based on a copper-based metal organic framework (Cu-MOF) nanofibers film has been developed for the evaluation of the freshness of chilled seafood (Wang et al. 2019b). Hypoxanthine and xanthine can be used as an indicator of the freshness of chilled seafood because these compounds occur during adenosine 5'-triphosphate (ATP) degradation process. Cu-MOF nanofibers were used to immobilize xanthine oxidase (XOD) and their suspension with sodium alginate was drop cast onto a glassy carbon electrode (GCE). The developed XOD-electrochemical biosensor had a wide linear range from 0.01 to 10 μM and low limit of detections (0.0023 and 0.0064 μM) for hypoxanthine and xanthine. The hypoxanthine and xanthine in chilled squid and large yellow croaker were detected sensitively with good recovery rates using the proposed biosensor.

Glucose biosensors are some of the most used electrochemical biosensors in the market (Hammond et al. 2016). Their practical applications from biomedical to the pharmaceutical field have been well developed (Mason et al. 2016). However, glucose biosensors also have many potential applications in food industry. For instance, real-time monitoring of carbohydrate levels can be useful for controlling the fermentation process. Mason et al. (2016) designed an electrochemical biosensor based on naylon nanofibers (NFM) and multi-walled carbon nanotubes (CNT) composites. A glassy carbon electrode was coated with a composite of nylon fibrous membrane and CNT. Glucose oxidase (GOX) was immobilized on the surface of electrode. The resulting electrode was used as a working electrode. Electrochemical characterization was performed by cyclic voltammetry and chronoamperometry. The developed sensor was used to determine glucose during the beer brewing process. It has been reported that the binding of GOX around the functionalized NFM greatly enhanced the electron transfer and the resulting current response of biosensor increased five-fold compared to NFM without CNT. The limit of quantification was 20 μM with the sensitivity of 1.2 μAmM^{-1}. The results of this study showed that the applicability of the biosensor may be limited only when the concentrations of other redox species such as ascorbic acid are high.

A maltose sensor based on the carbon paste electrode (CPE) modified with an electrospun nickel doped cobalt (II, III) nanofibers (Ni-Co3O4) has been developed (Xu et al. 2015). The composite nanofibers were produced by electrospinning of nickel acetate, cobalt acetate, and polyvinyl pyrolidone solutions. The calcination process was applied to the resulting fibers to obtain Ni-Co3O4 electrospun fibers. The suspension of Ni-Co3O4 electrospun fibers was casted on the surface carbon paste electrode and dried with an infrared lamp. Cyclic voltammetry and amperometry were employed in maltose determination. Malt syrup was used to test the proposed sensor. The sensor responded to maltose over a wide concentration range (0.2–100 μM) with a low detection limit (32.5 nM). The results demonstrated that the sensor can be used for practical analysis. In a recent study, titanium carbide (TiC)-carbon nanofibers (CNFs) film (TCNFs) has been used to improve the performance of the glucose biosensor (Guo et al. 2018). TCNFs were produced by electrospinning followed by carbothermal reduction. A glassy carbon electrode was coated with the suspension of the TCNFs with glucose oxidase (GOD) and used as a working electrode. The detection range of glucose sensor was from 0.013 to 10.5 mM with a detection limit of 3.7 μM. The use of both CNFs and TiC nanoparticles resulted in an increase in surface area and an improvement of electrical conductivity with an enhancement of electrocatalytic activity of TCNFs.

Zhang et al. (2015) fabricated a dual-purpose fluorescent and electrochemical biosensor based on the graphene quantum dots (GQDs) with polyvinyl alcohol (PVA) nanofibers for H_2O_2 and glucose determination. The PVA/GQD nanofibers were produced by electrospinning and collected on the glass slide coated with indium tin oxide (ITO) for fluorescent biosensing of H_2O_2 and glucose but coated on the glass carbon electrode (GCE) for electrochemical biosensing of H_2O_2. The immobilization of glucose oxidase on the surface of the PVA/GQD fibrous membrane was based on the physical adsorption. The fluorescent biosensor responded in wide concentration ranges of 0.05–35 and 0.25–24 mM and detection limits of 1.0 and 10.0 μM for both H_2O_2 and glucose. In the case of the electrochemical biosensor, the H_2O_2 detection range was 0.1–200 mM and a detection limit of 0.53 μM.

An electrochemical non-enzymatic glucose biosensor based on Ni/CoO loaded carbon nanofiber (Ni-CoO/CNF) has been developed by Mei et al. (2019). The electrospun fibers were prepared by electrospinning of polyacrylonitrile solutions containing nickel acetate and cobaltous acetate with different ratios of sodium dodecyl sulfate (SDS). The carbon nanofibers with Ni-CoO were obtained by the pyrolysis of the resulting electrospun fibers. SDS was used as a surfactant to improve the dispersion of metals (Ni, Co) on the carbon nanofibers surface to form unique shape (Fig. 13.6). Carbon nanofibers with acicular and spherical shaped materials grown on the surface were obtained when the amount of SDS was at 3% (Fig. 13.6C). Glass carbon electrode (GCE) with Ni-CoO/CNF was used as a working electrode. The glucose concentration ranged from 0.25 to 600 μM with a low limit of detection at 0.03 μM. The glucose concentration in human serum was determined to test the applicability of the developed sensor. The results demonstrated that the developed sensor can be used for glucose detection in real sample.

Recently, Baek et al. (2020) fabricated an electrochemical biosensor based on graphene oxide nanofibers (GO NFs) modified with gold nanoparticles (AuNPs) and copper (Cu)-nanoflowers for glucose detection. The mixture of poly(vinyl alcohol) (PVA) and GO solution was used to obtain GO NFs by electrospinning and the resulting electrospun nanofibers were deposited on the surface of the gold chip (Fig. 13.7a). After calcination of nanofibers, GO NFs were modified with gold nanoparticles (AuNPs) by electrostatic interaction to enhance the conductivity of NFs (Fig. 13.7b). Cu-nanoflowers (Fig. 13.7c) were synthesized using glucose oxidase (GOx), horseradish peroxidase (HRP), and $CuSO_4$ and they were casted on AuNPs-GO NFs (Cu-nanoflower@AuNPs-GO NFs). The prepared electrode was used as working electrode for glucose detection. The selectivity of developed glucose sensor was tested in the presence of some interfering substances (such as ascorbic acid, saccharose, urea, etc.). The results showed that the use of GO, AuNPs, and Cu-nanoflowers with nanofibers improved the electrochemical properties of the developed biosensor. The glucose detection limit of biosensor was found to be 0.018 μM with a linear range of 0.001–0.1 mM. These results suggested that the fabricated biosensor could be used for determining glucose in biofluids for clinical application.

Chauhan et al. (2018) developed an electrochemical immunosensor for Vitamin-D_3 detection. Electrospun polyacrylonitrile nanofibers (PANnFs) incorporated with

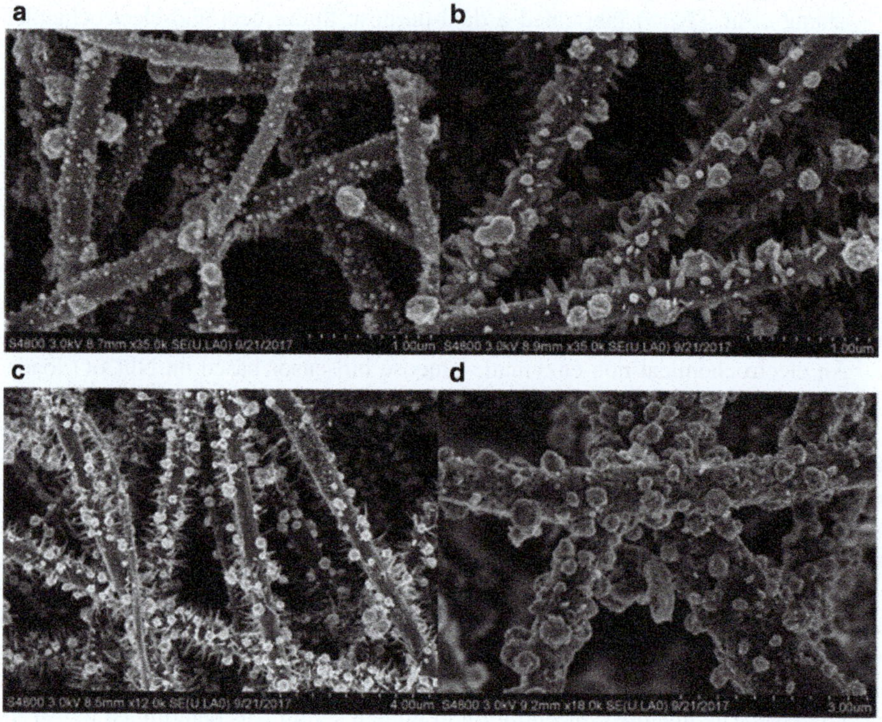

Fig. 13.6 SEM images of Ni-CoO/CNF with different mass ratio of SDS. (A) 1%, (B) 2%, (C) 3%, (D) 4% (Reprinted from Mei et al. 2019. Copyright with permission from Elsevier)

magnetite nanoparticles (Fe_3O_4 NPs) were fabricated using electrospinning technique (Fig. 13.8). The composite nanofibers (Fe_3O_4-PANnFs) were collected directly onto glass electrodes coated with indium tin oxide (ITO). Nafion solution was casted on the surface of electrode to ensure the adhesion of nanofibers. The fabricated electrode was hydrolyzed with NaOH solution to partially convert the nitrile group to carboxyl and amine functional groups. The electrode was further modified with crosslinking agents 1-ethyl-3-(3-dimethyllaminopropyl) carbodiimide (EDC) and N-hydroxysuccinimide (NHS) for immobilization of monoclonal antibody specific to Vitamin-D_3 (Anti-VD). Nonspecific binding sites on the surface of the immunoelectrode were blocked using BSA. The fabricated biosensor detected Vit-D_3 at a concentration range from 10 to 100 ng mL^{-1} with a low detection limit of 0.12 ng mL^{-1}. The developed immunosensor has been proposed to potential use in clinical applications for rapid detection of Vit-D_3. However, the application of this sensor can be extended to other fields such as food industry. The rapid and sensitive detections of vitamins are important for the purpose of labelling food products for their vitamin content present.

Recently, electrospun nanofibers have been used to produce artificial enzyme, which have shown many applications in various fields such as biosensing,

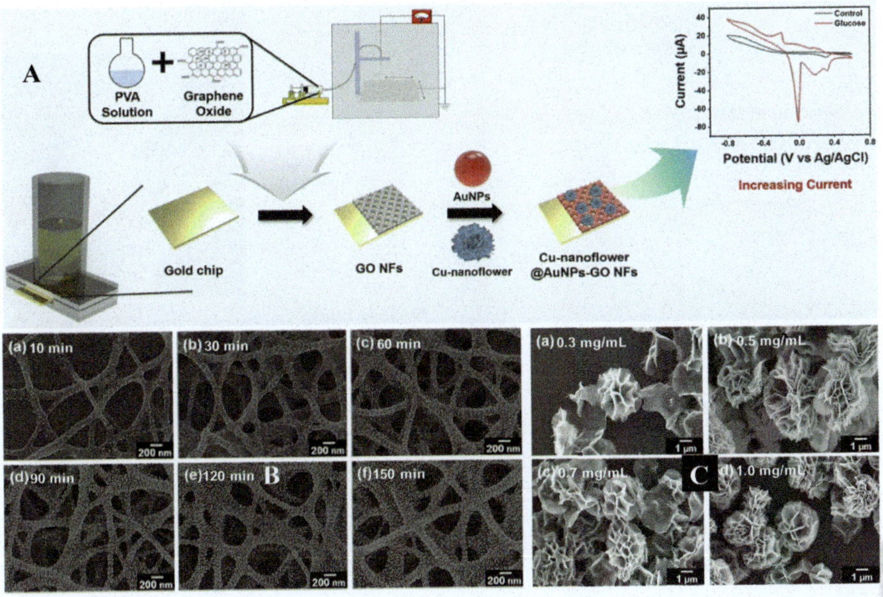

Fig. 13.7 **a** Schematic illustration for the fabrication of Cu-nanoflower@AuNPs-GO NFs-based electrochemical glucose nanobiosensor. **b** FE-SEM images of AuNPs-GO NFs at different reaction times (10–150 min). (C) Cu-nanoflowers at different GOx concentrations (0.3–1.0 mg/mL) (Reprinted from Baek et al. 2020. Copyright with permission from Elsevier)

immunoassays, and environmental monitoring (Gao et al. 2017). To fabricate oxidase mimics, carbon nanofibers (CNFs) were first produced by calcination of electrospun polyacrylonitrile nanofibers. The composite of CNFs with manganese cobalt oxides was prepared by electrochemical deposition of $MnCo_2O_{4.5}$ nanosheets on the surface of CNFs followed by annealing. The oxidase-like activities of CNFs/$MnCo_2O_{4.5}$ nanofibers were explored against to oxidation of 3,3',5,5'-tetramethylbenzidine (TMB), 2,2'-azino-bis(3-ethylbenzothiazoline-6-sulfonic acid) diammonium (ABTS), and o-pheynlenediamine (OPD). The prepared oxidase mimics were used to detect sulfite and ascorbic acid. The results demonstrated that an improved intrinsic oxidase-like activity was obtained with the metal oxide-carbon nanofibers compared to CNFs and $MnCo_2O_{4.5}$ alone. The prepared sensing platform had a sulfite detection limit of 15.9 nM and a linear concentration range of 0–1 μM, while the detection limit of ascorbic acid was 50 nM with a concentration range of 0–40 μM.

Adabi et al. (2015) fabricated carbon nanofibers (CNFs)-based electrodes produced from polyacrylonitrile nanofibers for potential use in electrochemical sensors and biosensors applications. Electrospinning technique was used to obtain electrospun PAN nanofibers. The effects of process parameters such as applied voltage, flow rate, and distance between needle tip and collector on morphology and diameter of PAN fibers were investigated. PAN nanofibers were carbonized to

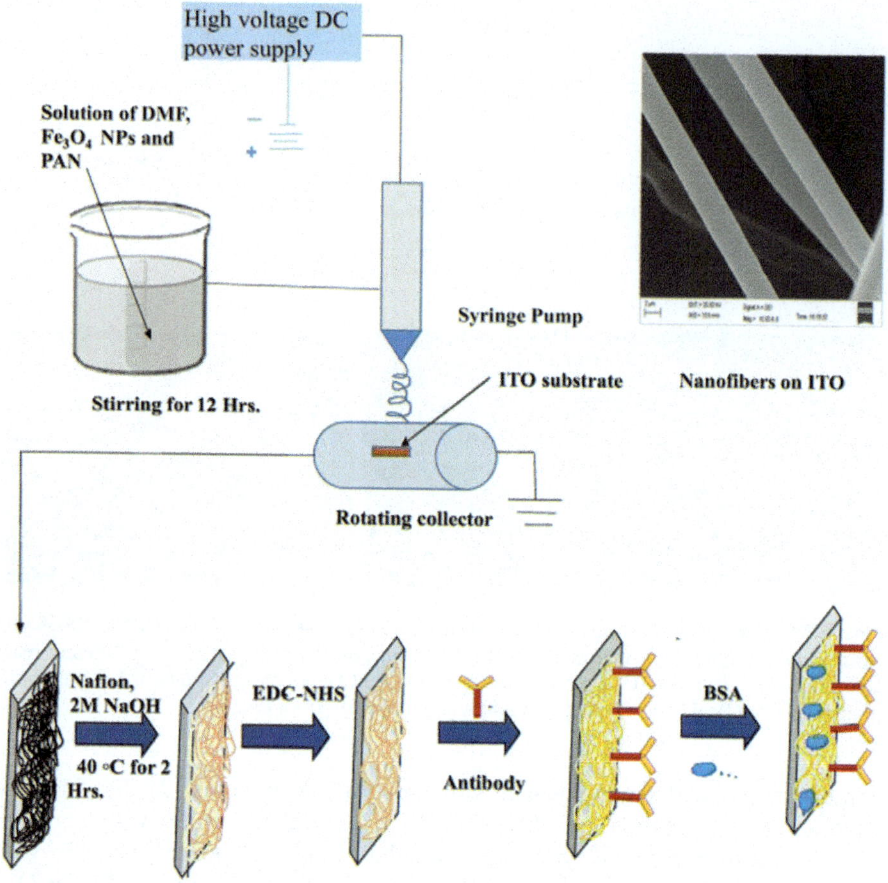

Fig. 13.8 Schematic illustration of the stepwise fabrication of electrospun nanofibers-based BSA/Anti-VD/Fe3O4-PANnFs/ITO immunoelectrode (Reprinted from Chauhan et al. 2018. Copyright with permission from Elsevier)

obtain CNFs and the resulting carbon nanofibers were used as electrode. Electrochemical characterization of electrode was evaluated by cyclic voltammetry. The diameter of PAN nanofibers was significantly affected by polymer concentration and electrospinning processing parameters except for feed flow rate. It was shown that the diameter of electrospun CNFs used as nanoelectrode ranged from 75 to 80 nm. The cyclic voltammetric response in CNFs electrode decreased as the diameter of the CNFs or the presence of beads increased.

The rapid and accurate determination of lactic acid or lactate plays an important role in monitoring product quality and controlling manufacturing processes because many foods, beverages, and fermented products contain lactic acid. Lactate levels are measured as an indicator during the fermentation process and provide information about freshness, stability, and storage quality of food products. Therefore, food and beverage industries need fast, easy, and affordable methods for the determination of lactate. Loaiza et al. (2015) fabricated a new lactate biosensor aimed at detecting lactate in wines and ciders. A composite nanomaterial composed of graphitized carbon nanofibers (GCNF) and platinum nanoparticles (PtNps) was used to prepare the biosensor. The dispersion of the PtNps/GCNFs was casted on the screen-printed carbon electrode and lactate oxidase (LOx) was immobilized on the modified electrode. The proposed sensor was used to determine lactate in Basque ciders and certified wines. The lactate biosensor concentration ranged from 10 to 2000 μM, while the limit of detection was 6.9 μM. The developed sensor exhibited long term stability, maintaining 90 and 95% of the initial response after storage at room temperature for 3 months and 18 months of storage at −20 °C. The use of the lactate sensor for L-lactate detection in wines and cider has been shown to be suitable when sample is sufficiently diluted for analysis.

13.4 Concluding Remarks and Future Perspectives

The versatility of electrospinning allows the production of high-performance and functional nanofibers from various polymers that can be used in different applications. Electrospun materials have many advantages such as large specific surface area, suitable porosity, tunable fiber diameters, and tailored morphology. Such characteristic features of electrospun nanofibers make them an excellent candidate for their use in biosensors. They can provide multiple sites for interaction or attachment of enzymes and the more efficient transport of analytes to the sensor receptor via nanofibers, which is important for efficiency, sensitivity, and rapid response of biosensors. Even though the particular recent interest in electrospun fibers has focused on encapsulation and food packaging, another important application area where electrospun nanofibers can be used in the future is in biosensors. Recent studies show that electrospun fibers have high potential for use in the biosensor platform in many fields such as clinical, biomedical, and environmental applications, but there are limited number of studies for their use in food applications. This chapter summarizes the recent developments within past five years regarding the application of electrospun based electrochemical biosensors for detection of various compounds with interest in food industry. In addition, some applications in other fields are also included because the detected analytes are also important in food analysis, process, and quality control. Nanofibers fabricated from a variety of materials, including polymers only or doped with nanomaterials (e.g., metal nanoparticles, carbon nanotubes, and quantum dots), metal oxides, carbon or conducting polymers (e.g., polyaniline and polypyrrole), have been explored to fabricate biosensing platforms. Biosensors developed based on these electrospun nanofibers have demonstrated a good analytic performance such as high selectivity, fast response time, and low analyte detection limits. Although the utilization of electrospun nanofibers in biosensing is very promising, there are some limitations and challenges that need to be solved for further development of nanofiber-based biosensors. Conventional electrospinning mostly produces randomly oriented electrospun nanofibers which limit the repeatability of the final structures and therefore a better control of the nanofiber production and arrangement on the transducers is needed. Electrospun nanofibers have been functionalized by dispersing nanomaterials in the nanofibers or on the surface of nanofibrous membrane to enhance the performance of biosensors. While dispersing nanomaterials in the nanofibers, the uniform distribution of nanomaterials without aggregation with the spinnable polymer solution is crucial for the electrospinning process. On the other hand, the binding and assembling of nanomaterials onto electrospun nanofibers should be optimized in post-processing. In addition, the composite nanofibers with different structures such as hollow, porous, and core-shell can be used to increase surface area and immobilization sites for target biomolecules. The increasing demand on rapid, simple, low-cost, and accurate monitoring of food quality and safety will push forward the development of biosensors. Electrospun-based biosensors are of great potential, but studies on food applications are limited and more research is needed to develop nanofibers-based biosensors applied to real sample analysis. In the future, it is expected that the development of electrospun nanofibers-based platforms will result in the production of highly sensitive and selective biosensors for fulfilling the needs of food industry.

References

Adabi M, Saber R, Faridi-Majidi R, Faridbod F (2015) Performance of electrodes synthesized with polyacrylonitrile-based carbon nanofibers for application in electrochemical sensors and biosensors. Mater Sci Eng, C 48:673–678

Asefa T, Duncan CT, Sharma KK (2009) Recent advances in nanostructured chemosensors and biosensors. Analyst 134(10):1980–1990

Baek SH, Roh J, Park CY, Kim MW, Shi R, Kailasa SK, Park TJ (2020) Cu-nanoflower decorated gold nanoparticles-graphene oxide nanofiber as electrochemical biosensor for glucose detection. Mater Sci Eng, C 107:110273

Baji A, Mai Y-W, Wong S-C, Abtahi M, Chen P (2010) Electrospinning of polymer nanofibers: effects on oriented morphology, structures, and tensile properties. Compos Sci Technol 70:703–718

Bao J, Hou C, Dong Q, Ma X, Chen J, Huo D, Yang M, El Galil KHA, Chen W, Lei Y (2016) ELP-OPH/BSA/TiO_2 nanofibers/c-MWCNTs based biosensor for sensitive and selective determination of *p-nitrophenyl* substituted organophosphate pesticides in aqueous system. Biosens Bioelectro 85:935–942

Bhardwaj N, Kundu SC (2010) Electrospinning: a fascinating fiber fabrication technique. Biotech Adv 28(3):325–347

Bhushani JA, Anandharamakrishnan C (2014) Electrospinning and electrospraying techniques: potential food based applications. Trends Food Sci Technol 38:21–33

Caon T, Martelli SM, Fakhouri FM (2017) New trends in the food industry: application of nanosensors in food packaging. In: Grumezescu AM (ed) Nanotechnology in the agri-food industry, vol 8, Nanobiosensors. Academic Press, an imprint of Elsevier, Cambridge, pp 773–804

Casper CL, Stephens JS, Tassi NG, Chase DB, Rabolt JF (2004) Controlling surface morphology of electrospun polystyrene fibers: effect of humidity and molecular weight in the electrospinning process. Macromolecules 37(2):573–578

Chauhan D, Gupta PK, Solanki PR (2018) Electrochemical immunosensor based on magnetite nanoparticles incorporated electrospun polyacrylonitrile nanofibers for Vitamin-D_3 detection. Mater Sci Eng, C 93:145–156

Chen A, Shah B (2013) Electrochemical sensing and biosensing based on square wave voltammetry. Anal Methods 5(9):2158–2173

Chen H-M, Yu D-G (2010) An elevated temperature electrospinning process for preparing acyclovir-loaded PAN ultrafine fibers. J Mater Process Tech 210(12):1551–1555

Chen X, Li D, Li G, Luo L, Ullah N, Wei Q, Huang F (2015) Facile fabrication of gold nanoparticle on zein ultrafine fibers and their application for catechol biosensor. Appl Surf Sci 328:444–452

De Cesare F, Macagnano A (2015) Electrospinning-based nanobiosensors. In: Macagnano A, Zampetti E, Kny E (eds) Electrospinning for high performance sensors. Springer, Switzerland p, pp 225–279

Deitzel JM, Kleinmeyer J, Harris D, Beck Tan NC (2001) The effect of processing variables on the morphology of electrospun nanofibers and textiles. Polymer 42:261–272

De Vrieze S, Van Camp T, Nelvig A, Hagström B, Westbroek P, De Clerck K (2008) The effect of temperature and humidity on electrospinning. J Mater Sci 44(5):1357–1362

Ding B, Wang M, Wang X, Yu J, Sun G (2010) Electrospun nanomaterials for ultrasensitive sensors. Mater Today 13(11):16–27

D'Souza AA, Kumari D, Banerjee R (2017) Nanocomposite biosensors for point-of-care—evaluation of food quality and safety. In: Grumezescu AM (ed) Nanotechnology in the agri-food industry, vol 8, Nanobiosensors. Academic Press, an imprint of Elsevier, Cambridge, pp 629–676

Echegoyen Y, Fabra MJ, Castro-Mayorga JL, Cherpinski A, Lagaron JM (2017) High throughput electro-hydrodynamic processing in food encapsulation and food packaging applications: viewpoint. Trends Food Sci Tech 60:71–79

El-Moghazy AY, Soliman EA, Ibrahim HZ, Marty J-L, Istamboulie G, Noguer T (2016) Biosensor based on electrospun blended chitosan-poly (vinyl alcohol) nanofibrous enzymatically sensitized membranes for pirimiphos-methyl detection in olive oil. Talanta 155:258–264

Gao M, Lu X, Nie G, Chi M, Wang C (2017) Hierarchical CNFs/MnCo2O4.5 nanofibers as a highly active oxidase mimetic and its application in biosensing. Nanotechnology 28(48):485708

Ghorani B, Tucker N (2015) Fundamentals of electrospinning as a novel delivery vehicle for bioactive compounds in food nanotechnology. Food Hydrocoll 51:227–240

Gunasekaran S (2014) Nanotechnology for food: principles and selected applications. In: Clark S, Jung S, Lamsal B (eds) Food processing: principles and applications, 2nd edn. Wiley Blackwell, Chichester, pp 171–205

Guo Q, Liu L, Wu T, Wang Q, Wang H, Liang J, Chen S (2018) Flexible and conductive titanium carbide–carbon nanofibers for high-performance glucose biosensing. Electrochim Acta 281:517–524

Haghi AK, Akbari M (2007) Trends in electrospinning of natural nanofibers. Phys Stat Sol (a) 204:1830–1834

Hammond JL, Formisano N, Estrela P, Carrara S, Tkac J (2016) Electrochemical biosensors and nanobiosensors. Essays Biochem 60(1):69–80

Hsu C-L, Chang K-S, Kuo J-C (2008) Determination of hydrogen peroxide residues in aseptically packaged beverages using an amperometric sensor based on a palladium electrode. Food Control 19(3):223–230

Jiang D, Ge P, Wang L, Jiang H, Yang M, Yuan L, Ge Q, Fang W, Ju X (2019) A novel electrochemical mast cell-based paper biosensor for the rapid detection of milk allergen casein. Biosens Bioelectro 130:299–306

Kim SG, Lee JS, Jun J, Shin DH, Jang J (2016) Ultrasensitive bisphenol A field-effect transistor sensor using an aptamer-modified multichannel carbon nanofiber transducer. ACS Appl Mater Interfaces 8(10):6602–6610

Kress-Rogers E (1998) Chemosensors, biosensors and immune sensors. In: Kress-Rodgers E (ed) Instrumentation and sensors for the food industry. Woodhead Publishing, Cambridge, pp 581–669

Liu T, Guo Y, Zhang Z, Miao Z, Zhang X, Su Z (2019) Fabrication of hollow CuO/PANI hybrid nanofibers for non-enzymatic electrochemical detection of H2O2 and glucose. Sens Actuator B Chem 286:370–376

Loaiza OA, Lamas-Ardisana PJ, Añorga L, Jubete E, Ruiz V, Borghei M, Cabañero G, Grande HJ (2015) Graphitized carbon nanofiber—Pt nanoparticle hybrids as sensitive tool for preparation of screen printing biosensors: Detection of lactate in wines and ciders. Bioelectrochem 101:58–65

Maftoonazad N, Ramaswamy H (2018) Novel techniques in food processing: bionanocomposites. Curr Opin Food Sci 23:49–56

Malhotra BD, Ali MA (2018) Nanomaterials in biosensors: fundamentals and applications. In: Malhotra BD, Ali MA (eds) Nanomaterials for biosensors: fundamentals and applications, 1st edn. Elsevier, Amsterdam, pp 1–74

Mason M, Longo E, Scampicchio M (2016) Monitoring of glucose in beer brewing by a carbon nanotubes based nylon nanofibrous biosensor. J Nanomater: 1–11

Matlock-Colangelo L, Baeumner AJ (2012) Recent progress in the design of nanofiber-based biosensing devices. Lab Chip 12(15):2612–2620

Mazoochi T, Jabbari V (2011) Chitosan nanofibrous scaffold fabricated via electrospinning: the effect of processing parameters on the nanofiber morphology. Int J Polymer Anal Char 16(5):277–289

Megelski S, Stephens JS, Chase DB, Rabolt JF (2002) Micro- and nanostructured surface morphology on electrospun polymer fibers. Macromolecules 35(22):8456–8466

Mei Q, Fu R, Ding Y, Li L, Wang A, Duan D, Ye D (2019) Electrospinning of highly dispersed Ni/CoO carbon nanofiber and its application in glucose electrochemical sensor. J Electroanal Chem 847:113075

Mercante LA, Scagion VP, Migliorini FL, Mattoso LHC, Correa DS (2017) Electrospinning-based (bio)sensors for food and agricultural applications: a review. Trends Anal Chem 91:91–103

Migliorini FL, Sanfelice RC, Mercante LA, Andre RS, Mattoso LHC, Correa DS (2018) Urea impedimetric biosensing using electrospun nanofibers modified with zinc oxide nanoparticles. Appl Surf Sci 443:18–23

Mit-uppatham C, Nithitanakul M, Supaphol P (2004) Ultrafine electrospun polyamide-6 fibers: effect of solution conditions on morphology and average fiber diameter. Macromol Chem Phys 205(17):2327–2338

Mo XM, Xu CY, Kotaki M, Ramakrishna S (2004) Electrospun P(LLA-CL) nanofiber: a biomimetic extracellular matrix for smooth muscle cell and endothelial cell proliferation. Biomaterials 25(10):1883–1890

Mondal K, Ali MA, Srivastava S, Malhotra BD, Sharma A (2016) Electrospun functional micro/nanochannels embedded in porous carbon electrodes for microfluidic biosensing. Sens Actuators B: Chem 229:82–91

Mondal K, Sharma A (2016) Recent advances in electrospun metal-oxide nanofiber-based interfaces for electrochemical biosensing. RSC Adv 6:94595–94616

Moradi S, Azizi-Lalabadi M, Bagheri V, Sadeghi E (2020) Fabrication of electrospun sensor based on a synthesized component doped into PAN (polyacrylonitrile) nanofibers for electrochemical detection of zearalenone mycotoxin in foods simulant. Sens Biosensing Res 28:100321

Neethirajan S, Weng X, Tah A, Cordero JO, Ragavan KV (2018) Nano-biosensor platforms for detecting food allergens—new trends. Sens Biosensing Res 18:13–30

Neo YP, Ray S, Easteal AJ, Nikolaidis MG, Quek SY (2012) Influence of solution and processing parameters towards the fabrication of electrospun zein fibers with sub-micron diameter. J Food Eng 109(4):645–651

Nikoleli G-P, Nikolelis DP, Methenitis C (2010) Construction of a simple optical sensor based on air stable lipid film with incorporated urease for the rapid detection of urea in milk. Anal Chim Acta 675(1):58–63

Pérez-López B, Merkoçi A (2011) Nanomaterials based biosensors for food analysis applications. Trends Food Sci Tech 22(11):625–639

Pilolli R, Monaci L, Visconti A (2013) Advances in biosensor development based on integrating nanotechnology and applied to food-allergen management. Trend Anal Chem 47:12–26
Quirós J, Boltes K, Rosal R (2016) Bioactive applications for electrospun fibers. Polym Rev 56(4):631–667
Rathee K, Dhull V, Dhull R, Singh S (2016) Biosensors based on electrochemical lactate detection: a comprehensive review. Biochem Biophys Rep 5:35–54
Sahota S, Singh J, Kumar P, Suresh S, Pandey KM (2017) Nano devices for contaminant detection. In: Oprea AE, Grumezescu AM (eds) Nanotechnology applications in food: flavor, stability, nutrition and safety. Academic Press, an imprint of Elsevier, London, pp 335–344
Sarika C, Shivakumar MS, Shivakumara C, Krishnamurthy G, Narasimha Murthy B, Lekshmi IC (2017) A novel amperometric catechol biosensor based on α-Fe_2O_3 nanocrystals-modified carbon paste electrode. Artif Cells, Nanomed Biotechnol 45(3):625–634
Shaibani PM, Etayash H, Jiang K, Sohrabi A, Hassanpourfard M, Naicker S, Sadrzadeh M, Thundat T (2018) Portable nanofiber-light addressable potentiometric sensor for rapid Escherichia coli detection in orange juice. ACS Sens 3(4):815–822
Shenoy SL, Bates WD, Frisch HL, Wnek GE (2005) Role of chain entanglements on fiber formation during electrospinning of polymer solutions: good solvent, non-specific polymer–polymer interaction limit. Polymer 46(10):3372–3384
Supraja P, Singh V, Vanjari SRK, Govind Singh S (2020) Electrospun CNT embedded ZnO nanofiber based biosensor for electrochemical detection of Atrazine: a step closure to single molecule detection. Microsyst Nanoeng 6:3
Tan S-H, Inai R, Kotaki M, Ramakrishna S (2005) Systematic parameter study for ultra-fine fiber fabrication via electrospinning process. Polymer 46(16):6128–6134
Teepoo S, Dawan P, Barnthip N (2017) Electrospun chitosan-gelatin biopolymer composite nanofibers for horseradish peroxidase immobilization in a hydrogen peroxide biosensor. Biosensors 7(4):47
Thévenot DR, Tóth K, Durst RA, Wilson GS (1999) Electrochemical biosensors: recommended definitions and classification. Pure Appl Chem 71(12):2333–2348
Thoppey NM, Bochinski JR, Clarke LI, Gorga RE (2010) Unconfined fluid electrospun into high quality nanofibers from a plate edge. Polymer 51(21):4928–4936
Thoppey NM, Bochinski JR, Clarke LI, Gorga RE (2011) Edge electrospinning for high throughput production of quality nanofibers. Nanotech 22(34):345301
Topuz F, Uyar T (2020) Antioxidant, antibacterial and antifungal electrospun nanofibers for food packaging applications. Food Res Int 130:108927
Tran TB, Son SJ, Min J (2016) Nanomaterials in label-free impedimetric biosensor: Current process and future perspectives. Biochip J 10(4):318–330
Tripathy S, Bhandari V, Sharma P, Vanjari SRK, Singh SG (2019) Chemiresistive DNA hybridization sensor with electrospun nanofibers: a method to minimize inter-device variability. Biosens Bioelectro 133:24–31
US Food and Drug Administration (FDA) (2018) Potential hazards associated with the manufacturing, processing, packing, and holding of human food. Chapter 3, United States, pp 1–42
Viswanathan S, Radecka H, Radecki J (2009) Electrochemical biosensors for food analysis. Monatshefte Chem. Mon 140:891–899
Wahab MA, Darain F (2017) Nanostructure-modified electrodes for food sensors. In: Ahmed MU, Zourob M, Tamiya E (eds) Food chemistry, function and analysis 1: Food biosensors. Royal Society of Chemistry, Cambridge, pp 299–326
Wang Z, Ma B, Shen C, Cheong L-Z (2019a) Direct, selective, and ultrasensitive electrochemical biosensing of methyl parathion in vegetables using burkholderia cepacia lipase@MOF nanofibers-based biosensor. Talanta 197:356–362
Wang Z, Ma B, Shen C, Lai O-M, Tan C-P, Cheong L-Z (2019b) Electrochemical biosensing of chilled seafood freshness by xanthine oxidase immobilized on copper-based metal–organic framework nanofiber film. Food Anal Methods 12:1715–1724

Wei L, Sun R, Liu C, Xiong J, Qin X (2019) Mass production of nanofibers from needleless electrospinning by a novel annular spinneret. Mater Des 179:107885
Wen P, Zong M-H, Linhardt RJ, Feng K, Wu H (2017) Electrospinning: a novel nano-encapsulation approach for bioactive compounds. Trends Food Sci Tech 70:56–68
Xu D, Gu S, Ding Y, Wang B (2015) Synthesis and characterization of electrospun nickel doped cobalt (II, III) nanofibers with application to maltose determination. Analy Lett 48(2):269–280
Xu L, Zhang Z, Zhang Q, Li P (2016) Mycotoxin determination in foods using advanced sensors based on antibodies or aptamers. Toxins 8(8):239
Yam KL, Takhistov PT, Miltz J (2005) Intelligent packaging: concepts and applications. J Food Sci 70:1–10
Yılmaz M, Altan A (2017) Applications of electrospinning-based nanosensors in food field. Gıda 42:708–725
Yuan X, Zhang Y, Dong C, Sheng J (2004) Morphology of ultrafine polysulfone fibers prepared by electrospinning. Polym Int 53(11):1704–1710
Zaid MHM, Saidykhan J, Abdullah J (2019) Nanosensors based detection of foodborne pathogens. In: Siddiquee S, Melvin GJH, Rahman MM (eds) Nanotechnology: applications in energy, drug and food. Springer Nature, Switzerland, pp 377–422
Zaikov GE (2017) Nanostructured fibers via electrospinning (Part I). In: Kodolov VI, Zaikov GE, Haghi AK (eds) Applied nanotechnology: materials and applications. Apple Academic Press Inc, Waretown, NJ, pp 253–280
Zhang C, Li Y, Wang P, Zhang H (2020) Electrospinning of nanofibers: potentials and perspectives for active food packaging. Compr Rev Food Sci Food Saf 19:479–502
Zhang M, Zhao X, Zhang G, Wei G, Su Z (2017) Electrospinning design of functional nanostructures for biosensor applications. J Mater Chem B 5(9):1699–1711
Zhang P, Zhao X, Ji Y, Ouyang Z, Wen X, Li J, Su Z, Wei G (2015) Electrospinning graphene quantum dots into a nanofibrous membrane for dual-purpose fluorescent and electrochemical biosensors. J Mater Chem B 3(12):2487–2496
Zhang S, Jia Z, Liu T, Wei G, Su Z (2019) Electrospinning nanoparticles-based materials interfaces for sensor applications. Sensors 19(18):3977
Zhao L, Duan G, Zhang G, Yang H, He S, Jiang S (2020) Electrospun functional materials toward food packaging applications: a review. Nanomaterials 10(1):150
Zhao S, Wu X, Wang L, Huang Y (2004) Electrospinning of ethyl-cyanoethyl cellulose/tetrafuran solutions. J Appl Polym Sci 91:242–246
Zong X, Kim K, Fang D, Ran S, Hsiao BS, Chu B (2002) Structure and process relationship of electrospun bioabsorbable nanofiber membranes. Polymer 43(16):4403–4412

Chapter 14
Current Trends of Electrochemical Sensing for Mycotoxins

Ruchika Chauhan, Rashi Bhardwaj, Sheetal K. Bharadwaj, Ajit Kaushik, Rajshekhar Karpoormath, and Tinku Basu

Abstract The presence of mycotoxins in food materials is difficult to control, and some mycotoxins are carcinogenic even at very low concentration. Therefore, there is a demand for fast, sensitive, user friendly, and stable method for the detection of mycotoxins. A portable smart biosensor is in a recent appeal in industries and daily life in place of tedious and time taking conventional techniques. In the present chapter, we analyze the current scenario for mycotoxin detection by immunosensors based on labeled and label-free methods, aptasensors. Some other recent trends such as mimotope and nanobody as a recognition agent for mycotoxin sensing are also discussed. We also discussed the new trends in nanomaterials, or matrix of electrode and transuding techniques. The current challenges, and technological aspects which experience during the development of biosensors are also discussed, at the end future research aspects mentioned.

Keywords Labeled and label-free immunosensor · Aptamers · Mycotoxins · Mimotopes · Molecular imprinting polymers · Nanobodies

R. Chauhan · R. Bhardwaj · T. Basu (✉)
Amity Centre for Nanomedicine, Amity University Uttar Pradesh, Noida 201303, UP, India
e-mail: tbasu@amity.edu

R. Chauhan · R. Karpoormath
Department of Pharmaceutical Chemistry, College of Health Sciences, University of KwaZulu-Natal, Westville Campus, Durban 4000, South Africa

S. K. Bharadwaj
University of Amsterdam, Amsterdam, The Netherlands

A. Kaushik
Florida Polytechnique University, Lakeland, FL, USA

R. N. Pudake et al. (eds.), *Biosensors in Agriculture: Recent Trends and Future Perspectives*, Concepts and Strategies in Plant Sciences,
https://doi.org/10.1007/978-3-030-66165-6_14

14.1 Introduction

The contamination of food materials by mycotoxins is becoming a serious problem worldwide. These mycotoxins are secondary metabolites of fungi, particularly from *Aspergillus* spp., *Fusarium* spp., and *Penicillium* spp. These species can contaminate during all stages of the crop, pre- and post-harvesting, or throughout the storage (Robbins et al. 2000). About 400 specific mycotoxins are known till now; among these mycotoxins, predominant are; aflatoxins (AFB), ochratoxin A (OTA), and zearalenone (ZEA) (Liew and Mohd-Redzwan 2018), which occurs in wheat, coffee, fruit juices, fermented food products, and even in animal feed (Reddy et al. 2010). Owing to low molecular weight, and heat resistance, they are capable to persist in the food chain. Among all mycotoxins, aflatoxin B1 has been categorized as a potent carcinogen (in Group 1 by International Agency for Research on Cancer [IARC]) and cause liver cancer in human (Henry et al. 1999). Also, ochratoxin A (OTA), and fumonisins are considered possible carcinogen (Group 2B) (IARC 2002). Consequently, the European Commission set the level of maximum acceptance of these Group A and Group B mycotoxins in food materials (Table 14.1). These mycotoxins are extremely hazardous than food additives and residues of pesticides present in foods (van Egmond et al. 2007). Therefore, mycotoxin-contaminated food consumption causes a harsh effect on human as well as animal health (Zheng et al. 2006; Rodrigues et al. 2011). The lowermost limits of mycotoxins in food materials are in part per billion (ppb) while for baby foods are below (0.05 ppb) which has been set authorized by the European Union (EU). The analysis of mycotoxins (molecular weight ~300–400 g/mol) in very small concentrations is the root of, major restriction to develop a standard method for analysis (Tamerler et al. 2006). Internationally, major research has been done by different research groups is designed to identify and quantify the mycotoxins and assessment of their effects on the health of live stocks (Zain 2011). The major concern group of fungi and their mycotoxins are listed in Table 14.1.

14.2 Traditional Methods for Mycotoxins Detection

Several traditional analytical methods including high-performance liquid chromatography (HPLC) (Lippolis et al. 2008), thin-layer chromatography (TLC), gas chromatography (GC) (Visconti et al. 2005), and enzyme-linked immunoassay (ELISA) (Zheng et al. 2005) are commonly exploited for mycotoxins detection. However, these methods are acknowledged for their accuracy and precision in the field of mycotoxin detection in food samples, despite the costly equipment, needed skilled personal, sample pretreatment, and time taking (Goryacheva et al. 2007). Consequently, a fast, highly specific, and sensitive technique is required for the rapid and accurate detection of mycotoxins. In this chapter, we would like to cover major research carried out in process of development of labeled and label-free immunosensors and aptasensors

Table 14.1 Occurrence, restricted limit as defined by the European Commission along with IARC classification of some hazardous mycotoxins

Sl. No.	Name of Fungi	Mycotoxin	Occurrence in food and feeds	Restricted limit (μg Kg^{-1})	IARC classification
1	*Aspergillus* spp.	Aflatoxins **B1** B2, G1 G2 M1, M2	Groundnuts, nuts, spices maize, rice, figs, cocoa, crude vegetable oils, (peanut, soybean, sunflower, cotton, chili pepper, coriander, turmeric, ginger) Milk and milk products	2 5-10 0.05	1 2B
2	*Aspergillus* spp., *Penicillium* spp.	Ochratoxins A B C	Dried vine fruit (currants, raisins, and sultanas), Wine, Grape juice, Wheat, Coffee, beans, Processed cereal	0.5-10	2B
3	*Fusarium* spp.	Fumonisins B1 B2	Mainly in maize, other cereals also wheat, barley	50-1000	2B
4	*Fusarium* spp.	Zearalenone	In wheat, maize, sorghum, and barley	100-400	-
5	*Fusarium* spp.	HT-2, T-2	Oats and oat products, barley, rice, maize etc.	50-100	-
6	*Fusarium* spp.	Deoxynivalenol	Cereal crops such as wheat, maize, or barley	200-2000	-

and additionally recent technology for rapid quantification of mycotoxins in daily consumable food items.

14.3 Current Trends of Mycotoxins Detection

Compared to conventional mycotoxin detection methods, biosensors attain more attention due to their simplicity, low cost, faster in response, easy-to-handle, and portability. Biosensor converts biological responses into the signal, in this, an analyte interacts with their bioreceptor, and this interaction measures by transducer detector. The basic components of biosensors include biorecognition agents (bioreceptor), transducer, and display. The biorecognition agent identified for mycotoxin sensing are antibodies, DNA, cell receptors, enzymes, etc. In addition, few engineered chemicals such as aptamers, MIP (molecularly impregnated polymer, mimotopes, etc.) are also exploited as a recognition agent. Mycotoxins are small and neutral organic molecules, so direct sensing through bio affinity-based receptor develops too weak signal to be detected. Reliable sensing of mycotoxin demands signal amplification through appropriate labeling like nanoparticles, enzymes, redox moieties, fluorophores, etc. As a result, the mycotoxin biosensors are broadly classified into two categories, first is labeled while second is label-free biosensors based on detection strategy, and those are further subdivided into two, (i) competitive (ii) non-competitive. Labeled sensor majorly dependent on label moiety and the response received by virtue of labels (like enzymes, Quantum Dots, Fluorescence nanomaterials, etc.), while label-free sensors are independent, and response receive due to direct interaction of antibody and analyte. Significant research efforts are devoted to developing a suitable mycotoxin biosensor for rapid and reliable sensing with respect to transducing techniques (electrochemical, optical, and piezoelectric), immobilizing matrix, bioreceptor, and label selection. The present chapter will discuss about the material composites, the function of labeled moieties, and the principle of both labeled and label-free sensor. A schematic representation of labeled and label-free biosensors, i.e., Fig. 14.1a, b, respectively, are given below:

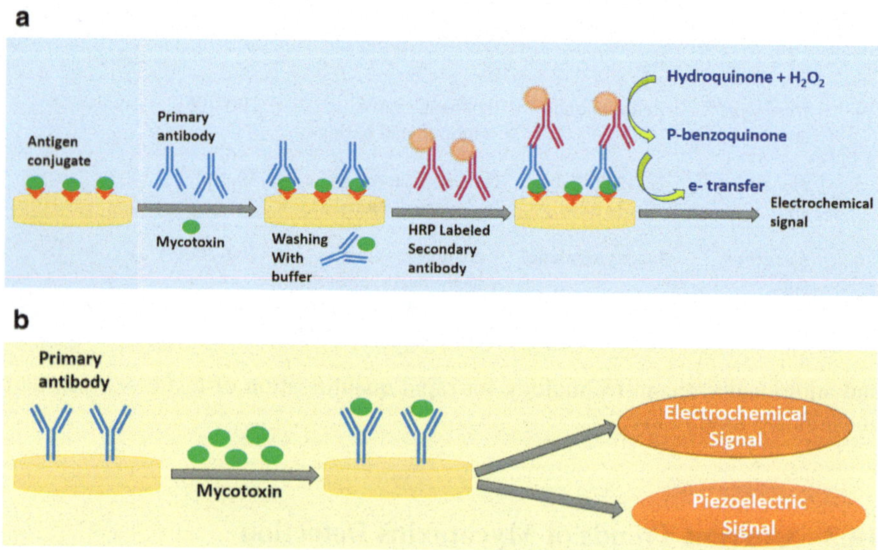

Fig. 14.1 **a** Schematic representation of labeled mycotoxin biosensor. **b** Schematic representation of label-free mycotoxin biosensor

14.3.1 Labeled Mycotoxin Biosensor

14.3.1.1 Immunosensor for Detection of Mycotoxins

For small molecules like mycotoxin, the competitive immunoassay is more reliable than a non-competitive one. In the competitive mechanism, competition occurs between the free mycotoxin and labeled mycotoxin or protein conjugated mycotoxin for limited antibodies present in the reaction mixture or surface of the electrode. When concentration of mycotoxin increases in the reaction mixture, more labeled mycotoxin or protein-mycotoxin conjugates will be displaced from the surface of transducer (Pemberton et al. 2006; Prieto-Simón et al. 2012; Piermarini et al. 2007). In a competitive mechanism, there are two types of approaches. In the first approach, the optimum concentration of protein conjugated or labeled mycotoxin was tagged with the optimized antibody concentration. When the analyte (mycotoxin) is added over the electrode surface, competition occurs between loaded mycotoxin and the sample mycotoxin for the optimized antibody, this method is named as "indirect method." The BSA-mycotoxin conjugate is extensively used. Sometimes, this competition causes a steric effect. To avoid steric effect, spacer molecules (like polyethylene glycol [PEG]) were introduced (Yuan et al. 2009) over the electrode in between the mycotoxin and protein conjugate. Alkaline phosphatase (AP) was studied as label in an indirect competitive immunosensing of Aflatoxin B1(AFB1), and the electrochemical signal produced by AP catalysis in which α naphthyl phosphate is hydrolyzed (Zhang et al. 2016).

High electron transfer rate (ETR) of the nanomaterial modified electrode surface amplifies the response signal of interaction between mycotoxin and their antibody (Bonel et al. 2010; Liu et al. 2009). The sandwiched immunosensor of AuNPs labeled secondary antibody amplified the surface plasmon resonance (SPR) signal for OTA detection by 10-fold with an extremely low sensing limit of (60 pg mL^{-1}), in comparison to AuNPs labeled primary anti-OTA (Urusov et al. 2011).

The cadmium telluride labeled antibody over the Fe-GO matrix was used for the detection of aflatoxin M1 with the 0.3 pg mL^{-1} of LOD (Gan et al. 2013). For concurrent detection of two or more mycotoxins in foods, upconversion nanoparticles (Wu et al. 2011) were successfully tried. Feng et al. have fabricated (Feng et al. 2018) a photo electrotypes competitive immunosensor for quantitative detection of AFB_1. The silver ion modified $BiVO_4$ nano-polyhedron ($BiVO_4@Ag^+$) was employed as a label of anti-AFB1. A nonenzymatic immunosensor was developed with photoactive material (Mn^{2+} incorporated zinc complex) for AFB_1 detection in foodstuff (Lin et al. 2017). The proposed photo active electrochemical immunosensor attained satisfactory detection range 0.5 pg mL^{-1}–10 ng mL^{-1}, and LOD as 0.3 pg mL^{-1} by using a competitive strategy. Here, dopamine loaded liposome was exploited as an electron donor, which significantly enhanced the photocurrent response of the photoelectrode.

Another phenomenon is called "direct method," in this format, competition occurs between the free analyte (mycotoxin) and the absorbed labeled mycotoxin for the optimized concentration of antibody loaded over the electrode surface. Initially, the signal is obtained from saturated concentration of labeled mycotoxin, later, alteration in transducer signal in the presence of sample mycotoxin reflects the interaction of analyte (mycotoxin) with their specific antibodies. It was found that the indirect method is less effective in comparison to the direct method (Alarcón et al. 2006).

The quartz crystal microbalance (QCM) is a highly sensitive technique to analyze the small molecule at the scale of the nano dimension. For mycotoxin, researchers used a sandwiched format for the more precision in results. For the signal amplification, relied on QCM plinth, HRP labeled secondary antibodies for the oxidation of 4-chloro-1-naphthol into insoluble benzo-4-chlorohexadienone and AuNP decorated secondary antibody were used as a mass enhancer on QCM surface which enhance the sensitivity of the QCM electrode (Jin et al. 2009a, b). Unlike the conventional QCM technique, these formats attained a low detection range of 0.01–10.0 ng mL^{-1}.

The magnetic nanoparticles have great advantages because of their inherent property of magnetism therefore these magnetic nanoparticles were used for regeneration procedure (Aguilar-Arteaga et al. 2010; Vidal et al. 2012). The almost ~100% efficiency of magnetic nanoparticles were reported for the separation of interfering substances from the reaction mixture (Prieto-Simón et al. 2008). By following this Perrotta et al. (2012) determined the OTA in red wine and achieved 0.008 ppb LOD. The simultaneous detection of mycotoxins was accomplished with a combination of competitive and sandwich immunoassay on a single array biosensor (Sapsford et al. 2006).

The non-competitive labeled immunoassays are mostly based on sandwiched format, in this, signal obtained through labeled secondary antibodies, as the number of labeled antibodies increases over the electrode surface, increase in signal is observed. A sandwiched immunosensor was designed by using catechol as a label for the detection of AFB1 and in this study Au–Fe nanoparticle played a role in the regeneration of sensor. The external horse magnet was used for the regeneration of immunosensor. When catechol oxidized into o-quinone signal read via a transducer and broad response from 0.6 to 110 ng mL^{-1} for AFB1 detection with a low detection limit of 0.2 ng mL^{-1} was observed (Masoomi et al. 2013), though the whole system of the biosensor was typical and chances of false signals. Another study of multiple labels was carried out for zearalenone detection with HRP tagged secondary antibodies (Feng et al. 2013a). They used sodium montmorillonites NPs and thionine for achieving LOD of 3 pg mL^{-1} for zearalenone detection in real samples.

14.3.1.2 Labeled Aptasensor for Detection of Mycotoxin

In aptasensor, aptamer is an important component, which is synthesized from 10 or more variable bases. These synthetic sequences of oligonucleotides (DNA/RNA) are specific for the target because of their specific binding affinity (Willner and Zayats 2007; Cho et al. 2009). Aptamers have plenty of advantages over the antibodies, as the synthesis and handling of aptamers are easy and other factors like cost effect, stability, and ability for further functionalization are enough to take over the antibody (Yang et al. 2013b; Schmidt et al. 2004). Aptamers are single-stranded oligonucleotides with three-dimensional structures, which can help ligands bind strongly with their complementary strands (Hermann and Patel 2000; Cruz-Aguado and Penner 2008a). The aptamers are highly stable, it can resist denaturation and active toward their targets (like peptides, proteins, toxins, drugs, whole cells, etc.) with strong affinity. Thus, aptamers are deliberated as a recovering substitute in many research studies instead of antibodies. Some reported sequences of aptamers for mycotoxins are listed in Table 14.2.

The mechanism of aptasensors is orchestrated on the interaction of aptamer with the analyte (mycotoxins) in the reaction mixture. After the interaction, the signal received by the transducer, which may be due to the enhancement of signal or decrement in signal depends upon the assay format called as "signal on" and "signal off," respectively. In labeled aptasensor signal receives to the transducer via labels (nanoparticles, enzyme, fluorescence material, etc.) either on the aptamer or on complementary DNA (Loo et al. 2015).

Labeled Signal On

"Signal On" reaction mechanism is shown in Fig. 14.2a. In this phenomenon, the signal increases after the interaction of aptamer and analyte because labeled moiety attached to the sequence aptamer of the analyte. For this phenomenon, mostly fluorescence materials have been used with the combination of quencher, as carbon-based nanomaterials are good quencher (CNTs, graphene) (Guo et al. 2011; Cruz-Aguado and Penner 2008b). Chen and coauthors achieved dynamic range for the OTA detection from 1 to 100 ng mL^{-1} via fluorescence-quencher mechanism (Chen et al. 2012). In fluorescence-quencher mechanism, the fluorescence material of aptamer interacts or quenched via "$\pi-\pi$" interaction between fluorescence material and carbon material. As the analyte (mycotoxin) enters, the aptamers change to their conformation due to more affinity toward the analyte (Zhu et al. 2010). The polymeric coating on carbon material like polyvinyl pyrrolidone coated on graphene oxide prevents the nonspecific adsorption over the electrode, which helps in enhancement in sensitivity (Sheng et al. 2011).

The upconversion nanoparticle (UCNP) in combination with rare earth metal (Tm, Er, Ho) gives multicolor. This property of UCNP has advantages over the other nanoparticles for the multi detection of analyte at the single electrode. Forster resonance energy transfer (FRET) is the best tool for multiple analyte detection using UCNP. The simultaneous detection of OTA and FB1 (Fumonisen B1) was reported by Wu et al. (Wu et al. 2012b) via "signal on" format. The AuNPs modified with luminol derivative-based aptasensor produced a 10-fold low detection limit (0.06 ng mL^{-1}) and linear range (0.2–5.0 ng mL^{-1}) for OTA (Wang et al. 2010b) in comparison to bare Au electrode.

The nucleic acids act as catalysts and receptors called as functional nuclic acid (DNAzyme or RNAzyme) (Li and Lu 2009). In aptasensor, DNAzyme plays an important catalytic role in HRP mimicking caused by hemin and G-quadruplexes interaction (Li et al. 2009; Cheng et al. 2009). Yang et al. (2013a) have designed colorimetric aptasensor for OTA detection, the "signal ON" phenomenon occurs

Table 14.2 Aptamer sequences for different mycotoxin detection in aptasensor

Sl. No.	Aptamer sequence	Toxin	Ref.
1	5′-ATGCTCCCTTTACGCCACGATCGGGTGTGGGTGGCGTAAAGGGAGCATCGGACA-3′	OTA	Chen et al. (2014)
2	5′-GATCGGGTGTGGGTGGCGTAAAGGGAGCATCGGACA-C6-biotin-3′	OTA	Zhang et al. (2013b)
3	5′-TGGGTAGGGCGGGTTGGGAAAGATCGGGTGTGGGTGGCGTAAAGGGAGCATCGGACA-3′	OTA	Yang et al. (2013a)
4	5′-MB-GATCGGGTGTGGGTGGCGTAAAGGGAGCATCGGACA-SH-3′	OTA	Wu et al. (2012b)
5	5′- AAAGATCGGGTGTGGGTGGCGTAAAGGGAGCATCGGACA-3′	OTA	Tong et al. (2011)
6	5′-GATCGGGTGTGGGTGGCGTAAAGGGAGCATCGGACA-FAM-3′	OTA	Chen et al. (2012)
7	5′-AGCAGCACAGAGGTCAGATG-N40-CCTATGCGTGCTACCGTGAA-3′	ZEN	Chen et al. (2013)
8	5 end-labeled by fluorescein (F-Apt05′-F-GATCGGGTGTGGGTGGCGTAAAGGGAGCATGGACA-3′)	OTA	Galarreta et al. (2013)

(continued)

Table 14.2 (continued)

Sl. No.	Aptamer sequence	Toxin	Ref.
9	5′-GATCGGGTGTGGGTGGCGTAAAGGGAGCATCGGACA-3′	OTA	Kuang et al. (2010), De Girolamo et al. (2011), Wu et al. (2012a), Rhouati et al. (2013), Hun et al. (2013), Lv et al. (2014)
10	OTA's aptamer (5′-GATCGGGTGTGGGTGGCGTAAAGGGAGCATCGGACA-FAM-3′)	OTA	Guo et al. (2011)
11	5-FAM-GATCGGGTGTGGGTGGCGTAAAGGGAGCATCGGACA-3′	OTA	Sheng et al. (2011)
12	5′-SH-C6-ATCTACGGGGCACGTTTATCCGTCCCTCCTAGTGGCGTGCCCC-3′	OTA	Wang et al. (2009, 2011)
13	5′-biotin-(CH_2) 6-ATACCAGCTTATTCAATTAATCGCATTACCTTATACCAGCTTATTCAATTACGTCTGCACATACCAGCTTATTCAATTAGATAGTAAGTGCAATCT-3′	Fumonisin	Wu et al. (2012b)

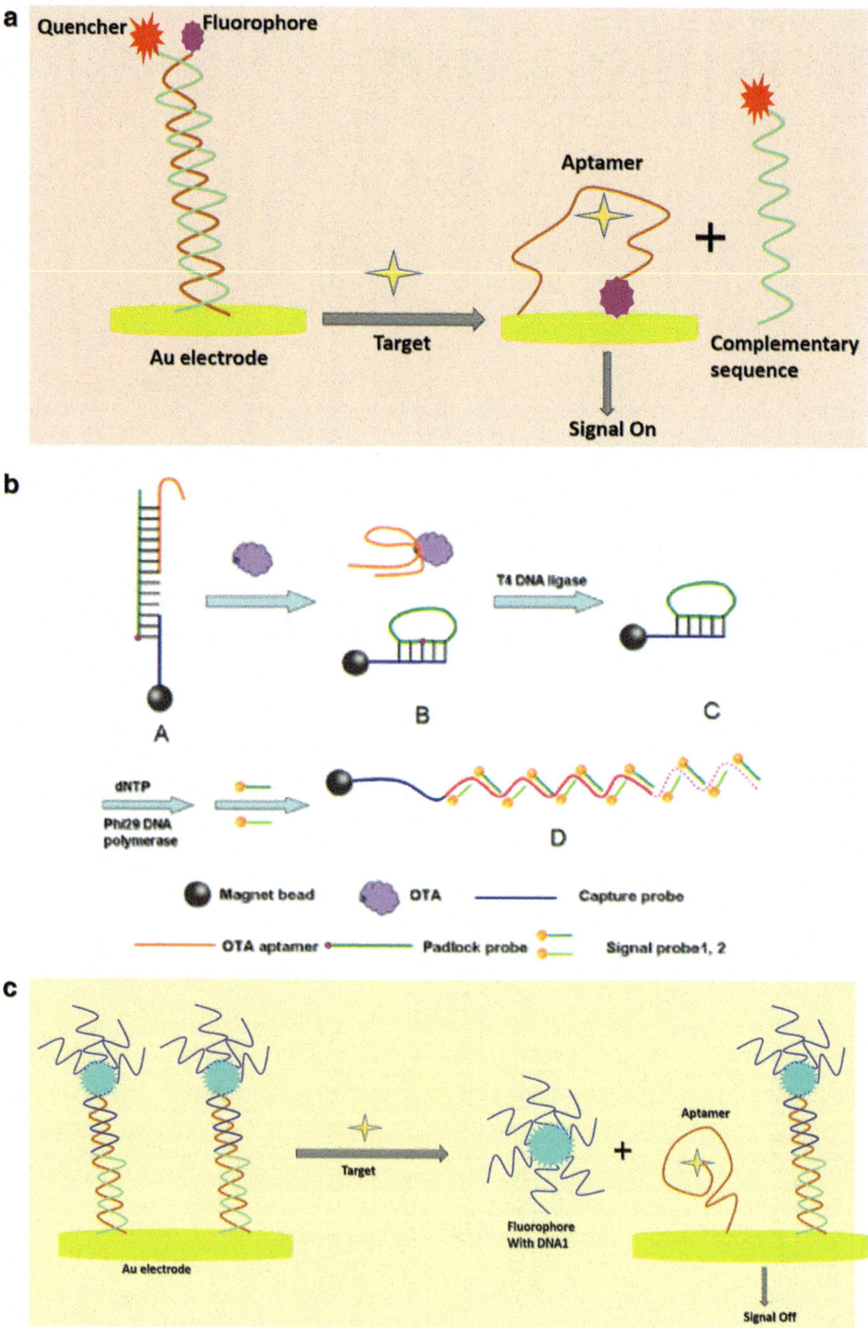

Fig. 14.2 **a** Schematic diagram for "signal on" aptasensor. **b** Scheme for rolling cycling amplification (RCA) for the detection of OTA (Tong et al. 2012). **c** Schematic diagram for labeled "signal off" mode for aptasensor

when OTA interacts with complex and hemin was displaced from G-quadruplex. The oxidation reaction was completed through H_2O_2 and tetramethylbenzidine (TMB). The multiple HRP-DNAzymes attached with AFB1 aptamer was used for signal amplification in chemiluminescence-based aptamers (Shim et al. 2014). Chen and group designed DNA probes based on biotinylated hairpin and labeled with streptavidin coated AuNPs for enzyme-free catalytic detection of AFB1 (Chen et al. 2016). In this AuNPs(red) aggregated into a cross-linked network via biotin-streptavidin coupling, on aggregation color changed from red to blue.

Based on peroxidase mimicking DNAzyme activity, Seok et al. (2015) have recognized AFB1 in ground corn samples. In this the aptamer-DNAzyme complex distorted and the DNAzyme halves split and generated a homogeneous colorimetric sensor with LOD of 0.1 ng mL^{-1}.

Aptasensor has limitations due to the equal ratio of receptor and analyte which causes the major issue of sensitivity (Tong et al. 2011). There are several signal amplification methods which expand the sensitivity of aptasensor, like rolling circle amplification (RCA) (Li et al. 2010), polymerase chain reaction (PCR) (Schmittgen and Livak 2008), nicking endonuclease amplification (Hun et al. 2010), and loop-mediated isothermal amplification (LAMP) (Wang et al. 2010a). A sandwiched aptasensor designed for OTA, in which DNAzyme quenched the chemiluminescent light from the O_2 and RecJf endonuclease recycle the assembly which enhanced the sensitivity up to 75 fg mL^{-1} (Chen et al. 2014).

Tong and group have electrochemically detected OTA using rolling cycling amplification, which amplified the signal. On the addition of T4 ligase capture DNA and padlock DNA formed circular template. Later this template DNA initiate cycling amplification reaction in the presence of phi2 DNA polymerase and dNTPs and generated a long tandem repeated sequence (Tong et al. 2012). Another circular amplification enzyme—nicking endonuclease, was used for the sensitive detection of OTA mycotoxins for sensitive detection that cleaved and released the OTA template strand (Hun et al. 2013). Hetero-enzyme-based two-round signal amplification strategy has been used for trace detection of aflatoxin B1 using an electrochemical aptasensor (Zheng et al. 2016).

Labeled Signal Off

The labeled signal off aptasensor based on a decrease in signal due to the formation of complex in between the aptamer and mycotoxin (Fig. 14.2c). Or in other words, the presence of labels (like enzyme, fluorophore QDs) in sensor show the maximum signal which suppose to reduce or off after the target interaction with their complementary aptamer, and aptamer with label moiety removes from the electrode. In signal off mode, the effect of nonspecific DNA adsorption reduces therefore, it has been found that the signal off mode is more sensitive and gives positive results rather than the signal on. The redox probe of AuNP/methylene blue used for the detection of OTA, on interaction with OTA to the probe, the transducer signal was reduced (Kuang et al. 2010).

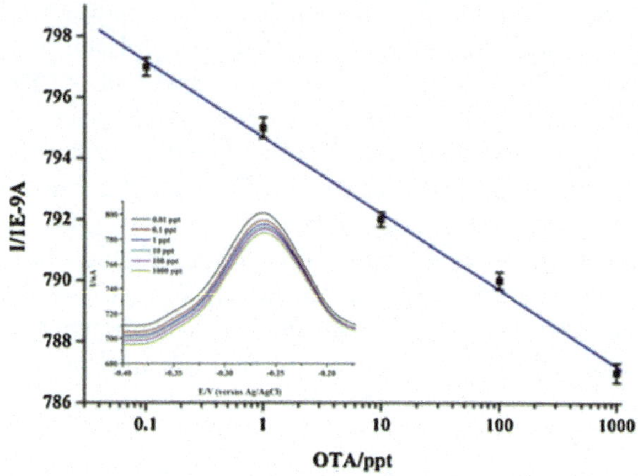

Fig. 14.3 Calibration curves for electrochemical aptasensor for detection of OTA (Wu et al. 2012a)

Those aptamers functionalized with magnetic bead have benefits of regeneration of aptasensor and easy to immobilize over the electrode with an external magnetic field (Rhouati et al. 2013). The multiple steps of processing of aptasensor produced the false results with real sample analysis. Though one step Au electrode-based aptasensor was designed with a dual label (thiol- and methylene blue) for the detection of OTA. The broad response range (0.1 pg mL^{-1}–1000 pg mL^{-1}) and 0.095 pg mL^{-1} LOD was obtained (Wu et al. 2012a) (Fig. 14.3).

14.3.2 Label-Free Mycotoxin Biosensor

14.3.2.1 Immunosensor for Label-Free Detection of Mycotoxin

The label-free immunosensors had advantages over the labeled immunosensor because of their simplicity and one step analysis. Like the labeled immunosensor it was also categorized into two (i) non-competitive and (ii) competitive. Sometimes label-free immunosensor are unable to reach the required sensitivity specially for small molecules like mycotoxin. The non-competitive immunosensors are fast and simple in comparison to competitive immunosensor, though they are less sensitive because of nonspecific binding and receive false positive results.

The mechanism of label-free non-competitive immunosensor is based on direct interaction between immobilized antibodies and mycotoxin. Researchers used different transducers and varying the matrix of electrode (carbon nanocomposites, CNTs, graphene oxide) or conducting polymer or metal, and metal oxide nanoparticles (Fe_3O_4, ZnO, TiO_2, etc.) to enhance the sensitivity of label-free immunosensor (Sharma et al. 2010; Solanki et al. 2009; Zhang et al. 2016; Díaz Nieto et al. 2019; Srivastava et al. 2013). An immunoassay based on lateral flow was developed by controlling the concentration of antibody and size of magnetic beads. It exhibited a cutoff value of 0.02 mg L^{-1} of AFM1 in milk samples (Liu et al. 2015a). Using electro-spinning technique, a polyacrylonitrile (PAN) nanofiber-modified pencil-graphite electrode was developed for Zearalenone (ZEN) detection in the liquid foods, including dairy, juices, and other liquid foods (Moradi et al. 2020).

Among the electrochemical spectroscopic studies impedance spectroscopy is more sensitive and reliable because in this interface of electrode studies with electrode surface reactions (Radi et al. 2009; Zamfir et al. 2011). A label free impedemetric immunosensor for OTA has been developed (Malvano et al. 2016). Yu and group developed an impedimetric immunosensor for the determination of AFB1 based on MWCNTs-ionic liquid modified electrode and acquired the LOD 0.03 ng mL^{-1}. The MWCNTs in ionic liquid exists as a three-dimensional (3D) network and cross-linked due to cation π and π–π interactions between the imidazolium cations and CNTs. Here the ionic liquid delivers excellent biocompatible microenvironment to uphold the activity of antibody and used as an exceptional conductive binder to immobilize antibodies and MWCNTs onto the electrode (Yu et al. 2015).

The self-assembled monolayer of 11-mercaptoundacanoic acid was assembled on the silver wire for dual detection of AFB1 and AFM1 in nuts and milk, respectively, using EIS immunosensing technique (Bacher et al. 2012). The online flow system with EIS technique is more persuasive then lateral flow system and shows enhanced sensitivity and lower LOD. On the coupling of these two systems Kanungo and group have detected AFM1 and AFM2 in beverages with the LOD of 1 pg mL^{-1} (Kanungo et al. 2014).

The fragment antibody of AFB1 showed 10-fold higher sensitivity than the whole antibody. Li and coauthors used fluorescence resonance energy transfer technique with tryptophan (Trp) residues-based fragment antibody, they followed a label-free non-competitive format, where tryptophan (Trp) residues played a role of donor wherein fragment comes in contact of AFB1 mycotoxin (Li et al. 2013). The optical sensing technique was found more compatible with label-free immunosensing because the reflection of light can read minor changes on the surface of the electrode (Nabok et al. 2005, 2007). The SPR technique based on the reflection of light from the gold surface of the electrode, therefore gold-coated protein G were more compatible rather than thiol or ProG modified electrode surface (Tamerler et al. 2006; Park et al. 2014). The simultaneous detection of AFB1 and OTA was done by optical waveguide light spectroscopy (OWLS). The OWLS technique compatible with competitive immunosensor and show highly sensitive response in compare to non-competitive immunosensor (Adanyi et al. 2007). Later study applied OWLS for the detection of AFB1 in competitive immunoassay with AuNPs of different sizes and origins (obtained by chemical or biotechnological synthesis) (Adányi et al. 2018). The addition of nanomaterial to the optical label-free immunosensor may reduce the challenge related to the sensitivity (Ah et al. 2012). The gold nanorods were applied for AFB1 detection on an optical electrode which followed the label-free competitive format and obtained the LOD of 0.16 ng mL^{-1} for AFB1 (Xu et al. 2013). Though optical immunosensors are sensitive they have some drawbacks like, expensive equipment, regeneration of sensor chip, and nonspecific adsorption. Over the other optical technologies, White Light Reflectance Spectroscopy (WLRS) has certain advantages like free-space interference concepts, the limited effect from the change in the refractive index of the sample under analysis and provides immunity to temperature variations. Moreover, the illumination of the sensing area is straightforward during the bioreaction monitoring. Additionally, biochips are easy to be fabricated and of low cost. The principle of WLRS depends on the change in the interference spectrum, which occurs after the interaction of analyte over the SiO_2 fabricated silicon chip. With a slight modification a new approach, applied by Anastasiadis and coworkers (Anastasiadis et al. 2020) in which they have designed the WLRS electrode surface for multiple detections of mycotoxins using the variable thickness of SiO_2.

The QCM transducer use is not much widespread for the detection of mycotoxins, because of its small size (Tsai and Hsieh 2007; Cheng et al. 2012; Vidal et al. 2009). However, label-free detection of an analyte using the QCM transducer has advantages over the other transducer. The mass-based cantilever was used for multiple detections of mycotoxins (Ricciardi et al. 2013). The coupling of QCM with electrochemical transducer showed a high sensitivity for label-free detection of mycotoxins. The electrochemical quartz crystal microbalance (EQCM)-based immunosensor was developed, in this thiol self-assembled monolayer modified with AuNPs (Fig. 14.4a) for the detection of AFB1 (Chauhan et al. 2015b; Chauhan et al. 2016). They developed a reusable label-free immunosensor (Fig. 14.4b). This immunosensor format was competitive and sandwiched (Au coated Fe_3O_4 nanocomposites with secondary antibodies) and detect AFB1 in wine samples (Chauhan et al. 2015a). A list of developed label-free and labeled immunosensors for mycotoxin given in Table 14.3.

A competitive multiplexed imaging surface plasmon resonance (iSPR) biosensor assays were corroborated for DON, ZEA, and T-2 toxin (Hossain and Maragos 2018). In this, AuNP was used for signal amplification conjugated with secondary antibodies. One of the advantages of iSPR is the ability to use the sensor multiple times, however, it is a challenging task to regulate the optimal conditions at the time of binding and regeneration for multiple antigens, and antibodies. Lu et al. (2019) designed a microfluidic immunosensor for simultaneous detection of FB1 and DON. This capillary-driven microfluidic system achieved LODs of 97 pg mL^{-1} and 35 pg mL^{-1}, and their corresponding linear ranges of detection are 0.3–140 ppb and 0.2–60 ppb, respectively, for FB1 and DON (Lu et al. 2019). Liu and group developed a label-free immunosensor based on magnetic $CdSeFe_3O_4$ nanocomposites for AFB_1 detection (Liu et al. 2015b). Anti-AFB_1 was immobilized on the $CdSeFe_3O_4$ modified screen-printed electrode (SPE). The determination of AFB_1 concentration was measured by a decrease in photocurrent, caused by the steric hindrances due to the immunocomplex. Bhardwaj et al. develop an AuNPs-based microfluidic SPR biosensor for AFB1 detection relied on modified AuNPs grafted over a SAM-linked Au chip surface via an amine linkage (Bhardwaj et al. 2020).

Xia and coworkers designed a cell-based impedance sensor for DON, ZEA, and AFB1 detection (Xia et al. 2017). A luminescent bacterium, *Photobacterium phosphoreum* was employed used for the multi mycotoxin detection under discrete and flow-through analysis (Senko et al. 2019). Though the method is simple, convenient, and exhibits a quick response rate incubation time was found to be more than 12 h

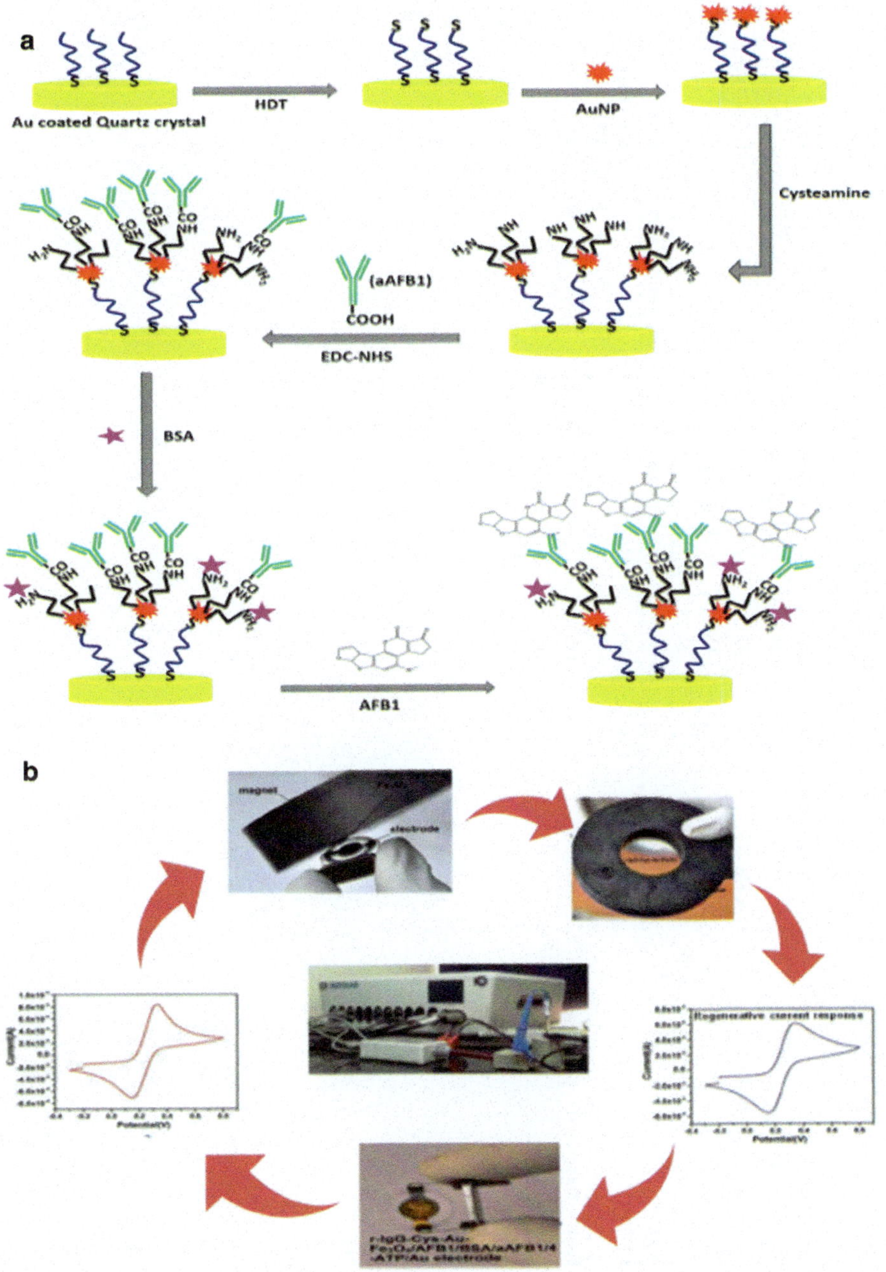

Fig. 14.4 **a** A diagrammatic representation of label-free detection of AFB1. **b** A pictorial representation of the regeneration of immuno-electrode

Table 14.3 The list of label-free and labeled immunosensors developed for detection of mycotoxins

Sl. No.	Matrix	Electrode	Type of Ab	Mode of detection	Label	Technique
1	BSA/ab/CeO_2–CS/GO–CS/SPE	CeO_2–CS/GO-CS/SPE.	Ab	Non-competitive	Label-free	DPV
2	AFB1/BSA/aAFB1/4-ATP/Au	–	mAb	Competitive	–	EQCM
3	CS–PtCo–Ab/Graphene-Thionine/GCE	GCE	mAb	NC	–	Electrochemical
4	Graphite/ silver	SPCE	mAb	Competitive	OTA-ALP	DPV
5	Graphite based ink	SPCE	pAb	Competitive	Au-labeled ALP-Sec Ab	DPV
6	1,6-Hexanedithiol	Au	mAb	Competitive	ALP	Electrochemical (DPV)
7	Magnetic beads with Protein G	SPCE	mAb	Competitive	HRP	SWV
8	–	SPE	mAb	Competitive	OTA	–

(continued)

Table 14.3 (continued)

Sl. No.	Matrix	Electrode	Type of Ab	Mode of detection	Label	Technique
9	GNRs	–	mAb	Competitive	–	Optical
10	16-MHDA	Au	pAb	Competitive	Au NPs	QCM
11	Magnetic Beads	SPCE	mAb	Competitive	HRP	DPV
12	11-MUOH	Au	mAb	Competitive	Au NPs	SPR
13			mAb	Competitive	–	EIS
14	MPA-BSA		mAb	Competitive	HRP	QCM
15	1-dodecanethiol (DDT)	Au	pAb	NC	–	Electrochemical
16	Au NPs	–	–	Competitive	–	SPR
17	GCE/PTH/AuNP AFB1- BSA-polythionine (PTH)/gold nanoparticles	GCE	pAb	Competitive	HRP	Electrochemical
18			pAb	Competitive	ALP	Intermittent pulse amperometry (IPA)

(continued)

Table 14.3 (continued)

Sl. No.	Matrix	Electrode	Type of Ab	Mode of detection	Label	Technique
19	magnetic nanoparticles (MNPs)	–	mAb	Competitive	–	Fluorescent
20	Au/3-Mercaptopropionic acid (MPA)		mAb	Competitive	Au NPs	Piezoelectric
21	Au/DDT SAM	Polycrystaline gold electrode	mAb	NC	–	TIRE and QCM
22	DSP (3,3′-Dithiodipropionic-acid-di-N-hydroxysuccinimide ester)	Au	mAb	NC	–	QCM
23	Fe3O4/SiO2 composite nanoparticles	Au	mAb	NC	–	QCM
24	Fe3O4-graphene oxides	SPCE	mAb	Non-competitive	CdTe-CNT QDs Conjugate	ECL
25	AET- Au NPs	Au interdigitated electrodes (150 nm thick)	mAb	Non-competitive	HRP	conductometric
26	GCE/chitosan/Au nanoparticle (AuNP)/anti-Aflatoxin B1 (a AFB1),	GCE	pAb	Non-competitive	catechol–Au–Fe3 O4 NPs	Electrochemical
27	nanoporous gold films (NPG)	GCE	mAb	Non-competitive (NC)	Na-Mont-TH-HRP-Ab2	Electrochemical
28	Chitosan-iron oxide	ITO	pAb	NC	–	DPV

(continued)

Table 14.3 (continued)

Sl. No.	Matrix	Electrode	Type of Ab	Mode of detection	Label	Technique
29	Nano CeO_2 – CH	ITO	pAb	NC	–	DPV
30	Chitosan- PANI	ITO	pAb	NC		EIS
31	4-carboxyphenyl SAM	Au	–	NC	–	EIS
32	BSA/AO-IgG/AUT/Au (SAM of AUT)	Au	pAb	NC	–	EIS
33	16-MHDA	Au	pAb	NC	–	QCM
34	Au/TA/Glu/BSA/MNP-Ab	Au	mAb	NC	–	EIS, CV
35	BSA/aAFB1-C-AuNP/MBA/Au	Au	mAb	NC	–	Volta metric
36	anti- AFB1/MWCNTs/ITO	ITO	mAb	Non-competitive	–	Volta metric
37	SAM of 11-MUA	Silver Wire	mAb	Non-competitive	–	Impedimetric
38	BSA/AbAFB1/n-Sm_2O_3/ITO	ITO	mAb	Non-competitive	–	Volta metric

(continued)

Table 14.3 (continued)

Sl. No.	Matrix	Electrode	Type of Ab	Mode of detection	Label	Technique
39	–	–	Fragment Ab	Non-competitive	–	FRET based Immunoassay
40	Nano-sized gold hollow balls (NGB) with dendritic surface	Au		NC	–	QCM
41	Silica gel-ionic liquid-antibody	GCE	pAb	NC	–	EIS
42	APTES, Silane, Protein G	–	mAb	NC	–	Microcantilever
43	RGO/ITO	ITO	mAb	NC	–	Electrochemical
44	MWCNTs/RTIL/Ab	glassy carbon electrodes	mAb	Competitive		EIS
46	Au/TGA/BSA-Ag/anti-OTA McAb	Au	Ab	competitive	–	DPV
47	anti-aflatoxin antibody/Au/rGO/ITO	AU/RGO/ITO	Ab	–	–	CV

Sl. No.	Toxin	Response range	LOD	Sensitivity	Relative Standard Deviation (RSD)in %	Coefficient of variation (CV)in %	Ref.
1	AFM1	0.01–1 μgL^{-1}	0.009 μgL^{-1}	–	2.7–3.5	–	An et al. (2020)
2	AFB 1	0.1–4.0 ng mL – 1	0.16 ng mL^{-1},	532.7 mA ng – 1 mL cm – 2	5.13	–	Chauhan et al. (2015b)
3	ZON	0.05–5.0 ng mL^{-1}	13 pg mL^{-1}	–	4.5	–	Feng et al. (2013b)
4	OTA	0.05–2.5 ng mL^{-1} 0.1–7.5 ng mL^{-1}	60 ng mL^{-1} 0.4 ng mL^{-1} 100 ng mL^{-1}	–	8–16	7	Alarcón et al. (2006)
5	OTA	0.9–9 ng mL^{-1} 0.3–8.5 ng mL^{-1}	0.86 ng mL^{-1} 0.20 ng mL^{-1}	–	8, 10	6	Bonel et al. (2010)
6	OTA	10 pg mL^{-1}- 100 ng mL^{-1}	8.2 pg mL^{-1}	–	–	–	Liu et al. (2009)
7	OTA	0.01–20 ppb	0.008 ppb	–	5.56	–	Perrotta et al. (2012)
8	OTA	0.3 ngL^{-1}- μgL^{-1}	–	–	20, 36	–	Prieto-Simón et al. (2012)
9	AFB 1	0.5–20 ng mL^{-1}	0.16 ng mL^{-1},	–	–	–	Xu et al. (2013)
10	OTA	10–128 ng mL^{-1} ng mL^{-1}	8 ng mL^{-1}	–	–	8.96	Vidal et al. (2009)
11	OTA	0.1 pg mL^{-1}–1 μg mL^{-1}	10 pg L^{-1}	–	9	–	Vidal et al. (2012)

(continued)

Table 14.3 (continued)

Sl. No.	Toxin	Response range	LOD	Sensitivity	Relative Standard Deviation (RSD)in %	Coefficient of variation (CV)in %	Ref.
12	OTA	–	1.5 ng mL^{-1} Au–0.042 ng mL^{-1}	–	–	–	Yuan et al. (2009)
13	AFM 1	1–100 pg mL^{-1}	1pgmL^{-1}		–	–	Kanungo et al. (2014)
14	AFB 1	0.01–10.0 ng mL^{-1}	0.01 ng mL^{-1}	–	4–7	–	Jin et al. (2009b)
15	altertoxin I (ATX-I)	–	4×10^{-8} moldm^{-3} (14 ppb)	–	7	–	Moressi et al. (2007)
16	OTA	0.1 ng mL^{-1}–0.1 μg mL^{-1}	60 pg mL^{-1}	–	–	–	Urusov et al. (2011)
17	AFB 1	0.6–2.4 ng mL^{-1}	0.07 ng mL^{-1}	–	2–5	–	Owino et al. (2008)
18	AFB 1	0.05 and 2 ng mL^{-1}	–	–	7–10	–	Piermarini et al. (2007)
19	AFB 1 OTA	0.01–10 ng mL^{-1}	0.01ngmL^{-1}	–	5.19 2.14	–	Wu et al. (2011)
20	AFB 1	0.1–100ngmL^{-1}	0.01 ng mL^{-1}	–	4–7	–	Jin et al. (2009a)
21	AFB1	100 μg mL^{-1} to 0.15 ng mL^{-1}		–	–	–	Nabok et al. (2007)

(continued)

Table 14.3 (continued)

Sl. No.	Toxin	Response range	LOD	Sensitivity	Relative Standard Deviation (RSD)in %	Coefficient of variation (CV)in %	Ref.
22	AFB 1	0.5–10 ppb	–	–	–	–	Spinella (2013)
23	AFB 1	0.3–7.0 ng mL^{-1}	0.3 ng mL^{-1}	–	8.0	6.3	Wang et al. (2009)
24	AFM 1	1.0–1.0 × 10^5 pg mL^{-1}	0.3 pg mL^{-1}	–	3.5	–	Gan et al. (2013)
25	AFB 1	0.5–10 ng mL^{-1}	0.1 ng mL^{-1} at 3δ.		7	7.1	Liu et al. (2006)
26	AFB 1	0.6–110 ng mL^{-1}	0.2 ng mL^{-1}	–	–	6.2, 5.7	Masoomi et al. (2013)
27	ZON	0.01–12 ng mL^{-1}	3 pg mL^{-1}	–	2.46–5.17	4.2–5.2	Feng et al. (2013a)
28	OTA		0.5 ng dL^{-1}	36μA$ng^{-1}dL^{-1}$	–	–	Kaushik et al. (2008)
29	OTA	0.5–6 ng dL^{-1}	0.25 ng dL^{-1}	1.27 μA ng^{-1} dl^{-1} cm^{-2}	–	–	Kaushik et al. (2009)
30	OTA	Up to 10 ng mL^{-1}	1.0 ng mL^{-1}	–	–	–	Khan and Dhayal. (2009)
31	OTA	1–20 ng mL^{-1}	0.5 ng mL^{-1}	–	3.7–5.5	–	Radi et al. (2009)

(continued)

Table 14.3 (continued)

Sl. No.	Toxin	Response range	LOD	Sensitivity	Relative Standard Deviation (RSD)in %	Coefficient of variation (CV)in %	Ref.
32	OTA	0.5–6 ng dL^{-1}	0.08 ng dL^{-1}	36.86 O $ng^{-1}dL^{-1}$	–	–	Solanki et al. (2010)
33	OTA	50–1000 ng mL^{-1}	16.1 ng mL^{-1}		–	64	Tsai and Hsieh (2007)
34	OTA	0.01–5 ng mL^{-1}	0.01 ng mL^{-1}	–	4.7	–	Zamfir et al. (2011)
35	AFM 1	10–100 ng dL^{-1}	17.90 ng dL^{-1}	0.45 μAng^{-1} dL	–	–	Sharma et al. (2010)
36	AFB 1	0.25–1.375 ng mL^{-1}	0.08 ng mL^{-1}	(95.2μAng^{-1} mLcm^{-2})	–	–	Singh et al. (2013a)
37	AFM 1	6.25–100pgmL^{-1}	1 pg mL^{-1}	6.25 pg mL^{-1}	–	–	Bacher et al. (2012)
38	AFB 1	10–700 pg mL^{-1}	57.82 pg mL^{-1} cm^{-2},	48.39 μApg^{-1}mL cm^{-2}	0.39	–	Singh et al. (2013b)
39	AFB 1	–	0.85 and 0.09 ng mL^{-1}	–	–	–	Li et al. (2013)
40	AFB 1		0.05 ng mL^{-1}		2.3, 5.8	3.8–5.1	Liao (2007)
41	AFB1	0.1–10 ng mL^{-1}	0.01 ng mL^{-1}		1.7	–	Zaijun et al. (2010)
42	Aflatoxins OTA 1	Aflatoxins; 3 ngmL^{-1} OTA 1: 6 ng mL^{-1}	–	–	–	–	Ricciardi et al. (2013)
43	AFB 1	–	0.12 ng mL^{-1}	68 μAng^{-1}mL cm^{-2}	–	–	Srivastava et al. (2013)
44	AFB 1	0.1–10 ng mL^{-1}	0.03 ng mL^{-1}	–	2.7–4.3		Yu et al. (2015)
46	OTA	0.1–1.0 ng mL^{-1}	0.08 ng mL^{-1}	–	<4.67	–	Sun et al. (2020)
47	AFB1	5 ng mL^{-1} to 1 pg mL^{-1}	6.9 pg mL^{-1}	–	–	–	Althagafi et al. (2019)

and the Hep G2 cell or any whole cell-based sensors are not specific for the targeted analyte, therefore there are more probabilities of false results.

The detection of mycotoxins by aptamers without using any label is called label-free aptasensor, where the change in signal in transducer for aptamer and mycotoxin interaction is read without a label (Schöning and Poghossian 2002). Researchers have modified the electrode surface via nanoparticles, polymers, oxides of metal nanoparticles, etc., for electrochemical detection of OTA (Cai et al. 2018; Khan et al. 2011). Some research group used click chemistry for uniform immobilization of aptamer, which provides the better surface to the interaction of mycotoxins, and trigger to enhance the sensitivity (Hayat et al. 2013b; Ran et al. 2011). The alteration in aptamer's conformation during the mycotoxin interaction reduces the background current in aptasensor. Hayat and group designed polyethylene glycol–aptamer-based screen-printed carbon electrode (SPCE). Consequently, macromolecules were formed through long tunnels, at the tunnel OTA aptamer was present for detection of OTA (Hayat et al. 2013a). The signal on label-free aptasensor formed G-quadruplex for the detection of OTA and Tb^{3+} introduce to enhance the sensitivity (Fig. 14.5). The LOD was noticed to be as 20 pg mL^{-1} (Zhang et al. 2013a). Castillo and coworkers have proposed an electrochemical aptasensor for detection of AFB1 using cystamine-dendrimers (Cys-PAMAM) layers on gold electrode. The limit of detection attained by this sensor is 0.40 ± 0.03 nM, in this dendrimer enhances the sensitivity. The dendrimers were positively charged due to this property of dendrimer the aptasensor possesses less negative charge over the electrode surface. This effect of dendrimer on the electrode surface helps in electron transfer and also effective for redox reaction (Castillo et al. 2015).

The reversible ligand grafted aptasensor was developed for wave optical detection of OTA. The real samples of food were analyzed, where the electrode contained glutaraldehyde and ethylenediamine as linker and the OTA attached with covalent bonding over optical fiber. The binding energy was measured on the bases of tied ligand and free OTA aptamer in the reaction mixture. Using a competitive detection

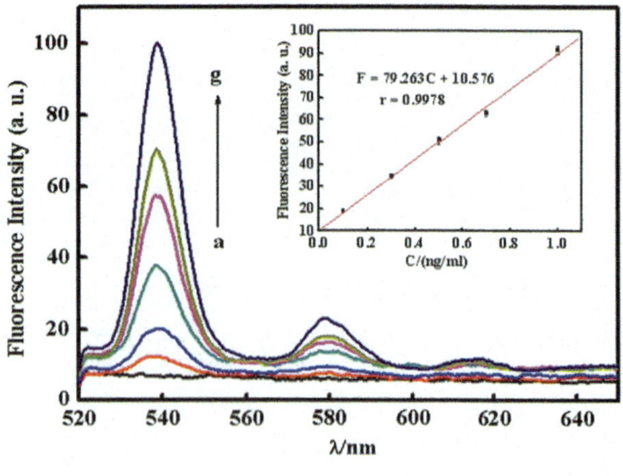

Fig. 14.5 The response analysis of aptasensor for OTA (from 0 to 1 ngmL^{-1}) in fluorimeter. Inset: shown calibration curve between different concentrations of OTA and fluorescence intensity (Zhang et al. 2013b)

mode, they attained the linear range from 0.73 μg L^{-1} to 12.50 μg L^{-1} and LOD of 0.39 μg L^{-1} (Liu et al. 2015a).

In optical resonance techniques, surface plasmon resonance spectroscopy(SPR) is also valuable for the detection of low molecular weight molecules. In some studies, it was observed that in SPR the signal obtained with the change of mycotoxin concentration was very low in respective of the antibody (Wang et al. 2009). The acoustic technique for label-free aptasensor was applied for the detection of OTA in real sample, the thickness-shear mode (TSM) transducer was applied (Lamberti et al. 2011). The other technique like Raman spectroscopy (SERS) was also reported (Galarreta et al. 2011). A Canada based Biotech company first time patented specific aptamers for AFB1 and ZAE in 2010. Guo and coworkers reported real time for PCR-based detection of AFB1. The real-time PCR is highly sensitive, quantitative technique, which can detect AFB1 at femtogram level. The biotinylated functionalized aptamer was used and obtained the LOD of 25 fg mL^{-1} for AFB1 (Guo et al. 2014). The list of developed aptasensor for mycotoxins is given in Table 14.4.

14.3.3 Molecular Imprinting Polymer

The synthetically produced polymer which possesses affinity toward the target molecule or has a frame to fix the target molecule is called molecularly imprinted polymers (MIPs). These are synthesized by polymerization of monomer around the target molecule and link via cross-linker to make solid frame of a respective target. On the complete formation of MIP there is a cavity in replacement of target molecule, which provides complimentary outline and functionality to the MIPS. However, very few works were reported on MIP-based detection of mycotoxins (Piletsky et al. 2001). Yu and team reported OTA detection based on MIPs, they electrochemically polymerized polypyrrole with chloride (PPy-Cl) on the gold surface (Yu and Lai 2005). The metal ions in MIP enhanced the chelation property. Gao and coworkers detected T-2 mycotoxin using Fe^{3+} introduced MIPs and attained selective and sensitive MIP-based detection (Gao et al. 2014).

14.3.4 Enzymatic Inhibitor

Mycotoxins inhibit enzymatic activity which includes cholinesterase, acetylcholinesterase (AChE), glucose oxidase, urease, etc. The AChE enzyme is sensitive toward AFB1 detection; therefore, reversible SPR sensor was used for real-time detection of AFB1(Puiu et al. 2012). Further, because of the inherent property of AChE toward AFB1, researcher exploited AChE for the detection of AFB1in real samples (Moscone et al. 2011; Ben Rejeb et al. 2009). The mycotoxins bound non-covalently and reversibly with an enzyme (Stepurska et al. 2015). Among the mycotoxins, AFB1 has maximum sensitivity (Soldatkin et al. 2013).

Table 14.4 The list of developed label-free and labeled aptasensor for mycotoxins

Sl. No.	Matrix	Electrode	Type of Aptamer	Mode of detection	Label	Technique
1	Magnetic Beads	–	Aptamer	Competitive	ALP	DPV
2	Paramagnetic microparticle beads	SPCE	Aptamer	Competitive	HRP	DPV
3	SH-Apta	Au	Aptamer	Competitive	Fluoresceinaptamer	Microfluidic channel SERS
4	Magnetic beads-based DNA (ssDNA) aptamers		DNA (ssDNA) aptamers	Non-competitive	–	Fluorometer
5	6-mercapto-1-hexanol/captured probe (MCH)/(CP)/AuE	Au	Aptamer	Competitive	–	Electrochemiluminescent
6	Carboxyfluorescein	SWCNT	Aptamer	Competitive	–	Fluorescence
7		magnetic bead (MB).	sensing DNA		ABEI-labeled carboxylic silica nanoparticles	chemiluminescence,
8	–	–		Competitive	Au NP -DNA	Electrochemical
9	–	–				Fluorescent
10	Magnetic beads	SPCE		Competitive	Biotin-labeled OTA, ALP	Automated flow electrochemical aptasensor
11	Graphene oxide			Competitive	FAM (carboxyfluorescein)	
12	Graphene				UCNPs	FRET
13	Phi 29 DNA	Magnetic bead	Aptamer	–	QDs	Electrochemical
14	Au NPs	Au	Aptamer	–	DNA 1–Au NPs DNA 2–ABEI	Electrochemi-luminescence
15	PGE/DNA	PGE	fsDNA	–	Label-free	EIS
16	OTA chemisorbed on Au	–	Aptamer 4 sequences with one Dimer	–	–	EIS
17	SA–Au NPs@Cd-MOF-74/S1/MCH/S2/Au NPs/Chit/GCE	MOF/GCE	cDNA	–	labeled	DPV

(continued)

Table 14.4 (continued)

Sl. No.	Matrix	Electrode	Type of Aptamer	Mode of detection	Label	Technique
18	OTA-APT:OTA-cAPT	Aptamer	RNA-based aptasensor	–	labeled OTA-RNA	Fluorescence
19	GCE/modified/aptamer/AFB1/rGO	GCE/rGO	Apt	–	Label-free	DPV, CV, EIS
20	Ab/BSA/Au-PANI/ITO	ITO	Ab	–	Label-free	EIS
21	AuNP/GQD/GSPE	GSPE		direct	Label-free	LSV
22	GO modified PAA/AFB1/Apt	GO/PAA	Apt	–	–	Amperometry
23	GO/AuNW/dsDNA/SPCE	SPE/GO	Aptamer	–	Label-free	DPV
24	BSA/anti-OTA/PdNPs/CF	PdNPs/CF	Aptamer	Non-competitive	Label-free	DPV
25			Aptamer			Electrochemical
26	Tb3þ, structure-switching aptamer, and magnetic beads (MBs)				Signal on	Fluorescent
27	label-free for rapid and high-throughput detection of OchratoxinA		DNAzyme–aptamer conjugate			colorimetric
28	Ferrocene-labeled DNA probe	Au	Aptamer	–	–	DPV
29	Au NPs and DNA probe	Gold	ssHSDNA	Non-competitive	–	Impedimetric
30	Biotin-Streptavidin	–	AFB1 Aptamer	Non-competitive	–	Electrochemical
31	(Fe_3O_4/PANI) film	Micro- IDE of Pt	21-mer aptamer	Non-competitive	–	Electrochemical-based aptasensor
32	GCE/rMoS_2-Au/APs/BSA/L-CPs	GCE	Aptamer	–	Label	DPV
33	Zn-SnS_2/N-GQDs	ITO	Aptamer			Photocurrent response
34	Au/PS/DNAapt/MB	Au/PS	Aptamer	–	Label-free	CV, DPV
35	cystamine, PAMAM G4 dendrimers, and DNA	Au	DNA	Competitive		EIS, CV
36	aptamer/G-quadruplex DNAzyme and Fe_3O_4@oMWCNTs	Fe_3O_4@MWCNTs	Aptamer	–	Fluorescence	Fluorescence spectra
37	SPCE/FGO/HMDA/Apt	SPCE	MB-tagged aptamer	Competitive	Labeled	DPV
38	Aptamer and DNA	–	HRP mimicking DNAzyme	–	Label-free	HCR, Gel electrophoresis

(continued)

Table 14.4 (continued)

Sl. No.	Matrix	Electrode	Type of Aptamer	Mode of detection	Label	Technique
39	SPCE/4-CP/act/Apt/OTA	SPCE		–	Label-free	EIS
40	PPy/Au	Au	Aptamer	competitive	Label-free	SPR, EIS
41	PEI-MoS_2-MWCNTs and Tb	Au	ZEN Aptamer	–	Label-free	
42	SDA and DNA G-quadruplex	–	cDNA	competitive	Fluoresce	AIE assay
43	Janus/COOH-GN/GCE	GCE	OTA aptamer	Non-competitive	Label-free	DPV
44	Gold nanoparticles decorated MoS2 nanosheets	Au	MB-Aptamer	–	label-free	
45	SPCE/4-CP/aptamer	SPCE	Aptamer	–	–	EIS
46	HRP/BSA/AFB1	Microplates	HRP-linked aptamer	Competitive	label	Chemiluminescence
47	anti-M13-HRP/phage/BSA/anti-OTA-McAb/Mal-PEG-NH2/L-Cysteine/MPA/AuNps/GCE.	GCE	mcAb	Competitive	label	CV, EIS and SWV
48	Apt-PDMS-AuNPs-SPCE	SPCE	Aptamer	–	label-free	EIS, CV
49	Au/cDNA	Au	dsDNA		Label-free	CV
50	MnO2/ACP	MnO_2	biotin-labeled OTA aptamer	Competitive	labeled	colorimetric
51	MB-cDNA/apt-SiO2@CdTe and MB-cDNA/apt-SiO2@PbS	SiO2	Aptamer	–	label	SWV
52	Streptavidin/AFM1 aptamer	–	ssDNA	–	fluorescence	RT-qPCR
53	dendritic-DNA/aptamer/AMNPs	Amino modified magnetic nanoparticles	DNA	–	Fluorescence probe	PAPDI-HCR amplification strategy
54	SPCE/PTH/IrO2NPs/aptamer/EA	SPCE	Aptamer		Label-Free	EIS
55	AFB1–QD–GO Quencher	GO	Aptamer	Competitive	Labeled	PL spectroscopy
56	LMOF-241	MOF	–	–	Photoluminescence	UV spectroscopy
57	Exo I enzyme/SYBR Gold/aptamer	SYBRAu	fluorescence	–	label-free	fluorescence measurement

(continued)

Table 14.4 (continued)

Sl. No.	Matrix	Electrode	Type of Aptamer	Mode of detection	Label	Technique
58	TS-AuNP-c-DNA	Au	cDNA	Non-Competitive	label-free	SWV
59	SPCE/4-CP/Apt/AFB1	SPCE	Aptamer	Non-Competitive	label-free	EIS
60	CAs-cDNA/apt/AuE	Au	Aptamer	Non-Competitive	label-free	DPV
61	3DOM MoS_2 – AuNPs	Au	H1/HRP/AuNPs-SiO@Fe_3O_4	–	–	DPV
62	poly-NR/GCE	GCE	Aptamer	–	mNR	EIS and DC voltammetry
63	pβ-CD/AuNPs/GC	GCE	Fc-cDNA	–	label	DPV and EIS
64	Anti-AFB1/MB/Au	Au	MB-labeled aptamer	Competitive	label	SWV
65	SiO_2@PbS/Fe_3O_4@SiO_2/CdTe	CdTe and PbS QDs	DNA	–	label	SWV
66	mSiO_2@Au-DNA1/apt	rGO/AuNP/CdTe/DNA2	–	–	label-free	SWV
67	RGO/MoS_2/PANI@AuNPs/Apt	Au	Aptamer	–	label-free	DPV
68	OTA-aptamer-ZnPPIX	ZnPPIX	Fluorescence	–	label-free	Fluorescence spectra
69	Apt-modified SPGE with CS-modified AuNPs	SPGE	cDNA	Non-Competitive	–	DPV
70	Fe3O4@Au-Apt	Fe3O4@Au	Aptamer	–	label-free	EIS
71	2D–MoS2 NPs	Au	cDNA	Competitive	label-free	chronoamperometry
72	Au@Au- Ag NNSs coupled Fe3O4 MNPs	Au@Au–Ag NNSs	cDNA	–	–	Surface-enhanced Raman
73	SPCE/PAM/PA/PDA/Apt	SPCE	Aptamer	–	–	DPV
74	SPCE/Apt	SPE	cApt	–	label	DPV
75	DNA-AuNPs-HRP nanoprobes	GCE	cDNA	–	–	DPV
76	AgNPs/PDANS	Au	cDNA	competitive	label	LSV
77	AuNP@CuCoPBA	Au	Aptamer		Label-free	EIS
78	Apt2-AuNRs-Fc/Apt1-AuNRs-Th/cDNA/Au	Au	Y-shaped DNA and AuNRs	–	–	DPV

(continued)

Table 14.4 (continued)

Sl. No.	Matrix	Electrode	Type of Aptamer	Mode of detection	Label	Technique
79	PANI-aptamer-PANI/GCE	GCE	Aptamer	Non-competitive	Label-free	CV, EIS
80	Apt/PtNP/MIL–101(Fe)/GCE	GCE	Aptamer	–	label–free	EIS

Sl. No.	Toxin	Response range	LOD	Sensitivity	Relative Standard Deviation (RSD)in %	Coefficient of variation (CV) in %	Reference
1	OTA	–	0.11 ng mL^{-1}	–	–	–	Barthelmebs et al. (2011)
2	OTA	0.78–8.74 ng mL^{-1}	0.07 ng mL^{-1}	–	–	–	(Bonel et al. (2011)
3	OTA	0.05–4 μM.			–	–	Galarreta et al. (2013)
4	ZON	3.14×10^{-9} to 3.14×10^{-5} M	7.85×10^{-10} M.	–	–	–	Chen et al. (2013)
5	OTA	0.2 pg mL^{-1} to 1 ng mL^{-1}	75 fg mL^{-1}		9.1, 6.0	–	Chen et al. (2014)
6	OTA	25 nM to 200 nM	24.1 nM		–	–	Guo et al. (2011)
7	OTA	1.0×10^{-12} to 5.0×10^{-8} g mL^{-1}	3.0×10^{-13} g mL^{-1}				Hun et al. (2013)
8	OTA	0.1–20 ng mL^{-1}.	30 pg mL^{-1},		–	10	Kuang et al. (2010)
9	OTA	1 ng mL^{-1}–100 mg mL^{-1}	1 ng mL^{-1}		–	–	Lv et al. (2014)
10	OTA	1 μL to 3μL	0.05 μgL^{-1}		4–5	5	Rhouati et al. (2013)
11	OTA	2 μM to 35 μM.	21.8 nM		–	–	Sheng et al. (2011)

(continued)

Table 14.4 (continued)

Sl. No.	Toxin	Response range	LOD	Sensitivity	Relative Standard Deviation (RSD)in %	Coefficient of variation (CV) in %	Reference
12	OTA, FB1	0.05–100 ng·mL^{-1} for OTA and 0.1–500 ng·mL^{-1} for FB1	OTA 0.02 ng·mL^{-1} and FB1 0.1 ng·mL^{-1}.		4.74, 5.86	–	Wu et al. (2012b)
13	OTA	0.5 pg mL^{-1}–10 ng mL^{-1}	0.2 pg mL^{-1}	–	–	–	Tong et al. (2012)
14	OTA	0.02–3 ng mL^{-1}	0.007 ng mL^{-1}	–	3.8	–	Wang et al. (2010b)
15	FB1	2.5 μgmL^{-1} to 12.5 gmL^{-1}	3.69 ng mL^{-1}	–	–	–	Kesici and Erdem (2019)
16	OTA	0.1–100 nM	0.12–0.4 nM	–	0.2–3.5	–	Castillo et al. (2012)
17	OTA	0.05–100 ng mL^{-1}	10 pg mL^{-1}	–	1.4–7.4	–	Li et al. (2018)
18	OTA	0.4–20 ng mL^{-1}	0.08 ng mL^{-1}	–	–	–	Wu et al. (2019a)
19	AFB1	0.5 nM-4 μM	0.07 nM	–	2.9	–	Beheshti-Marnani et al. (2019)
20	AFB1	0.1–100 ng mL^{-1}	0.05 ng mL^{-1}	–	3.6 to 5.3	–	Yagati et al. (2018)
21	AFB1	0.10–100 nmol L^{-1}	0.47 nmolL^{-1}	92.4 A mol L^{-1}	3.1	–	Gevaerd et al. (2020)
22	AFB1	1–20 ng mL^{-1}	0.13 ng mL^{-1}	–	–	–	Mo et al. (2018)
23	AFB1	5–750 pM	1.4 pM	–	5.2	–	Nodoushan et al. (2019)
24	OTA	0.5 – 20 ng mL^{-1}	0.096 ng mL^{-1}	–	5.6	–	Kunene et al. (2020)
25	OTA	0.1pgmL^{-1}–1000pgmL^{-1}	0.095 pg mL^{-1}	–	4.9	–	Wu et al. (2012a)
26	OTA		20 pg mL^{-1} OTA with high specificity	–	1.67	–	Zhang et al. (2013b)
27	OTA	upto30nM	4 nM	–	–	–	Yang et al. (2013a)
28	OTA	5pgmL^{-1}–10 ng mL^{-1}	1 pg mL^{-1}	–	–	–	Tong et al. (2011)

(continued)

Table 14.4 (continued)

Sl. No.	Toxin	Response range	LOD	Sensitivity	Relative Standard Deviation (RSD)in %	Coefficient of variation (CV) in %	Reference
29	AFM 1	1–14 ng mL^{-1}	–	–	–	6.2	Dinckaya et al. (2011)
30	AFB 1	5.0×10^{-5} to 5.0 ng mL^{-1}	25 fg mL^{-1}	–	2	–	Guo et al. (2014)
31	AFM 1	6–60 ngL^{-1}	1.98 ngL^{-1}.	–	5	–	Nguyen et al. (2013)
32	ZEN and FB1	1×10^{-3}–10 ng mL^{-1} (ZEN) 1×10^{-3}–1×10^{2} ng mL^{-1} (FB1)	5×10^{-4} ng mL^{-1}	–	1.6–3.8	–	Han et al. (2020)
33	AFB1	0.01 ng mL^{-1}–20 ng mL^{-1}	3 pg mL^{-1}	–	2.0		Kong et al. (2019)
34	OTA	0.1–300 ng mL^{-1}	78.3 pg mL^{-1}	–	–	–	Mazaafrianto et al. (2018)
35	AFB 1	0.1–10 nM	0.40 ± 0.03 nM	–	1.73	–	Castillo et al. (2015)
36	AFB1	0.5–15 ng mL^{-1}	0.02 ng mL^{-1}	–	2.8–4.5	–	Wang et al. (2019c)
37	AFB 1	0.05–6.0 ng mL^{-1}	0.05 ng mL^{-1}	–	3.7	–	Goud et al. (2017)
38	OTA	0.01–0.32 nM	0.01 nM	–	–	–	Wang et al. (2015)
39	Ochratoxin A	0.15–2.5 ng mL^{-1}	0.15 ng mL^{-1}		4.8		Mishra et al. (2015)
40	OTA	up to 5 mgL^{-1}	2 mgL^{-1}	0.1 kΩ/ng	–	–	Mejri-Omrani et al. (2016)
41	zearalenone	5.0×10^{-4} ~ 50 ng mL^{-1}	1.7×10^{-4}	–	3.23 and 9.96	–	Ma et al. (2019)
42	patulin	0.001–100 ng mL^{-1}	0.042 pg mL^{-1}		1.1–4.4	–	Zhang et al. (2020)
43	OTA	1×10^{-5} nM to 10 nM	3.3×10^{-3} pM	–	–	–	Yang et al. (2019)

(continued)

Table 14.4 (continued)

Sl. No.	Toxin	Response range	LOD	Sensitivity	Relative Standard Deviation (RSD)in %	Coefficient of variation (CV) in %	Reference
44	OTA	0.1–50.0 nM	0.06 nM	–	6.5	–	Wang et al. (2018)
45	AFM1	2–150 ngL^{-1}	1.15 ngL^{-1}	–	–	–	Istamboulié et al. (2016)
46	AFB1	0.01–0.2 nM	0.01 nM	–	–	–	Sun and Zhao (2018)
47	OTA	7.17–548.76 $fgmL^{-1}$	2.04 $fgmL^{-1}$	–	–	–	Hou et al. (2019)
48	fumonisin B1	10 pg mL^{-1} to 50 ng mL^{-1}	3.4 pg mL^{-1}		4.9	–	Lu et al. (2016)
49	OTA	0.2–500 nM	0.073 nM		2.8	–	Zhu et al. (2018)
50	OTA	1.25–250 nM	0.069 nM	–	–	–	Tian et al. (2019)
51	OTA and FB1	0.05 – 50 ng mL^{-1} (FB1) and 0.01 – 10 ng mL^{-1} (OTA)	20 pg mL (FB1) and 5 pg/mL (OTA)	Less than 3 pg/mL	6.1 for OTA and 5.3 for FB1	–	Wang et al. (2017)
52	OTA	1.0×10^{-4}–2×10^{-2} nM	0.03 ngL^{-1}	–	5	–	Guo et al. (2016)
53	OTA	1.0–20 pM	0.10 pM	–	2.9–4.3	–	Wang et al. (2016)
54	OTA	0.01 – 100 nM	14 pM	–	17	–	Rivas et al. (2015)
55	AFB1	6 nM–600 mM	4 ngL^{-1}	–	–	–	Kumar et al. (2018)
56	AFB1	1–16.7 nM	46 ppb	–	–	–	Hu et al. (2015)
57	OTA	20–500 nM	16.5 nM	–	2.5	–	Lv et al. (2017)
58	AFB1	0.001 ppt to 100 ppt	0.6×10^{-4} ppt	–	–	–	Zhang et al. (2016)
59	AFB1	0.125–16.0 ng mL^{-1}	0.125 ng mL^{-1}	0.271 ng mL^{-1}	2.72–4.83	0.25 ng mL^{-1}	Yugender Goud et al. (2016)
60	OTA	1.0×10^{-4}–500 ng mL^{-1}	1.0×10^{-4} ng mL^{-1}	–	6.7	–	Wei and Zhang (2018)
61	AFB1	0.1 $fgmL^{-1}$–0.1 μgmL^{-1}	0.01 $fgmL^{-1}$	–	4.2–6.5	–	Peng et al. (2018)

(continued)

Table 14.4 (continued)

Sl. No.	Toxin	Response range	LOD	Sensitivity	Relative Standard Deviation (RSD)in %	Coefficient of variation (CV) in %	Reference
62	AFM1	5–120 ngL^{-1}	0.5–1 ngL^{-1}	–	2.3	–	Smolko et al. (2018)
63	AFB1	0.1 pg mL^{-1}–10 ng mL^{-1}	0.0491 pg mL^{-1}	–	lower than 11.51%	–	Wu et al. (2019b)
64	AFB1	8 nM to 4 μM	2 nM	–	less than 4	–	Wang et al. (2019a)
65	AFB1	5 pg mL^{-1}–50 ng mL^{-1}	4.5 pg mL^{-1}	–	4.7	–	Wang et al. (2019b)
66	OTA	0.2 pg mL^{-1-4} ng mL^{-1}	0.07 pg mL^{-1}	–	–	–	Hao et al. (2016)
67	AFB1	0.01–1.0 $fgmL^{-1}$	0.002 $fgmL^{-1}$	–	5.3	–	Geleta et al. (2018a)
68	OTA	0.1 ~ 1.2 nM	0.03 nM	–	6.1	–	Liu et al. (2018)
69	AFM1	2–600 ngL^{-1}	0.9 ngL^{-1}	–	3.9–8.1	–	Jalalian et al. (2018)
70	AFB1	20 pg mL^{-1-5}0 ng mL^{-1}	15 pg mL^{-1}	–	6.2	–	Wang et al. (2018)
71	OTA	0.0005–10 ng mL^{-1}	0.23 pg mL^{-1}	–	8.8–9.0	–	Tang et al. (2018a)
72	OTA	0.01–50 ng mL^{-1}	0.004 ng mL^{-1}	–	–	–	Shao et al. (2018)
73	AFB2	0.1 pg mL^{-1} to 100 ng mL^{-1}	0.10 pg mL^{-1}		3.9	–	Geleta et al. (2018b)
74	OTA	0.15–5 ng mL^{-1}	0.07 ng mL^{-1}	–	3.7	–	Mishra et al. (2016)
75	AFB1	10^{-3} to 200 ng mL^{-1}	3.3 × 10 – 4 ng mL^{-1}	–	3.07	–	Hui et al. (2020)
76		1.0×10^{-6}–1.0×10^{-2} μM	0.57 pM	–	2.9	–	Zhang et al. (2019)
77	OTA	50 $fgmL^{-1}$ to 10 ng mL^{-1}	5.2 fg mL^{-1}	–	<4.8	–	Gu et al. (2019)
78	OTA and FB1	1×10^{-3} –100 ng mL^{-1}	0.47×10^{-3} ng mL^{-1} (OTA) 0.26×10^{-3} ng mL^{-1} (FB1)	–	6.17 and 5.38		Wei et al. (2020)
79	AFM1	3–90 ngL^{-1}	1–5 ngL^{-1}	–	–	–	Kulikova et al. (2019)

(continued)

Table 14.4 (continued)

Sl. No.	Toxin	Response range	LOD	Sensitivity	Relative Standard Deviation (RSD)in %	Coefficient of variation (CV) in %	Reference
80	AFM1	1.0×10^{-2} to 80.0 ng mL^{-1}	$2.0 \times 10 - 3$ ng mL^{-1}	–	6.4	–	Jahangiri-Dehaghani et al. (2020)

14.3.5 Mimotope for Mycotoxin

The mycotoxins conjugates need pure mycotoxin, which is highly risky to the manufacturer due to its carcinogenic nature, and these are hazardous to the environment, due to its persistence at high temperature. Therefore, for the environment and laboratory safety purpose, it is important to avoid the use of mycotoxins and choose the alternate. The formation of a nontoxic immunochemical compound, i.e., mimotope has advantages in the replacement of mycotoxin conjugate. The mimotopes are synthesized from peptides with the respective sequences of the peptide that have a good affinity toward mycotoxin. The peptide sequence selected from random phage-displayed peptide libraries. The mimotopes of OTA, ZEA, and DON were developed and applied for their respective mycotoxin's detection (Bazin et al. 2013; Lai et al. 2009; He et al. 2013)

14.3.6 Nanobody (VHHs Antibody)

All types of antibodies and labeled antibodies have their distinctive possessions and affinity toward the target (mycotoxins) on biosensor's surface (Regiart et al. 2018; Solanki et al. 2017; Karczmarczyk et al. 2017). The whole process of synthesis of antibodies and their conjugates with labels is expensive, time taking and depends on vaccination. Sometimes the enzyme chemically linked with antibodies which may cause false results because of instability and arbitrarily link in between the molecules. The aforesaid problems are not inconsequential, therefore, researchers have worked on synthetically engineered antibodies like (recombinant antibody, single domain antibody or nanobodies, and single chain fragment variable (scFv), to solve these problems. The nanobodies are designed from a variable part of the heavy chain of antibodies. The llamas, camels, and dromedaries' animals' antibodies have been used for the fragment and the cartilaginous fishes VNAR fragment was used (De Groof et al. 2019; De Meyer et al. 2014). Nbs have discrete advantages over the conventional antibodies, as they are high in yield and stability, easy to modify genetically, and refolding tendency (Steeland et al. 2016; Könning et al. 2017). Liu and their group have synthesized Nbs for the detection of OTA. The Nbs were derived from the alpaca VHH library and were analyzed OTA from PCR and ELISA methods (Liu et al. 2017). A one-step qualitative detection of OTA using Nbs has been conducted. Alkaline phosphatase linked Nbs were used on polyvinylidene fluoride (PVDF) membrane immunoassay. And this immunoassay reached up to the LOD of 0.3125 ng/mL^{-1} in 6 min for OTA in cereals (Tang et al. 2018b). Pan et al. (2018) fabricated direct immunosensor for AFB1, the signal received by the interaction of Nbs and carbon materials. The proposed immunosensor showed LOD of 3.3 pg mL^{-1} (S/N = 3). The immunosensor with Nbs was 10-fold more sensitive for the detection of AFB1 than the immunosensor with complete monoclonal antibody. Though the selection of Nbs which will interact with mycotoxins is very complexing, because each has their specific epitopes from the phage display Nbs range, and this is one of the disadvantages of wide applications of Nbs in mycotoxin.

14.4 Conclusions

The concept of the biosensor is demonstrating a compatible alternative to replace previous techniques which are tedious, costly, and time taking. In this regard there is an immense revolution in field of transducers (electrochemical, optical, and cantilever), nanomaterials synthesis (quantum dots, metal oxides, and metal nanoparticles), and in the selection of receptors (peptides, MIP, nucleotides, nanobody, etc.). These changes lead to enhancement in the linear range, shelf life, and LOD for the developed biosensor. The developed sensor can apply to detect the contaminated food sample. However, false-positive results observed due to the following reasons (i) cross-reactions between the receptors and mimic molecules (ii) nonspecific reaction on the electrode surface (iii) physical parameters temperature, pH, viscosity, and ionic strength, etc. Though everyday research steps on the pathway of new science and technology, innovate new materials and techniques, still they are far away from the practical applicability of these techniques in daily life. Therefore, there is a need of enormous advancements in mycotoxin detection available to the public and field sector.

14.5 Future Trends

The invention of nanotechnology is boon to the science and technology field. The researchers are paying attention to the advancement in technology for the development of mycotoxin detection devices. The success rate of the development of a smart, fast, reliable sensing device is very slow. The mycotoxin detection device was not developed for the field trial. The contamination of mycotoxins in food is not uniform and therefore it needs a huge number of samples for detection, researchers follow the set levels established by the European Commission and other international agencies. The conventional techniques (like PCR, chromatography, and ELISA techniques) are time consuming, costly, skilled technicians required for the detection of mycotoxins. The biosensor for mycotoxin detection is promising tool though there are some challenges affecting it, can be addressed judiciously such as (i) replacement of monoclonal antibody to the synthetically prepared receptors like aptamers, MIP, etc. (ii) label-free system adaptation for detection (iii) incorporation of contemporary technologies, e.g., microfluidic (iv) finally the adaptation of innovation to practical application. Future research can be detecting multi-analyte on a chip platform at a reasonable cost.

Acknowledgments Prof. Tinku Basu, Director, Centre for Nanomedicine is thankful to Funding Agency, BIRAC for financial support (BT/AIR0416/PACE-I 4/18). R. Chauhan acknowledges the postdoctoral fellowship from the University of KwaZulu Natal, Westville, South Africa.

References

Adanyi N, Levkovets IA, Rodriguez-Gil S, Ronald A, Varadi M, Szendro I (2007) Development of immunosensor based on OWLS technique for determining Aflatoxin B1 and Ochratoxin A. Biosens Bioelectron 22:797–802. https://doi.org/10.1016/j.bios.2006.02.015

Adányi N et al (2018) Sensitivity enhancement for mycotoxin determination by optical waveguide lightmode spectroscopy using gold nanoparticles of different size and origin. Food Chem 267:10–14. https://doi.org/10.1016/j.foodchem.2018.04.089

Aguilar-Arteaga K, Rodriguez JA, Barrado E (2010) Magnetic solids in analytical chemistry: A review. Anal Chim Acta 674:157–165. https://doi.org/10.1016/j.aca.2010.06.043

Ah CS et al (2012) Detection of uncharged or feebly charged small molecules by field-effect transistor biosensors. Biosens Bioelectron 33:233–240. https://doi.org/10.1016/j.bios.2012.01.010

Alarcón SH, Palleschi G, Compagnone D, Pascale M, Visconti A, Barna-Vetró I (2006) Monoclonal antibody based electrochemical immunosensor for the determination of ochratoxin A in wheat. Talanta 69:1031–1037

Althagafi II, Ahmed SA, El-Said WA (2019) Fabrication of gold/graphene nanostructures modified ITO electrode as highly sensitive electrochemical detection of Aflatoxin B1. PLoS ONE 14:e0210652. https://doi.org/10.1371/journal.pone.0210652

An X et al (2020) An electrochemical immunosensor based on a combined amplification strategy with the GO–CS/CeO2–CS nanocomposite for the detection of aflatoxin M1. New J Chem 44:1362–1370. https://doi.org/10.1039/c9nj04804a

Anastasiadis V et al (2020) Multiplexed mycotoxins determination employing white light reflectance spectroscopy and silicon chips with silicon oxide areas of different thickness. Biosens Bioelectron 153:112035. https://doi.org/10.1016/j.bios.2020.112035

Bacher G, Pal S, Kanungo L, Bhand S (2012) A label-free silver wire based impedimetric immunosensor for detection of aflatoxin M1 in milk. Sens Actuators B: Chem 168:223–230

Barthelmebs L, Jonca J, Hayat A, Prieto-Simon B, Marty J-L (2011) Enzyme-Linked Aptamer Assays (ELAAs), based on a competition format for a rapid and sensitive detection of Ochratoxin A in wine. Food Control 22:737–743. https://doi.org/10.1016/j.foodcont.2010.11.005

Bazin I, Andreotti N, Hassine AIH, De Waard M, Sabatier J-M, Gonzalez C (2013) Peptide binding to ochratoxin A mycotoxin: a new approach in conception of biosensors. Biosens Bioelectron 40:240–246

Beheshti-Marnani A, Hatefi-Mehrjardi A, Es'haghi Z (2019) A sensitive biosensing method for detecting of ultra-trace amounts of AFB1 based on "Aptamer/reduced graphene oxide" nano-bio interaction. Colloids Surf B Biointerfaces 175:98–105. https://doi.org/10.1016/j.colsurfb.2018.11.087

Ben Rejeb I et al. (2009) Development of a bio-electrochemical assay for AFB1 detection in olive oil. Biosens Bioelectron 24:1962–1968. https://doi.org/10.1016/j.bios.2008.10.002

Bhardwaj H, Sumana G, Marquette CA (2020) A label-free ultrasensitive microfluidic surface Plasmon resonance biosensor for Aflatoxin B1 detection using nanoparticles integrated gold chip. Food chemistry 307:125530. https://doi.org/10.1016/j.foodchem.2019.125530

Bonel L, Vidal JC, Duato P, Castillo JR (2010) Ochratoxin A nanostructured electrochemical immunosensors based on polyclonal antibodies and gold nanoparticles coupled to the antigen. Anal Methods 2:335–341. https://doi.org/10.1039/B9AY00297A

Bonel L, Vidal JC, Duato P, Castillo JR (2011) An electrochemical competitive biosensor for ochratoxin A based on a DNA biotinylated aptamer. Biosens Bioelectron 26:3254–3259. https://doi.org/10.1016/j.bios.2010.12.036
Cai J, Hao C, Sun M, Ma W, Xu C, Kuang H (2018) Chiral shell core-satellite nanostructures for ultrasensitive detection of mycotoxin. Small 14:1703931. https://doi.org/10.1002/smll.201703931
Castillo G, Lamberti I, Mosiello L, Hianik T (2012) Impedimetric DNA aptasensor for sensitive detection of ochratoxin A in food. Electroanalysis 24:512–520. https://doi.org/10.1002/elan.201100485
Castillo G, Spinella K, Poturnayová A, Šnejdárková M, Mosiello L, Hianik T (2015) Detection of aflatoxin B1 by aptamer-based biosensor using PAMAM dendrimers as immobilization platform. Food Control 52:9–18. https://doi.org/10.1016/j.foodcont.2014.12.008
Chauhan R, Singh J, Solanki PR, Basu T, O'Kennedy R, Malhotra B (2015a) Electrochemical piezoelectric reusable immunosensor for aflatoxin B1 detection. Biochem Eng J 103:103–113
Chauhan R, Solanki PR, Singh J, Mukherjee I, Basu T, Malhotra BD (2015b) A novel electrochemical piezoelectric label free immunosensor for aflatoxin B1 detection in groundnut. Food Control 52:60–70. https://doi.org/10.1016/j.foodcont.2014.12.009
Chauhan R, Singh J, Solanki PR, Manaka T, Iwamoto M, Basu T, Malhotra B (2016) Label-free piezoelectric immunosensor decorated with gold nanoparticles: Kinetic analysis and biosensing application. Sens Actuators B: Chem 222:804–814
Chen J, Fang Z, Liu J, Zeng L (2012) A simple and rapid biosensor for ochratoxin A based on a structure-switching signaling aptamer. Food Control 25:555–560. https://doi.org/10.1016/j.foodcont.2011.11.039
Chen X et al (2013) Selection and identification of ssDNA aptamers recognizing zearalenone. Anal Bioanal Chem 405:6573–6581. https://doi.org/10.1007/s00216-013-7085-9
Chen Y, Yang M, Xiang Y, Yuan R, Chai Y (2014) Binding-induced autonomous disassembly of aptamer–DNAzyme supersandwich nanostructures for sensitive electrochemiluminescence turn-on detection of ochratoxin A. Nanoscale 6:1099–1104. https://doi.org/10.1039/c3nr05499c
Chen J, Wen J, Zhuang L, Zhou S (2016) An enzyme-free catalytic DNA circuit for amplified detection of aflatoxin B1 using gold nanoparticles as colorimetric indicators. Nanoscale 8:9791–9797. https://doi.org/10.1039/c6nr01381c
Cheng X, Liu X, Bing T, Cao Z, Shangguan D (2009) General peroxidase activity of G-quadruplex-hemin complexes and its application in ligand screening. Biochemistry 48:7817–7823. https://doi.org/10.1021/bi9006786
Cheng CI, Chang Y-P, Chu Y-H (2012) Biomolecular interactions and tools for their recognition: focus on the quartz crystal microbalance and its diverse surface chemistries and applications. Chem Soc Rev 41:1947–1971. https://doi.org/10.1039/C1CS15168A
Cho EJ, Lee J-W, Ellington AD (2009) Applications of aptamers as sensors. Annu Rev Anal Chem 2:241–264. https://doi.org/10.1146/annurev.anchem.1.031207.112851
Cruz-Aguado JA, Penner G (2008a) Determination of ochratoxin A with a DNA aptamer. J Agric Food Chem 56:10456–10461. https://doi.org/10.1021/jf801957h
Cruz-Aguado JA, Penner G (2008b) Fluorescence polarization based displacement assay for the determination of small molecules with aptamers. Anal Chem 80:8853–8855. https://doi.org/10.1021/ac8017058
De Girolamo A, McKeague M, Miller JD, DeRosa MC, Visconti A (2011) Determination of ochratoxin A in wheat after clean-up through a DNA aptamer-based solid phase extraction column. Food Chem 127:1378–1384. https://doi.org/10.1016/j.foodchem.2011.01.107
De Groof TWM, Bobkov V, Heukers R, Smit MJ (2019) Nanobodies: new avenues for imaging, stabilizing and modulating GPCRs. Mol Cell Endocrinol 484:15–24. https://doi.org/10.1016/j.mce.2019.01.021
De Meyer T, Muyldermans S, Depicker A (2014) Nanobody-based products as research and diagnostic tools. Trends Biotechnol 32:263–270. https://doi.org/10.1016/j.tibtech.2014.03.001

Díaz Nieto CH, Granero AM, Garcia D, Nesci A, Barros G, Zon MA, Fernández H (2019) Development of a third-generation biosensor to determine sterigmatocystin mycotoxin: An early warning system to detect aflatoxin B1. Talanta 194:253–258. https://doi.org/10.1016/j.talanta.2018.10.032

Dinckaya E, Kinik O, Sezginturk MK, Altug C, Akkoca A (2011) Development of an impedimetric aflatoxin M1 biosensor based on a DNA probe and gold nanoparticles. Biosens Bioelectron 26:3806–3811. https://doi.org/10.1016/j.bios.2011.02.038

Feng J et al (2018) A competitive-type photoelectrochemical immunosensor for aflatoxin B1 detection based on flower-like WO3 as matrix and Ag2S-enhanced BiVO4 for signal amplification. Sens Actuators B Chem 270:104–111. https://doi.org/10.1016/j.snb.2018.05.015

Feng R et al (2013a) Ultrasensitive electrochemical immunosensor for zeranol detection based on signal amplification strategy of nanoporous gold films and nano-montmorillonite as labels. Anal Chim Acta 758:72–79

Feng R et al (2013b) Nanoporous PtCo-based ultrasensitive enzyme-free immunosensor for zeranol detection. Biosens Bioelectron 42:367–372. https://doi.org/10.1016/j.bios.2012.10.031

Galarreta BC, Norton PR, Lagugne-Labarthet F (2011) SERS detection of streptavidin/biotin monolayer assemblies. Langmuir 27:1494–1498

Galarreta BC, Tabatabaei M, Guieu V, Peyrin E, Lagugné-Labarthet F (2013) Microfluidic channel with embedded SERS 2D platform for the aptamer detection of ochratoxin A. Anal Bioanal Chem 405:1613–1621. https://doi.org/10.1007/s00216-012-6557-7

Gan N, Zhou J, Xiong P, Hu F, Cao Y, Li T, Jiang Q (2013) An ultrasensitive electrochemiluminescent immunoassay for aflatoxin M1 in milk, based on extraction by magnetic graphene and detection by antibody-labeled CdTe quantumn dots-carbon nanotubes nanocomposite. Toxins 5:865–883

Gao X, Cao W, Chen M, Xiong H, Zhang X, Wang S (2014) A high sensitivity electrochemical sensor based on Fe3+-ion molecularly imprinted film for the detection of T-2 toxin. Electroanalysis 26:2739–2746

Geleta GS, Zhao Z, Wang Z (2018a) A novel reduced graphene oxide/molybdenum disulfide/polyaniline nanocomposite-based electrochemical aptasensor for detection of aflatoxin B1. Analyst 143:1644–1649. https://doi.org/10.1039/c7an02050c

Geleta GS, Zhao Z, Wang Z (2018b) A sensitive electrochemical aptasensor for detection of Aflatoxin B2 based on a polyacrylamide/phytic acid/polydopamine hydrogel modified screen printed carbon electrode. Anal Methods 10:4689–4694. https://doi.org/10.1039/c8ay01675e

Gevaerd A, Banks CE, Bergamini MF, Marcolino-Junior LH (2020) Nanomodified screen-printed electrode for direct determination of Aflatoxin B1 in malted barley samples. Sens Actuators B: Chem 307:127547. https://doi.org/10.1016/j.snb.2019.127547

Goud KY, Hayat A, Catanante G, M S, Gobi KV, Marty JL (2017) An electrochemical aptasensor based on functionalized graphene oxide assisted electrocatalytic signal amplification of methylene blue for aflatoxin B1 detection. Electrochimica Acta 244:96–103. https://doi.org/10.1016/j.electacta.2017.05.089

Goryacheva IY, De Saeger S, Nesterenko IS, Eremin SA, Van Peteghem C (2007) Rapid all-in-one three-step immunoassay for non-instrumental detection of ochratoxin A in high-coloured herbs and spices. Talanta 72:1230–1234. https://doi.org/10.1016/j.talanta.2006.12.049

Gu C, Yang L, Wang M, Zhou N, He L, Zhang Z, Du M (2019) A bimetallic (Cu-Co) Prussian Blue analogue loaded with gold nanoparticles for impedimetric aptasensing of ochratoxin a. Mikrochim Acta 186:343. https://doi.org/10.1007/s00604-019-3479-5

Guo X et al (2014) Development of an ultrasensitive aptasensor for the detection of aflatoxin B1. Biosens Bioelectron 56:340–344. https://doi.org/10.1016/j.bios.2014.01.045

Guo Z, Ren J, Wang J, Wang E (2011) Single-walled carbon nanotubes based quenching of free FAM-aptamer for selective determination of ochratoxin A. Talanta 85:2517–2521. https://doi.org/10.1016/j.talanta.2011.08.015

Guo X, Wen F, Zheng N, Li S, Fauconnier ML, Wang J (2016) A qPCR aptasensor for sensitive detection of aflatoxin M1. Anal Bioanal Chem 408:5577–5584. https://doi.org/10.1007/s00216-016-9656-z

Han Z, Tang Z, Jiang K, Huang Q, Meng J, Nie D, Zhao Z (2020) Dual-target electrochemical aptasensor based on co-reduced molybdenum disulfide and Au NPs (rMoS2-Au) for multiplex detection of mycotoxins. Biosens Bioelectron 150:111894. https://doi.org/10.1016/j.bios.2019.111894

Hao N, Jiang L, Qian J, Wang K (2016) Ultrasensitive electrochemical Ochratoxin A aptasensor based on CdTe quantum dots functionalized graphene/Au nanocomposites and magnetic separation. J Electroanal Chem 781:332–338. https://doi.org/10.1016/j.jelechem.2016.09.053

Hayat A, Andreescu S, Marty J-L (2013a) Design of PEG-aptamer two piece macromolecules as convenient and integrated sensing platform: application to the label free detection of small size molecules. Biosens Bioelectron 45:168–173

Hayat A, Sassolas A, Marty J-L, Radi A-E (2013b) Highly sensitive ochratoxin A impedimetric aptasensor based on the immobilization of azido-aptamer onto electrografted binary film via click chemistry. Talanta 103:14–19. https://doi.org/10.1016/j.talanta.2012.09.048

He Z-y et al (2013) Ochratoxin A mimotope from second-generation peptide library and its application in immunoassay. Anal Chem 85:10304–10311

Henry SH, Bosch FX, Troxell TC, Bolger PM (1999) Reducing liver cancer—global control of aflatoxin science 286:2453–2454. https://doi.org/10.1126/science.286.5449.2453

Hermann T, Patel DJ (2000) Adaptive recognition by nucleic acid aptamers. Science 287:820–825. https://doi.org/10.1126/science.287.5454.820

Hossain MZ, Maragos CM (2018) Gold nanoparticle-enhanced multiplexed imaging surface plasmon resonance (iSPR) detection of Fusarium mycotoxins in wheat. Biosens Bioelectron 101:245–252. https://doi.org/10.1016/j.bios.2017.10.033

Hou S-l, Ma Z-e (2019) Meng H, Xu Y, He Q-h. Ultrasensitive and green electrochemical immunosensor for mycotoxin ochratoxin A based on phage displayed mimotope peptide Talanta 194:919–924. https://doi.org/10.1016/j.talanta.2018.10.081

Hu Z et al (2015) Effective detection of mycotoxins by a highly luminescent metal-organic framework. J Am Chem Soc 137:16209–16215. https://doi.org/10.1021/jacs.5b10308

Hui Y, Wang B, Ren R, Zhao A, Zhang F, Song S, He Y (2020) An electrochemical aptasensor based on DNA-AuNPs-HRP nanoprobes and exonuclease-assisted signal amplification for detection of aflatoxin B1. Food Control 109:106902. https://doi.org/10.1016/j.foodcont.2019.106902

Hun X, Chen H, Wang W (2010) Design of ultrasensitive chemiluminescence detection of lysozyme in cancer cells based on nicking endonuclease signal amplification technology. Biosens Bioelectron 26:248–254. https://doi.org/10.1016/j.bios.2010.06.042

Hun X, Liu F, Mei Z, Ma L, Wang Z, Luo X (2013) Signal amplified strategy based on target-induced strand release coupling cleavage of nicking endonuclease for the ultrasensitive detection of ochratoxin A. Biosens Bioelectron 39:145–151

IARC (2002) IARC Monographs on the evaluation of carcinogenic risks to humans 82

Istamboulié G, Paniel N, Zara L, Granados LR, Barthelmebs L, Noguer T (2016) Development of an impedimetric aptasensor for the determination of aflatoxin M1 in milk. Talanta 146:464–469. https://doi.org/10.1016/j.talanta.2015.09.012

Jahangiri-Dehaghani F, Zare HR, Shekari Z (2020) Measurement of aflatoxin M1 in powder and pasteurized milk samples by using a label-free electrochemical aptasensor based on platinum nanoparticles loaded on Fe-based metal-organic frameworks. Food Chem 310:125820. https://doi.org/10.1016/j.foodchem.2019.125820

Jalalian SH, Ramezani M, Danesh NM, Alibolandi M, Abnous K, Taghdisi SM (2018) A novel electrochemical aptasensor for detection of aflatoxin M1 based on target-induced immobilization of gold nanoparticles on the surface of electrode. Biosens Bioelectron 117:487–492. https://doi.org/10.1016/j.bios.2018.06.055

Jin X, Jin X, Chen L, Jiang J, Shen G, Yu R (2009a) Piezoelectric immunosensor with gold nanoparticles enhanced competitive immunoreaction technique for quantification of aflatoxin B1. Biosens Bioelectron 24:2580–2585. https://doi.org/10.1016/j.bios.2009.01.014

Jin X, Jin X, Liu X, Chen L, Jiang J, Shen G, Yu R (2009b) Biocatalyzed deposition amplification for detection of aflatoxin B1 based on quartz crystal microbalance. Anal Chim Acta 645:92–97. https://doi.org/10.1016/j.aca.2009.04.041
Kanungo L, Bacher G, Bhand S (2014) Flow-based impedimetric immunosensor for aflatoxin analysis in milk products. Appl Biochem Biotechnol 174:1157–1165
Karczmarczyk A, Baeumner AJ, Feller K-H (2017) Rapid and sensitive inhibition-based assay for the electrochemical detection of Ochratoxin A and Aflatoxin M1 in red wine and milk. Electrochimica Acta 243:82–89. https://doi.org/10.1016/j.electacta.2017.05.046
Kaushik A, Solanki PR, Ansari AA, Ahmad S, Malhotra BD (2008) Chitosan–iron oxide nanobiocomposite based immunosensor for ochratoxin-A. Electrochem Commun 10:1364–1368. https://doi.org/10.1016/j.elecom.2008.07.007
Kaushik A, Solanki PR, Ansari AA, Ahmad S, Malhotra BD (2009) A nanostructured cerium oxide film-based immunosensor for mycotoxin detection. Nanotechnology 20:055105. https://doi.org/10.1088/0957-4484/20/5/055105
Kesici E, Erdem A (2019) Impedimetric detection of Fumonisin B1 and its biointeraction with fsDNA. Int J Biol Macromol 139:1117–1122. https://doi.org/10.1016/j.ijbiomac.2019.08.024
Khan R, Dhayal M (2009) Chitosan/polyaniline hybrid conducting biopolymer base impedimetric immunosensor to detect Ochratoxin-A. Biosens Bioelectron 24:1700–1705. https://doi.org/10.1016/j.bios.2008.08.046
Khan R, Dey NC, Hazarika AK, Saini KK, Dhayal M (2011) Mycotoxin detection on antibody-immobilized conducting polymer-supported electrochemically polymerized acacia gum. Anal Biochem 410:185–190
Kong L, Chen J, Chen Z, Feng J, Fan D, Wei Q (2019) A label-free photoelectrochemical aptasensor based on N-GQDs sensitized Zn-SnS2 for aflatoxin B1 detection. IEEE Sens J 19:1633–1639. https://doi.org/10.1109/jsen.2018.2883430
Könning D et al (2017) Camelid and shark single domain antibodies: structural features and therapeutic potential. Curr Opin Struct Biol 45:10–16. https://doi.org/10.1016/j.sbi.2016.10.019
Kuang H et al (2010) Fabricated aptamer-based electrochemical "signal-off" sensor of ochratoxin A. Biosens Bioelectron 26:710–716. https://doi.org/10.1016/j.bios.2010.06.058
Kulikova TN, Porfireva AV, Evtugyn GA, Hianik T (2019) Electrochemical aptasensor with layer-by-layer deposited polyaniline for aflatoxin M1 voltammetric determination. Electroanalysis 31:1913–1924. https://doi.org/10.1002/elan.201900274
Kumar YVVA, Renuka RM, Achuth J, Mudili V, Poda S (2018) Development of a FRET-based fluorescence aptasensor for the detection of aflatoxin B1 in contaminated food grain samples. RSC Advances 8:10465–10473. https://doi.org/10.1039/C8RA00317C
Kunene K, Weber M, Sabela M, Voiry D, Kanchi S, Bisetty K, Bechelany M (2020) Highly-efficient electrochemical label-free immunosensor for the detection of ochratoxin A in coffee samples. Sens Actuators B: Chem 305:127438. https://doi.org/10.1016/j.snb.2019.127438
Lai W, Fung DY, Yang X, Renrong L, Xiong Y (2009) Development of a colloidal gold strip for rapid detection of ochratoxin A with mimotope peptide. Food Control 20:791–795
Lamberti I, Mosiello L, & Hianik, T (2011) Development of thickness shear mode biosensor based on DNA aptamers for detection ochratoxin. Chemical Sensors, Vol 1, pp 11–15, 2231–6035.
Li D-l, Zhang X, Ma Y, Deng Y, Hu R, Yang Y (2018) Preparation of an OTA aptasensor based on a metal-organic framework. Anal Methods 10:3273–3279. https://doi.org/10.1039/c8ay00758f
Li Y, Lu Y (2009) Functional nucleic acids for analytical applications. Springer, New York, pp 15–16
Li T, Dong S, Wang E (2009) Label-free colorimetric detection of aqueous mercury ion (Hg2+) using Hg2+-modulated G-quadruplex-based DNAzymes. Anal Chem 81:2144–2149. https://doi.org/10.1021/ac900188y
Li J, Deng T, Chu X, Yang R, Jiang J, Shen G, Yu R (2010) Rolling circle amplification combined with gold nanoparticle aggregates for highly sensitive identification of single-nucleotide polymorphisms. Anal Chem 82:2811–2816. https://doi.org/10.1021/ac100336n

Li T, Byun JY, Kim BB, Shin YB, Kim MG (2013) Label-free homogeneous FRET immunoassay for the detection of mycotoxins that utilizes quenching of the intrinsic fluorescence of antibodies. Biosens Bioelectron 42:403–408. https://doi.org/10.1016/j.bios.2012.10.085

Liao JY (2007) Construction of nanogold hollow balls with dendritic surface as immobilized affinity support for protein adsorption. Colloids Surf, B 57:75–80

Liew WPP, Mohd-Redzwan S (2018) Mycotoxin: its impact on gut health and microbiota. Front Cell Infect Microbiol 8:60. https://doi.org/10.3389/fcimb.2018.00060

Lin Y, Zhou Q, Tang D (2017) Dopamine-loaded liposomes for in-situ amplified Photoelectrochemical Immunoassay of AFB1 to enhance photocurrent of Mn2+-doped Zn3(OH)2V2O7 nanobelts. Anal Chem 89:11803–11810. https://doi.org/10.1021/acs.analchem.7b03451

Lippolis V, Pascale M, Maragos CM, Visconti A (2008) Improvement of detection sensitivity of T-2 and HT-2 toxins using different fluorescent labeling reagents by high-performance liquid chromatography. Talanta 74:1476–1483. https://doi.org/10.1016/j.talanta.2007.09.024

Liu Y, Qin Z, Wu X, Jiang H (2006) Immune-biosensor for aflatoxin B1 based bio-electrocatalytic reaction on micro-comb electrode. Biochem Eng J 32:211–217

Liu J et al (2009) Nanomaterial-assisted signal enhancement of hybridization for DNA biosensors: a review. Sensors 9:7343–7364

Liu LH, Zhou XH, Shi HC (2015a) Portable optical aptasensor for rapid detection of mycotoxin with a reversible ligand-grafted biosensing surface. Biosens Bioelectron 72:300–305. https://doi.org/10.1016/j.bios.2015.05.033

Liu X et al (2017) Nanobody-based enzyme immunoassay for ochratoxin A in cereal with high resistance to matrix interference. Talanta 164:154–158. https://doi.org/10.1016/j.talanta.2016.11.039

Liu Y, Yan T, Li Y, Cao W, Pang X, Wu D, Wei Q (2015b) A simple label-free photoelectrochemical immunosensor for highly sensitive detection of aflatoxin B1 based on Cds Fe3O4 magnetic nanocomposites. RSC Adv 5:19581–19586. https://doi.org/10.1039/C4RA15918G

Liu F, Ding A, Zheng J, Chen J, Wang B (2018) A label-free aptasensor for ochratoxin a detection based on the structure switch of aptamer. Sensors 18:1769

Loo AH, Bonanni A, Pumera M (2015) Mycotoxin aptasensing amplification by using inherently electroactive graphene-oxide nanoplatelet labels. ChemElectroChem 2:743–747. https://doi.org/10.1002/celc.201402403

Lu L et al (2019) Dual-channel ITO-microfluidic electrochemical immunosensor for simultaneous detection of two mycotoxins. Talanta v. 194:pp. 709–716-2019 v.2194. https://doi.org/10.1016/j.talanta.2018.10.091

Lu L, Seenivasan R, Wang YC, Yu JH, Gunasekaran S (2016) An electrochemical immunosensor for rapid and sensitive detection of mycotoxins fumonisin B1 and deoxynivalenol. Electrochim Acta 213:89–97

Lv Z et al (2014) A simple and sensitive approach for ochratoxin A detection using a label-free fluorescent aptasensor. Plos One 9:e85968–e85968. https://doi.org/10.1371/journal.pone.0085968

Lv L, Li D, Liu R, Cui C, Guo Z (2017) Label-free aptasensor for ochratoxin A detection using SYBR Gold as a probe. Sens Actuators B: Chem 246:647–652. https://doi.org/10.1016/j.snb.2017.02.143

Ma L, Bai L, Zhao M, Zhou J, Chen Y, Mu Z (2019) An electrochemical aptasensor for highly sensitive detection of zearalenone based on PEI-MoS2-MWCNTs nanocomposite for signal enhancement. Anal Chim Acta 1060:71–78. https://doi.org/10.1016/j.aca.2019.02.012

Malvano F, Albanese D, Pilloton R, Di Matteo M (2016) A highly sensitive impedimetric label free immunosensor for Ochratoxin measurement in cocoa beans. Food Chem 212:688–694. https://doi.org/10.1016/j.foodchem.2016.06.034

Masoomi L, Sadeghi O, Banitaba MH, Shahrjerdi A, Davarani SSH (2013) A non-enzymatic nanomagnetic electro-immunosensor for determination of Aflatoxin B1 as a model antigen. Sens Actuators B: Chem 177:1122–1127

Mazaafrianto DN, Ishida A, Maeki M, Tani H, Tokeshi M (2018) Label-free electrochemical sensor for ochratoxin A using a microfabricated electrode with immobilized aptamer ACS. Omega 3:16823–16830. https://doi.org/10.1021/acsomega.8b01996

Mejri-Omrani N, Miodek A, Zribi B, Marrakchi M, Hamdi M, Marty JL, Korri-Youssoufi H (2016) Direct detection of OTA by impedimetric aptasensor based on modified polypyrrole-dendrimers. Anal Chim Acta 920:37–46. https://doi.org/10.1016/j.aca.2016.03.038

Mishra RK, Hayat A, Catanante G, Ocana C, Marty JL (2015) A label free aptasensor for ochratoxin A detection in cocoa beans: an application to chocolate industries. Anal Chim Acta 889:106–112. https://doi.org/10.1016/j.aca.2015.06.052

Mishra RK, Hayat A, Catanante G, Istamboulie G, Marty JL (2016) Sensitive quantitation of ochratoxin A in cocoa beans using differential pulse voltammetry based aptasensor. Food Chem 192:799–804. https://doi.org/10.1016/j.foodchem.2015.07.080

Mo R et al (2018) A novel aflatoxin B1 biosensor based on a porous anodized alumina membrane modified with graphene oxide and an aflatoxin B1 aptamer. Electrochem Commun 95:9–13. https://doi.org/10.1016/j.elecom.2018.08.012

Moradi S, Azizi-Lalabadi M, Bagheri V, Sadeghi E (2020) Fabrication of electrospun sensor based on a synthesized component doped into PAN (polyacrylonitrile) nanofibers for electrochemical detection of zearalenone mycotoxin in foods simulant. Sens Bio-Sens Res 28:100321. https://doi.org/10.1016/j.sbsr.2020.100321

Moressi MB, Calvente JJ, Andreu R, Fernández H, Zón MA (2007) Electro-oxidation of altertoxin I (ATX-I) at gold electrodes modified by dodecanethiol self-assembled monolayers. J Electroanal Chem 605:118–124. https://doi.org/10.1016/j.jelechem.2007.03.024

Moscone D, Arduini F, Amine A (2011) A rapid enzymatic method for aflatoxin B detection. In: Microbial Toxins. Springer, pp 217–235

Nabok AV, Tsargorodskaya A, Hassan AK, Starodub NF (2005) Total internal reflection ellipsometry and SPR detection of low molecular weight environmental toxins. Appl Surf Sci 246:381–386. https://doi.org/10.1016/j.apsusc.2004.11.084

Nabok AV, Tsargorodskaya A, Holloway A, Starodub NF, Gojster O (2007) Registration of T-2 mycotoxin with total internal reflection ellipsometry and QCM impedance methods. Biosens Bioelectron 22:885–890. https://doi.org/10.1016/j.bios.2006.03.010

Nguyen BH, Tran LD, Do QP, Nguyen HL, Tran NH, Nguyen PX (2013) Label-free detection of aflatoxin M1 with electrochemical Fe3O4/polyaniline-based aptasensor. Mater Sci Eng C Mater Biol Appl 33:2229–2234. https://doi.org/10.1016/j.msec.2013.01.044

Nodoushan MS, Nasirizadeh N, Kachuei R, Imani Fooladi AA (2019) Electrochemical detection of aflatoxin B1: an aptasensor prepared using graphene oxide and gold nanowires. Anal Methods 11:6033–6042. https://doi.org/10.1039/c9ay01673b

Owino JHO et al (2008) Electrochemical immunosensor based on polythionine/gold nanoparticles for the determination of aflatoxin B1. Sensors (Basel) 8:8262–8274. https://doi.org/10.3390/s8128262

Pan D et al (2018) Direct immunoassay for facile and sensitive detection of small molecule aflatoxin B1 based on nanobody. Chem Eur J 24:9869–9876. https://doi.org/10.1002/chem.201801202

Park JH, Kim Y-P, Kim I-H, Ko S (2014) Rapid detection of aflatoxin B1 by a bifunctional protein crosslinker-based surface plasmon resonance biosensor. Food Control 36:183–190

Pemberton RM, Pittson R, Biddle N, Drago GA, Hart JP (2006) Studies towards the development of a screen-printed carbon electrochemical immunosensor array for mycotoxins: a sensor for aflatoxin B1. Anal Lett 39:1573–1586. https://doi.org/10.1080/00032710600713289

Peng G, Li X, Cui F, Qiu Q, Chen X, Huang H (2018) Aflatoxin B1 electrochemical aptasensor based on tetrahedral DNA nanostructures functionalized three dimensionally ordered macroporous MoS2–AuNPs film. ACS Appl Mater Interfaces 10:17551–17559. https://doi.org/10.1021/acsami.8b01693

Perrotta PR, Arévalo FJ, Vettorazzi NR, Zón MA, Fernández H (2012) Development of a very sensitive electrochemical magneto immunosensor for the direct determination of ochratoxin A in red wine. Sens Actuators B: Chem 162:327–333. https://doi.org/10.1016/j.snb.2011.12.089

Piermarini S, Micheli L, Ammida NHS, Palleschi G, Moscone D (2007) Electrochemical immunosensor array using a 96-well screen-printed microplate for aflatoxin B1 detection. Biosens Bioelectron 22:1434–1440. https://doi.org/10.1016/j.bios.2006.06.029

Piletsky SA, Subrahmanyam S, Turner AP (2001) Application of molecularly imprinted polymers in sensors for the environment and biotechnology. Sens Rev

Prieto-Simón B, Campàs M, Marty J-L, Noguer T (2008) Novel highly-performing immunosensor-based strategy for ochratoxin A detection in wine samples. Biosens Bioelectron 23:995–1002. https://doi.org/10.1016/j.bios.2007.10.002

Prieto-Simón B, Karube I, Saiki H (2012) Sensitive detection of ochratoxin A in wine and cereals using fluorescence-based immunosensing. Food Chem 135:1323–1329. https://doi.org/10.1016/j.foodchem.2012.05.060

Puiu M, Istrate O, Rotariu L, Bala C (2012) Kinetic approach of aflatoxin B1–acetylcholinesterase interaction: A tool for developing surface plasmon resonance biosensors. Anal Biochem 421:587–594

Radi AE, Munoz-Berbel X, Lates V, Marty JL (2009) Label-free impedimetric immunosensor for sensitive detection of ochratoxin A. Biosens Bioelectron 24:1888–1892. https://doi.org/10.1016/j.bios.2008.09.021

Ran Q, Peng R, Liang C, Ye S, Xian Y, Zhang W, Jin L (2011) Covalent immobilization of horseradish peroxidase via click chemistry and its direct electrochemistry. Talanta 83:1381–1385. https://doi.org/10.1016/j.talanta.2010.11.024

Reddy KRN, Salleh B, Saad B, Abbas HK, Abel CA, Shier WT (2010) An overview of mycotoxin contamination in foods and its implications for human health. Toxin Reviews 29:3–26. https://doi.org/10.3109/15569541003598553

Regiart M et al (2018) Mesoporous immunosensor applied to zearalenone determination in Amaranthus cruentus seeds. Microchem J 141:388–394. https://doi.org/10.1016/j.microc.2018.05.051

Rhouati A, Hayat A, Hernandez DB, Meraihi Z, Munoz R, Marty J-L (2013) Development of an automated flow-based electrochemical aptasensor for on-line detection of ochratoxin A. Sens Actuators B: Chem 176:1160–1166

Ricciardi C et al (2013) Development of a microcantilever-based immunosensing method for mycotoxin detection. Biosens Bioelectron 40:233–239

Rivas L, Mayorga-Martinez CC, Quesada-González D, Zamora-Gálvez A, de la Escosura-Muñiz A, Merkoçi A (2015) Label-free impedimetric aptasensor for ochratoxin-A detection using iridium oxide nanoparticles. Anal Chem 87:5167–5172. https://doi.org/10.1021/acs.analchem.5b00890

Robbins CA, Swenson LJ, Nealley ML, Gots RE, Kelman BJ (2000) Health effects of mycotoxins in indoor air: a critical review. Appl Occup Environ Hyg 15:773–784. https://doi.org/10.1080/10473220050129419

Rodrigues I, Handl J, Binder EM (2011) Mycotoxin occurrence in commodities, feeds and feed ingredients sourced in the Middle East and Africa. Food Addit Contam Part B Surveill 4:168–179. https://doi.org/10.1080/19393210.2011.589034

Sapsford KE, Ngundi MM, Moore MH, Lassman ME, Shriver-Lake LC, Taitt CR, Ligler FS (2006) Rapid detection of foodborne contaminants using an Array Biosensor. Sens Actuators B: Chem 113:599–607. https://doi.org/10.1016/j.snb.2005.07.008

Schmidt KS et al (2004) Application of locked nucleic acids to improve aptamer in vivo stability and targeting function. Nucleic Acids Res 32:5757–5765. https://doi.org/10.1093/nar/gkh862

Schmittgen TD, Livak KJ (2008) Analyzing real-time PCR data by the comparative CT method. Nat Protoc 3:1101–1108. https://doi.org/10.1038/nprot.2008.73

Schöning MJ, Poghossian A (2002) Recent advances in biologically sensitive field-effect transistors (BioFETs). Analyst 127:1137–1151

Senko O, Stepanov N, Maslova O, Akhundov R, Ismailov A, Efremenko E (2019) Immobilized luminescent bacteria for the detection of mycotoxins under discrete and flow-through conditions. Biosensors 9:63. https://doi.org/10.3390/bios9020063

Seok Y, Byun JY, Shim WB, Kim MG (2015) A structure-switchable aptasensor for aflatoxin B1 detection based on assembly of an aptamer/split DNAzyme. Anal Chim Acta 886:182–187. https://doi.org/10.1016/j.aca.2015.05.041

Shao B, Ma X, Zhao S, Lv Y, Hun X, Wang H, Wang Z (2018) Nanogapped Au(core) @ Au-Ag(shell) structures coupled with Fe3O4 magnetic nanoparticles for the detection of Ochratoxin A. Anal Chim Acta 1033:165–172. https://doi.org/10.1016/j.aca.2018.05.058

Sharma A, Matharu Z, Sumana G, Solanki PR, Kim C, Malhotra B (2010) Antibody immobilized cysteamine functionalized-gold nanoparticles for aflatoxin detection. Thin Solid Films 519:1213–1218

Sheng L, Ren J, Miao Y, Wang J, Wang E (2011) PVP-coated graphene oxide for selective determination of ochratoxin A via quenching fluorescence of free aptamer. Biosens Bioelectron 26:3494–3499. https://doi.org/10.1016/j.bios.2011.01.032

Shim W-B, Mun H, Joung H-A, Ofori JA, Chung D-H, Kim M-G (2014) Chemiluminescence competitive aptamer assay for the detection of aflatoxin B1 in corn samples. Food Control 36:30–35. https://doi.org/10.1016/j.foodcont.2013.07.042

Singh C et al (2013a) Carboxylated multiwalled carbon nanotubes based biosensor for aflatoxin detection. Sens Actuators B: Chem 185:258–264

Singh J, Roychoudhury A, Srivastava M, Solanki PR, Lee DW, Lee SH, Malhotra B (2013b) A highly efficient rare earth metal oxide nanorods based platform for aflatoxin detection. J Mater Chem B 1:4493–4503

Smolko V, Shurpik D, Porfireva A, Evtugyn G, Stoikov I, Hianik T (2018) Electrochemical aptasensor based on poly(neutral red) and carboxylated pillar[5]arene for sensitive determination of aflatoxin M1. Electroanalysis 30:486–496. https://doi.org/10.1002/elan.201700735

Solanki PR, Kaushik A, Ansari AA, Malhotra B (2009) Nanostructured zinc oxide platform for cholesterol sensor. Appl Phys Lett 94:143901

Solanki PR, Kaushik A, Manaka T, Pandey MK, Iwamoto M, Agrawal VV, Malhotra BD (2010) Self-assembled monolayer based impedimetric platform for food borne mycotoxin detection. Nanoscale 2:2811–2817. https://doi.org/10.1039/C0NR00289E

Solanki PR, Singh J, Rupavali B, Tiwari S, Malhotra BD (2017) Bismuth oxide nanorods based immunosensor for mycotoxin detection. Mater Sci Eng, C 70:564–571. https://doi.org/10.1016/j.msec.2016.09.027

Soldatkin OO, Burdak OS, Sergeyeva TA, Arkhypova VM, Dzyadevych SV, Soldatkin AP (2013) Acetylcholinesterase-based conductometric biosensor for determination of aflatoxin B1. Sens Actuators B: Chem 188:999–1003. https://doi.org/10.1016/j.snb.2013.06.107

Spinella K (2013) Development of a QCM (quartz crystal microbalance) biosensor to the detection of aflatoxin B1. Open J Appl Biosens 2:112

Srivastava S et al (2013) Electrophoretically deposited reduced graphene oxide platform for food toxin detection. Nanoscale 5:3043–3051. https://doi.org/10.1039/C3NR32242D

Steeland S, Vandenbroucke RE, Libert C (2016) Nanobodies as therapeutics: big opportunities for small antibodies. Drug Discovery Today 21:1076–1113. https://doi.org/10.1016/j.drudis.2016.04.003

Stepurska K, Soldatkin O, Kucherenko I, Arkhypova V, Dzyadevych S, Soldatkin A (2015) Feasibility of application of conductometric biosensor based on acetylcholinesterase for the inhibitory analysis of toxic compounds of different nature. Anal Chim Acta 854:161–168

Sun L, Zhao Q (2018) Competitive horseradish peroxidase-linked aptamer assay for sensitive detection of Aflatoxin B1. Talanta 179:344–349. https://doi.org/10.1016/j.talanta.2017.11.048

Sun C et al (2020) A self-assembled electrochemical immunosensor for ultra-sensitive detection of ochratoxin A in medicinal and edible malt. Food Chem 315:126289. https://doi.org/10.1016/j.foodchem.2020.126289

Tamerler C, Oren EE, Duman M, Venkatasubramanian E, Sarikaya M (2006) Adsorption kinetics of an engineered gold binding Peptide by surface plasmon resonance spectroscopy and a quartz crystal microbalance. Langmuir 22:7712–7718. https://doi.org/10.1021/la0606897

Tang J, Huang Y, Cheng Y, Huang L, Zhuang J, Tang D (2018a) Two-dimensional MoS2 as a nano-binder for ssDNA: ultrasensitive aptamer based amperometric detection of Ochratoxin A. Microchimica Acta 185:162. https://doi.org/10.1007/s00604-018-2706-9
Tang Z, Wang X, Lv J, Hu X, Liu X (2018b) One-step detection of ochratoxin A in cereal by dot immunoassay using a nanobody-alkaline phosphatase fusion protein. Food Control 92:430–436. https://doi.org/10.1016/j.foodcont.2018.05.013
Tian F, Zhou J, Jiao B, He Y (2019) A nanozyme-based cascade colorimetric aptasensor for amplified detection of ochratoxin A. Nanoscale 11:9547–9555. https://doi.org/10.1039/C9NR02872B
Tong P, Zhang L, Xu J-J, Chen H-Y (2011) Simply amplified electrochemical aptasensor of Ochratoxin A based on exonuclease-catalyzed target recycling. Biosens Bioelectron 29:97–101. https://doi.org/10.1016/j.bios.2011.07.075
Tong P, Zhao W-W, Zhang L, Xu J-J, Chen H-Y (2012) Double-probe signal enhancing strategy for toxin aptasensing based on rolling circle amplification. Biosens Bioelectron 33:146–151. https://doi.org/10.1016/j.bios.2011.12.042
Tsai WC, Hsieh CK (2007) QCM-based immunosensor for the determination of ochratoxin A. Anal Lett 40:1979–1991
Urusov AE, Kostenko SN, Sveshnikov PG, Zherdev AV, Dzantiev BB (2011) Ochratoxin A immunoassay with surface plasmon resonance registration: lowering limit of detection by the use of colloidal gold immunoconjugates. Sens Actuators B: Chem 156:343–349. https://doi.org/10.1016/j.snb.2011.04.044
van Egmond HP, Schothorst RC, Jonker MA (2007) Regulations relating to mycotoxins in food. Anal Bioanal Chem 389:147–157. https://doi.org/10.1007/s00216-007-1317-9
Vidal J, Duato P, Bonel L, Castillo J (2009) Use of polyclonal antibodies to ochratoxin A with a quartz–crystal microbalance for developing real-time mycotoxin piezoelectric immunosensors. Anal Bioanal Chem 394:575–582
Vidal JC, Bonel L, Ezquerra A, Duato P, Castillo JR (2012) An electrochemical immunosensor for ochratoxin A determination in wines based on a monoclonal antibody and paramagnetic microbeads. Anal Bioanal Chem 403:1585–1593. https://doi.org/10.1007/s00216-012-5951-5
Visconti A, Lattanzio VMT, Pascale M, Haidukowski M (2005) Analysis of T-2 and HT-2 toxins in cereal grains by immunoaffinity clean-up and liquid chromatography with fluorescence detection. J Chromatogr A 1075:151–158. https://doi.org/10.1016/j.chroma.2005.04.009
Wang C et al (2018) Fabrication of magnetically assembled aptasensing device for label-free determination of aflatoxin B1 based on EIS. Biosens Bioelectron 108:69–75. https://doi.org/10.1016/j.bios.2018.02.043
Wang J, Munir A, Zhou HS (2009) Au NPs-aptamer conjugates as a powerful competitive reagent for ultrasensitive detection of small molecules by surface plasmon resonance spectroscopy. Talanta 79:72–76
Wang H, Li J, Wang Y, Jin J, Yang R, Wang K, Tan W (2010a) Combination of DNA ligase reaction and gold nanoparticle-quenched fluorescent oligonucleotides: a simple and efficient approach for fluorescent assaying of single-nucleotide polymorphisms. Anal Chem 82:7684–7690. https://doi.org/10.1021/ac101503t
Wang Z, Duan N, Hun X, Wu S (2010b) Electrochemiluminescent aptamer biosensor for the determination of ochratoxin A at a gold-nanoparticles-modified gold electrode using N-(aminobutyl)-N-ethylisoluminol as a luminescent label. Anal Bioanal Chem 398:2125–2132. https://doi.org/10.1007/s00216-010-4146-1
Wang X et al (2011) An immunoarray for the simultaneous detection of two mycotoxins, ochratoxin a and fumonisin b1. J Food Saf 31:408–416. https://doi.org/10.1111/j.1745-4565.2011.00314.x
Wang C, Dong X, Liu Q, Wang K (2015) Label-free colorimetric aptasensor for sensitive detection of ochratoxin A utilizing hybridization chain reaction. Anal Chim Acta 860:83–88. https://doi.org/10.1016/j.aca.2014.12.031
Wang B, Wu Y, Chen Y, Weng B, Xu L, Li C (2016) A highly sensitive aptasensor for OTA detection based on hybridization chain reaction and fluorescent perylene probe. Biosens Bioelectron 81:125–130. https://doi.org/10.1016/j.bios.2016.02.062

Wang C et al (2017) Magneto-controlled aptasensor for simultaneous electrochemical detection of dual mycotoxins in maize using metal sulfide quantum dots coated silica as labels. Biosens Bioelectron 89:802–809. https://doi.org/10.1016/j.bios.2016.10.010

Wang C, Li Y, Zhao Q (2019a) A signal-on electrochemical aptasensor for rapid detection of aflatoxin B1 based on competition with complementary DNA. Biosens Bioelectron 144:111641. https://doi.org/10.1016/j.bios.2019.111641

Wang C, Qian J, An K, Lu X, Huang X (2019b) A semiconductor quantum dot-based ratiometric electrochemical aptasensor for the selective and reliable determination of aflatoxin B1. Analyst 144:4772–4780. https://doi.org/10.1039/c9an00825j

Wang L, Zhu F, Chen M, Zhu Y, Xiao J, Yang H, Chen X (2019c) Rapid and visual detection of aflatoxin B1 in foodstuffs using aptamer/G-quadruplex DNAzyme probe with low background noise. Food Chem 271:581–587. https://doi.org/10.1016/j.foodchem.2018.08.007

Wei M, Zhang W (2018) The determination of Ochratoxin A based on the electrochemical aptasensor by carbon aerogels and methylene blue assisted signal amplification. Chem Cent J 12:45. https://doi.org/10.1186/s13065-018-0415-4

Wei M, Xin L, Feng S et al (2020) Simultaneous electrochemical determination of ochratoxin A and fumonisin B1 with an aptasensor based on the use of a Y-shaped DNA structure on gold nanorods. Microchim Acta 187:102. https://doi.org/10.1007/s00604-019-4089-y

Willner I, Zayats M (2007) Electronic aptamer-based sensors. Angew Chem Int Ed Engl 46:6408–6418. https://doi.org/10.1002/anie.200604524

Wu S, Duan N, Zhu C, Ma X, Wang M, Wang Z (2011) Magnetic nanobead-based immunoassay for the simultaneous detection of aflatoxin B1 and ochratoxin A using upconversion nanoparticles as multicolor labels. Biosens Bioelectron 30:35–42. https://doi.org/10.1016/j.bios.2011.08.023

Wu J, Chu H, Mei Z, Deng Y, Xue F, Zheng L, Chen W (2012a) Ultrasensitive one-step rapid detection of ochratoxin A by the folding-based electrochemical aptasensor. Anal Chim Acta 753:27–31

Wu S, Duan N, Ma X, Xia Y, Wang H, Wang Z, Zhang Q (2012b) Multiplexed fluorescence resonance energy transfer aptasensor between upconversion nanoparticles and graphene oxide for the simultaneous determination of mycotoxins. Anal Chem 84:6263–6270. https://doi.org/10.1021/ac301534w

Wu K, Ma C, Zhao H, Chen M, Deng Z (2019a) Sensitive aptamer-based fluorescene assay for ochratoxin A based on RNase H signal amplification. Food Chem 277:273–278. https://doi.org/10.1016/j.foodchem.2018.10.130

Wu SS, Wei M, Wei W, Liu Y, Liu S (2019b) Electrochemical aptasensor for aflatoxin B1 based on smart host-guest recognition of β-cyclodextrin polymer. Biosens Bioelectron 129:58–63. https://doi.org/10.1016/j.bios.2019.01.022

Xia S, Zhu P, Pi F, Zhang Y, Li Y, Wang J, Sun X (2017) Development of a simple and convenient cell-based electrochemical biosensor for evaluating the individual and combined toxicity of DON, ZEN, and AFB1. Biosens Bioelectron 97:345–351. https://doi.org/10.1016/j.bios.2017.06.002

Xu X, Liu X, Li Y, Ying Y (2013) A simple and rapid optical biosensor for detection of aflatoxin B1 based on competitive dispersion of gold nanorods. Biosens Bioelectron 47:361–367. https://doi.org/10.1016/j.bios.2013.03.048

Yagati AK, Chavan SG, Baek C, Lee M-H, Min J (2018) Label-free impedance sensing of aflatoxin B1 with polyaniline nanofibers/Au nanoparticle electrode array. Sensors 18:1320

Yang C, Lates V, Prieto-Simón B, Marty J-L, Yang X (2013a) Rapid high-throughput analysis of ochratoxin A by the self-assembly of DNAzyme-aptamer conjugates in wine. Talanta 116:520–526. https://doi.org/10.1016/j.talanta.2013.07.011

Yang X-H, Kong W-J, Yang M-H, Zhao M, Ouyang Z (2013b) Application of aptamer identification technology in rapid analysis of mycotoxins. Chin J Anal Chem 41:297–306. https://doi.org/10.1016/S1872-2040(13)60630-1

Yang Y-J et al (2019) A Label-free aptasensor based on Aptamer/NH2 Janus particles for ultrasensitive electrochemical detection of Ochratoxin A. Talanta 199:310–316. https://doi.org/10.1016/j.talanta.2019.02.015

Yu JCC, Lai EPC (2005) Interaction of ochratoxin A with molecularly imprinted polypyrrole film on surface plasmon resonance sensor. React Funct Polym 63:171–176. https://doi.org/10.1016/j.reactfunctpolym.2005.03.001

Yu L, Zhang Y, Hu C, Wu H, Yang Y, Huang C, Jia N (2015) Highly sensitive electrochemical impedance spectroscopy immunosensor for the detection of AFB1 in olive oil. Food Chem 176:22–26

Yuan J, Deng D, Lauren DR, Aguilar M-I, Wu Y (2009) Surface plasmon resonance biosensor for the detection of ochratoxin A in cereals and beverages. Anal Chim Acta 656:63–71. https://doi.org/10.1016/j.aca.2009.10.003

Yugender Goud K, Catanante G, Hayat A, M S, Vengatajalabathy Gobi K, Marty JL (2016) Disposable and portable electrochemical aptasensor for label free detection of aflatoxin B1 in alcoholic beverages. Sens Actuators B: Chem 235:466–473. https://doi.org/10.1016/j.snb.2016.05.112

Zaijun L, Zhongyun W, Xiulan S, Yinjun F, Peipei C (2010) A sensitive and highly stable electrochemical impedance immunosensor based on the formation of silica gel-ionic liquid biocompatible film on the glassy carbon electrode for the determination of aflatoxin B1 in bee pollen. Talanta 80:1632–1637. https://doi.org/10.1016/j.talanta.2009.09.058

Zain ME (2011) Impact of mycotoxins on humans and animals. J Saudi Chem Soc 15:129–144. https://doi.org/10.1016/j.jscs.2010.06.006

Zamfir L-G, Geana I, Bourigua S, Rotariu L, Bala C, Errachid A, Jaffrezic-Renault N (2011) Highly sensitive label-free immunosensor for ochratoxin A based on functionalized magnetic nanoparticles and EIS/SPR detection. Sens Actuators B: Chem 159:178–184

Zhang J, Liu B, Liu H, Zhang X, Tan W (2013a) Aptamer-conjugated gold nanoparticles for bioanalysis. Nanomedicine 8:983–993

Zhang J, Zhang X, Yang G, Chen J, Wang S (2013b) A signal-on fluorescent aptasensor based on Tb3+ and structure-switching aptamer for label-free detection of ochratoxin A in wheat. Biosens Bioelectron 41:704–709

Zhang S, Shen Y, Shen G, Wang S, Shen G, Yu R (2016) Electrochemical immunosensor based on Pd–Au nanoparticles supported on functionalized PDDA-MWCNT nanocomposites for aflatoxin B1 detection. Anal Biochem 494:10–15. https://doi.org/10.1016/j.ab.2015.10.008

Zhang J, Yang K, Chen L (2019) Situ deposition of silver nanoparticles on polydopamine nanospheres for an ultrasensitive electrochemical aptasensor of ochratoxin A. J Electrochem Soc 166:H182–H186. https://doi.org/10.1149/2.0931906jes

Zhang M et al (2020) Ultrasensitive competitive detection of patulin toxin by using strand displacement amplification and DNA G-quadruplex with aggregation-induced emission. Anal Chim Acta 1106:161–167. https://doi.org/10.1016/j.aca.2020.01.064

Zheng Z, Humphrey CW, King RS, Richard JL (2005) Validation of an ELISA test kit for the detection of total aflatoxins in grain and grain products by comparison with HPLC. Mycopathologia 159:255–263. https://doi.org/10.1007/s11046-004-8666-0

Zheng MZ, Richard JL, Binder J (2006) A review of rapid methods for the analysis of mycotoxins. Mycopathologia 161:261–273. https://doi.org/10.1007/s11046-006-0215-6

Zheng W et al (2016) Hetero-enzyme-based two-round signal amplification strategy for trace detection of aflatoxin B1 using an electrochemical aptasensor. Biosens Bioelectron 80:574–581. https://doi.org/10.1016/j.bios.2016.01.091

Zhu Z, Yang R, You M, Zhang X, Wu Y, Tan W (2010) Single-walled carbon nanotube as an effective quencher. Anal Bioanal Chem 396:73–83. https://doi.org/10.1007/s00216-009-3192-z

Zhu X et al (2018) Label-free ochratoxin A electrochemical aptasensor based on target-induced noncovalent assembly of peroxidase-like graphitic carbon nitride nanosheet. Sen Actuators B: Chem 270:263–269. https://doi.org/10.1016/j.snb.2018.05.048

Chapter 15
Biosensors for Fruit Quality Monitoring

Vinita Hooda, Nidhi Chauhan, and Shringika Soni

Abstract The growing interest in development of reliable and sensitive detection methods for fruit quality and safety has been observed in scientific communities and food industry. For better management of existing analytical technologies and to bring next-generation sensing platform in realistic market, updated and accurate information is necessary for proper planning and decision-making. This chapter describes the current status of chemical and biological sensors available for fruit quality monitoring and smart packaging. These sensing designs should have features for measuring adulterants, freshness, allergens, microorganisms, and toxicants. On the other hand potential applications of smart packaging and e-nose are also explained in fruit quality and spoilage monitoring in logistic applications. The last section of the chapter also summarized commercially available sensing platform for fruit quality assessment and opportunities for future research discussed in brief.

Keywords Fruit quality · Sensors · Indicator sensing · RFID tag · Smart and intelligent packaging · Nanotechnology

15.1 Introduction

According to literature, approximately one-third of food produced across the globe is never consumed and from the last three decades quality and safety of food products are compromised (Huang et al. 2008; Jedermann et al. 2014). The World Health Organization (WHO) based report also evaluated contaminated food induced 600 million cases of illness and 420,000 deaths each year (WHO 2017). In addition, food poisoning, diarrhea, and various infections are major human diseases caused by consuming spoiled fruit products. The major share of these losses is linked with

V. Hooda (✉)
Department of Botany, Maharshi Dayanand University, Rohtak, Haryana, India
e-mail: vinitahooda@yahoo.co.in

N. Chauhan · S. Soni
Amity Institute of Nanotechnology, Amity University Uttar Pradesh, Noida, UP, India

R. N. Pudake et al. (eds.), *Biosensors in Agriculture: Recent Trends and Future Perspectives*, Concepts and Strategies in Plant Sciences,
https://doi.org/10.1007/978-3-030-66165-6_15

improper handling and mismanagement during pre- and post-harvesting operations. Besides, economic development and improvement in living standards increased fresh and nutritious fruit consumption and also the availability of fruits in all seasons. Yet fruit damage during transportation is an important factor of concern, because of mismanagement of controlled environmental variables and longer time taken in transportation.

The fruit quality measurement is defined by different variables such as, fruit health, nutritional values, consumer standards, stability, and various other country-wise factors (Yousefi et al. 2019), though consumers' primary perception of packaged fruit products depends upon visual appeal, aroma, flavor, ripeness, and date of packaging mentioned on the label. Conventionally spectroscopy, chromatography, and culture-based laboratory methods were applied in agricultural sector for qualitative evaluation of food products. Imaging and spectroscopy are important conventional optical technology where they provide spatial, chemical, and physical information about fruit products (Wang et al. 2015). But these techniques are time-consuming, expensive, laboratory-based, and requires heavy equipment and expertise for result analysis. Therefore, we require a rapid and reliable on-field and real-time monitoring system of food products.

In recent years, several new portable and remote wireless sensing technologies have come into the worldwide market. On-field sensors monitor crops, soil, and water samples for potential disease or pesticide contaminations, whereas smart packaging and wireless sensor systems show satisfactory solutions for real-time monitoring in food. On-field electrochemical or optical sensors are direct detection method to capture analyte and know the current status of food product, but smart packaging and wireless sensing applied for real-time monitoring on the basis of changes in pH, temperature, moisture, ripening, and sugar content.

15.2 Biosensor: A Miniaturized Tool for Detection

The term "Biosensor" is a very powerful and innovative analytical platform that involves biological sensing element as a recognition template with a broad range of applications in diagnosis, food safety, healthcare, defense, biomedicine, and environmental monitoring (Chauhan et al. 2020; Gu and Liu 2020). It works on the principle of biological and chemical reactions and output signals are directly proportional to the concentration of analyte (Chauhan et al. 2019; Gu and Liu 2020). Their purpose is to provide quick, reliable, real time, and accurate detection of analyte of interrogation, yet ideally biosensors are capable of continuous and reversible monitoring. After the first enzyme-based biosensor by Clark in 60's, huge progress has been made in sensing technologies by development in biorecognition of elements and artificial transducers (Liu et al. 2019; Coulet and Blum 2019). The most typical components of biosensors include suitable receptors such as enzymes, antibodies, oligonucleotides, cells, nanoparticles, and imprinting polymers along with pH, temperature, change in electrical and optical signals-based physiochemical transducers (Mehrotra 2016;

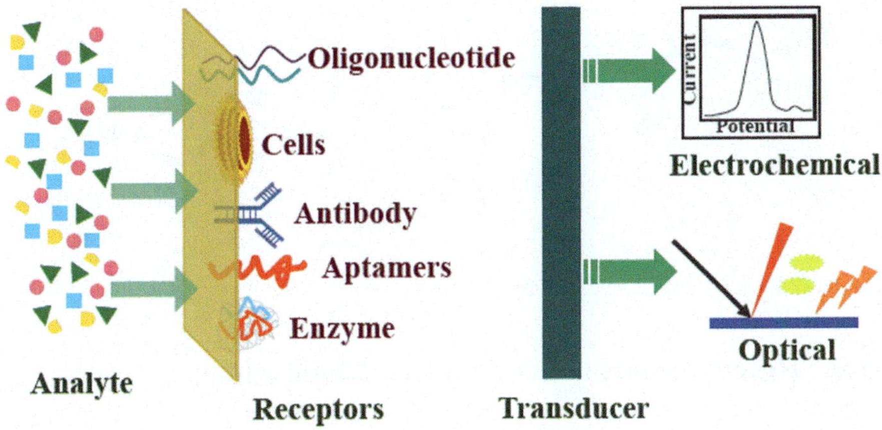

Fig. 15.1 Components and signaling of electrochemical and optical sensors

Davis and Altintas 2018). The basic configurations of biosensors are explained in Fig. 15.1. On the basis of biorecognition elements, the biosensors are classified into immunosensors, DNA or aptamer sensors, microbial sensors, enzymatic sensors, and colorimetric sensors. These sensors have benefits of higher selectivity, sensitivity, easy portability, quick response time, and no sample preprocessing requirements (Zhu et al. 2019; Cui et al. 2019; Jain et al. 2020). Further improvement and generation of next-generation sensing technology, nanotechnology-based sensing platform recently has been reported for application in analytical chemistry.

Recently electronic nose (e-nose) and radio frequency-based sensing techniques have been used in most of the agriculture sector, where nanomaterials assisted sensing improved detection capacity of biosensor. The e-nose is artificial olfactory sensor technology, which is a combination of biochemistry, electronics, and artificial intelligence to use in numerous applications (Burlachenko et al. 2016; Kodogiannis 2017; Wasilewski et al. 2019). The e-nose is basically a gas multisensor array that is designed to measure gas mixture as a whole instead of targeting individual chemical species. The sensing array-like electrochemical and optical sensors, data acquisition system as signal processing unit along with multivariate data processing tools, and a pattern recognition unit to identify and classify target on the basis of specific odors are the major components of e-nose, as shown in Fig. 15.2 (Karakaya et al. 2020; Ramgir 2013). Volatile chemicals originated from adulteration, microbial contamination, maturity, and other quality parameters of post-harvested or packaged food products, reacts with biorecognition molecules on e-nose multisensory array and pattern recognition unit identify the chemicals according to their database (Gliszczyńska-Świgło and Chmielewski 2017; Karakaya et al. 2020; Shi et al. 2018). Recently, nanoparticles namely, tin oxide, carbon nanotubes, silicon and titanium nanowires, and many more, were reported in e-nose based sensing due to high surface to volume ratio, microelectronic processing, and sensitivity toward target molecule (Chauhan et al. 2020; Soni et al. 2016).

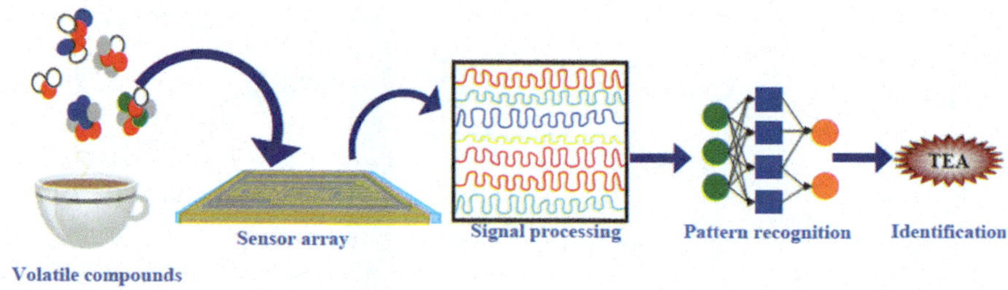

Fig. 15.2 Working mechanism of e-nose

15.3 Recent Trends in Fruit Quality Monitoring

In the last decades, farmers and manufacturers faced huge economical losses due to inaccessibility of cultivable land, depleting water resources, use of excessive pesticides, pathogen infection, and post-market commercialization in agricultural industries. For assurance of good quality agricultural output, on-site detection is very much required for quality control in the agricultural food sector. Due to increasing number of pollutants and hazardous chemicals in soil and water resources, scientists are looking forward to develop new methods in monitoring and controlling the quality of the food. Therefore, the diagnostic industry is flourishing in the agricultural sector after playing an important role in medical diagnosis. In recent years, countless sensing platforms have been developed for detection of fruit ripening, contamination, and real-time monitoring for packaged fruit quality (Satpute and Jagdale 2016; Mustafa and Andreescu 2018). Markets-based sensing platforms for agro-food-diagnostic are generally multianalyzer, bench-top portable equipment, and single-time use disposable sensors (Tothill 2001). In this chapter, we will be elaborating on emerging sensing technologies and recent trends in biosensor development for fruit quality monitoring.

15.3.1 Gas Detector for Fruit Quality Assessment

To avoid fruit spoilage during transport and maintain the quality in the market, fruit products are shipped in preservation chambers to delay the ripening process. These transported fruits are exposed to 2–3% oxygen, ~ 0 °C temperature, high humidity, and high CO_2 level (Jochum et al. 2016). Also, the application of ethylene (C_2H_4) in fruit ripening increases respiration rise, change in texture, aroma, color, and flavor along with autocatalytic production of C_2H_2 (Pech et al. 2018; Gao et al. 2020). Therefore, C_2H_4 concentration should be kept as low as possible. Additionally, ammonia (NH_3) monitoring is of particular interest because it is the commonly used coolant in fruit transportation yet leakage of NH_3 can endanger the ozone layer, effect human health at a bad scale, and spoil stored fruits (Jochum et al. 2016; Tzortzakis and Chrysargyris 2017; Yang and Jiao 2017).

Since last decades, lots of sensing platforms and electronic noses have been designed for C_2H_2 detection. Sklorz and colleagues designed an intelligent container for C_2H_4 measurement in fruit logistic applications (Sklorz et al. 2012a). They exhibited selective detection of C_2H_4 gas detection via miniaturized gas chromatograph (μGC) coupled with miniaturized preconcentrator device (μPCD) with absorbing material, Carbosieve S-II. The carbosieve S-II absorbed C_2H_4 and 16 times higher peak in chromatogram was achieved as compared to empty μPCD. Thus, this innovative technology could be very beneficial for fruit logistic applications. Similarly, another preconcentrator microfluidic sensing platform has been previously developed by Germany-based research team, where they used Carboxen 1000 fabricated microchannel to detect C_2H_4 and 100-fold enhancement was achieved (Dow et al. 2011). The Carbosieve S-II and Carboxen 1000 are carbon molecular sieve absorbent materials with large surface area and small pore size therefore ideal for trapping small and volatile organics.

A recent study reported infrared (IR) thermal emission-based sensing platform for C_2H_4 detection in banana fruit samples (Kathirvelan and Vijayaraghavan 2017). This efficient alternate approach, combined IR thermal emitter with tunable wavelength with silicon temperature detector and enclosed the system in a gas test cell. When released C_2H_4 introduced in the test cell, the analyte absorbed IR radiation and decreased the temperature of the system. Though this method has C_2H_4 detectability of 3.3% during fruit ripening process that was higher than previously developed IR-based thermopile, 0.2% (Sklorz et al. 2012b) and fiber optic sensing, 0.25% (Wei et al. 2015) methods. But due to limitation of silicon temperature detector, the sensitivity was achieved at 5 ppm, while sensing ppb range is desirable for higher accuracy in long-term preserved fruit samples.

For further advancement in fruit ripening detection, Agarwal and colleagues have used dielectric property of SnO_2 nanoparticles (NPs) for the detection of C_2H_4 in fruit samples (Agarwal et al. 2012). The n-type SnO_2 with Pd/Pt NPs exhibited higher conductivity in the presence of target and change in depletion region also presented by reaction between O^- and Pd /Pt nanoparticle (NPs). The developed sensing platform showed decrease in capacitance with increasing C_2H_4 concentration from 0 to 100 ppm, which has the potential for usage in RFID tags for fruit quality monitoring. Similarly, USA-based research team also used highly resistive copper (I) complex 1 conjugated single-walled carbon nanotubes (SWNTs) chemo-resistive sensing of C_2H_4 in banana, orange, apple, pear, and avocado fruit samples (Esser et al. 2012). When 1-SWNT fabricated field-effect transistor (FET) exposed to target C_2H_4 gas, the target bound to 1-SWNT in associative fashion without breaking up compound 1 with SWNT and a linear change in resistivity was observed in the range of 0.5–50 ppm C_2H_4. Similarly, a recent study utilized MoS_2, SWNTs, and Cu(I)–tris (mercaptoimidazolyl) borate complexes (Cu-Tm) fabricated sensing platform for direct detection of C_2H_4 in fruit samples (Chen, et al. 2020). The conductometric property of MoS_2 and Cu(I) compounds as C_2H_4 receptors followed the similar concept as in the previous study, but in the presence of analyte charge transfer coupling between MoS_2 and Cu-TM lowered the conductance and output recorded in the range of 0.1–1.0 ppm. In continuation of nanotechnology application in C_2H_4 detection in fruit ripening, Kathirvelan and colleagues have also designed TiO_2-WO_3 nanocomposite-based chemo-resistive sensor which showed detectability of 46.2% in 200 ppm C_2H_4 and reported limit of detection (LOD) of 8 ppm in real fruit samples (Kathirvelan et al. 2017). Similarly, $AgBF_4$/polyvinylpyrrolidone polymer composite-based quartz crystal microbalance (QCM) (Tolentino et al. 2018), ZnO nanosheets (Wang et al. 2019), and MnO_2 NPs (Bigiani et al., 2020)-based sensing platforms are recently developed for effective quantification of C_2H_4 released during fruit ripening.

In another study, poly(3-hexylthiophene) (P3HT) and polystyrene (PS) fabricated organic field-effect transistor (OFET) were developed for ammonia (NH_3) detection (Han et al. 2016). The dual-sensing possibility of the platform worked on the basis of reduced subthreshold slope and negative threshold voltage due to positivity of trapped NH_3. On the other hand, when lone pair of diffused NH_3 interacted with P3HT/PS layer there was decrement in drain current and reduced net positive charge was observed, therefore, decrement in current was obtained with increasing NH_3 concentration in the range of 0–50 ppm. This poly/insulator blended OFET device stable for 40 days, thus they can be utilized in transportation and logistics applications. Similarly, USA-based research team recently developed N-(tert-Butoxycarbonyloxy)-phthalimide and palladium particles blended P3HT fabricated OFET sensor for direct detection of C_2H_4 (Besar et al. 2017). When analyte exposed to sensing platform palladium particles acted as C_2H_4 receptor and detection principle was similar as for ammonia in the previous study with sensitivity of 25 ppm V/V. The catalytic property of Pd was also used by Li and colleagues where they used Pd NPs and rGO modified porous α-Fe_2O_3 in C_2H_4 detection in fruit samples (Li et al. 2019). The Pd NPs assisted catalysis of O_2 molecule not only formed depletion layer onto α-Fe_2O_3 to increase air resistance, it also promoted $C = C$ bond cleavage to allow interaction between C_2H_4 and absorbed O^- to produce ethylene oxide and free electrons. These free electrons caused resistance drop while rGO acted as an excellent electrical site for rapid transmission of electrons between O^- and α-Fe_2O_3. Therefore, sensitivity of the developed platform significantly increased with increasing C_2H_4 concentration in the range of 100–1000 ppm and LOD of 10 ppb were reported in 18 s.

In commercially developed portable electronic nose (PEN 2) e-nose-based sensing technology for C_2H_4 detection, a principal component analysis (PCA) was performed on sensor data (Benedetti et al. 2008). This system discriminated quality of peach fruit on the basis of increasing days of shelf life and avoided aroma-based fruit quality monitoring, which means consistent low presence of C_2H_4 for a long time period in stored fruit. Similarly, China-based research team also designed FOX 4000 e-nose-based sensing system to discriminate a variety of peaches on the basis of respiration rate (CO_2) and C_2H_4 (Su et al. 2017). Whereas India-based scientists reported enhancement in conductivity when ripped fruits release C_2H_4 and reacted with GO (Mallick and Das 2017; Mallick et al. 2018), thus low cost and easily synthesized GO can be used in smart packaging or e-nose for fruit monitoring. In another study, nano zeolite-molybdate was also used as a pH indicator for avocado ripening where C_2H_4 reacted with nanosystem and change in color from yellow to blue was stable till 10 days in storage condition (Putri et al. 2019).

In continuation of this, scientists incorporated more and more artificial intelligence (AI) in prediction modeling and pattern recognition. In this scenario, Du and colleagues designed metal oxides-based gas sensors to predict different ripening stages of kiwifruit (Du et al. 2019). They incorporated different chemometrics pattern recognition algorithms, such as linear discriminant analysis (LDA) to recognize ripeness of kiwifruit at different time points, whereas compared partial least squares regression (PLSR), random forest (RF), and support vector machine (SVM) determined complete ripeness, firmness, and soluble solids content. The regression outcome supported RF-based algorithm for predicting ripeness indexes in post-harvested kiwifruits. In another study, China-based research team used similar chemometrics algorithms to predict the quality of peaches where they studied the after-damage quality of packaged peaches and confirmed accuracy of 93.33% after 24 h of fruit spoilage (Yang et al. 2020).

15.3.2 pH-Based Detection in Fruit Quality Measurement

The pH and titratable acidity (TA) are considered as important quality attributes for fruit processing and ripening, where pH gradually increases with increasing fruit ripening and TA is significantly high at the beginning of fruit ripening (Anthon et al. 2011). This acidity of fruit is maintained by oligosaccharides such as glucose and fructose and organic acids, namely, citrate, malate, and tartaric acids (Batista-Silva et al. 2018). Malic acid is dominant in apple, pear, and loquat whereas citric acid is generally found in citrus fruits such as pineapple (Etienne et al. 2013). The metabolic tri-citric acid (TCA) cycle intermediates play important role in fruit ripening and changes organic acids or oligosaccharides level in fruits, thus enzyme-based sensing platform can be used to determine the qualitative changes in fruit product.

In recent years, several dye and optical nanoparticles-based colorimetric sensing technology came into market for fruit quality assessment. Jawaheer and colleagues have developed an enzymatic electrochemical sensing platform for detection of β-D-glucose, ascorbic acid, total D-glucose, and sucrose in fruit samples (Jawaheer et al. 2003). They further used cellulose acetate (CA) membrane onto developed enzymatic sensor where enzymatic reaction generates H_2O_2 and CA improved the sensitivity by avoiding binding of other interfering agents of sensing platform. On the other hand, Malaysia-based research team developed optical fiber sensor for quality detection of *Averrhoa carambola* (star fruit) on the basis of pH and firmness (Omar and Matjafri 2013). At the peak of 635 nm, the optical fiber red system (OF-RS) obtained excellent accuracy in pH and firmness determination. For further improvement in sensing platform, a paper-based microfluidic device (μPAD) was designed for glucose determination in fruit samples (Li et al. 2017). When horseradish peroxidase/glucose oxidase fabricated microchannel were exposed to fruits pulp samples, the color of the working electrode gradually changed to dark brown due to HRP assisted metabolic reaction which resulted into I_2 (brown color). To incorporate the nanotechnology in fruit quality sensing system, date fruit was exposed to polyvinyl alcohol (PVA) matrix-based nanomat in a study, where anthocyanin (AC) induced pH changes was determined by variable colorimetric images and good quality fruit further consumed (Maftoonazad and Ramaswamy 2019).

For the pH-based e-nose or smart packaging development, several studies have been performed. In this direction Kantor and colleagues have designed an electronic tongue (ET) for suitable apricot measurement (Kantor et al. 2008). They studied the effect of 1-methylcyclopropene (1-MCP) on stored apricot ripening, which confirmed consistent pH and increment in glucose content throughout the four weeks of storage at 1 °C and 95% relative humidity. Thus, 1-MCP treated apricot showed reduced ripening and sensing can be further used in long-term fruit storage. On other hand, China-based research team established a quality model index to analyze peach quality in controlled conditions (Zhang and Wang 2009). They developed PCA and LDA-based algorithms determined ripening monitoring whereas multiple linear regression (MLR) and artificial neural network (ANN) prediction models helped in differentiating the right quality of peaches. The sensing system reported an increase in pH, sugar content, and at last 9.91% growth of peaches within 14-days time period. Song and colleagues designed MLR, partial least square regression (PLS), and back-propagation (BP) network-based e-nose system for quality prediction of kiwifruit (Song et al. 2014). According to the study, when the firmness, soluble solid content, and pH of the kiwifruit in cold storage were reduced, the change in volatile aromatic compounds was also observed. In continuation of this, China-based scientists team used PEN2 based e-nose and integrated it with metal oxide sensor to detect five different quality parameters in strawberry juice (Qiu et al. 2015). They also used e-tongue to check the acidity and basicity on the basis of glucose, citrate, caffeine, and salt presence and significant accuracy of 98% and 99% for e-tongue and e-nose were obtained. Similarly, metal oxide sensing platform integrated LDA system was used to established quality indices for stored banana fruit (Sanaeifar et al. 2016). They discriminate the banana fruit on the basis of TA, pH, and total soluble solids at ripening and senescence stage and demonstrated decrease and increase in pH in both stages respectively and correlate the pH change with respect to malic acid. Zhang and colleagues designed extreme learning machine (ELM) and partial least squares (PLS) based e-nose to determine firmness, pH, and total soluble solids in stored cherry tomato fruit (Feng et al. 2018). The ELM model system showed significant discrimination property in which firmness of the fruits steadily decreased till 25 days, whereas pH and total soluble solids increase till 5 days then gradually decreased. Thus, eatable, and non-eatable quality of stored cherry tomato fruit can be easily determined.

For the real-time monitoring of grape quality in logistics chain, Indonesia-based research team developed a colorimetric pH sensor-based package indicator label (Kuswandi and Murdyaningsih 2017). When grapes exposed to chlorophenol red (CPR) fabricated sensing film, the sensitive CPR confirmed the changes in pH, TA, and total soluble solids in ripening grapefruit and change in color directly observed by naked eyes. Recently, Kurnianto and colleagues used betacyanin extracts from dragon fruit's (*Caesalpina sappan L.*) peel as a pH indicator to monitor banana deterioration (Kurnianto et al. 2020). The sensing platform was sensitive toward pH, texture and respiration rate, CO_2, and colorimetric change from yellow to dark red determined the spoilage of banana. Although, Indonesia-based research team has previously designed bromophenol blue (BPB) tagged bacterial cellulose film-based smart packaging to monitor mangoes Arummanis (*Mangifera indica L.* var. Arummanisa) (Dirpan et al. 2018). The study evaluated changes in pH, TA, and total soluble solids and colorimetric changes of tag from blue to green confirmed the gradual ripening of mangoes.

15.3.3 Aroma or Flavor Sensing in Fruit Quality Assessment

Aroma or flavor is the perception of aromatic volatile compounds by olfactory system or taste buds in which phenolic, few aldehydes, and triterpenes compounds provide bitterness whereas soluble sugar and organic acids are responsible for sweet and sour taste (Sánchez-Rodríguez et al. 2019; Zhu et al. 2020). Been as a precursor of volatile compounds (VOC), fatty acids are indirectly linked to fruit quality traits (Zhu et al. 2020), whereas amino acids induced polyphenol oxidases reaction causes fruit browning (Mertens et al. 2019). Despite several analytical advantages, aroma-based sensing is difficult to study because of the mixture of volatile compounds and chemical sensations. Yet recent studies on new sensor or e-nose development have improved the fruit quality monitoring.

For on-field sensing, Malaysia-based research team designed MIP-based QCM sensor for detection of α-pinene volatile compound for mango ripening confirmation (Hawari et al. 2013). They evaluated the freshness of mango fruit in 27 ± 2 °C and $55 \pm 10\%$ humidity for seven weeks and significant results were obtained with higher sensitivity and selectivity toward target volatile molecules. Similarly, Ghatak and colleagues have developed molecularly imprinted polymer (MIP) fabricated QCM sensing platform and reported 0.16 Hz/ppm sensitivity of the sensor toward d-limonene from ripped mango (Ghatak et al. 2019a). This research team has also designed MIP-based (3-Aminopropyl) trimethoxysilane (APTMS) fabricated QCM sensing platform for selective detection of furaneol from stored mango (Ghatak et al. 2019b). This sensor exhibited 79.70% selectivity and 1.23 ppm resolution in the range of 100–1000 ppm concentration of furaneol.

In another study, China-based research team developed four different enzymatic sensors for volatile compound detection in real fruit samples (Hou et al. 2014). They fabricated QCM sensing platform with ethyl cellulose, cellulose acetate, 1,2-dioleoyl-sn-glycero-3-[phosphor-L-serine], and galactosylceramide for detection of isobutyl alcohol, ethyl acetate, and ethylene VOCs. The result indicated linear change in singling in the range of 5–25 ppm. In another study, Ali and colleagues have also designed QCM sensing platform for efficient detection of β- Caryophyllene (β-CF) for mango quality evaluation (Ali et al. 2018). They fabricated QCM platform with embedded 2-mercaptobenzothiazole as receptor for β-CF and interaction between ring structures of thiolated sensing matrix and terpene caused change in resonant frequency in the range of 5–1000 ppm analyte concentration and LOD of 0.32 ppm. This research team has previously designed PDMS fabricated QCM sensor 3-carene gas detection from ripped mango fruit (Ali et al. 2016). The hydrophilic property of CH_3 group on -$OSi(CH_3)_2$- of PDMS and hydrophobic nature of 3-carene allows the absorption-based sensing in the range of 10–1000 ppm with sensitivity of 0.06 Hz/ppm. For further improvement, they also studied absorptive binding of ocimene (OCM) onto large molecular voids containing mustard-based QCM sensing platform (Ghatak et al. 2019c). The developed sensor showed excellent sensitivity of 0.276 Hz/ppm with LOD of 1.04 ppm in mango samples. Not stopping these studies, Ali and colleagues developed ethyl cellulose fabricated QCM sensor for β-myrcene VOC detection in mango quality monitoring (Ali et al. 2019). They evaluated sensitivity of 0.1 Hz/ppm due to the strong binding of ethyl ether groups from ethyl cellulose with mango specific β-myrcene in fruit quality monitoring.

To develop smart packaging material, Brazil-based research team used low-cost polyaniline film as a sensing material gas sensor to detect different aromas from apples, grapes, and strawberry fruit samples (Tiggemann et al. 2016). The HCl doped polyaniline layer allowed transferring from or to the active species of VOC with sensitivities in 75–125% range. The sensing film reported LOD of 0.081 ppm, 0.46 ppm, and 0.621 ppm for strawberries, grapes, and apples in the linear range of 40–200 ppm.

Till date several researches have been conducted on developing e-nose and smart packaging for real-time fruit quality monitoring and aroma-based sensing is proved as excellent target analyte. In this direction Chile-based research team designed "single point" algorithm (SACMI)-based e-nose for evaluation of peaches stored in cold storage (Infante et al. 2008). They compared four types of peaches on the basis of TA, total soluble solid, and aroma at 0 °C and 90% RH; and reported decline in TA/total soluble solid ratio and aroma till 42 days. In another study, Lu and colleagues have generated sensitive e-nose for 1-MCP and calcium coated apple fruit (Lu et al. 2018). The combination of 1-MCP and calcium are in use to improve volatile aroma and lowers fruit ripening, therefore scientists used PEN3 portable e-nose consisting 10 different sensing arrays and resulted in high-quality aroma detection till ~ 90 days. Similarly, a study was conducted in which several pattern recognition system modeling was combined to discriminate good quality saffron, obtained from *Crocus sativus L.* (Iridaceae) (Kiani et al. 2016). The PCA model revealed 11 different groups of aroma associated with 11 distinct saffron samples and this further classified into five quality classes on the basis of hierarchical clustering analysis (HCA) and MLP model system. Thus, the system was 100% successful in differentiating saffron samples and proposed to utilize as aroma quality control system.

Recent studies also combined e-nose with sensors to improve sensitivity and reliability toward aroma detection. In this direction, China-based research integrated e-nose with surface acoustic wave resonator (SAWR) for quality evaluation of kiwifruit (*Actinidia chinensis*) (Wei and Guohua 2015). The PCA-based method discriminated kiwifruits with different storage time on the basis of aroma-induced quality and SAWR frequency analysis modeling system exhibited excellent qualitative discrimination from all time zoned groups with high accuracy and high regression coefficient.

15.4 Commercially Available Sensors in Fruit Monitoring

Biosensors have been grown from a tiny niche laboratory discipline since the 70's to multidollar industrial market in today's world. However, this growth represents in a very small fraction of agricultural and industries level, yet there are few effective seining platforms available in market. Few smartphone-based techniques, such as SCiO was developed by Consumer Physics Ltd., Israel, which work on the basis of infrared technology and provide ripeness and nutritional values of fruits ("This Pocket-Sized Sensor Will Tell You When Fruit Is Ripe" 2014; "What Happened When We Took the SCiO Food Analyzer Grocery Shopping" 2017). The SCiO is car-clicker-sized tiny optical sensor that captures the molecular footprint when the fruit was scanned and SCiO's in-house mobile app provides barcode-like readout. Clarifruit and Fresh Fruit Detector—Check Fruits Quality by Dragster Production are the other smartphone-based application for real-time fruit quality assessment on the basis of texture and soluble solid contents. A smart ripening label, namely, RipeSense® is also proposed by RipeSense, Auckland, New Zealand that worked on ethylene sensing to communicate the degree of ripening by change in color of label.

15.5 Outlook and Conclusions

With current economic status and improved lifestyle, consumers are aware of hands-on fruit quality monitoring and prefer nutritional values and healthy fruit consumption. The next-generation trends in commercialization also entered daily life for effective and quick quality assessment system. Unfortunately, effective portable sensors and smart packaging or e-nose are very limited in market and compromise consumer goods. The major reasons for this shortcoming can be additional costs and acceptance of sensing tags by dealers or manufacturer. Though, nobody can deny the advantages of these sensing systems in fruit quality monitoring.

The success of fruit quality assessment approaches critically relies on sensitivity and selectivity of the sensing platform. In this scenario, e-noses have futuristic position for real time monitoring compared to on-field static sensing technology. Conventionally used sensing approaches measure the quality of fruits in bulk therefore intrinsic variability of fruit will be negotiable for quality assessment, hence individual quality measurement can be performed with e-nose and smart/intelligent packaging.

In order to fulfill the requirement, next-generation RFID tags and nanotechnology-based sensing platform are generated which are linked with multimodel pattern recognition models.

The practical implementation of reliable sensors and indicator labels in active and intelligent packaging, along with the development of next-generation sensing methods for freshness investigation of fruits, is growing field of study. As we discussed in this chapter, significant progresses have been made till date including time–temperature indicator (TTI), pH, and aroma-based indictors from on-site to consumer end and on-field fruit quality assessment. But there are still major efforts need to make to apply the functionality of these sensing devices in realistic settings. The major challenges in this scenario are the molecular complexity of fruits and difficulty to identify accurate markers for fruit ripening or spoilage. On other hand, on-field fruit quality monitoring still requires fruit sample pre-treatment. Most successful approach of fruit quality monitoring till now is the assessment of VOCs such as ethylene and amines, however no effective and hurdle-free approach has been selected for on-field assessment.

Future development in fruit quality monitoring needs to be directed toward lowering detection limit, for say ppm and improving the possibility of accurate marker detection upon simple contact with fruit sample. The futuristic benefits of polyaniline due to higher stability as sensing element also need to be promoted in e-nose and smart packaging material. The higher stability and low reactivity with fruits of packaging material will be very helpful in fruit quality assessment in stored packaged fruit for long time. Maintaining variable environmental parameters in fruit logistics application is another challenge that needed to be addressed. Recently, the use of antimicrobial agents and nanomaterials in sensing and packaging technology geared up the fruit quality assessment approaches and positive impact has been achieved. However, toxicity and safety concern of nanomaterials limit their application in food industry. These sensing technologies are still in initial proof of concept stage and significant improvement and efforts should be made for worldwide launch and acceptability of sensing packaging material and indicator tags into a marketable product to meet positive health concerns. Additionally, effective safety regulations need to be issued by governmental health authorities and manufacturers should strictly follow the use of these sensing platforms in their fruit products.

Acknowledgements Dr. Vinita Hooda is thankful to the Department of Science and Technology (DST), Ministry of Science and Technology, Government of India grant under its "Fund for Improvement of Science & Technology Infrastructure (FIST)" scheme (No. SR/FST/LS1-529/2012(C)) (http://www.fist-dst.org/) and Haryana State Council for Science, Innovation, and Technology (HSCSIT, Endst. No. HSCSIT/R&D/2020/474) for providing financial support to work in the field of Agriculture Nanotechnology.

References

Agarwal M, Balachandran MD, Shrestha S, Varahramyan K (2012) SnO_2 nanoparticle-based passive capacitive sensor for ethylene detection. Journal of Nanomaterials, 2012, 145406

Ali SB, Ghatak B, Debabhuti N, Sharma P, Ghosh A, Tudu B, Bhattacharya N, Bandyopadhyay R (2018) Detection of β-caryophyllene in mango using a quartz crystal microbalance sensor. Sensors and Actuators B: Chemical, 255, 3064–3073

Ali SB, Ghatak B, Debabhuti N, Pal S, Sharma P, Tudu B, Bandyopadhyay R (2019) Sensitive detection of β-Myrcene in mango using ethyl cellulose modified quartz crystal microbalance sensor. Materials Today: Proceedings 18:1025–1032

Ali SB, Ghatak B, Gupta SD, Debabhuti N, Chakraborty P, Sharma P, Ghosh A, Tudu B, Mitra S, Sarkar MP, Bhattacharyya N (2016) Detection of 3-carene in mango using a quartz crystal microbalance sensor. Sensors and Actuators B: Chemical 230:791–800

Anthon GE, LeStrange M, Barrett DM (2011) Changes in pH, acids, sugars and other quality parameters during extended vine holding of ripe processing tomatoes. J Sci Food Agric 91(7):1175–1181

Batista-Silva W, Nascimento VL, Medeiros DB, Nunes-Nesi A, Ribeiro DM, Zsögön A, Araújo WL (2018) Modifications in organic acid profiles during fruit development and ripening: correlation or causation? Frontiers in Plant Science 9:1689

Benedetti S, Buratti S, Spinardi A, Mannino S, Mignani I (2008) Electronic nose as a non-destructive tool to characterise peach cultivars and to monitor their ripening stage during shelf-life. Postharvest Biol Technol 47(2):181–188

Besar K, Dailey J, Katz HE (2017) Ethylene detection based on organic field-effect transistors with porogen and palladium particle receptor enhancements. ACS Appl Mater Interfaces 9(2):1173–1177

Bigiani L, Zappa D, Comini E, Maccato C, Gasparotto A, Barreca D (2020) Manganese oxide nanoarchitectures as chemoresistive gas sensors to monitor fruit ripening. J Nanosci Nanotechnol 20(5):3025–3030

Burlachenko J, Kruglenko I, Snopok B, Persaud K (2016) Sample handling for electronic nose technology: State of the art and future trends. TrAC Trends Anal Chem 82:222–236

Chauhan N, Jain U, Soni S (2019) Sensors for Food Quality Monitoring. In Nanoscience for sustainable agriculture (pp 601–626). Springer, Cham

Chauhan N, Soni S, Agrawal P, Balhara YPS, Jain U (2020) Recent advancement in nanosensors for neurotransmitters detection: Present and future perspective. Process Biochem 91:241–259

Chen WY, Yermembetova A, Washer BM, Jiang X, Shuvo SN, Peroulis D, Wei A, Stanciu L (2020) Selective detection of Ethylene by MoS_2–Carbon nanotube networks coated with Cu (I)–Pincer complexes. ACS Sensors 5(6):1699–1706

Coulet PR, Blum LJ (Eds.) (2019) Biosensor principles and applications. CRC Press

Cui Y, Lai B, Tang X (2019) Microbial fuel cell-based biosensors. Biosensors 9(3):92

Davis F, Altintas Z (2018) General introduction to biosensors and recognition receptors. Applications in Health Care Diagnostics, Biosensors and Nanotechnology, pp 1–15

Dirpan A, Latief R, Syarifuddin A, Rahman ANF, Putra RP, Hidayat SH (2018) The use of colour indicator as a smart packaging system for evaluating mangoes Arummanis (*Mangifera indica* L. var. Arummanisa) freshness IOP Conf. Ser. Earth Environ. Sci, 157, 012031

Dow ABA, Sklorz A, Lang W (2011) A microfluidic preconcentrator for enhanced monitoring of ethylene gas. Sens Actuators, A 167(2):226–230

Du D, Wang J, Wang B, Zhu L, Hong X (2019) Ripeness prediction of postharvest kiwifruit using a MOS e-nose combined with chemometrics. Sensors 19(2):419

Esser B, Schnorr JM, Swager TM (2012) Selective detection of ethylene gas using carbon nanotube-based devices: Utility in determination of fruit ripeness. AngewandteChemie International Edition 51(23):5752–5756

Etienne A, Génard M, Lobit P, Mbeguié-A-Mbéguié D, Bugaud C (2013) What controls fleshy fruit acidity? A review of malate and citrate accumulation in fruit cells. J Exp Bot 64(6):1451–1469

Feng L, Zhang M, Bhandari B, Guo Z (2018) A novel method using MOS electronic nose and ELM for predicting postharvest quality of cherry tomato fruit treated with high pressure argon. Comput Electron Agric 154:411–419

Gao J, Zhang Y, Li Z, Liu M (2020) Role of ethylene response factors (ERFs) in fruit ripening. Food Quality and Safety 4(1):15–20

Ghatak B, Ali SB, Naskar H, Tudu B, Pramanik P, Mukherji S, Bandyopadhyay R (2019a) Selective and sensitive detection of limonene in mango using molecularly imprinted polymer based quartz crystal microbalance sensor. In 2019 IEEE International Symposium on Olfaction and Electronic Nose (ISOEN) (pp 1–3). IEEE

Ghatak B, Naskar H, Ali SB, Tudu B, Pramanik P, Mukherji S, Bandyopadhyay R (2019b) Development of furaneol imprinted polymer based QCM sensor for discrimination of artificially and naturally ripened mango. In 2019 IEEE International Symposium on Olfaction and Electronic Nose (ISOEN) (pp 1–3). IEEE

Ghatak B, Ali SB, Tudu B, Pramanik P, Mukherji S, Bandyopadhyay R (2019c) Detecting Ocimene in mango using mustard oil based quartz crystal microbalance sensor. Sensors and Actuators B: Chemical 284:514–524.

Gliszczyńska-Świgło A, Chmielewski J (2017) Electronic nose as a tool for monitoring the authenticity of food. A review. Food Analytical Methods 10(6):1800–1816

Gu N, Liu S (2020) Introduction to biosensors. Journal of Materials Chemistry B 8(16):3168–3170

Han S, Zhuang X, Shi W, Yang X, Li L, Yu J (2016) Poly (3-hexylthiophene)/polystyrene (P3HT/PS) blends based organic field-effect transistor ammonia gas sensor. Sensors and Actuators B: Chemical 225:10–15

Hawari HF, Samsudin NM, Shakaff AM, Wahab Y, Hashim U, Zakaria A, Ghani SA, Ahmad MN (2013) Highly selective molecular imprinted polymer (MIP) based sensor array using interdigitated electrode (IDE) platform for detection of mango ripeness. Sensors and Actuators B: Chemical 187:434–444

Huang H, Yu H, Xu H, Ying Y (2008) Near infrared spectroscopy for on/in-line monitoring of quality in foods and beverages: A review. J Food Eng 87(3):303–313

Infante R, Farcuh M, Meneses C (2008) Monitoring the sensorial quality and aroma through an electronic nose in peaches during cold storage. J Sci Food Agric 88(12):2073–2078

Jain U, Soni S, Balhara YPS, Khanuja M, Chauhan N (2020) Dual-layered nanomaterial-based molecular pattering on polymer surface biomimetic impedimetric sensing of a bliss molecule. Anandamide Neurotransmitter. ACS omega 5(19):10750–10758

Jawaheer S, White SF, Rughooputh SDDV, Cullen DC (2003) Development of a common biosensor format for an enzyme based biosensor array to monitor fruit quality. Biosens Bioelectron 18(12):1429–1437

Jedermann R, Pötsch T, Lloyd C (2014) Communication techniques and challenges for wireless food quality monitoring. Philosophical Transactions of the Royal Society A: Mathematical, Physical and Engineering Sciences 372(2017):20130304

Jochum T, Rahal L, Suckert RJ, Popp J, Frosch T (2016) All-in-one: A versatile gas sensor based on fiber enhanced Raman spectroscopy for monitoring postharvest fruit conservation and ripening. Analyst 141(6):2023–2029

Juncai H, Yaohua H, Kangquan G (2014) Application of modified quartz crystal sensors: detection of isobutyl alcohol, ethyl acetate and ethylene. International Journal of Agricultural and Biological Engineering 7(5):71–77

Kantor DB, Hitka G, Fekete A, Balla C (2008) Electronic tongue for sensing taste changes with apricots during storage. Sensors and Actuators B: Chemical 131(1):43–47

Karakaya D, Ulucan O, Turkan M (2020) Electronic nose and its applications: A survey. International Journal of Automation and Computing 17:1–31

Kathirvelan J, Vijayaraghavan R (2017) An infrared based sensor system for the detection of ethylene for the discrimination of fruit ripening. Infrared Phys Technol 85:403–409

Kathirvelan J, Vijayaraghavan R, Thomas A (2017) Ethylene detection using TiO2–WO3 composite sensor for fruit ripening applications. Sensor Review 37(2):147–154

Kiani S, Minaei S, Ghasemi-Varnamkhasti M (2016) A portable electronic nose as an expert system for aroma-based classification of saffron. Chemometr Intell Lab Syst 156:148–156

Kodogiannis VS (2017) Application of an electronic nose coupled with fuzzy-wavelet network for the detection of meat spoilage. Food Bioprocess Technol 10(4):730–749

Kurnianto MF, Poerwanto B, Wahyono A, Apriliyanti MW, Lestari IP (2020) Monitoring of banana deteriorations using intelligent-packaging containing brazilien extract (*Caesalpina sappan* L.). In IOP conference series: Earth and environmental science (Vol 411, No 1, p 012043). IOP Publishing

Kuswandi B, Murdyaningsih EA (2017) Simple on package indicator label for monitoring of grape ripening process using colorimetric pH sensor. Journal of Food Measurement and Characterization 11(4):2180–2194

Li B, Li M, Meng F, Liu J (2019) Highly sensitive ethylene sensors using Pd nanoparticles and rGO modified flower-like hierarchical porous α-Fe_2O_3. Sensors and Actuators B: Chemical 290:396–405

Li Z, Zhu Y, Zhang W, Xu C, Pan Y, Zhao Y (2017) A low-cost and high sensitive paper-based microfluidic device for rapid detection of glucose in fruit. Food Anal Methods 10(3):666–674

Liu H, Ge J, Ma E, Yang L (2019) Advanced biomaterials for biosensor and theranostics. In Biomaterials in translational medicine (pp 213–255). Academic Press

Lu X, Meng G, Jin W, Gao H (2018) Effects of 1-MCP in combination with Ca application on aroma volatiles production and softening of 'Fuji'apple fruit. Sci Hortic 229:91–98

Maftoonazad N, Ramaswamy H (2019) Design and testing of an electrospun nanofiber mat as a pH biosensor and monitor the pH associated quality in fresh date fruit (Rutab). Polym Testing 75:76–84

Mallick M, Das J (2017) Graphene oxide based ethylene gas sensor for e-nose application. Invertis Journal of Science & Technology 10(4):178–182

Mallick M, Hossain SM, Das J (2018) Graphene oxide based fruit ripeness sensing e-nose. Materials Today: Proceedings 5(3):9866–9870

Mehrotra P (2016) Biosensors and their applications–A review. Journal of Oral Biology And Craniofacial Research 6(2):153–159

Mertens N, Mai F, Glomb MA (2019) Influence of nucleophilic amino acids on enzymatic browning systems. J Agric Food Chem 67(6):1719–1725

Mustafa F, Andreescu S (2018) Chemical and biological sensors for food-quality monitoring and smart packaging. Foods 7(10):168

Omar AF, Matjafri MZ (2013) Specialized optical fiber sensor for nondestructive intrinsic quality measurement of *Averrhoa Carambola*. Photonic Sensors 3(3):272–282

Pech JC, Purgatto E, Bouzayen M, Latché A (2018) Ethylene and fruit ripening. Annual Plant Reviews online, 275–304

Putri VJ, Warsiki E, Syamsu K, Iskandar A (2019) Application nano zeolite-molybdate for avocado ripeness indicator. In IOP Conference Series: Earth and Environmental Science (Vol 347, No 1, p 012063). IOP Publishing

Qiu S, Wang J, Gao L (2015) Qualification and quantisation of processed strawberry juice based on electronic nose and tongue. LWT-Food Science and Technology 60(1):115–123

Ramgir NS (2013) Electronic nose based on nanomaterials: Issues, challenges, and prospects. International Scholarly Research Notices, 2013, Article ID 941581, 21 pages

Sanaeifar A, Mohtasebi SS, Ghasemi-Varnamkhasti M, Ahmadi H (2016) Application of MOS based electronic nose for the prediction of banana quality properties. Measurement 82:105–114

Sánchez-Rodríguez L, Ali NS, Cano-Lamadrid M, Noguera-Artiaga L, Lipan L, Carbonell-Barrachina ÁA, Sendra E (2019) Flavors and Aromas. Postharvest Physiology and Biochemistry of Fruits and Vegetables, 385–404

Satpute MR, Jagdale SM (2016) Automatic fruit quality inspection system. In 2016 International Conference on Inventive Computation Technologies (ICICT) (Vol 1, pp 1–4). IEEE

Shi H, Zhang M, Adhikari B (2018) Advances of electronic nose and its application in fresh foods: A review. Crit Rev Food Sci Nutr 58(16):2700–2710

Sklorz A, Janßen S, Lang W (2012a) Gas chromatograph based on packed μGC-columns and μ-preconcentrator devices for ethylene detection in fruit logistic applications. Procedia Engineering 47:486–489
Sklorz A, Schafer A, Lang W (2012b) Merging ethylene NDIR gas sensors with preconcentrator-devices for sensitivity enhancement. Sensors and Actuators B: Chemical 170:21–27
Song XQ, Ren YM, Zhang YY, Li Y, Peng GY, Ma T (2014) Prediction of kiwifruit quality during cold storage by electronic nose. Food Sci 35(20):230–235
Soni S, Ruhela RK, Medhi B (2016) Nanomedicine in central nervous system (CNS) disorders: A present and future prospective. Advanced Pharmaceutical Bulletin 6(3):319
Su M, Ye Z, Zhang B, Chen K (2017) Ripening season, ethylene production and respiration rate are related to fruit non-destructively-analyzed volatiles measured by an electronic nose in 57 peach (*Prunus persica* L.) samples. Emirates Journal of Food and Agriculture, 807–814
This Pocket-Sized Sensor Will Tell You When Fruit Is Ripe. 02 July, 2014. By: Lauren Rothman. Available on: [accessed on 29 June, 2020] https://modernfarmer.com/2014/07/handheld-device-might-indispensible-shopping-future/
Tiggemann L, Ballen S, Bocalon C, Graboski AM, Manzoli A, de Paula Herrmann PS, Steffens J, Valduga E, Steffens C (2016) Low-cost gas sensors with polyaniline film for aroma detection. J Food Eng 180:16–21
Tolentino MAKP, Albano DRB, Sevilla FB III (2018) Piezoelectric sensor for ethylene based on silver (I)/polymer composite. Sensors and Actuators B: Chemical 254:299–306
Tothill IE (2001) Biosensors developments and potential applications in the agricultural diagnosis sector. Comput Electron Agric 30(1–3):205–218
Tzortzakis N, Chrysargyris A (2017) Postharvest ozone application for the preservation of fruits and vegetables. Food Reviews International 33(3):270–315
Wang H, Peng J, Xie C, Bao Y, He Y (2015) Fruit quality evaluation using spectroscopy technology: A review. Sensors 15(5):11889–11927. https://doi.org/10.3390/s150511889
Wang LP, Jin Z, Luo T, Ding Y, Liu JH, Wang XF, Li MQ (2019) The detection of ethylene using porous ZnO nanosheets: utility in the determination of fruit ripeness. New J Chem 43(8):3619–3624
Wasilewski T, Migoń D, Gębicki J, Kamysz W (2019) Critical review of electronic nose and tongue instruments prospects in pharmaceutical analysis. Anal Chim Acta 1077:14–29
Wei L, Guohua H (2015) Kiwi fruit (*Actinidia chinensis*) quality determination based on surface acoustic wave resonator combined with electronic nose. Bioengineered 6(1):53–61
Wei Y, Chang J, Lian J, Liu T (2015) A coal mine multi-point fiber ethylene gas concentration sensor. Photonic Sensors 5(1):67–71
What Happened When We Took the SCiO Food Analyzer Grocery Shopping. 14 Mar 2017. By Tekla S. Perry. Available on: [accessed on 29 June, 2020] https://spectrum.ieee.org/view-from-the-valley/at-work/start-ups/israeli-startup-consumer-physics-says-its-scio-food-analyzer-is-finally-ready-for-prime-timeso-we-took-it-grocery-shopping
WHO (2017) Food Safety; https://www.who.int/news-room/fact-sheets/detail/food-safety
Yang WZ, Jiao Y (2017) The international research progress of Ammonia (NH3) emissions and emissions reduction technology in farmland ecosystem. In IOP Conf. Series: Earth and Environmental Science (Vol 59, p 012024)
Yang X, Chen J, Jia L, Yu W, Wang D, Wei W, Li S, Tian S, Wu D (2020) Rapid and non-destructive detection of compression damage of yellow peach using an electronic nose and chemometrics. Sensors 20(7):1866
Yousefi H, Su HM, Imani SM, Alkhaldi KM, Filipe CD, Didar TF (2019) Intelligent food packaging: A review of smart sensing technologies for monitoring food quality. ACS Sensors 4(4):808–821
Zhang H, Wang J (2009) Evaluation of peach quality attribute using an electronic nose. Sens. Mater 21:419–431
Zhu C, Lu Q, Xianyan Z, Li J, Yue J, Wang Z, Pan S (2020) Metabolic variations of organic acids, amino acids, fatty acids, and aroma compounds in the pulp of different pummelo varieties. LWT, 109445

Zhu YC, Mei LP, Ruan YF, Zhang N, Zhao WW, Xu JJ, Chen HY (2019) Enzyme-based biosensors and their applications. In Advances in Enzyme Technology (pp 201–223). Elsevier

Chapter 16
Lateral Flow Assays for Food Authentication

Despina P. Kalogianni

Abstract Foodstuffs with a particularly high nutritional and economic value are often adulterated with lower value products. Products of protected geographical origin are also included in the list of the most prone to fraud products, having a negative impact on the national economy. This contravenes the rights of consumers and producers, while it often jeopardizes public health. Due to these reasons, rapid and reliable tests, as well as state control mechanisms, should be developed to detect food adulteration and protect consumers and local producers from fraud. Lateral flow assays (strips) have played an important role in this direction and for on-site applications. This chapter reports all the DNA and protein-based lateral flow assays that have been constructed so far for food adulteration detection. The advantages of these assays are the rapid and one-step analysis (within a few minutes), the simplicity, the high detectability and reproducibility and the portability, maintaining also a low-cost per sample analysis.

Keywords Strip · Biosensor · Dipstick · Adulteration · Authenticity · Nanoparticles

16.1 Introduction

Food adulteration has been a global concern in current times for all the stakeholders involved in the food industry (Zhao et al. 2020a; Hong et al. 2017). The term adulteration refers mostly (95%) to the substitution of a food ingredient with another ingredient of a much lower nutritional or quality value and of lower cost, for economical profit. Usually the substitution is difficult to be recognized by the consumers or routine analytical techniques. Apart from the financial fraud, food adulteration may have a great impact on public health, such as severe allergic reactions (Magiati et al. 2019; Böhme et al. 2019; Abbas et al. 2018). For example, substitution of mustard oil with that of argemone causes oxidative stress and ultimate death of red blood

D. P. Kalogianni (✉)
Department of Chemistry, University of Patras, Rio, Patras 26504, Greece
e-mail: kalogian@upatras.gr

R. N. Pudake et al. (eds.), *Biosensors in Agriculture: Recent Trends and Future Perspectives*, Concepts and Strategies in Plant Sciences,
https://doi.org/10.1007/978-3-030-66165-6_16

cells, resulting in serious health problems (Lo and Shaw 2018). For these reasons, people have the right to be protected from fraud, health risks, and producers from illicit antagonism.

The most vulnerable to adulteration are products of high nutritional and economic value such as meat and fish, olive oil, dairy products, honey, spices and several herbs, grain-based foods and organic foods, fruit juices, wine and other alcoholic beverages, coffee and special varieties of tea (Hong et al. 2017).

Sensitive and robust analytical methods as well as biosensors have to be developed for food authenticity testing of varieties of food products checking safety, quality, and religious issues. Moreover, reducing unfair competition is also a concern, while identification of the geographical origin of special food products has also gained the interest of the consumers (Wadood et al. 2020). Traditional methods used for food adulteration detection include gas/liquid chromatography, single or in combination with mass spectrometry, nuclear magnetic resonance spectroscopy, infrared and mid-infrared spectroscopy, Raman spectroscopy, inductively coupled plasma-atomic emission (elemental analysis) and fluorescence spectroscopy, electrophoresis, isotope ratio analysis, and immunoassays. Mass spectrometry has been extensively used in most of the categories of adulterated products, while the advantages of some spectroscopic techniques are the fast and non-invasive analysis with no sample preparation (Hong et al. 2017; Wadood et al. 2020). The drawbacks of these techniques are mainly the long analysis time, the extensive and laborious protocols, and the expensive instruments required (Bansal et al. 2017). Some of these techniques use different chemical or metabolic profiles for species discrimination. However, chemical profiles are strongly affected by environmental changes, harvest time, storage conditions, and different processing procedures (Woolfe and Primrose 2004). Moreover, proteins are not stable during thermal food processing or in case of plants, different proteins are secreted in different parts of the plant. For these reasons, DNA-based techniques are preferred, because DNA has high thermal stability, it is present in all cells of the organisms and not affected by environmental conditions, therefore it overcomes many limitations of the aforementioned techniques (Zhao et al. 2020a; Böhme et al. 2019). Molecular authentication is also a powerful tool for the most specific discrimination of closed relative species or cultivars of different geographical origin. However, DNA-based techniques are also affected by some factors including DNA degradation and PCR inhibitors. The researchers are trying to overcome these limitations by introducing more effective DNA isolation protocols, while amplification targets small-size DNA sequences (Lo and Shaw 2018).

16.2 Lateral Low Assays Used in Food Authentication (LFAs)

Among other methods used for food authenticity testing, lateral flow assays (LFAs) have been constructed for the identification and authentication of various food products, providing simplicity, rapidness, one-step analysis, user-friendly format, long-term stability, high specificity, and detectability. LFAs are also cost-effective, as they do not require expensive instruments, disposable, and portable in some cases (Sajid et al. 2015). Some DNA-based LFAs are also universal and can be applied for the detection of any DNA sequence, because the molecular recognition is performed prior to the application to the LFA. Both DNA-based LFAs and lateral flow immunoassays have been reported for food authentication (Fig. 16.1). LFAs are divided into DNA-based and protein-based analytical systems, whereas the former possessing the greater part of the applications.

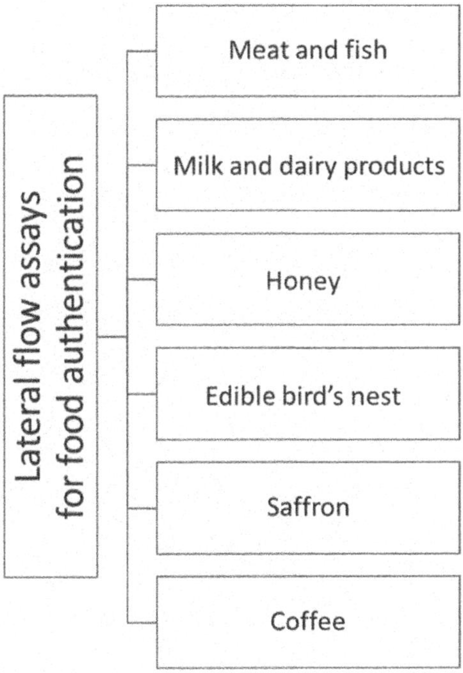

Fig. 16.1 The lateral flow assays developed for the detection of adulteration in specific food products

16.2.1 Principle of the Lateral Flow Assays

Lateral flow assays are constructed in a strip-based format and consist of four discrete parts: an immersion pad, a conjugate pad, a diagnostic membrane, and an adsorbent pad (Magiati et al. 2018; Spyrou et al. 2016; Sajid et al. 2015). A schematic illustration of a lateral flow assay is presented in Fig. 16.2. The four parts of the strip are summarized below:

Immersion pad: It is usually made of cellulose and dipped into the appropriate developing solution. It enables the smooth, continuous, and homogeneous flow of the developing solution.

Conjugate pad: It is made of glass fibers, where the reporters (usually colored nanoparticles conjugated to biorecognition molecules) and the sample are deposited on. This material must have no interaction with the sample and the reporters, while it should be immediately releasing the sample and reagents upon contact with the developing solution. For the preparation of dry-reagent strip format, reporters are

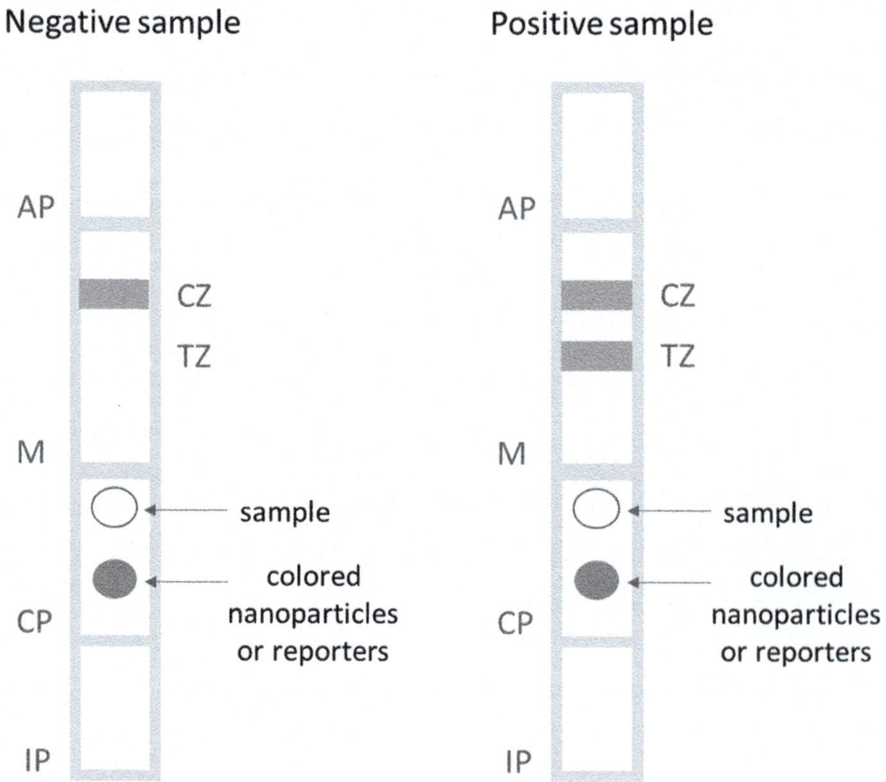

Fig. 16.2 Lateral flow assay schematic configuration in case of a negative (left) and a positive (right) sample. IP: immersion pad, CP: conjugate pad, M: membrane, AP: adsorbent pad, TZ: test zone, and CZ: control zone

dispensed and let to dry on the conjugate pad. The reagents must have long-term stability on the strip. Other materials used for conjugate pad are cellulose and polyesters.

Diagnostic membrane: It is usually made of nitrocellulose, available in different pore sizes. It is crucial for the sensitivity and detectability of the assay, providing good binding ability for the immobilization of biomolecules and minimizing non-specific binding. A colored test line and a control line are formed onto the membrane by immobilizing specific biorecognition molecules at these two areas. Again, biomolecules are sprayed and dried onto the membrane to construct the test and the control zone of the membrane.

Adsorbent pad: It consists of an absorbent material and enables the continuous flow of the developing solution and reagents, avoiding the backflow of the sample.

The four parts are attached on a plastic backing pad that serves as a solid support, with overlapping ends to enable the continuous flow on the strip. The reporters are usually conjugated with specific recognition biomolecules and deposited onto the conjugate pad of the strip. The sample is also added on the conjugate pad of the strip just above the reporters. The strip is then dipped into an appropriate developing solution. The solution is moving upwards due to capillary forces and removes the reporters and the sample in the same direction. The reporters interact with the sample and the complex is captured at the test line of the strip by other immobilized recognition molecules, through a label carried by the sample, forming a visual (colored) line, due to the accumulation of colored nanoparticles. The excess of the reporters is finally captured by specific immobilized molecules at the control line of the strip forming a second visual line and ensuring that the lateral flow assay works perfectly. So, in case of a positive result, both visual lines are formed and in the absence of the analyte, only the control line of the strip is shaped (Fig. 16.2). The concentration of all reagents, the constitution of the developing solution, as well as the molar ratio of coupling agents have to be carefully optimized in order to get the highest sensitivity and specificity of the assay, along with reliable and accurate results. Surface surfactants and denaturing agents are also used in the developing solution to minimize cross-reactivities, and to avoid false-positive outcomes. The developing solution must contain also the optimum salt concentration to enable hybridization in DNA-based LFAs and liquids with high viscosity, such as glycerol and sucrose, in order to gain the optimum mobility through the strip and to provide enough time for successful interaction between the reagents. Finally, diagnostic membranes with different pore sizes provide different liquid mobilities and have to be carefully selected.

LFAs provide qualitative or semi-quantitative analysis. The visualization of the lines is accomplished either by the naked eye when colored nanoparticles are used or by a digital camera or a smartphone in case of a fluorescent or a chemiluminescent signal. The images of the strips with the colored lines can be also obtained using a regular camera or a smartphone, while all captured images can be further processed by an image processing software. The intensity of the color of the test line is measured and compared to standards to get semi-quantitative results. Optical strip-readers are also available to get more accurate quantitative results, by also measuring the intensity of the colored line and using suitable image processing software (Kalogianni et al. 2006; Sajid et al. 2015; Dzantiev et al. 2014).

16.2.2 Sensing Systems and Labels

LFAs are available in different formats using different biorecognition molecules, labels, and detection systems. Sensitive and selective sensing systems have been exploited in LFAs for the detection of specific analytes. These systems are based on the strong and extremely specific and stable interaction between the following molecules: antigen (analyte)–antibody, streptavidin–biotin, avidin–biotin, anti-biotin antibody–biotin, anti-FITC antibody–FITC (fluorescein isothiocyanate), anti-fluorescein antibody–fluorescein, anti-digoxigenin antibody–digoxigenin, and anti-Cy5 antibody–Cy5 dye.

16.2.3 Nanoparticles and Nanoparticles Conjugates Used as Reporters in LFAs for Food Authentication Applications

Metallic nanoparticles, in general, have gained the great interest of the researchers because of their visual and electrical properties for applications in food analysis. Their characteristic properties are mostly dependent on size and shape. As their synthesis has become easier and of lower cost, they have been further used as a new analytical tool and applied to food authenticity testing. Gold nanoparticles (AuNPs) are the main nanoparticles used in LFAs developed for food authentication, as they are considered the key tool for visual detection. Gold nanoparticles are well known for their unique optical characteristics, having a bright red color that enables their visual detection even at low concentrations, as well as for their good stability, large surface to volume ratio, and capability of facile conjugation to several different biomolecules. The optical properties of gold nanoparticles have increased the detectability of LFAs, due to their extremely high molar absorption coefficient due to surface plasmon resonance. They have been extensively used in numerous applications, providing high sensitivity and detectability (Subara and Jaswir 2018; Sajid et al. 2015).

Gold nanoparticles have been conjugated to specific biorecognition molecules, in order to be applied for the detection of specific adulterants by LFAs. The substances that have been used for bioconjugation to the gold nanoparticles in LFAs for authentication purposes are the following:

- *Streptavidin*
- *Anti-biotin antibody*
- *Anti- digoxigenin antibody*
- *Anti-fluorescein (FAM) antibody*

- *Anti-fluorescein isothiocyanate (FITC) antibody*
- *Species-specific antibodies*
- *Species-specific DNA probes or other oligonucleotides.*

16.3 DNA-Based Lateral Flow Assays

As discussed above, DNA-based LFAs are the most preferred one especially because of their higher discrimination between close-related species compared to protein-based analysis. Most of the DNA-based techniques applied for food adulteration detection employ the amplification of genomic material using species-specific primers. Apart from real-time PCR that enables the detection and quantification of the PCR products in real time during each cycle, all other methods need the post-PCR identification of the amplified products. However, real-time PCR requires special and expensive instrumentation. LFAs are considered one of the most suitable alternatives for post-PCR analysis and have successfully integrated with amplification techniques for food authenticity assessments (Zhao et al. 2020a). DNA-based LFAs have been applied for the authenticity assessment of meat, fish, milk, saffron, and coffee.

16.3.1 Meat Authentication

Most of the developed DNA-based assays have been applied for the detection of meat adulteration, as meat is largely consumed and is one of the most susceptible to adulteration food products worldwide (Magiati et al. 2019; Rahmati et al. 2016). Meat adulteration mainly includes the adulteration of meat with other meat species of lower value and quality, meat from other geographic origin, processed meat, or non-meat ingredients, such as water or vegetable proteins (Hong et al. 2017). Meat adulteration arises substantial threats to public health and ethical issues in some cases (Böhme et al. 2019). Recently, Zhao and coworkers developed a lateral flow assay for the identification of turkey meat in meat products. Turkey-specific DNA sequences were amplified using specific primers, while the PCR products were labeled with biotin at one end and with fluorescein at the other end. The PCR products were directly applied onto the conjugate pad, along with gold nanoparticles as reporters that were conjugated to anti-fluorescein polyclonal antibodies. The strip was then immersed into the developing solution. A red line was formed at the test zone of the strip due to the immobilized streptavidin at this zone via biotin–streptavidin interaction, while the gold nanoparticles conjugates were attached to the PCR products via the interaction between the anti-fluorescein antibody and the fluorescein label. The excess of the gold nanoparticles conjugates was captured by a second antibody specific to anti-fluorescein antibody immobilized at the control zone of the strip forming a second line and ensuring the right strip function (Fig. 16.3). The LFA was specific

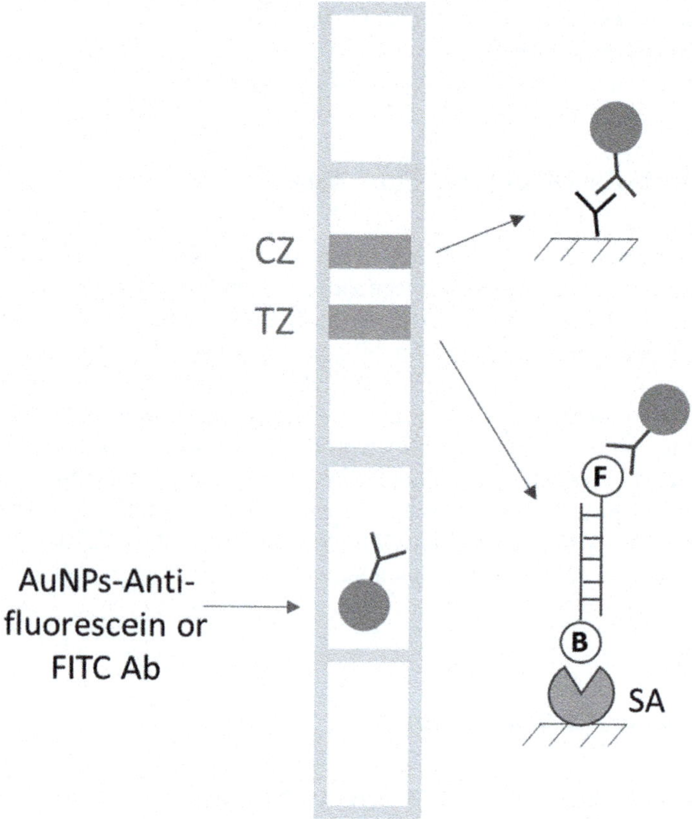

Fig. 16.3 The detection of species-specific meat products using PCR-amplified DNA sequences labeled with biotin and fluorescein (FAM) or fluorescein isothiocyanate (FITC) and gold nanoparticles coupled to anti-FAM or anti-FITC antibodies as reporters. AuNPs: gold nanoparticles, SA: streptavidin, B: biotin, F: fluorescein or fluorescein isothiocyanate, TZ: test zone, and CZ: control zone

to turkey meat among nineteen other meat samples and two plant species, while the detectability of the method was set at 0.1% of turkey meat in mixtures with beef meat samples. The detectability of the method was not affected by heat treatment of the meat mixtures in the same concentration. The analysis was completed in 5 min with good reproducibility evaluated by both intra- and inter-laboratory tests (Zhao et al. 2020a). The same configuration was also used for the identification of donkey meat in food samples with a limit of detection (LOD) of 0.001% w/w for raw and 0.01% w/w for cooked donkey meat in beef meat (Zhao et al. 2019a). The same assay was exploited as well for camel species identification with a limit of detection of 0.1% w/w of camel meat in beef for both the raw and cooked meat samples (Zhao et al. 2020b) and for 0.01% w/w of goose meat in uncooked binary mixtures and 0.1% w/w in cooked binary mixtures (Zhao et al. 2020c). In some of these approaches, fluorescein isothiocyanate (FITC) was used as label instead of fluorescein (FAM) along with the corresponding antibodies.

Another report involves the detection of pork meat in meat mixtures using PCR combined with LFA. Pork-specific sequences were amplified by PCR using a primer labeled with fluorescein. The PCR products were then heat-denatured and let to hybridize with a complementary probe labeled with biotin. The hybrids were loaded onto the conjugate pad along with streptavidin-gold nanoparticles conjugates. The strip was then dipped into the developing solution. A first red line was formed as the hybrids are captured by immobilized anti-fluorescein antibody at the test zone of the strip, while gold nanoparticles were accumulated there through streptavidin–biotin interaction. The excess of the gold nanoparticles conjugates was captured by biotinylated bovine serum albumin (BSA), producing a second red line at the control zone of the strip. The authors achieved a limit of detection of 10 fg of pork-specific DNA sequence and 0.01 of pork meat in meat mixtures of mutton, beef, chicken, goose, duck, horsemeat, and rabbit meats in equal quantities. The result was obtained within 3 min (Yin et al. 2020).

An LFA was developed for the on-site detection of horse and donkey meat. This LFA involved gold nanoparticles and tag-labeled multiplex loop-mediated isothermal amplification (LAMP) in order to simplify amplification and detection. The advantages of the LAMP technique were the rapid amplification of the DNA sequences, along with high specificity and sensitivity, performed in one step and at a constant temperature. Three LAMP products were produced carrying one biotin moiety at their 5' end. The three products were specific to donkey meat, to horse meat, and to an endogenous reference gene present in all mammals. The donkey-specific products carried a digoxigenin label, the horse-specific products a fluorescein isothiocyanate (FITC) label, and the endogenous gene a Cy5 label at their 3' end to allow discrimination with the LFA. The LAMP products were added onto the conjugate pad of the strip after amplification, and the strips were immersed again into the appropriate solution. Then, three red test lines were formed on the test zone of three single strips by immobilizing antibodies specific to digoxigenin, FITC, and Cy5, respectively, while the control zone was constructed in all strips by immobilized goat anti-mouse IgG. The captured LAMP products were detected by gold nanoparticles attached to mouse anti-biotin antibodies through their biotin moieties, while the excess of the gold nanoparticles was captured at the control line of the strip through interaction with the IgG molecules to ensure the proper function of the assay. The authors could detect as low as 40 pg or 15 copies of DNA sequences corresponding to horse and donkey species. The method was finally applied to various processed foods and mixed meat products. The whole procedure was very fast, and the final result was obtained within 40 min (Zhang et al. 2019a). A similar approach was used by Xu et al. using the same endogenous gene for the on-site detection of mammalian-derived meat samples. The PCR products were again labeled with biotin and FITC and detected at the test zone by anti-FITC antibody and anti-biotin antibody—gold nanoparticles conjugates. The control line was formed when the excess of the gold nanoparticles conjugates was captivated by a secondary antibody. The LOD of the method was 10 pg, which was equivalent to 3 ~ 5 copies in mammals. Finally, the method was successfully applied to various meat materials and processed foods (Xu et al. 2017). Again, LAMP was combined with LFA for the detection of duck-derived ingredients in adulterated meat. LAMP primers were labeled with FITC and biotin and detected using LFA and gold nanoparticles. An amount of 3 pg of duck DNA, as well as 0.01% of duck meat in beef meat were detectd by this approach. The LOD of 3 pg is 10 times lower than that of quantitative real-time PCR (Shi et al. 2017).

PCR products specific to duck species were labeled, as above, with biotin and FITC and applied onto a new LFA. The amplified products were detected by anti-FITC antibody immobilized to the test zone and visualized by streptavidin-gold nanoparticles conjugates, in order to check the adulteration of beef meat samples. Beef species were also amplified in the same reaction but were dually labeled with biotin. The beef-specific DNA sequences were then captured at the control line of the strip by immobilized streptavidin to give a positive control signal, in the presence or absence of duck meat in the sample. The LFA was finally applied for the detection of adulteration in raw and processed meat samples. The whole protocol had a duration of 2 h (Qin et al. 2019).

In another study, an LFA was combined with recombinase polymerase amplification (RPA) for authentication of mutton slices or heat-processed shish kebabs. RPA is an isothermal amplification technology that can be completed within 20 min, and is the method of choice for on-site detection compared with other isothermal amplification methods, such as loop-mediated isothermal amplification (LAMP), strand displacement amplification (SDA), and rolling circle amplification (RCA). The RPA reaction was performed in low temperature (37–42 °C) and finished within 15 min, while the DNA extraction step and the subsequent detection with the LFA both ended up in 5 min. The sheep-specific PCR products were labeled with biotin at one end and digoxigenin at the other end and were immediately applied onto the strip. The amplified products were detected at the test zone of the strip by immobilized avidin and gold nanoparticles conjugated to anti-digoxigenin antibody through avidin–biotin and anti-digoxigenin antibody–digoxigenin interactions. The excess of the gold nanoparticles conjugates was captured finally by goat anti-mouse antibodies to form the control line of the strip. The LOD of this LFA was assessed to 200 fg of sheep DNA (Li et al. 2019). Mutton has also been detected by an LFA combined this time with a cross-priming amplification reaction (CPA). This reaction was performed at 63 °C for 60 min. The products were labeled at both ends with biotin and fluorescein instead of digoxigenin and detected as described above using anti-fluorescein antibody—gold nanoparticles conjugates. The LOD of the method was 1% of mutton in thermal treated meat mixtures, while the detection with the LFA was completed within 5 min (Feng et al. 2018). Sheep-specific PCR products were also detected by another LFA for the authentic identification of raw and heat-treated mutton. The strip format was the same as the one described by Zhao et al. (2020a). This LFA had an LOD of 0.01% of adulterated meat in mixtures with six other meats, as well as 0.01 pg of sheep DNA (Yin et al. 2016).

Magiati et al. developed an LFA for the detection of 0.01% of adulteration of beef and horse meat with pork and horse meat, respectively, in binary mixtures. Four species-specific DNA sequences were PCR-amplified. The PCR products carried biotin at one end and allowed to hybridize to species-specific oligonucleotide probes that had a poly(dA) tail at their 5' end. The hybrids were then applied to the conjugate pad of the strip, next to gold nanoparticles coupled to streptavidin. A first red line was formed at the test zone of the strip when the hybrids were captured by immobilized poly(dT) sequences through dA/dT hybridization and gold nanoparticles conjugates were accumulated there via streptavidin–biotin interaction. The excess of gold nanoparticles was captured finally at the control zone of the strip to form a second red line to ensure that the strip functions perfectly. The LFA was completed within 25–30 min (Magiati et al. 2019).

Finally, a multiplex DNA LFA kit has been commercially produced. Ha et al. have used this kit for the simultaneous detection of as low as 0.1% of adulteration of seven meat species: beef, pork, lamb, goat, horse, chicken, and turkey. Species-specific DNA sequences were amplified by a multiplex PCR. The PCR products were then detected simultaneously on the strip by immobilized species-specific oligonucleotide probes and gold nanoparticles (Ha et al. 2019).

16.3.2 Fish Authentication

Four fish species were also detected by a PCR-based LFA. Four sets of specific primers were used for the amplification of each species. The PCR products were dually labeled with biotin and fluorescein that allowed their visual detection with the LFA through immobilized biotin ligands at the test zone of the strip and anti-fluorescein gold nanoparticles. The excess of the gold nanoparticles was captured by immobilized anti-rabbit antibodies at the control zone of the strip. The LOD of this assay was 50 pg for *Gadus morhua* and *Molva molva* species and 500 pg for *Gadus chalcogrammus* and *Gadus macrocephalus* fish species, respectively. Finally, the LFA was successfully applied for the identification of the four species in 31 commercial fish-derived samples (Taboada et al. 2017).

16.3.3 Milk Authentication

The adulteration of milk-based products derived from goat, sheep, buffalos, and other animal species with cow milk that has lower quality and cost is a serious concern for many years now. Apart from health risks that are discussed above, many Protected Designation of Origin (PDO) councils have made great efforts in developing specific and reliable tests, to assure the quality and the genuineness of such dairy products (Galan-Malo et al. 2018).

Lateral flow assays were also exploited for milk authentication in dairy products. An LFA was developed for the detection of yak milk. This LFA was based on rapid isothermal recombinant polymerase amplification (RPA). Double-stranded amplified products were obtained containing a DNA tail at both ends that was attached to the primers through a special abasic C3 spacer. One DNA tail was used for the direct binding to gold nanoparticles labeled with complementary detection probes, while the final complex was captured by immobilized specific DNA sequences at the test zone of the strip through hybridization to the second DNA tail. The assay was completed within 40 min, allowing detection down to 5% yak milk in mixed cow milk samples (Wang et al. 2020).

Another LFA was developed for the detection of adulteration of sheep and goat milk with cow milk in yogurt samples. Three animal-specific DNA sequences were amplified by PCR. The products, biotinylated at one end, were let to hybridize to species-specific probes that contained a poly(dA) tail and applied to the strip next to gold nanoparticles conjugated to streptavidin. Then a red spot formed at the test area of the strip when the hybrids were captured by immobilized carboxylated beads, 2 μm in diameter, attached to an amino-poly(dT) probe through dA/dT hybridization, while the detection was accomplished through the streptavidin-gold nanoparticles conjugates and streptavidin–biotin interaction. The excess of the nanoparticle conjugates was captured at the control area of the strips by carboxylated beads that coupled to biotinylated DNA probe (Fig. 16.4). The authors could detect as low as 1.6 fmol of cow and goat, and 3.1 fmol of sheep PCR product, respectively. Also, adulteration of 0.01% of cow species in binary yogurt samples could be detected by this LFA (Bougadi and Kalogianni 2020).

16.3.4 Crocus Sativus *(Saffron)* Adulteration

Saffron is derived from the dried stigmas of the flowers of the plant *Crocus sativus* and is one of the most expensive herbs cultivated mainly in India, China, Greece, Spain, Italy, France, and Turkey. Its special flavor, along with some nutritional characteristics, makes saffron one of the most valuable spices, and at the same time vulnerable to adulteration. Some adulterants used are carrot, marigold, safflower, and others with a similar appearance that makes the discrimination by the consumers extremely difficult.

Recombinase polymerase amplification (RPA) in combination with a universal LFA was reported for the adulteration of saffron. The RPA products were labeled with fluorescein and biotin and were detected by the LFA through immobilized anti-biotin substances at the test zone of the strip, while the visualization was accomplished through conjugates of gold nanoparticles with anti-fluorescein antibody. The whole assay was ended up within 15 min at 37 °C and the authors could detect as low as 1 pg of genomic DNA with high specificity among common saffron adulterants. (Zhao et al. 2019b).

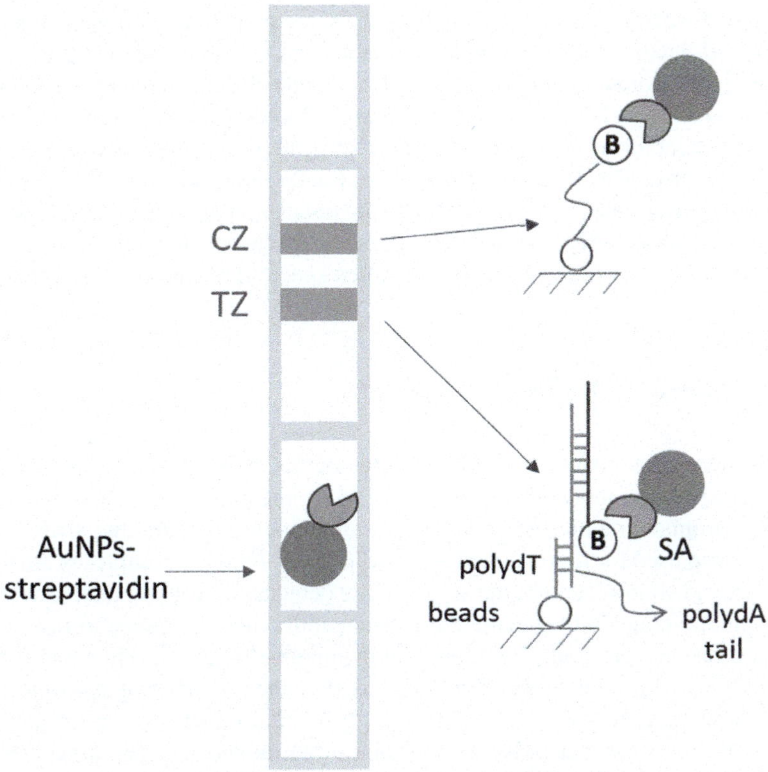

Fig. 16.4 Detection of adulteration of goat- and sheep-derived milk samples with cow milk using an LFA. The visual detection is based on species-specific PCR products hybridized to species-specific DNA probes carrying a poly(dA) tail and gold nanoparticles conjugated to streptavidin. AuNPs: gold nanoparticles, SA: streptavidin, B: biotin, TZ: test zone, CZ: control zone

16.3.5 Coffee Adulteration

Coffee authenticity assessment was accomplished through another LFA that used gold nanoparticles as reporters. Coffee is one of the most used food products daily with increasing global consumption. Two different coffee varieties-species are commercially available worldwide: *Coffea arabica* and *Coffea robusta. C. arabica* is preferred by the consumers, because it has a better and more intense taste, better quality, and lower caffeine content, therefore it is two–three times more expensive than *C. robusta.* The discrimination of the two coffee species was based on the detection of a unique single-nucleotide polymorphisms (SNP) in the chloroplastic trnL(UAA)—trnF(GAA) intraspacer region of the plant genome. DNA was isolated from coffee beans from the two species and amplified by PCR using a common primer pair. Then, the discrimination of the two alleles that correspond to the two coffee species was accomplished through a 15-min primer extension reaction (PEXT). Two allele-specific primers were used that have the same sequence, containing a poly(dA) tail at 5' end and differ at the 3' end that contained the analyzed SNP. The primers were extended, afterwards, by the DNA polymerase only if they were perfectly complementary to the target DNA. During the extension, biotin molecules were incorporated to enable the capturing of the PEXT products by immobilized streptavidin at the test zone of the strip. The PEXT products were finally detected by gold nanoparticles conjugated to poly(dT) sequences through the poly(dA) tail of the primer. The detection was finished in a few minutes, and *Robusta* species could be detected as low as 5% of by this LFA, in the presence of *Arabica* species (Trantakis et al. 2012)

16.4 Lateral Flow Immunoassays

Lateral flow immunoassays exhibit similar or better detectability than common enzyme-linked immunosorbent assay (ELISA) methods. The most widespread immunochromatographic assembly is the immunochromatographic strip test that has been also used for food authentication purposes. Immune complexes are formed at the test and control zone of the strip and are detectable by gold nanoparticles, that are covalently coupled to specific recognition antibodies. Immunochromatographic strips are easy to use, simple and rapid, allowing on-site testing. The sensitivity and specificity of the immunoassay is mainly based on the immunoreagents (antibodies) used. These reagents must have high affinity for the target (protein or antibody), along with low cross-reactivity with similar target molecules. The kinetic binding constant is also a factor that affects the detectability of the strip assay. Moreover, these strip tests usually include no sample pre-treatment, so interfering substances are not removed from the sample leading to high matrix effect and resulting in lower detection limit. In some cases, however, purification of analytes is required. Finally, the gold nanoparticles are again the most popular reporters in lateral flow immunoassays, due to their unique optical characteristics (Dzantiev et al. 2014).

16.4.1 Meat Authentication

Horse meat residues were detected in raw and cooked samples by an LFA, based on both a competitive and a sandwich-type immunoassay. Horse serum albumin (HSA) was used as the target in raw meat, while the horse thermal-stable meat protein (H-TSMP) was used as the target in cooked meat. Anti-HSA and anti-H-TSMP polyclonal antibodies produced by immunization protocols were conjugated to gold nanoparticles and used for the preparation of the sandwich-type test zones of the strip. Standard HSA and H-TSMP targets were used for the formation of the competitive format test lines. Chicken anti-goat antibodies were immobilized at the control line of the strip to capture the excess of the gold nanoparticles conjugates, ensuring the validity of the LFA. This LFA had a detection limit (LOD) of 0.01% and 1.0% for raw and cooked horse meat in beef meat, respectively. The final results were observed within 35 min with very good specificity and reproducibility (Masiri et al. 2017). A similar approach was developed by Masiri et al. (2016) for pork identification. The results were obtained after 15 min with an LOD of 0.01% of pork meat in beef meat (Masiri et al. 2016).

A rapid immunochromatographic strip was also developed for the identification of pork adulteration on beef cooked meatballs. Gold nanoparticles were coated with anti-swine IgG polyclonal antibody, while the same polyclonal antibodies were used to construct the test zone of the strip and a sandwich format immunoassay. The excess of gold nanoparticles was captured by goat anti-mouse IgG at the control zone of the strip (Fig. 16.5). The detection limit of the immunoassay was 0.1% (w/w) of pork meat, while the response was visualized within 5 min (Kuswandi et al. 2017).

Moreover, bovine authentication was based on the detection of bovine fat by a lateral flow immunoassay. This assay was based on the detection of heat-stable muscle protein. The protein was extracted from meat samples and the protein extracts were deposited onto the conjugate pad of the strip along with gold nanoparticles conjugated to anti-muscle protein antibodies, forming an immunocomplex. The strip was then dipped into the appropriate solution. A first red line formed at the test zone of the

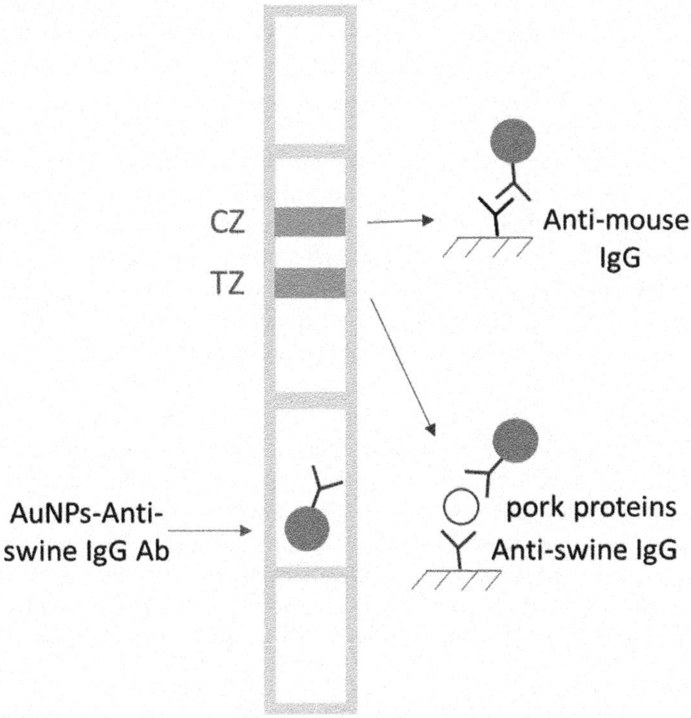

Fig. 16.5 A sandwich-type immunochromatographic strip for the detection of pork meat. AuNPs: gold nanoparticles, TZ: test zone, and CZ: control zone

strip, as the immunocomplex was captured by immobilized secondary antibodies specific to muscle protein. A second visual line formed, when the excess of the gold nanoparticles conjugates was accumulated by a third antibody immobilized at this area. This LFA could detect 2% bovine fat-in-pork fat, 1% bovine fat-in-porcine meat-and-bone meal, and 0.5% bovine fat-in-soy meal mixtures. Bovine fat could also be detected at high temperatures up to 213 °C (Hsieh and Gajewski 2016). The same format was used by another research team in 2007 for the detection of beef or ovine meat in beef-in-chicken, beef-in-turkey, and lamb-in-pork meat mixtures, with an LOD of 0.5%. The method was successfully applied in raw and cooked meat mixtures (Rao and Hsieh 2007).

16.4.2 Milk Authentication

Lateral flow immunoassays have been also developed for the identification of animal-specific proteins for the authentication of milk and milk-based products. A new competitive lateral flow immunoassay was reported for the detection of adulteration of goat milk with cow milk. Gold nanoparticles were coupled to monoclonal antibodies specific to the cow milk protein casein and used as reporters for visual detection. The limit of detection of this immunoassay was as low as 0.07% of cow milk in milk mixtures with goat milk. The assay offered equal specificity and higher detectability than commercial ELISA assay (Liu et al. 2019).

Another rapid immunochromatographic strip test was developed for the adulteration of milk from goat, sheep, and buffalo with cow milk based on the detection of bovine immunoglobulins (IgG). A validation study was also conducted including evaluation of the specificity, analyzing 146 negative samples, detectability, reproducibility, and robustness by studying the effect of several factors such as different batches of the strip, sample dilution, assay time, high level of fat, protein, and somatic cell counts. The detectability of the assay was as low as 0.5% of cow milk in milk mixtures, while a strip reader could be useful for interpretation of the results (Galan-Malo et al. 2018).

Along with these, a competitive lateral flow immunoassay was developed for the detection of soy milk in cow's milk. The strip test was based on the detection of soy milk protein through rabbit anti-soy protein antibodies conjugated to gold nanoparticles. Standard soy milk protein was immobilized at the test line of the strip competing the protein present in soy milk sample. At the control line of the strip, anti-rabbit immunoglobulin was immobilized to ensure the properties of antibody—conjugated gold nanoparticles (Fig. 16.6). The detection limit of soy milk was 1.75% in whole bovine milk, while the results were available in 5 min (Gautam et al. 2017).

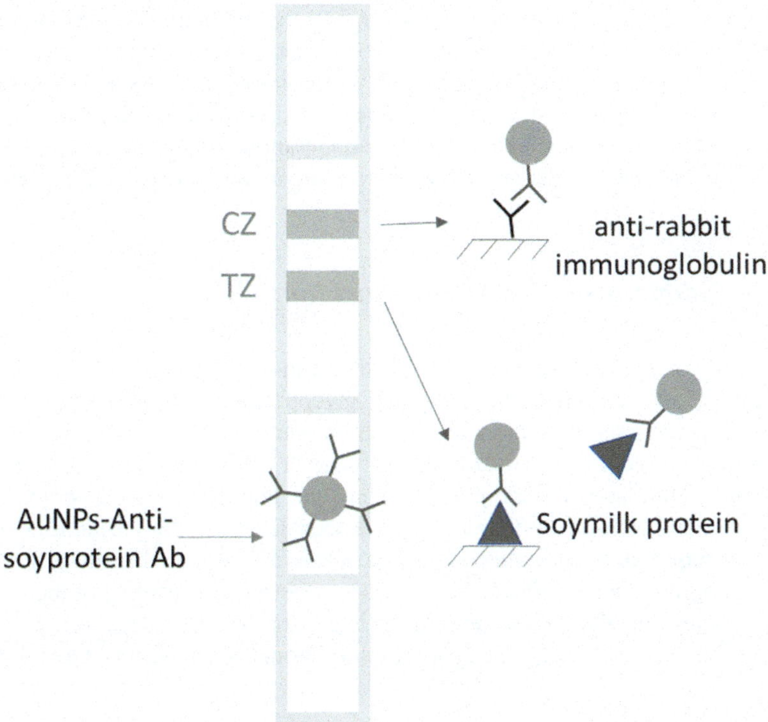

Fig. 16.6 Detection of adulteration of bovine milk with soy milk by a lateral flow competitive immunoassay targeting the soy milk protein. AuNPs: gold nanoparticles, Ab: antibody, TZ: test zone, and CZ: control zone

16.4.3 Honey Authentication

Honey is a natural sweet source that is considered to have health-promoting and nutritional characteristics, such as antioxidant, antibacterial, antimutagenic, and anti-inflammatory properties, as well as healing properties. Honey is rich in enzymatic and non-enzymatic substances, including polyphenols, ascorbic acid, α-tocopherol, carotenoids, nucleic acids, proteins, and others. Its antibacterial activity is mainly attributed to its high osmosis phenomena in combination with its low pH and the presence of hydrogen peroxide, bee defensin-1, and methylglyoxal. For the above reasons, honey is one of the favorite natural products worldwide. Its composition includes mainly carbohydrates and water that makes its adulteration with cheaper sugar syrups with similar composition very easily. Many analytical methods have been used for the detection of honey adulteration, but they use expensive instruments and require high analysis time (Zhang et al. 2019b; Pita-Calvo and Vázquez 2018).

An immunochromatographic strip test was constructed for the detection of the major royal jelly protein 1 (MRJP1) in honey. The assay was based on the use of novel polyclonal antibodies, a sandwich-type ELISA format, and gold nanoparticles conjugated to the polyclonal antibodies as reporters. For quality control of the strip test, goat anti-rabbit IgG was immobilized on the control zone of the strip to capture the excess of gold nanoparticles conjugates with the specific antibodies. The LOD of the method was 2 μg mL^{-1} for the analyzed protein. Also, the test line formed in case of a positive sample remained constant for adulterant mixtures that contained above 30% honey. The adulterants tested were rice and corn syrup (Zhang et al. 2019b).

16.4.4 Edible Bird's Nest Authentication

Edible bird's nest is produced by the saliva secretion of several kinds of birds and is considered as a health food, containing glycoproteins, carbohydrates, essential amino acids, minerals, and other elements. It is commercially available at high price; therefore, it is adulterated with substances of lower cost. These adulterants may are plant-exude, gum karaya, pork and fish skin, algae and the white part of the egg. A lateral flow sandwich-type immunoassay has been developed for the identification of bird's nest based on the detection of its glycoprotein. Polyclonal antibodies specific to the glycoprotein were produced and used to construct the test line of the strip and prepare conjugates with gold-silica (SiO_2) nanoparticles. The detection limit of the LFA was 7.02 ng mL^{-1} of purified glycoprotein, from nest, in buffer (Xu et al. 2019).

16.5 Conclusions and Future Perspectives

Food products with an extra added value are more and more often vulnerable to fraud. Food safety is a severe issue continuously investigated by the research community to prevent food fraud and counteract new "sophisticated" types of adulteration. Lateral flow assays have gained the interest of researches for a rapid and cheap alternative for food authenticity testing. DNA- and protein-based techniques in combination with LFAs have successfully applied for the detection of food adulteration and species identification in various food products such as meat, fish, and milk. DNA-based techniques are preferred due to higher DNA stability especially in processed food, higher sensitivity, and specificity among closely related species. Moreover, the integration of the nanotechnology (nanoparticles) in LFAs have increased the flexibility in the strip format configuration, which also provide very low detection limits, even in food mixtures and processed foodstuffs.

On the other hand, some issues have to be further addressed in order to improve the analytical performance of LFAs. New and different labels, as well as new detection systems are needed in order to increase the detectability and the multiplicity of these assays. The combination of LFAs with easy amplification systems, in case of DNA targets, could enable the on-site detection of food adulteration. Moreover, the use of portable recording devices along with the test strips could increase the possibility for more accurate quantitative results, enabling the data processing outside of the laboratory. This would increase the portability of these rapid tests that would greatly outperform modern technologies of food analysis in terms of speed, cost, simplicity, and efficacy. Finally, a strong effort has to be made in order to construct and produce universal strip tests that are not dedicated only to the detection of a single analyte but can be easily applied for the analysis of various different analytes. It is anticipated that the recognition is performed outside the strip, while the final detection is achieved with the strip test.

References

Abbas O, Zadravec M, Baeten V, Mikuš T, Lešić T, Vulić A, Prpić J, Jemeršić L, Pleadin J (2018) Analytical methods used for the authentication of food of animal origin. Food Chem 246:6–17

Bansal S, Singh A, Mangal M, Mangal AK, Kumar S (2017) Food adulteration: sources, health risks, and detection methods. Crit Rev Food Sci Nutr 57:1174–1189

Böhme K, Calo-Mata P, Barros-Velázquez J, Ortea I (2019) Review of recent DNA-based methods for main food-authentication topics. J Agric Food Chem 67:3854–3864

Bougadi ET, Kalogianni DP (2020) Paper-based DNA biosensor for food authenticity testing. Food Chem, 126758

Dzantiev BB, Byzova NA, Urusov AE, Zherdev AV (2014) Immunochromatographic methods in food analysis. TrAC—Trends Anal Chem 55:81–93

Feng T, Li S, Wang S, Pan J (2018) Cross priming amplification with nucleic acid test strip analysis of mutton in meat mixtures. Food Chem 245:641–645

Galan-Malo P, Mendiara I, Razquin P, Mata L (2018) Validation of a rapid lateral flow method for the detection of cows' milk in water buffalo, sheep or goat milk. Food Addit Contam—Part A Chem Anal Control Expo Risk Assess 35:599–604

Gautam PB, Sharma R, Lata K, Rajput YS, Mann B (2017) Construction of a lateral flow strip for detection of soymilk in milk. J Food Sci Technol 54:4213–4219

Ha Y, Thienes CP, Agapov AA, Laznicka AV, Han S, Nadala C, Samad Pour M (2019) Comparison of ELISA and DNA lateral flow assays for detection of pork, horse, beef, chicken, Turkey, and goat contamination in meat products. J AOAC Int 102:189–195

Hong E, Lee SY, Jeong JY, Park JM, Kim BH, Kwon K, Chun HS (2017) Modern analytical methods for the detection of food fraud and adulteration by food category. J Sci Food Agric 97:3877–3896

Hsieh YHP, Gajewski K (2016) Rapid detection of bovine adipose tissue using lateral flow strip assay. Food Sci Nutr 4:588–594

Kalogianni DP, Koraki T, Christopoulos TK, Ioannou PC (2006) Nanoparticle-based DNA biosensor for visual detection of genetically modified organisms. Biosens Bioelectron 21:1069–1076

Kuswandi B, Gani AA, Ahmad M (2017) Immuno strip test for detection of pork adulteration in cooked meatballs. Food Biosci 19:1–6

Li T, Jalbani YM, Zhang G, Zhao Z, Wang Z, Zhao Y, Zhao X, Chen A (2019) Rapid authentication of mutton products by recombinase polymerase amplification coupled with lateral flow dipsticks. Sens Actuators, B Chem 290:242–248

Liu B, Si J, Zhao F, Wang Q, Wang Y, Li J, Li C, Li T (2019) Rapid detection of cow milk adulteration/contamination in goat milk by a lateral flow colloidal gold immunoassay strip. J Dairy Res 86:94–97

Lo YT, Shaw PC (2018) DNA-based techniques for authentication of processed food and food supplements. Food Chem 240:767–774

Magiati M, Sevastou A, Kalogianni DP (2018) A fluorometric lateral flow assay for visual detection of nucleic acids using a digital camera readout. Microchim Acta 185:314

Magiati M, Myridaki VM, Christopoulos TK, Kalogianni DP (2019) Lateral flow test for meat authentication with visual detection. Food Chem 274:803–807

Masiri J, Benoit L, Barrios-Lopez B, Thienes C, Meshgi M, Agapov A, Dobritsa A, Nadala C, Samadpour M (2016) Development and validation of a rapid test system for detection of pork meat and collagen residues. Meat Sci 121:397–402

Masiri J, Benoit L, Thienes C, Kainrath C, Barrios-Lopez B, Agapov A, Dobritsa A, Nadala C, Sung SL, Samadpour M (2017) A rapid, semi-quantitative test for detection of raw and cooked horse meat residues. Food Control 76:102–107

Pita-Calvo C, Vázquez M (2018) Honeydew honeys: a review on the characterization and authentication of botanical and geographical origins. J Agric Food Chem 66:2523–2537

Qin P, Qiao D, Xu J, Song Q, Yao L, Lu J, Chen W (2019) Rapid visual sensing and quantitative identification of duck meat in adulterated beef with a lateral flow strip platform. Food Chem 294:224–230

Rahmati S, Julkapli NM, Yehye WA, Basirun WJ (2016) Identification of meat origin in food products-A review. Food Control 68:379–390

Rao Q, Hsieh YHP (2007) Evaluation of a commercial lateral flow feed test for rapid detection of beef and sheep content in raw and cooked meats. Meat Sci 76:489–494

Sajid M, Kawde AN, Daud M (2015) Designs, formats, and applications of lateral flow assay: a literature review. J Saudi Chem Soc 19:689–705

Shi Y, Feng Y, Xu C, Xu Z, Cheng D, Lu Y (2017) Loop-mediated isothermal amplification assays for the rapid identification of Duck-derived ingredients in adulterated meat. Food Anal Methods 10:2325–2331

Spyrou EM, Kalogianni DP, Tragoulias SS, Ioannou PC, Christopoulos TK (2016) Digital camera and smartphone as detectors in paper-based chemiluminometric genotyping of single nucleotide polymorphisms. Anal Bioanal Chem 408:7393–7402

Subara D, Jaswir I (2018) Gold nanoparticles: synthesis and application for Halal authentication in meat and meat products. Int J Adv Sci Eng Inf Technol 8:1633–1641

Taboada L, Sánchez A, Pérez-Martín RI, Sotelo CG (2017) A new method for the rapid detection of Atlantic cod (*Gadus morhua*), Pacific cod (Gadus macrocephalus), Alaska pollock (Gadus chalcogrammus) and ling (Molva molva) using a lateral flow dipstick assay. Food Chem 233:182–189

Trantakis IA, Spaniolas S, Kalaitzis P, Ioannou PC, Tucker GA, Christopoulos TK (2012) Dipstick test for DNA-based food authentication: application to coffee authenticity assessment. J Agric Food Chem 60:713–717

Wadood SA, Boli G, Xiaowen Z, Hussain I, Yimin W (2020) Recent development in the application of analytical techniques for the traceability and authenticity of food of plant origin. Microchem J 152:104295

Wang Z, Li T, Yu W, Qiao L, Yang S, Chen A (2020) A low-cost novel lateral flow nucleic acid assay (LFNAA) for yak milk authentication. Lwt 122:109038

Woolfe M, Primrose S (2004) Food forensics: using DNA technology to combat misdescription and fraud. Trends Biotechnol 22:222–226

Xu H, Zheng L, Xie Y, Zeng H, Fan Q, Zheng B, Zhang Y (2019) Identification and determination of glycoprotein of edible brid's nest by nanocomposites based lateral flow immunoassay. Food Control 102:214–220

Xu Y, Xiang W, Wang Q, Cheng N, Zhang L, Huang K, Xu W (2017) A smart sealed nucleic acid biosensor based on endogenous reference gene detection to screen and identify mammals on site. Sci Rep 7:1–10

Yin R, Sun Y, Yu S, Wang Y, Zhang M, Xu Y, Xue J, Xu N (2016) A validated strip-based lateral flow assay for the confirmation of sheep-specific PCR products for the authentication of meat. Food Control 60:146–150

Yin R, Sun Y, Wang K, Feng N, Zhang H, Xiao M (2020) Development of a PCR-based lateral flow strip assay for the simple, rapid, and accurate detection of pork in meat and meat products. Food Chem 318:126541

Zhang C, Zhang X, Liao G, Shang Y, Ge C, Chen R, Wang Y, Xu W (2019a) Species-specific TM-LAMP and Trident-like lateral flow biosensor for on-site authenticity detection of horse and donkey meat. Sens Actuators, B Chem 301:127039

Zhang Y, Chen Y, Cai Y, Cui Z, Zhang J, Wang X, Shen L (2019b) Novel polyclonal antibody-based rapid gold sandwich immunochromatographic strip for detecting the major royal jelly protein 1 (MRJP1) in honey. PLoS ONE 14:1–15

Zhao L, Hua MZ, Li S, Liu J, Zheng W, Lu X (2019a) Identification of donkey meat in foods using species-specific PCR combined with lateral flow immunoassay. RSC Adv 9:26552–26558

Zhao M, Wang B, Xiang L, Xiong C, Shi Y, Wu L, Meng X, Dong G, Xie Y, Sun W (2019b) A novel onsite and visual molecular technique to authenticate saffron (Crocus sativus) and its adulterants based on recombinase polymerase amplification. Food Control 100:117–121

Zhao L, Wang K, Yan C, Xiao J, Wu H, Zhang H, Zhang X, Zhang C, Hu Y, Lu X, Zheng W (2020a) A PCR-based lateral flow assay for the detection of Turkey ingredient in food products. Food Control 107:106774

Zhao L, Hu Y, Liu W, Wu H, Xiao J, Zhang C, Zhang H, Zhang X, Liu J, Lu X, Zheng W (2020b) Identification of camel species in food products by a polymerase chain reaction-lateral flow immunoassay. Food Chem 319:126538

Zhao L, Li S, Hua MZ, Liu J, Zhang H, Hu Y, Chen Y, Lu X, Zheng W (2020c) Development of a species-specific PCR coupled with lateral flow immunoassay for the identification of goose ingredient in foods. Food Control 114:107240

Chapter 17
Nanobiosensors in Agriculture and Foods: A Scientometric Review

Ozcan Konur

Abstract Nanobiosensors have been developed and increasingly used in agriculture and food to ensure the safety of the agricultural produces and foods. Thus, these advanced sensors have the utmost public importance. However, for the efficient development and use of these nanobiosensors, it is necessary to develop efficient incentive structures for the primary stakeholders and to inform these stakeholders about the research on nanobiosensors in the agricultural and food context. Scientometric analysis of the research offers ways to evaluate the research in a respective field. This method has been used to evaluate research in several research fields. There has been a limited number of pioneering scientometric studies on the various aspects of the nanobiosensors in recent years. This book chapter presents a study on the scientometric evaluation of the research on the nanobiosensing and nanobioanalysis of agricultural and food analytes using two data sets. The first data set includes 100 most-cited papers whilst the second set includes over 22,000 papers published between 1980 and 2019. The data on the indices, document types, authors, institutions, funding bodies, source titles, 'Web of Science' subject categories, key words, research fronts and citation impact are presented and discussed.

Keywords Nanobiosensors · Nanobiosensing · Nanobioanalysis · Scientometrics · Nanobiomaterials · Food · Agriculture · Analytes · Graphene · Nanoparticles

17.1 Introduction

Nanobiosensors are being developed and increasingly used in agriculture and food to ensure the safety of the agricultural produces and foods (Nam et al. 2003; Song et al. 2010; Dong et al. 2012; Shan et al. 2009; Srivastava et al. 2018; Lee et al. 2007). Thus, these advanced sensors have the utmost public importance (Carvalho 2006; Franz et al. 1999; Grunert 2005; McLaughlin et al. 1999; Shalaby 1996). However, for the efficient development and use of these nanobiosensors, it is necessary to

O. Konur (✉)
Formerly Ankara Yildirim Beyazit University, Ankara, Turkey
e-mail: okonur@hotmail.com

R. N. Pudake et al. (eds.), *Biosensors in Agriculture: Recent Trends and Future Perspectives*, Concepts and Strategies in Plant Sciences,
https://doi.org/10.1007/978-3-030-66165-6_17

develop efficient incentive structures for the primary stakeholders and to inform these stakeholders about the research on the nanobiosensors in the agricultural and food context (North 1990, 1991; Konur 2000, 2002a, b, c, 2006a, b, 2007a, b).

Scientometric analysis of the research offers ways to evaluate the research in a respective field (Garfield 1955, 1972). This method has been used to evaluate research in a number of research fields (Konur 2011, 2012a, b, c, d, e, f, g, h, i, j, k, l, m, n, 2016a, c, d, e, f, 2017a, c, d, f, 2018a, b, 2019a). Although there have been a number of studies on the nanotechnology in general (Porter et al. 2008; Schummer 2004), there has been a limited number of pioneering scientometric studies on the various aspects of the nanobiosensors in recent years (Konur 2016b, 2017b, e, 2019b).

This book chapter presents a study on the scientometric evaluation of the research on the nanobiosensing and nanobioanalysis of agricultural and food analytes using two data sets. The first data set includes 100 most-cited papers whilst the second set includes over 22,000 papers published between 1980 and 2019. The data on the indices, document types, authors, institutions, funding bodies, source titles, 'Web of Science' subject categories, key words, research fronts, and citation impact are presented and discussed.

17.2 Materials and Methodology

The search for the literature was carried out in the 'Web of Science' database in January 2020. This database includes 'Science Citation Index Expanded' (SCI-E), 'Social Sciences Citation Index' (SSCI), 'Book Citation Index–Science' (BCI-S), 'Conference Proceedings Citation Index-Science' (CPCI-S), 'Emerging Sources Citation Index' (ESCI), 'Book Citation Index-Social Sciences and Humanities' (BCI-SSH), 'Conference Proceedings Citation Index-Social Sciences and Humanities'(CPCI-SSH), and 'Arts and Humanities Citation Index' (A&HCI).

The keywords for the search of the literature are developed from the screening of abstract pages for the first 1,000 highly cited papers. Three sets of keywords were developed for food and agriculture, sensors and nanotechnology. These keyword sets are provided in the appendix.

Two data sets are used for this study. The most-cited 100 papers form the first data set (sample data set, $n = 100$ papers) whilst all the papers form the second data set (population data set, $n =$ over 22,400 papers). The data on the indices, document types, publication years, institutions, funding bodies, source titles, countries, 'Web of Science' subject categories, citation impact, keywords and research fronts are collated from these data sets.

The key findings are provided in the relevant tables and one figure, supplemented with explanatory notes in the text. These findings are discussed whilst a number of conclusions are drawn and a number of recommendations for further study are made.

Table 17.1 Document types

	Document type	Sample data set %	Population data set %	Difference %
1	Article	95	97.9	−2.9
2	Review	5	1.8	3.2
3	Book chapter	0	0.8	−0.8
4	Proc. paper	0	1.8	−1.8
5	Editorial mat.	0	0.2	−0.2
6	Letter	0	0.1	−0.1
7	Book	0	0.0	0.0
8	Note	0	0.1	−0.1

17.3 Results

17.3.1 Indices and Documents

There are over 24,500 papers related to nanobiosensors in the 'Web of Science' as of January 2020. This original population data set was refined for the document type (article, review, book chapter, book, editorial material, note, and letter) and language (English), resulting in over 22,400 papers comprising over 91% of the original population data set.

The primary index is SCI-E with 97.9% of this refined population data set. The papers on the social and humanitarian aspects of the nanobiosensors are relatively negligible with only 3 papers indexed by the SSCI and A&HCI. The BCI-S, CPCI-S, and ESCI form 0.8, 1.8, and 1.4% of this data set, respectively. On the other hand, the sole index for the sample data set is the SCI-E. The brief information on the document types for both data set is provided in Table 17.1. The key finding is that article and review types of documents are the primary documents and they are under-represented and over-represented in the sample papers by around 3%, respectively.

17.3.2 Authors

The brief information about the most-prolific 10 authors is provided in Table 17.2. The sample and population papers are authored by over 450 and 41,600 researchers, respectively. The most-prolific authors are 'Chad A. Mirkin' of Northwestern University of the US working primarily on protein, ion, and pathogen detection and 'Yuehe Lin' of Pacific Northwest National Laboratory. of the US working primarily on the glucose and pesticide detection with 4 sample papers, each.

The most-prolific institutions for these top authors are the 'Chinese Academy of Sciences' and 'National Taiwan University' with 3 and 2 authors from these institutes, respectively. In total, 7 institutions house these top authors. It is notable that 4 of

Table 17.2 Authors

	Authors	No. papers	Inst.	Country	Research front	HCR fields
1	Lin, Yuehe	4	Pacific NW Natl. Lab.	US	Glucose and pesticide detection	Chem.
2	Mirkin, Chad A.	4	Northwestern Univ.	US	Protein, ion and pathogen detection	Chem.
3	Chang, Huan-Tsung	3	Natl. Taiwan Univ.	Taiwan	Ion detection	
4	Huang, Chih-Ching	3	Natl. Taiwan Univ.	Taiwan	Ion detection	
5	Lei, Yu	3	Nanyang Technol. Univ.	Singapore	Glucose detection	
6	Li, Di	3	Chinese Acad. Sci.	China	Ion detection	
7	Liu, Guodong	3	New Mexico State Univ.	US	Protein, pesticide and DNA detection	
8	Qu, Xiaogang	3	Chinese Acad. Sci.	China	Glucose, ion, dopamine and thiol detection	Cross-field
9	Ren, Jinsong	3	Chinese Acad. Sci.	China	Ion, glucose, dopamine and thiol detection	Cross-field
10	Wang, Ying	3	Tsinghua Univ.	China	Dopamine, glucose and thiol detection	

*HCR: Highly Cited Researchers

these top researchers are listed in the 'Highly Cited Researchers' (HCR) list in 2019 (Docampo and Cram 2019).

The most-prolific countries for these top authors are China, the US, and Taiwan with 4, 3, and 2 authors, respectively. The other country is Singapore. The most-prolific research front for these top authors is 'metal ion detection' with 6 authors. The other prolific research fronts are 'glucose detection', 'dopamine detection', 'thiol detection', 'pesticide detection', and 'protein detection' with 5, 3, 3, 2 and 2 authors, respectively. It is further notable that there is a significant gender deficit amongst these top authors as only one researcher is female.

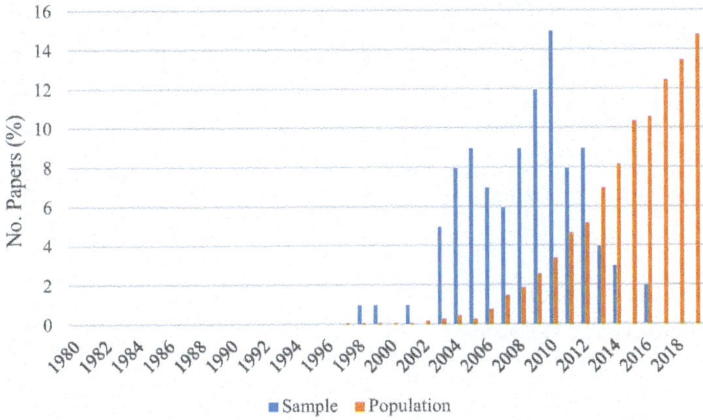

Fig. 17.1 The research output between 1980 and 2019

17.3.3 Publication Years

The information about the publication years for both data sets is provided in Fig. 17.1. This figure shows that whilst 57 and 41% of the sample papers are published in the 2000s and 2010s, respectively, over 90% of the population papers are published in the 2010s. The research output in the 1980s and 1990s are relatively negligible for both data sets. Similarly, the most-prolific publication years for the sample data set are 2009 and 2010 with 12 and 15 sample papers, respectively. On the other hand, the most-prolific publication years for the population data set are 2015, 2016, 2017, 2018 and 2019 with over 10% of the population papers each.

17.3.4 Institutions

The brief information on the top 8 institutions with at least 3 sample papers is provided in Table 17.3. In total, 136 and over 6,300 institutions contribute to the sample and population papers, respectively. These top institutions publish 48 and 12% of the sample and population papers, respectively. The top institution is 'Chinese Academy of Sciences' with 17 sample papers and with 10.5% publication surplus. All of these 8 institutions are significantly over-represented in the sample papers. The most-prolific countries for these top institutions are the US and China with 3 institutions each. Additionally, Taiwan and Singapore have 2 and one institutions, respectively.

Table 17.3 Institutions

	Institutions	Country	Sample no. papers %	Population no. papers %	Difference %
1	Chinese Acad. Sci.	China	17	6.5	10.5
2	Northwestern Univ.	US	7	0.8	6.2
3	Nanjing Univ.	China	4	1.6	2.4
4	Pacific NW Natl. Lab.	US	4	0.3	3.7
5	Tsinghua Univ.	China	4	0.7	3.3
6	Nanyang Technol. Univ.	Singapore	3	0.8	2.2
7	Natl. Hlth. Inst.	US	3	0.3	2.7
8	Natl. Taiwan Univ.	Taiwan	3	0.7	2.3
9	Natl. Taitung Univ.	Taiwan	3	0.3	2.7

17.3.5 Funding Bodies

The brief info about top 11 funding bodies is provided in Table 17.4. It is significant that only 58% of the sample papers declare any funding. This funding rate is less than the funding rate, 67%, reported in a study (Wang and Shapira 2011). In total, 140 and over 14,900 funding bodies all over the world fund the sample and population papers, respectively. The funding rate is higher for the population papers with 84% funding rate.

The top funding body is the 'National Natural Science Foundation' of China, funding 31 and 39.2% of the sample and population papers, respectively. The 'National Basic Research Program of China', 'National Institutes of Health', and 'US Department of Health Human Services' of the US further fund 12, 9, and 9% of the sample papers, respectively. The most-prolific country for these top funding bodies is China with 7 funding bodies. Three funding bodies are from the US whilst there is one Japanese funding body.

17.3.6 Source Titles

The brief information about the top source titles is provided in Table 17.5. In total, 32 and over 1200 source titles publish the sample and population papers, respectively. On the other hand, 11 top journals publish 79 and 12.8% of the sample and population papers, respectively. They are over-represented in the sample papers with 66.2% publication surplus.

The top journals are 'Analytical Chemistry' and 'Journal of the American Chemical Society' publishing 16 and 15 sample papers, with 13.1 and 14.7% publication surplus, respectively. The other top journals are 'Biosensors Bioelectronics',

Table 17.4 Funding bodies

	Institutions	Country	Sample no. papers %	Population no. papers %	Difference %
1	National Natural Science Foundation of China	China	31	39.2	−8.2
2	National Basic Research Program of China	China	12	4.6	7.4
3	National Institutes of Health	US	9	1.9	7.1
4	US Department of Health Human Services	US	9	1.9	7.1
5	Chinese Academy of Sciences	China	4	1.5	2.5
6	Ministry of Education Culture Sports Science and Technology Japan	Japan	4	1.0	3.0
7	National Science Foundation	US	4	1.9	2.1
8	Fundamental Research Funds for the Central Universities	China	3	4.3	−1.3
9	Ministry of Science and Technology China	China	3	1.1	1.9
10	Program for Changjiang Scholars Innovative Research Team in University	China	3	1.0	2.0
11	Program for New Century Excellent Talents in University	China	3	1.2	1.8

'Chemical Communications', 'Nano Letters', and 'ACS Nano'. The sample papers are indexed by 8 subject categories in total. The most-prolific subject categories are 'Chemistry Multidisciplinary', and 'Nanoscience Nanotechnology' with 6 and 5 sample papers, respectively. The other prolific categories are 'Materials Science Multidisciplinary', 'Physics Applied', 'Physics Condensed Matter' and 'Chemistry Analytics'.

Table 17.5 Source titles

	Source titles	WOS subject cat.	Sample no. papers %	Population no. papers %	Difference %
1	Analytical Chemistry	Chem. Analy.	16	2.9	13.1
2	Journal of the American Chemical Society	Chem. Mult.	15	0.3	14.7
3	Biosensors Bioelectronics	Biophys., Biot. Appl. Microb., Chem. Analy., Electrochem., Nanosci. Nanotechnol.	11	6.2	4.8
4	Chemical Communications	Chem. Mult.	8	1.3	6.7
5	Nano Letters	Chem. Mult., Chem. Phys., Nanosci. Nanotechnol., Mats. Sci. Mult., Phys. Appl., Phys. Cond. Matt.	7	0.3	6.7
6	ACS Nano	Chem. Mult., Chem. Phys., Nanosci. Nanotechnol., Mats. Sci. Mult.	6	0.5	5.5
7	Angewandte Chemie International Edition	Chem. Mult.	5	0.3	4.7
8	Proceedings of the National Academy of Sciences of the United States of America	Mult. Sci.	4	0.3	3.7
9	Applied Physics Letters	Phys. Appl.	3	0.3	2.7
10	Advanced Materials	Chem. Mult., Chem. Phys., Nanosci. Nanotechnol., Mats. Sci. Mult., Phys. Appl., Phys. Cond. Matt.	2	0.3	1.7
11	Nature Nanotechnology	Nanosci. Nanotechnol., Mats. Sci. Mult.	2	0.1	1.9

(continued)

Table 17.5 (continued)

	Source titles	WOS subject cat.	Sample no. papers %	Population no. papers %	Difference %
			79	12.8	66.2

17.3.7 Countries

The brief info about the top countries is provided in Table 17.6. In total 18 and 117 countries contribute to the sample and population papers, respectively. The top county is China publishing 45 and 51% of the sample and population papers, respectively with 6% publication deficit. The US follows China publishing 37 and 10% of the sample and population papers with 27% publication surplus. The other prolific countries are Singapore, Germany, South Korea, and Taiwan. The European and Asian countries represented in this table publish altogether 11 and 65% of the sample papers whilst they publish 4.6 and 74.5% of the population papers, respectively.

It is notable that the publication surplus for the US, these European and Asian countries is 27, 6.4 and −9.5%, respectively. It is further notable that the bulk of the Asian publication deficit originates from China and India with −6 and −7.4% publication deficit, respectively.

Table 17.6 Countries

	Countries	Sample no. papers %	Population no. papers %	Difference %
1	China	45	51.0	−6.0
2	US	37	10.0	27.0
3	Singapore	6	1.3	4.7
4	Germany	4	1.8	2.2
5	South Korea	4	5.8	−1.8
6	Taiwan	4	3.6	0.4
7	Unt. Kingd.	3	1.9	1.1
8	India	3	10.4	−7.4
9	Japan	3	2.3	0.7
10	Saudi Arabia	3	1.6	1.4
11	Finland	2	0.2	1.8
12	Israel	2	0.4	1.6
13	Sweden	2	0.8	1.2
14	Europe-4	11	4.6	6.4
15	Asia-5	65	74.5	−9.5

Table 17.7 Web of science subject categories

	Subjects	Sample no. papers %	Population no. papers %	Difference %
1	Chemistry Multidisciplinary	48	17.2	30.8
2	Chemistry Analytical	34	42.7	−8.7
3	Nanoscience Nanotechnology	30	19.4	10.6
4	Materials Science Multidisciplinary	21	16.4	4.6
5	Chemistry Physical	19	8.6	10.4
6	Physics Applied	14	9.7	4.3
7	Electrochemistry	13	25.4	−12.4
8	Biotechnology Applied Microbiology	12	7.7	4.3
9	Biophysics	11	7.3	3.7
10	Physics Condensed Matter	11	4.8	6.2
11	Multidisciplinary Sciences	6	1.4	4.6

17.3.8 Web of Science Subject Categories

The brief information about the top 'Web of Science' subject categories is provided in Table 17.7. The sample and population papers are indexed by 20 and 133 subject categories, respectively. The top subject category is 'Chemistry Multidisciplinary' publishing 48 and 17.2% of the sample and population papers, respectively with 30.8% publication surplus. Additionally, 'Chemistry Analytical' and 'Nanoscience Nanotechnology' index 34 and 30% of the sample papers, with −8.7 and 10.6% publication surpluses. The other prolific subject categories are 'Materials Science Multidisciplinary', 'Chemistry Physical', 'Physics Applied', and 'Electrochemistry' with 21, 19, 14, and 13 sample papers, respectively.

The subject categories with the large publication surplus are 'Chemistry Multidisciplinary', 'Nanoscience Nanotechnology', 'Chemistry Physical' and 'Physics Condensed Matter' with 30.8, 10.6, 10.4 and 6.2% publication surplus, respectively. On the other hand, the subjects with least impact are 'Chemistry Analytical' and 'Electrochemistry' with −8.7 and −12.4% publication deficit, respectively.

17.3.9 Citation Impact

These sample papers receive over 51,000 citations as of January 2020. Thus, the average number of citations per paper is over 500.

17.3.10 Keywords

Although several keywords are listed in the appendix for the data sets of food and agriculture, sensors and nanotechnology, some of them are more significant. For the data set of agriculture and food, the most-prolific keyword is 'glucose' with 30 occurrences. The other prolific keywords are 'protein*', 'ion*', 'mercur*', 'dopamine', 'dna', and 'hg + 2' with over 7 occurrences each at the first instance. The other keywords are 'hydrogen peroxide', 'ascorbic acid', 'virus*', 'food*', 'pathogen*', 'uric acid*', 'cysteine', 'pesticide', and 'nerve agent*' with over 2 occurrences each.

For the data set of sensors, the most-prolific keywords are 'detect*' and '*sensor*'with 57 and 47 occurrences, respectively. The other keyword is '*sensing' with 19 occurrences. For the data set of nanotechnologies, the most prolific keyword is '*nano*' with 93 occurrences. The other prolific keywords are 'graphene', 'single-molecule*' and 'metal-organic framework*' with 19, 2, and 2 occurrences, respectively.

17.3.11 Research Fronts

The brief information about the key research fronts is provided in Table 17.8. The most-prolific research fronts for the analytes are 'glucose detection', 'protein detection' and 'toxic metal ion detection' with 27, 27 and 23 sample papers, respectively. The other prolific keywords for the analytes are 'dopamine detection', 'hydrogen peroxide detection', 'virus detection', and 'thiol detection'.

Table 17.8 Research fronts

	Res. fronts-analytes	Sample no. papers %	Res. fronts-nanotechnology	Sample papers %
1	Glucose	27	Nanoparticles	30
2	Proteins	27	Graphene	19
3	Metal ions	23	Quantum dots	9
4	Dopamine	6	Nanotubes	9
5	Hydrogen peroxide	5	Nanorods	3
6	Viruses	4	Surface-enhanced Raman scattering	3
7	Thiols	4	Nanoclusters	3
8	Vitamins	3	Metal-organic framework	3
9	Uric acids	3	Nanocomposites	3
10	Pesticides	2	Nanowires	2
11	Melamine	2	Single molecules	2

The most-prolific research fronts for the nanomaterials and nanotechnology are 'nanoparticles' and 'graphene' with 30 and 19 sample papers, respectively. The other prolific keywords for in this area are 'quantum dots' and 'nanotubes' with 9 sample papers each.

17.4 Discussion

The number of papers on nanobiosensing of agricultural and food analytes has increased to over 22,000 papers, forming over 1.4% of the research papers on the nanotechnology. The research has developed more in the technological aspects of this field, rather than the social and humanitarian pathways as there are only 3 papers in these pathways in the 'Web of Science'.

The articles and reviews are under-represented and over-represented by about 3% each in the sample papers, respectively (Table 17.1). Thus, the contribution of reviews by 5% of the sample papers in this field is not highly exceptional (c.f. Konur 2016a, b, c, d, e, f, 2017a, b, c, d, e, 2019b).

Ten authors from 7 institutions have 3 or more sample papers (Table 17.2). These prolific authors are mainly from China, the US, and Taiwan. These authors focus on 'metal ion detection', 'glucose detection', 'dopamine detection', 'thiol detection', 'pesticide detection', and 'protein detection'. It is significant that there is ample 'gender deficit' amongst these top authors as only one researcher is female.

Whilst 57 and 41% of the sample papers are published in the 2000s and 2010s, respectively, over 90% of the population papers are published in the 2010s (Fig. 17.1). This finding suggests that the population papers have built up on the sample papers, primarily published in the 2000s and 2010s, in the 2010s. These data suggest that this field of research is an 'emerging research', booming in the 2010s (Rotolo et al. 2015; Wang 2018).

The engagement of the institutions on the research in the nanobiosensing of agricultural and food analytes at the global scale is significant as 130 and over 6,300 institutions contribute to the sample and population papers, respectively. Eight top institutions publish 48 and 12% of the sample and population papers, respectively (Table 17.3). The top institutions are 'Chinese Academy of Sciences' 'of China and 'Northwestern University' of the US with 17 and 7 sample papers, respectively as they are significantly over-represented in sample papers with 10.5 and 6.2% publication surplus. As in the case of the top authors, most-prolific countries for these top instructions are the US and China with 3 institutions each.

It is significant that only 58% of the sample papers declare any funding. The most-prolific country for these top funding bodies is China with 7 funding bodies (Table 17.4). The other 3 funding bodies are from the US. The top funding body is the 'National Natural Science Foundation of China' (NSFC) of China, funding 31% and 39.2% of the sample and population papers, respectively, and with 8.2% publication deficit. These findings are in line with the studies showing the heavy research funding in China and the NSFC is the primary funding agency in China (Wang et al. 2012).

The sample and population papers are published by 32 and over 1,200 journals, respectively. It is significant that 11 top journals publish 79 and 12.8% of the sample and population papers, respectively (Table 17.5). The top journals 'Analytical Chemistry', 'Journal of the American Chemical Society', and 'Biosensors Bioelectronics' publish together 42 sample papers with 32.6% publication surplus. It is significant that there are no journals related to 'Agriculture' and 'Food Science and Technology' in this top journal table. On the contrary, there are 3 journals related to 'Nanotechnology' in this top journal table. The most-prolific subject categories are 'Chemistry Multidisciplinary', and 'Nanoscience Nanotechnology'.

In total, 18 and 117 countries contribute to the sample and population papers, respectively. The top country is China publishing 45 and 51% of the sample and population papers, respectively with 6% publication deficit (Table 17.6). The US follows China publishing 37 and 10% of the sample and population papers with 27% publication surplus. This finding is in line with the studies arguing that the US is not losing ground completely in science and technology (Leydesdorff and Wagner 2009). The other prolific countries are Singapore, Germany, South Korea, and Taiwan. These findings are in contrast to the studies showing that European countries have superior publication performance in nanotechnology (Youtie et al. 2008).

The European and Asian countries represented in this table publish altogether 11 and 65% of the sample papers whilst they publish 4.6 and 74.5% of the population papers, respectively. It is notable that the publication surplus for the US, these European and Asian countries is 27, 6.4 and −9.5%, respectively. It is further notable that the bulk of the Asian publication deficit originates from China and India with −6 and −7.4% publication deficit, respectively. This finding is in contrast with China's and India's efforts to be a leading nation in science and technology (Zhou and Leydesdorff 2006) but it is in line with the findings of Guan and Ma (2007), Youtie et al. (2008), and Karpagam et al. (2011).

The sample and population papers are indexed by 20 and 133 subject categories, respectively. For the sample papers, the top subject categories are 'Chemistry Multidisciplinary', 'Chemistry Analytical' and 'Nanoscience Nanotechnology' with 48, 34 and 30 papers, respectively (Table 17.7). The other prolific subject categories are 'Materials Science Multidisciplinary', 'Chemistry Physical', 'Physics Applied' and 'Electrochemistry'. It is notable that the publication surplus is most significant for 'Chemistry Multidisciplinary' with 30.8% surplus. The other high-impact subjects are 'Nanoscience Nanotechnology', 'Chemistry Physical', and 'Physics Condensed Matter'. On the other hand, the subjects with least impact are 'Chemistry Analytical' and 'Electrochemistry' with 8.7 and 12.4% publication deficit, respectively. These sample papers receive over 51,000 citations as of January 2020. This is highly exceptional (Konur 2016a, b, c, d, e, f, 2017a, b, c, d, e, 2019b).

Although several keywords are listed in the appendix for the data sets of data sets of analytes, sensors and nanotechnology, some of them are more significant. For the data set of agriculture and food, the most-prolific keyword is 'glucose' with 30 occurrences. The other prolific keywords are 'protein*', 'ion*', 'mercur*', 'dopamine', 'dna' and 'hg + 2' with over 7 occurrences each at the first instance. The other keywords are 'hydrogen peroxide', 'ascorbic acid', 'virus*', 'food*', 'pathogen*', 'uric acid*', 'cysteine', 'pesticide' and 'nerve agent*' with over 2 occurrences each.

For the data set of sensors, the most-prolific keywords are 'detect*' and '*sensor*' with 57 and 47 occurrences, respectively. The other keyword is '*sensing' with 19 occurrences.

For the data set of nanotechnologies, the most prolific keyword is '*nano*' with 93 occurrences. The other prolific keywords are 'graphene', 'single-molecule*', and 'metal-organic framework*' with 19, 2 and 2 occurrences, respectively.

As expected, these keywords provide valuable information about the pathways of the research on the nanobiosensing of agricultural and food analytes. Pragmatically, 11 research fronts emerge from the examination of the sample papers for both the analytes and nanomaterials (Table 17.8).

The most-prolific research fronts for the analytes are 'glucose detection', 'protein detection', and 'toxic metal ion detection' with 27, 27 and 23 sample papers, respectively. The other prolific keywords for the analytes are 'dopamine detection', 'hydrogen peroxide detection' 'virus detection' and 'thiol detection'. The most-prolific research fronts for the nanomaterials and nanotechnology are 'nanoparticles' and 'graphene' with 30 and 19 sample papers, respectively. The other prolific keywords in this area are 'quantum dots' and 'nanotubes' with 9 sample papers each. The key emphasis in these research fronts is the exploration of the structure-processing-property relationships for these nanobiosensors for nanobiosensing agricultural and food analytes (Konur and Matthews 1989; Nam et al. 2003; Song et al. 2010; Dong et al. 2012; Shan et al. 2009; Lee et al. 2007).

17.5 Conclusion

This chapter maps the research on the nanobiosensing and nanobioanalysis of agricultural and food analytes using a scientometric method. The significant size of over 22,000 population papers shows the public importance of this interdisciplinary research field. However, it is significant that the research has developed more in the technological aspects of these issues, rather than the social and humanitarian pathways. It is also significant to note that the research in this field has peaked in the 2010s with over 90% of the population papers as an 'emerging research' field.

Articles and reviews dominate the sample papers, primarily published in the 2000s and 2010s. This is also relevant for the population papers, mostly published in the 2010s. The data presented in the tables and in one figure show that a small number of authors, institutions, funding bodies, journals, keywords, research fronts, subject categories, and countries have shaped the research in this field.

It is notable that the authors, institutions and funding bodies from China dominate the research in this field. Furthermore, China and other Asian countries represented in the top country table are dominant in both sample and population papers. These findings show the importance of the development of efficient incentive structures for the development of the research in this field as in other fields. It seems that the US and, to a lesser extent, European countries, have efficient incentive structures for the development of the research in this field, contrary to the Asian countries. It is likely that the key reason behind that surge in the number of population papers during the last decade has been the development of efficient incentive structures in this field. It is also likely that with the continuation of these incentive structures in this field, the number of papers would continue to rise, reaching to over 50,000 papers at the end of the 2020s.

However, it seems there is more to do to reduce significant gender deficit in this field as in other field of the biomedical research (Xie and Shauman 1998). It further seems that although the research funding is a significant element of these incentive structures, it might not be a sole solution for increasing the incentives for the research in this field as in the case of Asian countries.

The data on the research fronts, keywords, source titles and subject categories provides valuable evidence for the interdisciplinary nature of the research in this field. These findings are in line with studies arguing for the interdisciplinarity of the research in nanomaterials (Schummer 2004; Meyer and Persson 1998). As only 19.4% of the population papers are indexed by the subject category of the 'Nanoscience and Nanotechnology', there is ample justification for the broad search strategy employed in this study. It is also significant that the contribution of agriculture-related subject categories to the research in this field is relatively negligible with about 1% contribution. The contribution of 'Food Science and Technology' is 7.5%.

The search strategy employed in this study is in line with the search strategies employed for nanotechnology research in the relevant studies (Porter et al. 2008; Mogoutov and Kahane 2007; Arora et al. 2013). Pragmatically, 11 research fronts for both analytes and nanomaterials emerge from the examination of the sample papers. The most-prolific research fronts for the analytes are 'glucose detection', 'protein detection' and 'toxic metal ion detection' with 27, 27 and 23 sample papers, respectively. The other prolific keywords for the analytes are 'dopamine detection', 'hydrogen peroxide detection', 'virus detection' and 'thiol detection'.

The most-prolific research fronts for the nanomaterials and nanotechnology are 'nanoparticles' and 'graphene' with 30 and 19 sample papers, respectively. The other prolific keywords for in this area are 'quantum dots' and 'nanotubes' with 9 sample papers each.

It is recommended that further scientometric studies are carried out for each of these research fronts building on the pioneering studies in these fields. Such studies might also be carried out for special types of nanomaterials such as nanoparticles, graphene and quantum dots.

Acknowledgements The contribution of the highly cited researchers in the field of nanobiosensing and nanobioanalysis of agricultural and food analytes has been gratefully acknowledged.

Appendix

Syntax: I and II and III

I Food and agriculture-related keywords

(TI = (*food* or *agr* or *pathogen* or pesticid* or "nerve agent*" or *toxin* or contaminant* or meat* or nutr* or dairy or adulteran* or milk* or melamine* or feed* or plant* or escherichia or listeri* or salmonella or *bacter* or campylobacter or insecticid* or organophosphat* or organophosphor* or (enzyme and (sensing or detection)) or tyrosin* or (food* and (*toxic* or analysis or safety)) or "polybrominated diphenyl ether*" or carbohydrate* or carcin* or staphylococc* or "hydrogen peroxide" or *microb* or vibrio or h2o2 or bacillus or shigella or *carbamate* or *bisphenol or *phenol* or antibiotic* or *virus* or poultry or vegetable* or fruit* or protein* or chloramphenicol or sulfamethazine or sulfadimidine or "e-coli" or sudan or chicken or tartrazine or colourant* or "ascorbic acid*" or lovastatin or rice or chicken or glutathione or "sunset yellow" or colourant* or clenbuterol or "crystal violet" or "*malachite green" or dye* or sulfonamide* or acetaminophen or arabidopsis or streptomycin or "picric acid" or nitrophenol or tyramine or phosmet or thiabendazole or concanavalin or carbofuran or mercury or "hg2+" or zearalenone or enrofloxacin or estradiol or *dopamine or "uric acid" or *epinephrine or pork or goat* or sheep or carbaryl or maize or acetamiprid or apple* or fungicide or tocopherol or prion or corn or ellagic or gallic or punica or myrtus or itriphal or anthelmintic* or rhizosphere or amlodipine or hydrochlorothiazide or fish* or honey or *tetracycline or glyphosate or herb* or allerg* or *albumin or wheat or gliadin or rutin or *flavonoid* or amaranth* or *cysteine* or biothiol* or salbutamol or residue* or ions or phenylethanolamine or tomato* or cholesterol or "soft drink*" or carmoisine or arsenic or *xanthine or glucose or pyrophosphate or "tert-butylhydroquinone" or chlorpyrifos or theanine or "parathion-methyl" or chilli or tricyclazole or ricin or sulfaquinoxaline or ciprofloxacin or gelatin* or methyldopa or methamidophos or "lactic acid" or tocotrienol* or triclosan or diet* or orange or patulin or pyrimethanil or fumonisin or deoxynivalenol or kanamycin or carbendazim or "cr(iii)" or nitrofuran or juice or gallate or ponceat or dichlorvos or risperidone or brucella or paracetamol or crop* or parathion or pear* or apple* or fusarium or morphine or capsaicin or wine* or omethoate or peach* or thiabendazole or histamine or beverage* or digoxin or parathion or eggs or freshness or bleomycin or nitrate or nitrite or trimethylamine or "l-dopa" or acetylcysteine or lysine or honokiol or magnolol or saccharide* or carmine or phenylenediamine or curcum* or "methylene blue" or aurantifolia or tobramycin or benzimidazole or mucin or tryptophan or polymyxa or phosmet or linezolid or bean* or hypochlorous or folic or sesam*or iodide or sulphite or infant) or WC = (food* or agr* or nutr* or hort* or plant*) NOT TI = (sensory or lithium*)

II Sensor-related keywords

TI = (*sensor* or *sensing or interferomet* or ellipsomet* or *detect*)

III Nanotechnology-related keywords

(TI = (*nano* or "quantum dot*" or "quantum-dot*" or plasmonic* or triboelectric* or "carbon dot*" or skyrmion* *or dendrimer* or* "carbon onion*" or "exfoliated graphite*" or "atomic crystal*" or "langmuir-blodgett" or "c-dot*" or electrospinning or ((graphit* or carbon) and *tubule*) or semimetal* or "polymer dot") OR TI = (*fulleren* or pcbm or *c60* or *c61* or *c70* or *c120* or "*butyric acid methyl ester*" or pc70bm or cnt or swnt or mwnt or swcnt or mwcnt or "c-60" or "c-61" or "c-70" or "c-120" or fulleri* or fullero* or buckyball*) OR TI = (mos2 or *graphene* or "transition metal dichalcogenide*" or mose2 or ws2 or wse2 or "black phosphorus" or phosphorene or "two-dimensional material*" or "low-dimensional material*" or "molybdenum disulfide" or "boron nitride layer*" or "g-c3n4" or mxene* or "2d *material*" or "atom thick" or "*atomically thin*" *or* graphane* or "graphitic carbon nitride*" or "atomic layer*" or "two-dimensional carbon*" or tmdc* or "waals heterostructure*" or "two-dimensional titan*" or "two-dimensional carbide*" or "2d crystal*" or "two-dimensional crystal*") OR TI = ("inorganic-organic hybrid*" or "organic-inorganic hybrid*" or "metal-organic framework*" or mof* or mesoporous or microporous or "hybrid inorganic-organic" or "hybrid organic-inorganic" or "*coordination polymer*" or* "zeolitic imidazolate framework*" or "*molecular sieve*" or "metal-organic material*" or "covalent organic framework*" or cofs or mesopore* or "organometal halide perovskite*" or "hybrid halide perovskite*" or mesocrystal*) OR TI = ("atomic force microscop*" or afm or "scanning tunneling microscop*" or "fluorescence microscop*" or sted or "*stimulated emission depletion microscop*" or* "localized surface plasmon resonance" or lspr or "surface-enhanced ramanspectroscop*" or sers or "scanning probe microscop*" or "optical reconstruction microscop*" or "dynamic force spectroscop*") OR TI = ("**molecular machin*" or* "*molecular motor*" or "*molecular self-assembl*" or superlens or "quantum wire*" or "quantum well*" or "molecular photov*" or "*molecular assembl*" or "*molecular wire*" or "*molecular device*" or supramolecul* or "*molecular engineering" or "single atom*" or "single molecul*" or "dna machine*" or "molecular rotor*" or "molecular electronic device*" or "molecular biomimetic*" or "*molecular junction*" or "atomic scale" or "*sub nm*" *or* "single layer*" or "atomic sized" or *nems*) OR SO = ("ACS Nano" or "Advances in Nano*" or "Applied Nanoscience" or "Beilstein Journal of Nano*" or "Cancer Nano*" or Chem Nanomat or "Current Nano*" or "Environmental Science—Nano" or "Frontiers of Nano*" or Fullerenes* or "IEEE Transactions on Nano*" or "IET Nano*" or "International Journal of Nano*" or "Journal of Biomedical Nano*" or "Journal of Computational and Theoretical Nano*" or "Journal of Experimental Nano*" or "Journal of Nano*" or "Microporous and Mesoporous*" or Nano* or "Nature Nano*" or "NPJ 2d" or "Recent Patents on Nano*" or "RSC Nano*" or Small* or "Wiley Interdisciplinary Reviews—Nano*") or SO = ("ACS Applied Nano*" or "Advances in Natural Sciences—Nano*" or "European Journal of Nano*" or "IEEE Nano*" or "International Journal of Nano*" or "International Nano*" or "Journal of Nano*" or Nano* or "Proceedings of the Institution of Mechanical Engineers Part N-Journal of Nano*")) NOT TI = (nanog or nanogram* or "nano-gram*" or nanosecond* or "nano-second*" or nanoliter* or nanolitre* or nanokelvin or nanoamper* or "nano-amper*" or nano2 or nano3 or nanoplan* or nanoflag* or nanoprot* or nanobacter* or nanor or nanobody or nanostring or nanosatel*)

References

Arora SK, Porter AL, Youtie J, Shapira P (2013) Capturing new developments in an emerging technology: an updated search strategy for identifying nanotechnology research outputs. Scientometrics 95:351–370

Carvalho FP (2006) Agriculture, pesticides, food security and food safety. Environ Sci Policy 9:685–692

Docampo D, Cram L (2019) Highly cited researchers: a moving target. Scientometrics 118:1011–1025

Dong XC, Xu H, Wang XW et al (2012) 3D graphene-cobalt oxide electrode for high-performance supercapacitor and enzymeless glucose detection. ACS Nano 6:3206–3213

Franz CM, Holzapfel WH, Stiles ME (1999) Enterococci at the crossroads of food safety? Int J Food Microbiol 47:1–24

Garfield E (1955) Citation indexes for science. Science 122:108–111

Garfield E (1972) Citation analysis as a tool in journal evaluation. Science 178:471–479

Grunert KG (2005) Food quality and safety: consumer perception and demand. European Review of Agricultural Economics 32:369–391

Guan JC, MA N (2007) China's emerging presence in nanoscience and nanotechnology—a comparative bibliometric study of several nanoscience 'giants'. Res Policy 36:880–886

Karpagam R, Gopalakrishnan S, Natarajan M, Babu BR (2011) Mapping of nanoscience and nanotechnology research in India: a scientometric analysis, 1990–2009. Scientometrics 89:501

Konur O (2000) Creating enforceable civil rights for disabled students in higher education: an institutional theory perspective. Disabil Soc 15:1041–1063

Konur O (2002a) Access to nursing education by disabled students: rights AND duties of nursing programs. Nurse Educ Today 22:364–374

Konur O (2002b) Assessment of disabled students in higher education: current public policy issues. Assess Eval High Educ 27:131–152

Konur O (2002c) Access to employment by disabled people in the UK: is the disability discrimination act working? Int J Discrim Law 5:247–279

Konur O (2006a) Participation of children with dyslexia in compulsory education: current public policy issues. Dyslexia 12:51–67

Konur O (2006b) Teaching disabled students in higher education. Teach High Educ 11:351–363

Konur O (2007a) A judicial outcome analysis of the disability discrimination act: a windfall for the employers? Disabil Soc 22:187–204

Konur O (2007b) Computer-assisted teaching and assessment of disabled students in higher education: the interface between academic standards and disability rights. J Comput Assist Learn 23:207–219

Konur O (2011) The scientometric evaluation of the research on the algae and bio-energy. Appl Energy 88:3532–3540

Konur O (2012a) Evaluation of the research on the social sciences in Turkey: a scientometric approach. Energy Educ Sci Technol Part B: Soc Educ Stud 4:1893–1908

Konur O (2012b) Prof. Dr. Ayhan Demirbas' scientometric biography. Energy Educ Sci Technol Part A: Energy Sci Res 28:727–738

Konur O (2012c) The evaluation of the biogas research: a scientometric approach. Energy Educ Sci Technol Part A: Energy Sci Res 29:1277–1292

Konur O (2012d) The evaluation of the educational research: a scientometric approach. Energy Educ Sci Technol Part B: Soc and Educ Stud 4:1935–1948

Konur O (2012e) The evaluation of the global energy and fuels research: a scientometric approach. Energy Educ Sci Technol Part A: Energy Sci Res 30:613–628

Konur O (2012f) The evaluation of the research on the arts and humanities in Turkey: a scientometric approach. Energy Educ Sci Technol Part B: Soc Educ Stud 4:1603–1618

Konur O (2012g) The evaluation of the research on the biodiesel: a scientometric approach. Energy Educ Sci Technol Part A: Energy Sci Res 28:1003–1014

Konur O (2012h) The evaluation of the research on the bioethanol: a scientometric approach. Energy Educ Sci Technol Part A: Energy Sci Res 28:1051–1064

Konur O (2012i) The evaluation of the research on the biofuels: a scientometric approach. Energy Educ Sci Technol Part A: Energy Sci Res 28:903–916

Konur O (2012j) The evaluation of the research on the biohydrogen: a scientometric approach. Energy Educ Sci Technol Part A: Energy Sci Res 29:323–338

Konur O (2012k) The evaluation of the research on the microbial fuel cells: a scientometric approach. Energy Educ Sci Technol Part A: Energy Sci Res 29:309–322

Konur O (2012l). The scientometric evaluation of the research on the production of bioenergy from biomass. *Biomass Bioenergy* 47:504–515. SCI

Konur O (2012m) The scientometric evaluation of the research on the deaf students in higher education. Energy Educ Sci Technol Part B: Soc Educ Stud 4:1573–1588

Konur O (2012n) The scientometric evaluation of the research on the students with ADHD in higher education. Energy Educ Sci Technol Part B: Soc Educ Stud 4:1547–1562

Konur O (2016a) Scientometric overview in nanobiodrugs. In: Holban AM, Grumezescu AM (eds) Nanoarchitectonics for smart delivery and drug targeting. Elsevier, Amsterdam, pp 405–428

Konur O (2016b) Scientometric overview regarding nanoemulsions used in the food industry. In: Grumezescu AM (ed) Emulsions: nanotechnology in the agri-food industry. Elsevier, Amsterdam, pp 689–711

Konur O (2016c) Scientometric overview regarding the nanobiomaterials in antimicrobial therapy. In: Grumezescu AM (ed) Nanobiomaterials in antimicrobial therapy. Elsevier, Amsterdam, pp 511–535

Konur O (2016d) Scientometric overview regarding the nanobiomaterials in dentistry. In: Grumezescu AM (ed) Nanobiomaterials in dentistry. Elsevier, Amsterdam, pp 425–453

Konur O (2016e) Scientometric overview regarding the surface chemistry of nanobiomaterials. In: Grumezescu AM (ed) Surface chemistry of nanobiomaterials. Elsevier, Amsterdam, pp 463–486

Konur O (2016f). The scientometric overview in cancer targeting. In: Holban AM, Grumezescu A (eds) Nanoarchitectonics for smart delivery and drug targeting. Amsterdam, Elsevier, pp 871–895

Konur O (2017a) Recent citation classics in antimicrobial nanobiomaterials. In: Ficai A, Grumezescu AM (eds) Nanostructures for antimicrobial therapy. Elsevier, Amsterdam, pp 669–685

Konur O (2017b) Scientometric overview in nanopesticides. In: Grumezescu AM (ed) New pesticides and soil sensors. Elsevier, Amsterdam, pp 719–744

Konur O (2017c) Scientometric overview regarding oral cancer nanomedicine. In: Andronescu E, Grumezescu AM (eds) Nanostructures for oral medicine. Elsevier, Amsterdam, pp 939–962

Konur O (2017d) Scientometric overview regarding water nanopurification. In: Grumezescu AM (ed) Water purification. Elsevier, Amsterdam, pp 693–716

Konur O (2017e) Scientometric overview in food nanopreservation. In: Grumezescu AM (ed) Food preservation. Elsevier, Amsterdam, pp 703–729

Konur O (2017f) The top citation classics in alginates for biomedicine. In: Venkatesan J, Anil S, Kim SK (eds) Seaweed polysaccharides: isolation, biological and biomedical applications. Elsevier, Amsterdam, pp 223–249

Konur O (2018a). Scientometric evaluation of the global research in spine: an update on the pioneering study by Wei et al. Eur Spine J 27:525–529

Konur O (2018b) Bioenergy and biofuels science and technology: scientometric overview and citation classics. In: Konur O (ed) Bienergy Biofuels. CRC Press, Boca Raton, pp 3–63
Konur O (2019a) Cyanobacterial bioenergy and biofuels science and technology: a scientometric overview. In: Mishra AK, Tiwari DN, Rai AN (eds) Cyanobacteria: from basic science to applications. Elsevier, Amsterdam, pp 419–442
Konur O (2019b) Nanotechnology applications in food: a scientometric overview. In Pudake RN, Chauhan N, Kole C (eds) *Nanoscience for sustainable agriculture*. Springer, Cham, pp. 683–711
Konur O, Matthews FL (1989) Effect of the properties of the constituents on the fatigue performance of composites: a review. Composites 20:317–328
Lee JS, Han MS, Mirkin CA (2007) Colorimetric detection of mercuric ion (Hg2+) in aqueous media using DNA-functionalized gold nanoparticles. Angew Chem-Int Ed 46:4093–4096
Leydesdorff L, Wagner C (2009) Is the United States losing ground in science? a global perspective on the world science system. Scientometrics 78:23–36
McLaughlin MJ, Parker DR, Clarke JM (1999) Metals and micronutrients–food safety issues. Field Cr Res 60:143–163
Meyer M, Persson O (1998) Nanotechnology—interdisciplinarity, patterns of collaboration and differences in application. Scientometrics 42:195–205
Mogoutov A, Kahane B (2007) Data search strategy for science and technology emergence: a scalable and evolutionary query for nanotechnology tracking. Res Policy 36:893–903
Nam JM, Thaxton CS, Mirkin CA (2003) Nanoparticle-based bio-bar codes for the ultrasensitive detection of proteins. Science 301:1884–1886
North DC (1990) Institutions, institutional change and economic performance. Cambridge University Press, Cambridge, MA
North DC (1991) Institutions. J Econ Perspect 5:97–112
Porter AL, Youtie J, Shapira P, Schoeneck DJ (2008) Refining search terms for nanotechnology. J Nanopart Res 10:715–728
Rotolo D, Hicks D, Martin BR (2015) What is an emerging technology? Res Policy 44:1827–1843
Schummer J (2004) Multidisciplinarity, interdisciplinarity, and patterns of research collaboration in nanoscience and nanotechnology. Scientometrics 59:425–465
Shalaby AR (1996) Significance of biogenic amines to food safety and human health. Food Res Int 29:675–690
Shan CS, Yang HF, Song JF et al (2009) Direct electrochemistry of glucose oxidase and biosensing for glucose based on graphene. Anal Chem 81:2378–2382
Song YJ, Qu KG, Zhao C, Ren JS, Qu XG (2010) Graphene oxide: intrinsic peroxidase catalytic activity and its application to glucose detection. Adv Mater 22:2206–2210
Srivastava AK, Dev A, Karmakar S (2018) Nanosensors and nanobiosensors in food and agriculture. Environ Chem Lett 16:161–182
Wang J, Shapira P (2011) Funding acknowledgement analysis: an enhanced tool to investigate research sponsorship impacts: the case of nanotechnology. Scientometrics 87:563–586
Wang Q (2018) A bibliometric model for identifying emerging research topics. J Assoc Inf Sci Technol 69:290–304
Wang X, Liu D, Ding K, Wang X (2012) Science funding and research output: a study on 10 countries. Scientometrics 91:591–599
Xie Y, Shauman KA (1998). Sex differences in research productivity: new evidence about an old puzzle. Am Sociol Rev 63:847–870
Youtie J, Shapira P, Porter AL (2008) Nanotechnology publications and citations by leading countries and blocs. J Nanopart Res 10:981–986
Zhou P, Leydesdorff L (2006) The emergence of China as a leading nation in science. Res Policy 35:83–104

Part III
Biosensors in Animal and Fishery Sciences

Chapter 18
Biosensors: Modern Tools for Disease Diagnosis and Animal Health Monitoring

Anuj Tewari, Beenu Jain, Basanti Brar, Gaya Prasad, and Minakshi Prasad

Abstract Animals play a vital role in our lives as they provide milk, meat, and other by-products for daily consumption. At the same time, their health directly/indirectly affects human beings as we both share a common environment. The One Health concept makes it imperative that the animals should remain healthy as it will have an impact on our health as well as economy. Animal disease diagnosis, their health monitoring, prevention, and control of diseases are important from this aspect. Tools for rapid diagnosis of animal diseases are crucial for early diagnosis and imposing control measures. Molecular techniques such as, genome sequencing, restriction fragment length polymorphism (RFLP), DNA microarray, PCR, and real-time PCR are some of the rapid techniques, but they need expertise and sophisticated labs. Conventional methods such as isolation of the pathogen, serological techniques, are laborious and time taking process. Therefore, to overcome the said limitations of conventional and molecular tools, biosensors are the better alternatives as they are rapid and can be used as pen-side diagnostic tests. They can provide test results in a few minutes under field conditions. In this chapter, we give an elaborative description of various types of biosensors and their utility in animal disease diagnostics and health monitoring.

Keywords Biosensors · Animal disease diagnosis · Field-based tests · Rapid tests · Animal health monitoring

---

A. Tewari (✉)
Department of Veterinary Microbiology, College of Veterinary and Animal Sciences, Govind Ballabh Pant University of Agriculture and Technology, Pantnagar 263145, Uttarakhand, India
e-mail: anuj474@gmail.com

B. Jain
Department of Animal Husbandry, Lucknow, Uttar Pradesh, India

G. Prasad
International Institute of Veterinary Education and Research, Rohtak, Haryana, India

B. Brar · M. Prasad (✉)
Department of Animal Biotechnology, College of Veterinary Sciences, Lala Lajpat Rai University of Veterinary and Animal Sciences, Hisar, Haryana, India
e-mail: minakshi.abt@gmail.com

R. N. Pudake et al. (eds.), *Biosensors in Agriculture: Recent Trends and Future Perspectives*, Concepts and Strategies in Plant Sciences,
https://doi.org/10.1007/978-3-030-66165-6_18

18.1 Introduction

Globally, infectious diseases are the leading cause of mortality in man and animals. Infectious diseases of both domestic and wildlife pose a serious health risk to the human race as exhibited by the current ongoing COVID-19 pandemic. These infectious diseases can be of zoonotic origin affecting not only human health, but also, animal health and production. Such diseases pose a serious threat to a country's growth, economy, and health perspectives. To prevent the spread of infectious diseases it is imperative to develop rapid and sensitive tests for early detection of the causative agents. It aids in adopting suitable preventive measures, compartmentalization of the infected zones, and designing strategies to prevent further spread of the disease. The viral agents can be detected in the clinical samples by gold standard virus isolation in cell culture system. This method has well-known merits and demerits such as economic investment, expert manpower, and time-consuming lengthy protocols (from days to weeks). On the other hand, molecular detection methods such as conventional Polymerase Chain Reaction (PCR), Real-time PCR, although they are both sensitive and specific, cannot be employed for pen-side diagnosis due to the involvement of sophisticated instruments.

In such a scenario, the development of a rapid, sensitive, and specific pen-side test is the need of the hour and is a challenge for the scientific community to combat the emerging and re-emerging diseases. With the limited infrastructure for the diagnosis of the emerging and re-emerging infection, biosensors can be an effective tool to perform field-based diagnosis and animal health monitoring. Easy handling, user-friendly, and minimum processing endows biosensor an effective tool for disease diagnosis.

In the past few decades, a lot of work has been done on this aspect for the detection of both viral and bacterial pathogens. Various molecules like nucleic acids, antibodies, antigens, and proteins of animal origin are detected in the clinical samples in these biosensing elements.

18.2 Biosensors Principles and Types

According to The National Research Council (part of the US National Academy of Science), a biosensor is defined as a detection device that incorporates a living organism or product derived from a living system as the biorecognition element (BRE) or a bioreceptor and a transducer (electrochemical, optical, or mass-based) to convert a biological reaction into a quantifiable signal. The BRE can be enzymes, nucleic acids, antibodies, aptamers, bacterial cells, bacteriophages. The first biosensor was found by Clark and Lyon in 1962 for glucose detection.

The boost up in the field of biosensor came from the pioneering work done by a group of scientists in the year 1962–1969 (Clark and Lyons 1962; Guilbault et al.

1962; Updike and Hicks 1967; Guilbault and Montalvo 1969). Later on, an electrochemical biosensor had been described using ferrocene to detect electroactive species (Di Gleria et al. 1986). This invention led to the commercialization of a glucose pen by the company Medisens. After these reports, there has been a large amount of work done in the domain of biosensors which has an extensive application for environmental sensing and biological monitoring (Luong et al. 2008).

The principle of a biosensor is based on the detection of a biomarker molecule in the clinical sample (Fig. 18.1). This biomarker can be any protein, antigen, or antibody of the infectious agent. The biomarker molecule is detected by the bioreceptor which is immobilized on a chip or another base. This bioreceptor could be DNA, RNA, monoclonal antibody, protein, or any cell. However, it is important to choose the bioreceptor carefully as it is the deciding factor for the sensitivity and specificity of a biosensor. Although there are several types of bioreceptors, the most common types used are nucleic acids, enzymes, and antibodies. The interaction between the biomarker and bioreceptor generates signals via transductor which are read and interpreted. The signals generated give information regarding the presence or absence of the pathogen in the sample.

Biosensors can be classified on the basis of biorecognition elements or transducer, more details are discussed below.

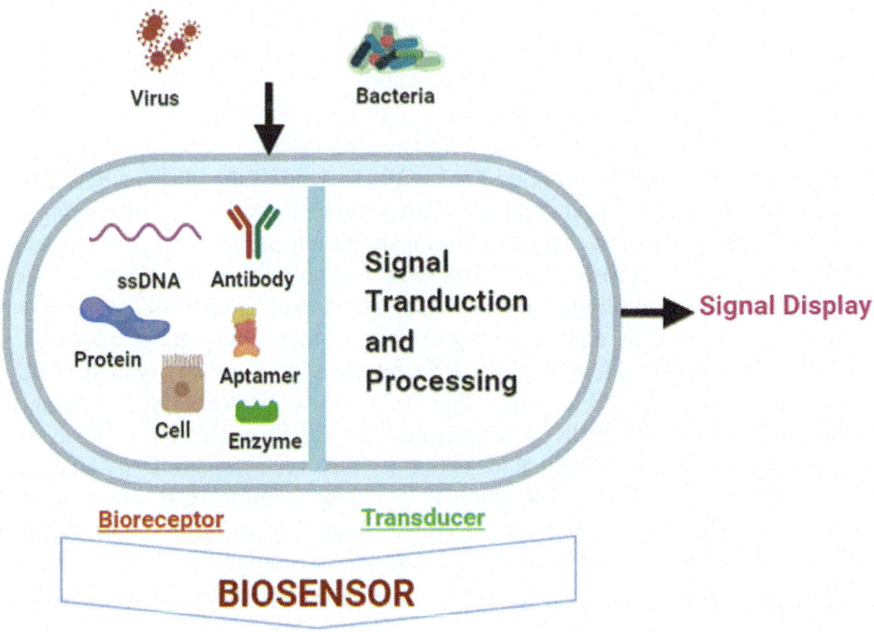

Fig. 18.1 Principle of biosensor for disease/pathogen detection

18.2.1 Biorecognition Element (BRE)

The type of BRE to be used depends on the target molecule to be recognized. Each of the BRE has its own advantage and disadvantage, hence, its selection depends on the application. Various types of BREs used are as follows:

18.2.1.1 Antibodies

Antibodies are the most commonly used BRE and their use does not require antigen purification during sample addition (Chambers et al. 2008). Monoclonal, polyclonal, and recombinant antibodies can be used. Monoclonal antibodies recognize a single antigenic epitope hence detects the antigen with less sensitivity in comparison to the polyclonal antibodies. Polyclonal antibodies recognize several epitopes and are able to detect the antigen with more sensitivity than the monoclonal antibodies. However, polyclonal antibodies show batch-to-batch variation in titre and reactivity in comparison to monoclonal antibodies which show unique specificity in all batches (Nelson et al. 2000). The third generation of antibodies known as recombinant antibodies are less expensive and time-consuming in production (Sharon et al. 2005; Haurum 2006; Chambers et al. 2008).

18.2.1.2 Nucleic Acids

Nucleic acids (NAs) can bind with the complementary sequence present in the sample through hybridization (Wetmur and Fresco 1991). One of the class of NAs that is used as BRE is Nucleic Acid Aptamers (NAAs) which are capable of binding two proteins (Ni et al. 2011). Such kind of biosensors which employ the use of NAAs are called as genosensors (Tosar et al. 2010; Martinkova et al. 2017). NAAs are high-affinity short stretch of oligonucleotides (Chambers et al. 2008). NAAs are easy to synthesize, are foldable, and easy to store, hence, they offer similar attributes as antibodies. Disadvantages of NAAs are degradation, cross-reactivity, and labeling costs (Chambers et al. 2008; Lakhin et al. 2013). Aptazymes are types of aptamers which have catalytic properties like enzymes (Chambers et al. 2008). Other types of aptamers are peptide nucleic acids (PNAs) which are synthetic DNA analogues having polyamide backbone (Chambers et al. 2008; Bonifazi et al. 2012; Briones and Moreno 2012; Gupta et al. 2017). Since, PNAs do not have any charge they have higher hybridization characteristics than NAs (Bonifazi et al. 2012; Briones and Moreno 2012).

18.2.1.3 Enzymes and Proteins

Enzymes are exclusively specific for their substrate hence are an excellent BRE for biosensors. Enzymes consist of protein components, cofactors, or prosthetic groups (Elnashar 2010). Enzymes are used extensively for biosensing purposes as they can measure a variety of reactions and the products generated thereof.

18.2.1.4 Cells

In certain cases, the immobilization of the enzymes may be difficult, or the isolation and purification of the enzymes may not be possible. In such a scenario, the whole cells offer a better alternative than enzymes as the enzymes inside the cell are stable for a long time. Also, the cells have multiple enzymes that can be used for several analytes in a single testing. The immobilized cells can also be genetically modified to express specific proteins to detect the analytes in the extracellular matrix (Fine et al. 2006; Jung et al. 2014; Vopálenská et al. 2015).

18.2.1.5 Biomimetics

In these kinds of biosensors, the non-biological receptors are used as BREs. These biosensors can mimic the original biological bioreceptor. The various forms of biomimetics are molecular imprinted polymers (MIP), membrane mimics, self-assembled monolayers (SAM), nanoparticles, nanostructured materials, and quantum dots (QDs). Molecular imprints aim to design artificial receptors like artificial antibodies. Nanoparticles and nanostructures along with other receptors can also act as biomimetics.

18.2.2 Classification of Biosensors on the Basis of Transducer

Transducers are the important components that determine the specificity of a biosensor as they are responsible to translate the biorecognition signals to measurable signals. Biosensors are mainly classified as Electrochemical, Mass-based, and Optical biosensors. These are further subclassified.

18.2.2.1 Electrochemical Biosensors

The electrochemical (EC) biosensors are the most common types of biosensors. They are inexpensive and have good bio-interaction. Due to the biochemical interaction on the sensor surface, the potential difference is measured as a recognizable signal. These biosensors have been further classified into conductometric, impedimetric, potentiometric, and amperometric types which are based on the conductance, impedance, potential, and current generation, respectively (Lazcka et al. 2007; Setterington and Alocilja 2012).

The EC biosensors are the most successful biosensors which have been used commercially. However, the disadvantage associated with the EC biosensor is the random orientation and their immobilization without getting denatured (Jung et al. 2008). Potentiometric sensors measure the change in the potential at the electrode happening after the recognition of the analyte. However, this method comes with a limitation that for large molecules (bacteria) it cannot provide sensitive and specific signals. Amperometric sensors are the kind of sensors in which current is generated after oxidation or reduction due to analyte and bioreceptor interaction. The most common example of this type of biosensor is the measurement of glucose which is based on the amperometric sensing of glucose. These kinds of sensors impart good sensitivity but low specificity. Impedimetric sensors are the kind of biosensors which are good for detection of whole bacteria. The sensors are label-free, of low cost, high sensitivity, and are not affected by other analytes present in the sample matrix. Moreover, these can be miniaturized easily, hence, easily translated to the point of care systems. Conductometric biosensors measure the ability of an analyte to conduct current between electrodes or the reference electrodes. Several enzymatic reactions are capable of changing the ionic strength of the sample which can then be measured by a conductometric instrument.

With the recent advances in nanotechnology, the incorporation of nanoparticles has improved the diagnostic performance of the electrochemical biosensors by increasing the signal transduction. The signal transduction, which can be increased by using the nanomaterials along with the BREs can increase the signal significantly (Fig. 18.2). Nanoparticles provide increased surface area for more binding of the analyte and also have high conductance which results in increased signal intensity. The nanomaterials used are gold nanoparticles, iron oxide, graphene, and carbon nanotubes.

18.2.2.2 Optical Biosensors

Optical biosensors are useful in attending high specificity, sensitivity, and rapidity. It has been further divided into label-free and real-time detection. Based on the mechanism of the signal generation they have been further categorized as colorimetric sensors, surface plasmon resonance (SPR), fluorescence-based sensors, surface-enhanced Raman spectroscopy (SERS), and chemiluminescence-based sensors.

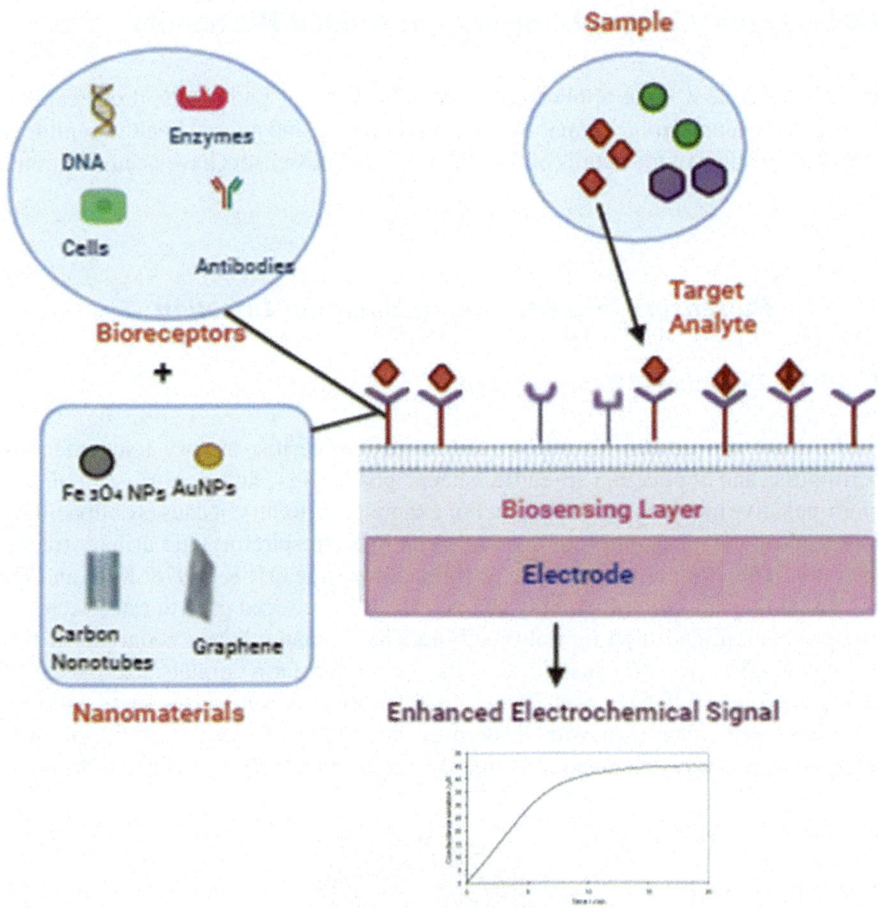

Fig. 18.2 Nanomaterials-based electrochemical biosensor

18.2.2.3 Mass-Based Biosensors

These biosensors are based on the principle of signal production which depend on the mass of chemicals interacting with a vibrating piezoelectric quartz crystal. Mass-based biosensors are label-free and offer increased sensitivity and specificity. The signal transduction is based on the minute change in the mass occurring on the sensor surface and is sensed by resonance phenomena (Turner and Zhang 2001). Various categories of mass-based biosensors are microcantilever-based (MCL) biosensors, surface acoustic wave (SAW) biosensors, and quartz crystal microbalances (QCM).

18.3 Applications of Biosensors in Animal Husbandry

Biosensors have a large application in the detection of pathogens, drug residues, food quality monitoring, animal location monitoring, and animal health monitoring (Fig. 18.3). Some of the details of the application of biosensors have been mentioned below.

18.3.1 Biosensors for Detection of Bacterial Infection

18.3.1.1 Detection of *Escherichia coli*

Escherichia coli causes disease related to gastroenteritis, urinary tract infection, peritonitis, and septicemia in cattle, sheep, goats, pigs, and poultry. *E. coli* is a gram-negative rod-shaped bacterium. For example, in poultry, it causes colibacillosis followed by the localization of this pathogen in the respiratory and urinary tract of the birds. The virulent strains causing colibacillosis are O1: K1, O78: K80, and O2: K1. In poultry, *E. coli* infection is characterized by reduced growth rate, decreases egg production, and high mortality subsequently causing heavy economic loss to the stakeholders. *E. coli* causes colibacillosis in other farm animals also like cattle, goats, and pigs (Bélanger et al. 2011). In newborn calves it causes septicemia and in milking cattle causes mastitis. Moreover, due to the excessive use of antibiotics, there is an increased incidence of antibiotic resistance in *E. coli*. Thus, it becomes

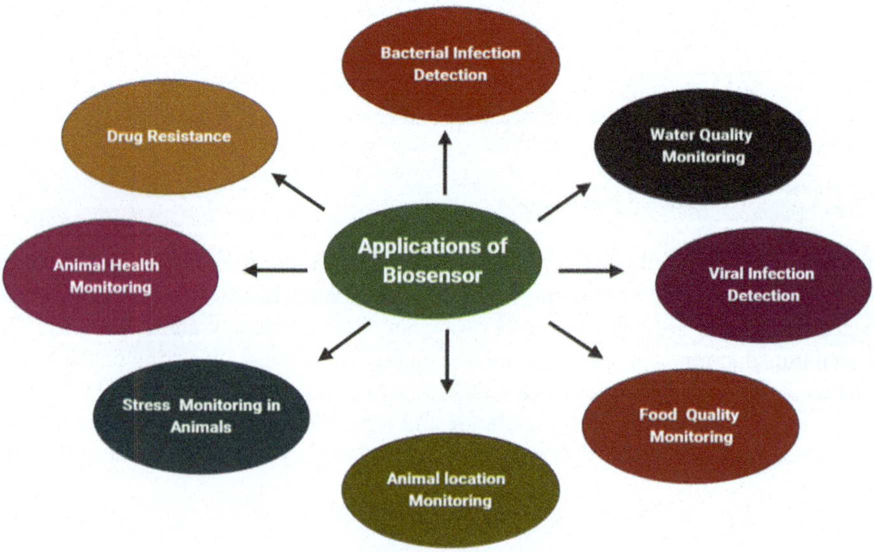

Fig. 18.3 Probable applications of biosensor in animal husbandry

imperative to detect the virulent strains of *E. coli* and to differentiate between them from the non-pathogenic strains.

The conventional and molecular methods used for the detection of *E. coli* infection are isolation of bacteria from clinical samples, ELISA, and PCR. However, these methods involve high cost, time, and manpower (Hobson et al. 1996; Fournier-Wirth et al. 2006). To overcome these hurdles advanced methods involving the use of biosensor have been explored. For detection of *E. coli*, the novel biosensors such as surface plasmon resonance (Yazgan et al. 2014), quartz crystal microbalance system (Jiang et al. 2011), chemiluminescence (Zhang et al. 2014), and electrochemistry (Kim et al. 2015a; Guo et al. 2016; Idil et al. 2017) have been explored. The *E. coli* cells are captured by specific antibodies directed against *E. coli* antigens like flagella. The signals generated are subsequently amplified by using suitable detectable labels such as fluorophores, enzymes, and biofunctionalized nanoparticles. The biosensors based on sandwich format have been developed for the rapid detection of *E. coli*. In an electrochemical biosensor (Jaffrezic-Renault et al. 2007) the *E. coli* cells are captured by anti-LPS antibodies coupled with magnetic nanoparticles which are conducted on the graphite ink electrode. The use of immunomagnetic beads aid in the detection of analyte present in the test sample.

The conductometric method can detect *E. coli* from the culture from 1 to 10^3 CFU mL^{-1}. The immunosensors based on conductometric method have been used successfully for detection of *Pseudomonas aeruginosa* and *Acinetobacter baumannii* from 10 to 10^3 CFU mL^{-1}. These pathogens were found to be undetectable using immunoblot methods. This test was also found to be specific as the gram-positive bacteria *Staphylococcus epidermidis* was found to be undetectable (El Ichi et al. 2014).

A point of care test using stacked paper membranes has also been proposed (Eltzov and Marks 2016, 2017) which can quantify *E.coli* in less than five minutes. When the infected clinical sample is added to the bottom layer of the membrane, the bacterial cells are pushed up to the top layer during which they bind with anti-*E. coli* antibodies conjugated with horseradish peroxidase (HRPO) which allows measurement of generated signals. In the case of the negative sample, the anti- *E. coli* antibodies are prevented to move to the topmost membrane due to their binding with the immobilized bacterial cells on the capture layer. HRPO specific substrate produces signals only with *E. coli* cells bound with HRPO labeled antibodies. This test is 1000 times more sensitive than ELISA and was able to quantify upto 100 cells mL^{-1}. Other added advantages were its rapidity and portability.

18.3.1.2 *Detection of* Clostridium Perfringens

C. perfringens is an important enteric pathogen and all domestic animals and poultry are infected by this pathogen. It is a gram-positive spore-forming bacteria with 17 different toxins (Gibert et al. 1997; Petit et al. 1999) and is also an important agent of foodborne illness. Cell-based biosensors are one of the emerging technologies which have mammalian cells as sensing elements to detect the pathogens or toxins in the clinical and food samples (Yoo and Lee 2016).

For this purpose, the mammalian cells fixed on 96 well plates and modified electrodes have been used (Banerjee and Bhunia 2010) to detect *C. perfringens* and their toxins in the samples, as each type of *C. perfringens* carries a unique sequence of virulence genes. In a microarray system, oligo probes against six toxin-producing sequences of *C. perfringens* have been designed and used successfully (Sergeev et al. 2004). The fluorescently labeled PCR amplicon is hybridized with the oligoprobes which is subsequently read to detect the presence of *C. perfringens* in the test sample. This platform enables the simultaneous detection of several bacterial strains and their toxins detected simultaneously (Volokhov et al. 2011).

Toxins of *C. perfringens* can also be detected by using specific antibodies in an immunosensor. It has been used successfully for the detection of epsilon toxin from *C. perfringens* (Palaniappan et al. 2013). The monoclonal antibody specific to the epsilon toxin was fixed on the carbon nanotube and was able to detect toxins with comparable sensitivity to ELISA.

18.3.1.3 Detection of *Campylobacter*

Campylobacter causes disease in wild animals, pets, domestic livestock, and human beings. *C. jejuni* and *C. coli* mainly infect cattle sheep, dogs, cats, pigs, and turkeys. *C. fetus* mainly infects cattle, goats, and sheep. *Campylobacter* causes enteritis with diarrhea, fever, vomiting, abortion, and infertility. To detect *Campylobacter*, various bioreceptor molecules like proteins, aptamers, nucleic acids, antibodies, and cells have been used. Out of these, DNA-based biosensors are very useful due to their low cost, specificity, and less time required to detect the pathogen.

An organic light-emitting diode (OLED) biosensor has been used for the detection of *Campylobacter* in meat samples (poultry) through capturing *Campylobacter* DNA via a labeled DNA probe on a glass slide. This captured DNA is detected by a secondary DNA probe labeled with a fluorophore. It allows detecting *Campylobacter* in the sample with a sensitivity of 0.37 $ng\mu L^{-1}$ DNA and 1.5×10^1 $CFUg^{-1}$ (Manzano et al. 2015).

A biosensor based on surface plasmon resonance (SPR) has also been proposed for the detection of *C. jejuni* (Wei et al. 2007); which can detect *C. jejuni* up to 10^3 CFU mL^{-1}. An optical biosensor based on SPR and diffraction optic technology has also been developed (Gnanaprakasa et al. 2011) targeting the hippuricase gene of *C. jejuni.* The complementary probe was thiolated and subsequently fixed on the gold surface of the chip. After hybridization of probe and hippuricase gene, it led to the transduction of an optical signal which could quantify *C. jejuni* up to 5 nM. In another study (Wadl et al. 2009) a lateral flow device was developed to detect *C. jejuni* or *C. coli* which can detect pathogen up to 7.3 log of CFU g^{-1}.

In recent study, functionalized magnetic particles with tripled silica layer and coated with anti-*C. jejuni* antibodies have been found to detect *C. jejuni* in water with a detection limit of 10^2 CFU mL^{-1}(Zhang et al. 2015).

18.3.1.4 Detection of *Salmonella*

Salmonella is a gram-negative rod-shaped bacterium which is associated with diarrhea and food poisoning in man and animals. An important reservoir of *Salmonella* is poultry. Several serotypes of *Salmonella* are present which can infect both man and animals (Alcaine et al. 2006). Infection arises through contaminated food and drinking water. Several workers have tried to develop different forms of biosensors for the detection of *Salmonella.* A wireless and mass sensitive biosensor for the detection of *S. typhimurium* has also been developed which was able to detect 1.6×10^2 CFU mL^{-1} (Chai et al. 2012). A lateral flow system developed by DuPont can detect *Salmonella* in 10 min. In this test, the samples are initially pre-enriched before sample addition to the strip which contains precoated antibodies against *Salmonella.*

A lateral flow immunochromatographic strip using the *invA* gene of the *Salmonella* as a target has also been developed (Bulut 2014). An aptamer-based lateral flow biosensor has been developed to detect *Salmonella* (Fang et al. 2014). It includes an amplification step which forms a single-stranded DNA followed by its deposition on a sensor membrane. This test has been found to have a sensitivity of 10^1 CFU mL^{-1}.

Further improvement in *Salmonella* detection has also been made by using microfluidic devices where efficient binding of *S. typhimurium* bacterial cells has been achieved by using anti-*Salmonella* antibodies. By using quantum dots conjugated to antibodies a portable fluorometer was also developed which could detect *Salmonella* cells up to 10^3 CFU mL^{-1} in chicken meat extract (Kim et al. 2015b).

18.3.2 Biosensors for Detection of Viral Infection

18.3.2.1 Detection of Avian Influenza (AI) Virus

Avian influenza virus (AIV) poses a serious threat to mankind and has been divided into highly pathogenic AIV and low pathogenic AIV. Highly pathogenic AIV tends to rapidly spread which may cause high mortality. Infection by low pathogenic AIV results in mild gastrointestinal and respiratory symptoms but may turn into a highly pathogenic type after multiple infections (Morse et al. 2012). A variety of tests which include haemagglutination inhibition test, haemagglutination test, virus neutralization test, ELISA, virus isolation in cell culture, virus propagation in embryonated chicken eggs are available to detect the virus infection. However, these processes require large cost input, specific maneuvers, and infrastructure are required for their conduction.

Moreover, certain strains of Avian influenza-like H5N1 kills the embryos very quickly, thus, the amplification obtained in eggs is not enough to proceed further. Similarly, molecular methods require RNA extraction and the requirement of high Biosecurity Levels equipment's. In such conditions, pen-side tests for detection provide a window for an early and rapid on-site detection (Chang et al. 2009; Chiou et al. 2010; Tseng et al. 2014). Some serological tests are also available to detect the serological response against AI. However, they are endowed with low sensitivity and false-negative results due to which it limits the application of serological tests.

The site of the binding of the influenza virus is different for both humans and birds. In the case of humans, the virus binds to the α-2, 6 sialic acid glycan of the upper respiratory tract. In the case of birds, it binds to the α-2, 3 sialic acid expressed in the intestine. Hence, based on this differentiating feature, biosensors have been developed to detect and differentiate between the human and avian influenza viruses.

A biosensor based on glycan-immobilized field-effect transistor has been demonstrated to differentiate between the avian (H5) and human (H1) influenza viruses with a sensitivity of up to 10^{-18} mol L^{-1} (Hideshima et al. 2013). Sialic acid bound to gold nanoparticles can bind with the HA of the virus, thus producing the signals for virus detection (Lee et al. 2013). The glycans have been printed in the glass slides to form a microarray (Dinh et al. 2014) which can detect various strains of influenza virus up to ten plaque-forming units. Portable lateral flow tests have also been developed for the diagnosis of influenza virus (Sajid et al. 2015). In these kinds of lateral flow assay, the sample to be tested is diluted in the detergent solution followed by its addition into a sample pad. Thus, the components in the sample start moving in the pad. If the sample is positive for the influenza virus, the virus particles bind with the anti-influenza antibodies conjugated with gold nanoparticles which are pre-embedded in the conjugate pad. The complex of virus-antibodies-GNP reaches the test lines where anti-influenza A and anti-influenza B antibodies are pre-coated. These antibodies are directed against the HA protein of influenza virus. In the case of positive samples, a visible red line is formed. The unbound GNPs reach the control line and bind with the secondary antibodies which subsequently form a red line. If the virus particles are absent in the test sample, only the GNP labeled antibodies will move and bind to the control line. However, these tests could not differentiate between the various subtypes of LPAIV and HPAIV.

This shortcoming can be overcomed by using aptamers against AIV or RNA/DNA probes on the test lines, which enables the differentiation of subtypes as well. These lateral flow assays are simple and easy to use, thus, it makes them indispensable. To amplify the signal generated silver nanoparticles have been used successfully and amplification of 1000 times has been achieved (Huang et al. 2015). By using quantum dots a limit of detection of 0.09 ng mL^{-1} was achieved for detection of AIV (Li et al. 2012) and a sensitivity of 100 times than ELISA was achieved.

For the quantitative detection of influenza A virus (subtypes H5 and H9) a quantum-dot-based lateral flow test has been developed (Wu et al. 2016). An electrochemical sensor employing the anti-M1 antibody has been found to detect all serotypes of influenza A virus (Nidzworski et al. 2014). After conjugation of the anti-M1 monoclonal antibodies with GNPs in a quartz crystal microbalance assay, a limit of detection up to 1×10^3 PFU mL^{-1} has been achieved (Hewa et al. 2009).

An electrochemical immunosensor that detects and quantifies PB1-F2 protein of the influenza virus has been found to be sensitive for detection of influenza virus (Miodek et al. 2014a, b) with a limit of detection of 0.42 nM. A surface plasma wave biosensor has also been developed for the detection of Influenza A virus (Su et al. 2012). Although this method is sensitive, accurate, and fast, however, it comes with a limitation that it can be used only for laboratory purposes.

Aptamers which are artificial nucleic acids with known 3D structures can differentiate between various serotypes of the influenza virus. These aptamers are synthesized by using systematic evolution of ligands by exponential enrichment (SELEX) method. Aptamers-based biosensors open newer insights; because they do not require any animal host for synthesis, are less expensive than antibodies, and consume less time. Moreover, they bind the target molecules in the quantity of picomolar which offers more sensitivity in the detection of the target molecule. Therefore, based on this concept aptasensors are developed for the detection of AI virus. These aptamer-based biosensors have been found to have target specificity to detect both the Avian influenza virus as well as its subtypes (H5N1 and H1N1). Due to the binding of aptamers with HA, it opens new opportunities for its use as an antiviral (Sung et al. 2004; Cheng et al. 2008).

For the multiplexed detection of influenza virus multiplexed RT-PCR microarray coupled with DNA microarray has been used for subtyping the influenza virus (Li et al. 2001). A microarray for rapid detection of various subtypes of influenza virus (H1N1, H3N2 in man and H5N1 from poultry) has also been developed (Townsend et al. 2006).

18.3.2.2 Detection of Bluetongue and Epizootic Hemorrhagic Disease Viruses

Bluetongue (BT) is a non-contagious infectious disease of domestic and wild ruminants caused by Bluetongue virus (BTV). It infects sheep, cattle, and deer. It is transmitted by biting midges belonging to the family *Ceratopogonidae.* The cattle may be subclinically infected and may act as reservoirs of the virus. Economic losses are caused due to reduced milk and meat production. Epizootic hemorrhagic disease is another disease that mainly affects sheep. It is caused by *Epizootic hemorrhagic disease virus* of family *Reoviridae* and genus *Orbivirus.* This virus is also transferred through the same arthropod that transfers BTV. However, this disease is exhibited by high mortality in contrast to BT which has low mortality, both having similar clinical signs and symptoms.

There is a wide array of tests like real-time PCR, DNA microarray, microsphere bead-based tests, and genotyping methods to detect the blue tongue-virus (Weis et al. 2015; Wilson et al. 2015). A lateral flow assay to detect the BTV antibodies has been employed and commercialized (Hanon et al. 2016). Magnetic modulation biosensing has also been developed using FRET assay which allows fluorescent detection of the virus using fluorescently labeled probes tagged to magnetic microspheres (Danielli et al. 2009).

18.3.2.3 Detection of Foot-and-Mouth Disease Virus

Foot-and-mouth disease (FMD) is caused by foot-and-mouth disease virus (FMDV) and is an economically important disease. It is a highly contagious disease of domestic and wild ruminants. The FMDV can be detected virus isolation, enzyme-linked immunosorbent assay (ELISA), and reverse-transcriptase polymerase chain reaction (RT-PCR) (King et al. 2006).

A pen-side test for detection of FMDV antigen has been developed (Reid et al. 2001; Ferris et al. 2010). In a GNP-IPCR (Gold nanoparticle improved immune PCR) the immune complex is formed by capturing the virus particles by polyclonal antibody followed by addition of gold nanoparticle conjugated oligonucleotides and FMDV specific monoclonal antibodies. The signal DNA is released by heating followed by an analysis by PCR/real-time PCR (Ding et al. 2011). This assay was found to detect 10 fg mL^{-1} of purified FMD particles with high sensitivity in clinical samples. A GNP biosensor using thiol-linked oligonucleotides has been designed for detection of FMDV (Hamdy et al. 2018). Recently, a GNP-based dot-blot assay has been developed for the detection of anti-FMDV antibodies (Jain et al. 2018).

18.3.2.4 Detection of Bovine Respiratory Syncytial Virus

Bovine respiratory syncytial virus (BRSV) is one of the major respiratory diseases of cattle and is present worldwide. It is a single-stranded negative-sense RNA of the family *Paramyxoviridae*. An electrochemical biosensor for the detection of mRNA of RSV has been developed (Cai et al. 2013). This kind of sensor has been developed by using molecular beacons which are oligonucleotides hybridization probes.

In another study enzymatic substrate reaction using HRPO and TMB have been used for the detection of RSV antigens (Rochelet et al. 2012). It was an electrochemical assay in which the RSV specific antibodies were immobilized on the polystyrene slide, followed by the addition of the antigen and then enzyme-labeled antibody. The signal generated by HRPO-TMB breakdown was transduced through an electrode. This test was found to be rapid which can be conducted within 25 min. An immuno-PCR has also been developed to detect the RSV particles which could detect virus particles up to 4.1 PFU mL^{-1} and showed a 10^3 higher limit of detection than ELISA and RT-PCR (Perez et al. 2011).

18.3.2.5 Detection of Bovine Viral Diarrhea Virus (BVDV)

An integrated nanowire-based immunosensor has been developed for the detection of BVDV in serum (Montrose et al. 2015). For the detection of anti-BVDV antibodies, an electrospun biosensor based on the principle of capillary separation and conductometric immunoassay has been developed (Luo et al. 2010). This biosensor has been found to have a detection limit of eight minutes and 10^3 CCID mL^{-1} (Luo et al. 2010). A microparticle immunoagglutination assay on a microfluidic chip has been used to detect BVDV particles (Heinze et al. 2009).

18.3.2.6 Detection of Porcine Reproductive and Respiratory Syndrome (PRRS) Virus

Several immunodetection-based biosensors have been developed for the detection of PRRSV. Infrared electrochemical luminescence biosensor (Shao et al. 2017), platinum nanotube-based fluorescent immunoassay (Chen et al. 2015), enzyme-linked aptamer antibody sandwich (Lee et al. 2013), Fluorescence Resonance Energy Transfer (FRET) based optical biosensor (Stringer et al. 2008) using gold nanoparticles and quantum dots have been developed for detection of PRRSV.

18.3.3 Biosensors for Detection of Mastitis Pathogens

Mastitis is the inflammation of the mammary gland and most of the time it is infectious in origin. Milk from mastitic animals is not suitable for purpose of human consumption because of the altered chemical composition, organoleptic properties (Seegers et al. 2003; Adkins and Middleton 2018; Ashraf and Imran 2018), and shelf life of the dairy products (Hogeveen et al. 2010). Mastitis can be classified as clinical or subclinical. The clinical form is manifested by visible features like the presence of clots, flakes, change in colour, swelling of the mammary gland. Subclinical mastitis does not manifest any clinical sign, but the milk quality and production are declined. Moreover, the excessive use of antibiotics to treat mastitis gives rise to the problem of antimicrobial resistance (Pol and Ruegg 2007). Therefore, it becomes imperative to detect mastitis at an early stage so that preventive measures can be applied as soon as possible (More 2009).

In this regard, there is an increasing demand for pen-side tests to confirm the infectious origin of mastitis (Adkins and Middleton 2018). Due to the persistence of the pathogen in the mammary gland, and the immunological response is generated. Therefore, for the detection of subclinical mastitis, several approaches have been suggested which are based on the testing for immunological modulators and chemical properties of milk (Viguier et al. 2009; Adkins and Middleton 2018; Ashraf and Imran 2018). Out of these increased somatic cell count (SCC), N-acetyl-b-D-glucosaminidase (NAGase) and lactate dehydrogenase (LDH) are the most commonly used markers (Viguier et al. 2009; Addis et al. 2016; Adkins and Middleton 2018; Ashraf and Imran 2018).

Both colorimetric and fluorometric assays have been developed for the detection of LDH and NAGase activity. A fluorometric assay has been developed for the estimation of the catalytic activity of NAGase on the substrate 4-MUAG (Hovinen et al. 2016). A portable spectrophotometer has also been developed to estimate the LDH activity (Hiss et al. 2007).

Electroconductivity (EC) metres for the measurement of increased EC of mastitic milk (due to increased sodium and chloride ions) has also been developed (Norberg et al. 2004; Lam et al. 2009; Viguier et al. 2009). However, they are found to vary between animals.

The pathogens associated with the causative agent of mastitis also need to be identified. Timely identification is essential to reduce the excessive use of administration. Most of the pathogens causing udder infection are derived from the environment. The most common pathogens associated with mastitis are *Streptococcus agalactiae, Staphylococcus aureus, Escherichia coli, Streptococcus dysgalactiae, Streptococcus uberis, Corynebacterium* sp., *Pseudomonas aeruginosa, Mycoplasma* sp., and *Kleb siella.* An electrochemical sensor based on a screen-printed carbon electrode has been devised to detect the NAGase enzyme in the mastitic milk with a limit of detection of 10 mU mL^{-1} (Pemberton et al. 2001a).

For the detection of *Mycoplasma bovis*, a single-stranded DNA aptamer has shown high specificity and binding affinity towards P48 protein of *M. bovis* (Fu et al. 2014). A biochip for the DNA amplification of genes specific for mastitis-causing pathogens could detect six pathogens and *M. bovis* with a limit of detection up to 10^3 CFU mL^{-1} (Lee et al. 2008). An electrochemical sensor that is based on electrochemical spectroscopy has been developed for the detection of pathogenic *Staphylococcus aureus* (Braiek et al. 2012), with a limit of detection of 10 CFU mL^{-1}. The assay was found to be specific as it did not detect *E. coli* and *S. epidermidis.* In another recent study, the electrochemical paper-based analytical device (EPAD) was developed. In this paper-based genosensors they used graphene nanodots (GNDs) and zeolite (Zeo) with the *S. aureus* specific DNA probe to detect the target (Mathur et al. 2018). This limit of detection (LOD) of this sensor was found to be 0.1 nM of ss-DNA in test samples.

18.3.4 Biosensors for Detection of Drug Residues in Meat and Dairy Products

The drug residues in food can give rise to antimicrobial resistance (Du and Liu 2012; Leibovici et al. 2016). There is a chance that via food (chicken, milk, egg, fish, meat, and honey) these antimicrobial-resistant strains may be transferred to human beings.

A label-free amperometric immunosensor was developed for the detection of kanamycin which showed a limit of detection of up to 5.74 pg mL^{-1} and was used to test kanamycin residues in animal-derived food samples (Wei et al. 2012). A label-free immunosensor designed using silver hybridized ferroferric oxide nanoparticles was found to detect kanamycin with a detection limit up to 15 pg mL^{-1} in pork meat samples (Yu et al. 2013).

Aptamer-based biosensors (Jiang and Yu 2008; Derbyshire et al. 2012; Citartan et al. 2012; Bai et al. 2014; Niu et al. 2014), have also been developed for the detection of kanamycin residues in food samples. For detection of chloramphenicol electrochemical sensors have been developed which are based on multiwalled carbon nanotubes (Yang and Zhao 2015). Similarly, an electrochemical aptasensor has also been developed for chloramphenicol detection which showed a detection limit of 0.059 ng mL^{-1} (Zhang et al. 2011).

An electrochemical sensor using three-dimensional reduced graphene oxide architectures has also been developed (Zhang et al. 2016) which was able to detect chloramphenicol up to 48.45 ng mL^{-1}. A molecularly imprinted polymer (MIP) has been synthesized for the electrochemical detection of streptomycin (Que et al. 2013). To detect Streptomycin, magnetic molecularly imprinted polymer nanospheres (mMIP) have been used (Liu et al. 2013) by assembling $AUCl_4$ ions on the magnetic bead surface. This was followed by the o-phenylenediamine polymerization on the surface of magnetic beads.

For the detection of Streptomycin an electrochemical aptasensor based on gold nanoparticle-functionalized magnetic multiwalled carbon nanotubes and nanoporous PtTi alloy has been developed (Yin et al. 2016). An electrochemical arc-shaped aptasensor has also been developed for the quantification of Streptomycin in milk (Danesh et al. 2016).

18.3.5 Biosensors for Measuring Physiological, Metabolic, and Biochemical Parameters of Livestock

Apart from disease diagnosis in livestock and poultry, biosensors have also been used to monitor other physiological, metabolic, and biochemical parameters like animal jaw movement, breath analysis, analysis of metabolites in perspiration, analyzing tears for glucose monitoring, analysis of stress in fish, and detection of ovulation. Jaw movements of animals are required to know the grazing behaviour of cattle. The rumination behaviour has been recorded by using a pressure sensor as a noseband along with a data logger which records the jaw movements via pressure (Braun et al. 2013). Acoustic analysis is used to estimate the biting and chewing movements for the analysis of food intake by the cattle (Laca and WallisDeVries 2000). Microphones have been used to record the jaw movement sounds of grazing animals (Ungar and Rutter 2006). Machine learning algorithms are used to analyze the signal pattern for the estimation of the intervals between the jaw movements, intensity, and duration (Navon et al. 2013). Accelerometer sensors have also been used to estimate the jaw movement and grazing behaviour of animals (Tani et al. 2013; Mattachini et al. 2016; Giovanetti et al. 2017). These sensors can convert the acceleration generated from movement or gravity into a suitable voltage output.

Some of the biosensors have also been developed to analyze the metabolites in sweat. These are the electrochemical sensors used for the measurement of sodium and lactate levels in sweat. They are also used to measure sweat as a measure of stress. Biosensors have also been used to analyze the infection by *M. bovis* by estimating the volatile organic compounds in breath (Ellis et al. 2014). A nanobiosensor and a strip-based lateral flow biosensor has been developed to detect *M. bovis* and *M. avium subspecies paratuberculosis*, respectively. An amperometric biosensor using immobilized glucose oxidase has also been used to detect glucose in the tear (Iguchi et al.

2007). It gives a measure of glucose concentration in blood. Other self-monitored biosensors are also in the stage of development and validation (La Belle et al. 2014).

Implanted biosensors inside the sclera of fish have also been used to monitor the stress biomarker as an indicator of the health of fish as well as the pollution status of the water (Hibi et al. 2012; Wu et al. 2015). Biosensors have also been implicated in determining ovulation in cattle. An amperometric progesterone biosensor has been developed to determine the ovulation in cattle. It consists of an anti-progesterone monoclonal antibody immobilized on a screen-printed carbon electrode (Pemberton et al. 2001b). Biosensors have also been developed to detect the progesterone in milk with software which lists the insemination time (Mazeris 2010) with more advancement integration of novel aptamer with a Surface Plasmon Resonance (SPR) imaging sensor (Zeidan et al. 2016). Intravaginal probes that can detect the conductivity and temperature inside the vagina have also been used to predict the ovulation of the cattle (Andersson et al. 2015, 2016).

Testing of saliva for various metabolites is an extremely useful non-invasive method (Bandodkar and Wang 2014) and is useful for disease diagnosis. The biomarkers found in saliva are useful in the early detection and diagnosis of diseases (Malon et al. 2014). Detection of corticosteroid hormones in saliva as a marker of stress has been used successfully. A cortisol immunosensor has been developed for the quantitative analysis of salivary cortisol (Yamaguchi et al. 2013). Similarly, α-amylase in the saliva is an indicator of psychosocial stress. An α-amylase biosensor has also been developed which is based on the hydrolysis of starch by α-amylase (Wu et al. 2007). The resonance of the starch gel immobilized sensor increases on the hydrolysis of starch by α-amylase. Salivary glucose biosensors have also been developed to measure glucose on the saliva of domestic pets and human beings (Stein and Greco 2002; Reusch et al. 2006). Sensors have also been developed for monitoring the physiological status of livestock. A livestock monitoring system integrated with a wireless sensor network for the collection of data on breathing rate, heart rate, and cattle movement has been developed (Park and Ha 2015).

Bioacoustics based system has been developed to detect the sex (Pereira et al. 2015), routine activities of poultry (Fontana et al. 2016), disease diagnosis like necrotic enteritis (Sadeghi et al. 2015). A sound detection system has been developed to differentiate between New Castle Disease, Infectious Bronchitis, and Avian Influenza (Banakar et al. 2016).

18.4 Conclusion

The Biosensors have a wide application in the field of veterinary sciences. Starting from the disease diagnosis biosensors have been used successfully in monitoring the metabolism and physiological parameters of animals and poultry. Animals' health plays an important role in any country's economy as directly and indirectly it affects the country's economy as well as human health. In the past few decades, there has been excessive use of antibiotics, the emergence of viral diseases, the emergence of new antibiotic-resistant bacterial strains, antibiotic residues in milk and other food commodities derived from animal sources. This imbalance is the outcome of the indiscriminate use of the available resources like antibiotics in feed/treatment of animal diseases, human activities like deforestation, mutation in viruses, and overlapping of human niche with that of wildlife. Excessive antibiotic use leads to antibiotic bacterial strains. Moreover, the infectious diseases in animals cause heavy economic losses and most of them are important from a zoonotic point of view. Therefore, with an increasing demand for rapid diagnosis of the viral/bacterial diseases, antibiotic residues, and monitoring the health status of animals, rapid pen-side tests need to be developed. The conventional diagnostic techniques take time, demands labour and sophisticated instruments for their interpretation. Biosensors can be the rapid pen-side tests that deliver the test results in a few minutes for early diagnosis and preventive measures required for disease control. Moreover, there is a requirement of multitargeting biosensors in the field to detect the target analyte with more confidence to overcome the impediment of false positive and false negative test outcomes. The nucleic acid-based biosensors are one of the biosensors for the sensitive detection of analytes. Biosensors powered with nanotechnologies and next-generation sequencing can offer more robust tools for the detection of pathogens and animal surveillance.

The layers on which the bioreceptors are immobilized also play an important role in the electrochemical signals and non-specific binding which depends on their thickness and surface charge. Moreover, choosing novel bioreceptors like bacteriophages, single-chain antibodies, non-antibody-binding proteins, and half antibodies can offer high specificity. Antibodies are the most widely used bioreceptors, however, their production and purification is costly. Moreover, the stability of antibodies on the sensor surface is not long-lasting and the binding capacity of these antibodies decreases with due time course. Hence, to counter these difficulties, some advancements in bioengineering have been implicated like peptoid nanosheets (Olivier et al. 2013). Such types of implications comprise thick sheets of 3–5 mm thickness formed by self-assembly of antibody mimetic peptoids. These thick sheets have surface loops expressing antigen-binding sites. The whole aggregation is stable and easy to produce. Some other alternatives for stable antibodies are, DARPins (Stumpp and Amstutz 2007), camel-derived heavy variable chain (VHH) (Hassanzadeh-Ghassabeh et al. 2013; Muyldermans 2013), single-chain antibodies expressed via yeast surface display (Richman et al. 2009), single-chain variable fragments (ScFv) (Ahmad et al. 2012), and adhirons (Tiede et al. 2014). These alternatives can be synthesized easily, are small, and can be produced in bacterial expression systems. Another important aspect is the regeneration of the sensor surface. Regeneration of the biosensor surface is cost-effective. Biosensor surface can be regenerated successfully after using the above bioreceptor molecules. These bioreceptors can withstand harsh buffers without affecting their binding capacity.

Nanomaterial-based electrochemical biosensors are also good biosensors due to their low cost, good selectivity, sensitivity. They need a precise choice of the transducer surface. The incorporation of the nanomaterials accelerates the sensitivity and specificity of the detection of the analyte. Electrochemical biosensors offer very low detection of analytes like as observed while detection of antibiotic residues of kanamycin, streptomycin, and tetracycline. Moreover, the portability and reusability of electrochemical biosensors add to its other features of rapid on-site detection. The sensitivity of the electrochemical biosensors can be increased further by replacing the enzyme-based labels by the nano labels.

Antibody-based biosensors can also offer high sensitivity for the target molecule depending on the selective binding of the target molecules with the immobilized antibodies and the stability of the antibody molecules on the sensor surface. There is also a need for international recognition and widespread use throughout the world these biosensors must be validated via intra-lab and inter-lab validation.

From future perspective, development, and validation of nanobiosensors can open new insights in animal disease diagnostics. With the recent advancements in nanotechnology, nanobiosensors will be an excellent platform with increased sensitivity for analyte detection.

References

Addis MF, Tedde V, Puggioni GMG et al (2016) Evaluation of milk cathelicidin for detection of bovine mastitis. J Dairy Sci 99:8250–8258. https://doi.org/10.3168/jds.2016-11407

Adkins PRF, Middleton JR (2018) Methods for diagnosing mastitis. Vet. Clin North Am—Food Anim Pract 34:479–491

Ahmad ZA, Yeap SK, Ali AM et al (2012) ScFv antibody: principles and clinical application. Clin Dev Immunol 2012:980250

Alcaine SD, Soyer Y, Warnick LD et al (2006) Multilocus sequence typing supports the hypothesis that cow- and human-associated Salmonella isolates represent distinct and overlapping populations. Appl Environ Microbiol 72:7575–7585. https://doi.org/10.1128/AEM.01174-06

Andersson LM, Okada H, Miura R et al (2016) Wearable wireless estrus detection sensor for cows. Comput Electron Agric 127:101–108. https://doi.org/10.1016/j.compag.2016.06.007

Andersson LM, Okada H, Zhang Y et al (2015) Wearable wireless sensor for estrus detection in cows by conductivity and temperature measurements. In: IEEE Sensors, Busan, pp 1–4. https://doi.org/10.1109/ICSENS.2015.7370219

Ashraf A, Imran M (2018) Diagnosis of bovine mastitis: from laboratory to farm. Trop Anim Health Prod 50:1193–1202

Bai X, Hou H, Zhang B, Tang J (2014) Label-free detection of kanamycin using aptamer-based cantilever array sensor. Biosens Bioelectron 56:112–116. https://doi.org/10.1016/j.bios.2013.12.068

Banakar A, Sadeghi M, Shushtari A (2016) An intelligent device for diagnosing avian diseases: newcastle, infectious bronchitis, avian influenza. Comput Electron Agric 127:744–753. https://doi.org/10.1016/j.compag.2016.08.006

Bandodkar AJ, Wang J (2014) Non-invasive wearable electrochemical sensors: a review. Trends Biotechnol 32(7):363–371

Banerjee P, Bhunia AK (2010) Cell-based biosensor for rapid screening of pathogens and toxins. Biosens Bioelectron 26:99–106. https://doi.org/10.1016/j.bios.2010.05.020

Bélanger L, Garenaux A, Harel J et al (2011) *Escherichia coli* from animal reservoirs as a potential source of human extraintestinal pathogenic *E. coli*. FEMS Immunol Med Microbiol 62:1–10

Bonifazi D, Carloni LE, Corvaglia V, Delforge A (2012) Peptide nucleic acids in materials science. Artificial Dna: Pna Xna 3(3):112–122

Braiek M, Rokbani KB, Chrouda A et al (2012) An electrochemical immunosensor for detection of Staphylococcus aureus bacteria based on immobilization of antibodies on self-assembled monolayers-functionalized gold electrode. Biosensors 2:417–426. https://doi.org/10.3390/bios2040417

Braun U, Trösch L, Nydegger F, Hässig M (2013) Evaluation of eating and rumination behaviour in cows using a noseband pressure sensor. BMC Vet Res 9:1–8. https://doi.org/10.1186/1746-6148-9-164

Briones C, Moreno M (2012) Applications of peptide nucleic acids (PNAs) and locked nucleic acids (LNAs) in biosensor development. Anal Bioanal Chem 402:3071–3089

Bulut O (2014) Development of nucleic acid based lateral flow immunochromatographic test platform for salmonella detection. Masters dissertation, Middle East Technical University

Cai Z, Song Y, Wu Y et al (2013) An electrochemical sensor based on label-free functional allosteric molecular beacons for detection target DNA/miRNA. Biosens Bioelectron 41:783–788. https://doi.org/10.1016/j.bios.2012.10.002

Chai Y, Li S, Horikawa S et al (2012) Rapid and sensitive detection of salmonella typhimurium on eggshells by using wireless biosensors. J Food Prot 75:631–636. https://doi.org/10.4315/0362-028X.JFP-11-339

Chambers J, Arulanandam B, Matta L et al (2008) Biosensor recognition elements. Curr Issues Mol Biol 10:1–12

Chang CM, Lebarbenchon C, Gauthier-Clerc M et al (2009) Molecular surveillance for avian influenza a virus in king penguins (Aptenodytes patagonicus). Polar Biol 32:663–665. https://doi.org/10.1007/s00300-009-0587-4

Chen L, Ye S, Cai K et al (2015) An aqueous platinum nanotube based fluorescent immuno-assay for porcine reproductive and respiratory syndrome virus detection. Talanta 144:324–328. https://doi.org/10.1016/j.talanta.2015.06.061

Cheng C, Dong J, Yao L et al (2008) Potent inhibition of human influenza H5N1 virus by oligonucleotides derived by SELEX. Biochem Biophys Res Commun 366:670–674. https://doi.org/10.1016/j.bbrc.2007.11.183

Chiou CC, Chen TL, Tsao KC et al (2010) Detection of pandemic (H1N1) 2009 influenza virus by allele discrimination. Clin Chim Acta 411:1080–1083. https://doi.org/10.1016/j.cca.2010.04.002

Citartan M, Gopinath SCB, Tominaga J et al (2012) Assays for aptamer-based platforms. Biosens Bioelectron 34:1–11

Clark LC, Lyons C (1962) Electrode systems for continuous monitoring in cardiovascular surgery. Ann N Y Acad Sci 102:29–45. https://doi.org/10.1111/j.1749-6632.1962.tb13623.x

Danesh NM, Ramezani M, Emrani AS et al (2016) A novel electrochemical aptasensor based on arch-shape structure of aptamer-complimentary strand conjugate and exonuclease I for sensitive detection of streptomycin. Biosens Bioelectron 75:123–128. https://doi.org/10.1016/j.bios.2015.08.017

Danielli A, Porat N, Arie A, Ehrlich M (2009) Rapid homogenous detection of the Ibaraki virus NS3 cDNA at picomolar concentrations by magnetic modulation. Biosens Bioelectron 25:858–863. https://doi.org/10.1016/j.bios.2009.08.047

Derbyshire N, White SJ, Bunka DHJ et al (2012) Toggled RNA aptamers against aminoglycosides allowing facile detection of antibiotics using gold nanoparticle assays. Anal Chem 84:6595–6602. https://doi.org/10.1021/ac300815c

Di Gleria K, Green MJ, Hill HAO, McNeil CJ (1986) Homogeneous ferrocene-mediated amperometric immunoassay. Anal Chem 58:1203–1205

Ding YZ, Liu YS, Zhou JH et al (2011) A highly sensitive detection for foot-and-mouth disease virus by gold nanopariticle improved immuno-PCR. Virol J 8. https://doi.org/10.1186/1743-422X-8-148

Dinh H, Zhang X, Sweeney J et al (2014) Glycan based detection and drug susceptibility of influenza virus. Anal Chem 86:8238–8244. https://doi.org/10.1021/ac501624v

Du L, Liu W (2012) Occurrence, fate, and ecotoxicity of antibiotics in agro-ecosystems: a review. Agron Sustain Dev 32:309–327

El Ichi S, Leon F, Vossier L et al (2014) Microconductometric immunosensor for label-free and sensitive detection of gram-negative bacteria. Biosens Bioelectron 54:378–384. https://doi.org/10.1016/j.bios.2013.11.016

Ellis CK, Stahl RS, Nol P et al (2014) A pilot study exploring the use of breath analysis to differentiate healthy cattle from cattle experimentally infected with Mycobacterium bovis. PLoS One 9. https://doi.org/10.1371/journal.pone.0089280

Elnashar M (2010) Low-cost foods and drugs using immobilized enzymes on biopolymers. In: BiopolymersMagdy Elnashar (Ed.), ISBN: 978-953-307-109-1. InTech. Available from http://www.intechopen.com/books/biopolymers/low-cost-foods-and-drugs-using-immobilized-enzymes-onbiopolymers

Eltzov E, Marks RS (2016) Miniaturized flow stacked immunoassay for detecting *Escherichia coli* in a single step. Anal Chem 88:6441–6449. https://doi.org/10.1021/acs.analchem.6b01034

Eltzov E, Marks RS (2017) Colorimetric stack pad immunoassay for bacterial identification. Biosens Bioelectron 87:572–578. https://doi.org/10.1016/j.bios.2016.08.044

Fang Z, Wu W, Lu X, Zeng L (2014) Lateral flow biosensor for DNA extraction-free detection of salmonella based on aptamer mediated strand displacement amplification. Biosens Bioelectron 56:192–197. https://doi.org/10.1016/j.bios.2014.01.015

Ferris NP, Nordengrahn A, Hutchings GH et al (2010) Development and laboratory validation of a lateral flow device for the detection of serotype SAT 2 foot-and-mouth disease viruses in clinical samples. J Virol Methods 163:474–476. https://doi.org/10.1016/j.jviromet.2009.09.022

Fine T, Leskinen P, Isobe T et al (2006) Luminescent yeast cells entrapped in hydrogels for estrogenic endocrine disrupting chemical biodetection. Biosens Bioelectron 21:2263–2269. https://doi.org/10.1016/j.bios.2005.11.004

Fontana I, Tullo E, Scrase A, Butterworth A (2016) Vocalisation sound pattern identification in young broiler chickens. Animal 10:1567–1574. https://doi.org/10.1017/S1751731115001408

Fournier-Wirth C, Deschaseaux M, Defer C et al (2006) Evaluation of the enhanced bacterial detection system for screening of contaminated platelets. Transfusion 46:220–224. https://doi.org/10.1111/j.1537-2995.2006.00704.x

Fu P, Sun Z, Yu Z et al (2014) Enzyme linked aptamer assay: based on a competition format for sensitive detection of antibodies to Mycoplasma bovis in serum. Anal Chem 86:1701–1709. https://doi.org/10.1021/ac4042203

Gibert M, Jolivet-Renaud C, Popoff MR (1997) Beta2 toxin, a novel toxin produced by Clostridium perfringens. Gene 203:65–73. https://doi.org/10.1016/S0378-1119(97)00493-9

Giovanetti V, Decandia M, Molle G et al (2017) Automatic classification system for grazing, ruminating and resting behaviour of dairy sheep using a tri-axial accelerometer. Livest Sci 196:42–48. https://doi.org/10.1016/j.livsci.2016.12.011

Gnanaprakasa TJ, Oyarzabal OA, Olsen EV et al (2011) Tethered DNA scaffolds on optical sensor platforms for detection of hipO gene from Campylobacter jejuni. Sens Actuators, B Chem 156:304–311. https://doi.org/10.1016/j.snb.2011.04.037

Guilbault G, Kramer D, Cannon PL (1962) Electrical determination of organophosphorous compoundse. Anal Chem 34:1437–1439

Guilbault GG, Montalvo JG (1969) A urea-specific enzyme electrode. J Am Chem Soc 91:2164–2165

Guo Y, Wang Y, Liu S et al (2016) Label-free and highly sensitive electrochemical detection of *E. coli* based on rolling circle amplifications coupled peroxidase-mimicking DNAzyme amplification. Biosens Bioelectron 75:315–319. https://doi.org/10.1016/j.bios.2015.08.031

Gupta A, Mishra A, Puri N (2017) Peptide nucleic acids: advanced tools for biomedical applications. J Biotechnol 259:148–159

Hamdy ME, Del Carlo M, Hussein HA et al (2018) Development of gold nanoparticles biosensor for ultrasensitive diagnosis of foot and mouth disease virus. J Nanobiotechnol. https://doi.org/10.1186/s12951-018-0374-x

Hanon JB, Vandenberge V, Deruelle M et al (2016) Inter-laboratory evaluation of the performance parameters of a Lateral Flow Test device for the detection of Bluetongue virus-specific antibodies. J Virol Methods 228:140–150. https://doi.org/10.1016/j.jviromet.2015.12.001
Hassanzadeh-Ghassabeh G, Devoogdt N, De Pauw P et al (2013) Nanobodies and their potential applications. Nanomedicine 8:1013–1026
Haurum JS (2006) Recombinant polyclonal antibodies: the next generation of antibody therapeutics? Drug Discov Today 11:655–660
Heinze BC, Song JY, Lee CH et al (2009) Microfluidic immunosensor for rapid and sensitive detection of bovine viral diarrhea virus. Sens Actuators, B Chem 138:491–496. https://doi.org/10.1016/j.snb.2009.02.058
Hewa TMP, Tannock GA, Mainwaring DE et al (2009) The detection of influenza A and B viruses in clinical specimens using a quartz crystal microbalance. J Virol Methods 162:14–21. https://doi.org/10.1016/j.jviromet.2009.07.001
Hibi K, Hatanaka K, Takase M et al (2012) Wireless biosensor system for Real-Time L-lactic acid monitoring in fish. Sensors (Switzerland) 12:6269–6281. https://doi.org/10.3390/s120506269
Hideshima S, Hinou H, Ebihara D et al (2013) Attomolar detection of influenza A virus hemagglutinin human H1 and avian H5 using glycan-blotted field effect transistor biosensor. Anal Chem 85:5641–5644. https://doi.org/10.1021/ac401085c
Hiss S, Mueller U, Neu-Zahren A, Sauerwein H (2007) Haptoglobin and lactate dehydrogenase measurements in milk for the identification of subclinically diseased udder quarters. Vet Med (Praha) 52:245–252. https://doi.org/10.17221/1879-VETMED
Hobson NS, Tothill I, Turner APF (1996) Microbial detection. Biosens Bioelectron 11:455–477
Hogeveen H, Huijps K, Halasa T (2010) Mastitis costs money: what's new? In: Proceedings of the 5th IDF mastitis conference on mastitis research into practice. Christchurch, New Zealand, 21–24 March 2010, pp 62–71
Hovinen M, Simojoki H, Pösö R et al (2016) N-acetyl-β-D-glucosaminidase activity in cow milk as an indicator of mastitis. J Dairy Res 83:219–227. https://doi.org/10.1017/S0022029916000224
Huang X, Aguilar ZP, Xu H et al (2015) Membrane-based lateral flow immunochromatographic strip with nanoparticles as reporters for detection: a review. Biosens Bioelectron 75:166–180
Idil N, Hedström M, Denizli A, Mattiasson B (2017) Whole cell based microcontact imprinted capacitive biosensor for the detection of Escherichia coli. Biosens Bioelectron 87:807–815. https://doi.org/10.1016/j.bios.2016.08.096
Iguchi S, Kudo H, Saito T et al (2007) A flexible and wearable biosensor for tear glucose measurement. Biomed Microdevices 9:603–609. https://doi.org/10.1007/s10544-007-9073-3
Jaffrezic-Renault N, Martelet C, Chevolot Y, Cloarec JP (2007) Biosensors and bio-bar code assays based on biofunctionalized magnetic microbeads. Sensors 7:589–614. https://doi.org/10.3390/s7040589
Jain B, Lambe U, Tewari A et al (2018) Development of a rapid test for detection of foot-and-mouth disease virus specific antibodies using gold nanoparticles. VirusDisease 29:192–198. https://doi.org/10.1007/s13337-018-0450-8
Jiang X, Wang R, Wang Y et al (2011) Evaluation of different micro/nanobeads used as amplifiers in QCM immunosensor for more sensitive detection of E. coli O157:H7. Biosens Bioelectron 29:23–28. https://doi.org/10.1016/j.bios.2011.07.059
Jiang XC, Yu AB (2008) Silver nanoplates: A highly sensitive material toward inorganic anions. Langmuir 24:4300–4309. https://doi.org/10.1021/la7032252
Jung I, Bin Seo H, Lee J-eun et al (2014) A dip-stick type biosensor using bioluminescent bacteria encapsulated in color-coded alginate microbeads for detection of water toxicity. Analyst 139:4696–4701. https://doi.org/10.1039/c4an00308j
Jung Y, Kang HJ, Lee JM et al (2008) Controlled antibody immobilization onto immunoanalytical platforms by synthetic peptide. Anal Biochem 374:99–105. https://doi.org/10.1016/j.ab.2007.10.022

Kim M, Jung T, Kim Y et al (2015a) A microfluidic device for label-free detection of *Escherichia coli* in drinking water using positive dielectrophoretic focusing, capturing, and impedance measurement. Biosens Bioelectron 74:1011–1015. https://doi.org/10.1016/j.bios.2015.07.059

Kim G, Moon JH, Moh CY, LimJG (2015b). A microfluidic nano-biosensor for the detection of pathogenic Salmonella. Biosens Bioelectron 67:243–247

King DP, Ferris NP, Shaw AE et al (2006) Detection of foot-and-mouth disease virus: Comparative diagnostic sensitivity of two independent real-time reverse transcription-polymerase chain reaction assays. J Vet Diagnostic Investig 18:93–97. https://doi.org/10.1177/104063870601800114

La Belle J T, Engelschall E, Lan K, Shah P, Saez N, Maxwell S, Adamson T, Abou-Eid M, McAferty K, Patel D R, Cook CB (2014) A disposable tear glucose biosensor-part 4: preliminary animal model study assessing efficacy, safety, and feasibility. J Diabetes Sci Technol 8:109–116.

Laca EA, WallisDeVries MF (2000) Acoustic measurement of intake and grazing behaviour of cattle. Grass Forage Sci 55:97–104. https://doi.org/10.1046/j.1365-2494.2000.00203.x

Lakhin AV, Tarantul VZ, Gening LV (2013) Aptamers: problems, solutions and prospects. Acta Naturae 5:34–43

Lam TJGM, Olde Riekerink RGM, Sampimon OC, Smith H (2009) Mastitis diagnostics and performance monitoring: a practical approach. Ir Vet J 62:34–39. https://doi.org/10.1186/2046-0481-62-S4-S34

Lazcka O, Del Campo FJ, Muñoz FX (2007) Pathogen detection: a perspective of traditional methods and biosensors. Biosens Bioelectron 22:1205–1217

Lee C, Gaston MA, Weiss AA, Zhang P (2013) Colorimetric viral detection based on sialic acid stabilized gold nanoparticles. Biosens Bioelectron 42:236–241. https://doi.org/10.1016/j.bios.2012.10.067

Lee KH, Lee JW, Wang SW et al (2008) Development of a novel biochip for rapid multiplex detection of seven mastitis-causing pathogens in bovine milk samples. J Vet Diagnostic Investig 20:463–471. https://doi.org/10.1177/104063870802000408

Leibovici L, Paul M, Garner P et al (2016) Addressing resistance to antibiotics in systematic reviews of antibiotic interventions. J Antimicrob Chemother 71:2367–2369

Li J, Chen S, Evans DH (2001) Typing and subtyping influenza virus using DNA microarrays and multiplex reverse transcriptase PCR. J Clin Microbiol 39:696–704. https://doi.org/10.1128/JCM.39.2.696-704.2001

Li X, Lu D, Sheng Z et al (2012) A fast and sensitive immunoassay of avian influenza virus based on label-free quantum dot probe and lateral flow test strip. Talanta 100:1–6. https://doi.org/10.1016/j.talanta.2012.08.041

Liu B, Tang D, Zhang B et al (2013) Au(III)-promoted magnetic molecularly imprinted polymer nanospheres for electrochemical determination of streptomycin residues in food. Biosens Bioelectron 41:551–556. https://doi.org/10.1016/j.bios.2012.09.021

Luo Y, Nartker S, Miller H et al (2010) Surface functionalization of electrospun nanofibers for detecting *E. coli* O157:H7 and BVDV cells in a direct-charge transfer biosensor. Biosens Bioelectron 26:1612–1617. https://doi.org/10.1016/j.bios.2010.08.028

Luong JHT, Male KB, Glennon JD (2008) Biosensor technology: technology push versus market pull. Biotechnol Adv 26:492–500

Malon RSP, Sadir S, Balakrishnan M, Córcoles EP (2014) Saliva-based biosensors: noninvasive monitoring tool for clinical diagnostics. Biomed Res Int:962903. https://doi.org/10.1155/2014/962903.

Manzano M, Cecchini F, Fontanot M et al (2015) OLED-based DNA biochip for *Campylobacter* spp. detection in poultry meat samples. Biosens Bioelectron 66:271–276. https://doi.org/10.1016/j.bios.2014.11.042

Martinkova P, Kostelnik A, Valek T, Pohanka M (2017) Main streams in the construction of biosensors and their applications. Int J Electrochem Sci 12: 7386–7403

Mathur A, Gupta R, Kondal S, Wadhwa S, Pudake RN, Kansal R, Pundir CS, Narang J (2018) A new tactics for the detection of *S. aureus* via paper based geno-interface incorporated with graphene nano dots and zeolites. Int J Biol Macromol 112:364–370

Mattachini G, Riva E, Perazzolo F et al (2016) Monitoring feeding behaviour of dairy cows using accelerometers. J Agric Eng 47:54–58. https://doi.org/10.4081/jae.2016.498

Mazeris F (2010) DeLaval herd navigator: proactive herd management. In: Proceedings of first North American conference on precision dairy management, pp 26–27

Miodek A, Sauriat-Dorizon H, Chevalier C et al (2014a) Direct electrochemical detection of PB1-F2 protein of influenza A virus in infected cells. Biosens Bioelectron 59:6–13. https://doi.org/10.1016/j.bios.2014.02.037

Miodek A, Vidic J, Sauriat-Dorizon H et al (2014b) Electrochemical detection of the oligomerization of PB1-F2 influenza a virus protein in infected cells. Anal Chem 86:9098–9105. https://doi.org/10.1021/ac5018056

Montrose A, Creedon N, Sayers R, Barry S, O'riordan A (2015) Novel single gold nanowire-based electrochemical immunosensor for rapid detection of bovine viral diarrhoea antibodies in serum. J Biosens Bioelectron 6. https://doi.org/10.4172/2155-6210.1000174

More S (2009) Global trends in milk quality: implications for the Irish dairy industry. Ir Vet J 62:5–14

Morse SS, Mazet JAK, Woolhouse M et al (2012) Prediction and prevention of the next pandemic zoonosis. Lancet 380:1956–1965

Muyldermans S (2013) Nanobodies: natural single-domain antibodies. Annu Rev Biochem 82:775–797. https://doi.org/10.1146/annurev-biochem-063011-092449

Navon S, Mizrach A, Hetzroni A, Ungar ED (2013) Automatic recognition of jaw movements in free-ranging cattle, goats and sheep, using acoustic monitoring. Biosyst Eng 114:474–483. https://doi.org/10.1016/j.biosystemseng.2012.08.005

Nelson PN, Reynolds GM, Waldron EE et al (2000) Demystified: monoclonal antibodies. J Clin Pathol—Mol Pathol 53:111–117

Ni X, Castanares M, Mukherjee A, Lupold SE (2011) Nucleic acid aptamers: clinical applications and promising new horizons. Curr Med Chem 18:4206–4214. https://doi.org/10.2174/092986711797189600

Nidzworski D, Pranszke P, Grudniewska M et al (2014) Universal biosensor for detection of influenza virus. Biosens Bioelectron 59:239–242. https://doi.org/10.1016/j.bios.2014.03.050

Niu S, Lv Z, Liu J, et al (2014) Colorimetric aptasensor using unmodified gold nanoparticles for homogeneous multiplex detection. PLoS One 9. https://doi.org/10.1371/journal.pone.0109263

Norberg E, Hogeveen H, Korsgaard IR et al (2004) Electrical conductivity of milk: ability to predict mastitis status. J Dairy Sci 87:1099–1107. https://doi.org/10.3168/jds.S0022-0302(04)73256-7

Olivier GK, Cho A, Sanii B et al (2013) Antibody-mimetic peptoid nanosheets for molecular recognition. ACS Nano 7:9276–9286. https://doi.org/10.1021/nn403899y

Palaniappan A, Goh WH, Fam DWH et al (2013) Label-free electronic detection of bio-toxins using aligned carbon nanotubes. Biosens Bioelectron 43:143–147. https://doi.org/10.1016/j.bios.2012.12.019

Park M-C, Ha O-K (2015) Development of effective cattle health monitoring system based on biosensors. Science & Engineering Research Support Society, pp 180–185

Pemberton RM, Hart JP, Mottram TT (2001a) An assay for the enzyme N-acetyl-β-D-glucosaminidase (NAGase) based on electrochemical detection using screen-printed carbon electrodes (SPCEs). Analyst 126:1866–1871. https://doi.org/10.1039/b104874k

Pemberton RM, Hart JP, Mottram TT (2001b) An electrochemical immunosensor for milk progesterone using a continuous flow system. Biosens Bioelectron 16(9–12):715–723

Pereira EM, Nääs IDA, Garcia RG (2015) Vocalization of broilers can be used to identify their sex and genetic strain. Eng Agric 35:192–196. https://doi.org/10.1590/1809-4430-Eng.Agric.v35n2p192-196/2015

Perez JW, Vargis EA, Russ PK et al (2011) Detection of respiratory syncytial virus using nanoparticle amplified immuno-polymerase chain reaction. Anal Biochem 410:141–148. https://doi.org/10.1016/j.ab.2010.11.033

Petit L, Gibert M, Popoff MR (1999) Clostridium perfringens: toxinotype and genotype. Trends Microbiol 7:104–110

Pol M, Ruegg PL (2007) Treatment practices and quantification of antimicrobial drug usage in conventional and organic dairy farms in Wisconsin. J Dairy Sci 90:249–261. https://doi.org/10.3168/jds.S0022-0302(07)72626-7

Que X, Liu B, Fu L et al (2013) Molecular imprint for electrochemical detection of streptomycin residues using enzyme signal amplification. Electroanalysis 25:531–537. https://doi.org/10.1002/elan.201200468

Reid SM, Ferris NP, Brüning A et al (2001) Development of a rapid chromatographic strip test for the pen-side detection of foot-and-mouth disease virus antigen. J Virol Methods 96:189–202. https://doi.org/10.1016/S0166-0934(01)00334-2

Reusch CE, Kley S, Casella M (2006) Home monitoring of the diabetic cat. J Feline Med Surg 8:119–127

Richman SA, Kranz DM, Stone JD (2009) Biosensor detection systems: engineering stable, high-affinity bioreceptors by yeast surface display. Methods Mol Biol 504:323–350. https://doi.org/10.1007/978-1-60327-569-9_19

Rochelet M, Solanas S, Grossiord C et al (2012) A thin layer-based amperometric enzyme immunoassay for the rapid and sensitive diagnosis of respiratory syncytial virus infections. Talanta 100:139–144. https://doi.org/10.1016/j.talanta.2012.07.088

Sadeghi M, Banakar A, Khazaee M, Soleimani MR (2015) An intelligent procedure for the detection and classification of chickens infected by clostridium perfringens based on their vocalization. Rev Bras Cienc Avic 17:537–544. https://doi.org/10.1590/1516-635x1704537-544

Sajid M, Kawde AN, Daud M (2015) Designs, formats and applications of lateral flow assay: a literature review. J Saudi Chem Soc 19:689–705. https://doi.org/10.1016/j.jscs.2014.09.001

Seegers H, Fourichon C, Beaudeau F (2003) Production effects related to mastitis and mastitis economics in dairy cattle herds. Vet Res 34:475–491

Sergeev N, Distler M, Courtney S et al (2004) Multipathogen oligonucleotide microarray for environmental and biodefense applications. Biosens Bioelectron 20:684–698. https://doi.org/10.1016/j.bios.2004.04.030

Setterington EB, Alocilja EC (2012) Electrochemical biosensor for rapid and sensitive detection of magnetically extracted bacterial pathogens. Biosensors 2:15–31. https://doi.org/10.3390/bios2010015

Shao K, Zhang C, Ye S et al (2017) Near–infrared electrochemiluminesence biosensor for high sensitive detection of porcine reproductive and respiratory syndrome virus based on cyclodextrin-grafted porous Au/PtAu nanotube. Sens Actuators, B Chem 240:586–594. https://doi.org/10.1016/j.snb.2016.08.162

Sharon J, Liebman MA, Williams BR (2005) Recombinant polyclonal antibodies for cancer therapy. J Cell Biochem 96(2):305–313. https://doi.org/10.1002/jcb.20536

Stein JE, Greco DS (2002) Portable blood glucose meters as a means of monitoring blood glucose concentrations in dogs and cats with diabetes mellitus. Clin Tech Small Anim Pract 17:70–72. https://doi.org/10.1053/svms.2002.33041

Stringer RC, Schommer S, Hoehn D, Grant SA (2008) Development of an optical biosensor using gold nanoparticles and quantum dots for the detection of Porcine Reproductive and Respiratory Syndrome Virus. Sens Actuators, B Chem 134:427–431. https://doi.org/10.1016/j.snb.2008.05.018

Stumpp MT, Amstutz P (2007) DARPins: a true alternative to antibodies. Curr Opin Drug Discov Dev 10:153–159

Su LC, Chang CM, Tseng YL et al (2012) Rapid and highly sensitive method for influenza A (H1N1) virus detection. Anal Chem 84:3914–3920. https://doi.org/10.1021/ac3002947

Sung HJ, Kayhan B, Ben-Yedidia T, Arnon R (2004) A DNA aptamer prevents influenza infection by blocking the receptor binding region of the viral hemagglutinin. J Biol Chem 279:48410–48419. https://doi.org/10.1074/jbc.M409059200

Tani Y, Yokota Y, Yayota M, Ohtani S (2013) Automatic recognition and classification of cattle chewing activity by an acoustic monitoring method with a single-axis acceleration sensor. Comput Electron Agric 92:54–65. https://doi.org/10.1016/j.compag.2013.01.001

Tiede C, Tang AAS, Deacon SE et al (2014) Adhiron: a stable and versatile peptide display scaffold for molecular recognition applications. Protein Eng Des Sel 27:145–155. https://doi.org/10.1093/protein/gzu007

Tosar JP, Brañas G, Laíz J (2010) Electrochemical DNA hybridization sensors applied to real and complex biological samples. Biosens Bioelectron 26:1205–1217

Townsend MB, Dawson ED, Mehlmann M et al (2006) Experimental evaluation of the FluChip diagnostic microarray for influenza virus surveillance. J Clin Microbiol 44:2863–2871. https://doi.org/10.1128/JCM.00134-06

Tseng CH, Tsai HJ, Chang CM (2014) A complete molecular diagnostic procedure for applications in surveillance and subtyping of avian influenza virus. Biomed Res Int 2014. https://doi.org/10.1155/2014/653056

Turner KL, Zhang W (2001) Design and analysis of a dynamic MEM chemical sensor. In: Proceedings of the 2001 American control conference, Arlington, VA, USA, pp 1214–1218

Ungar ED, Rutter SM (2006) Classifying cattle jaw movements: comparing IGER behaviour recorder and acoustic techniques. Appl Anim Behav Sci 98:11–27. https://doi.org/10.1016/j.applanim.2005.08.011

Updike SJ, Hicks GP (1967) The enzyme electrode. Nature 214:986–988. https://doi.org/10.1038/214986a0

Viguier C, Arora S, Gilmartin N et al (2009) Mastitis detection: current trends and future perspectives. Trends Biotechnol 27:486–493

Volokhov DV, Kong H, Herold K et al (2011) Oligonucleotide microarrays for identification of microbial pathogens and detection of their virulence-associated or drug-resistance determinants. Methods Mol Biol 671:55–94. https://doi.org/10.1007/978-1-59745-551-0_3

Vopálenská I, Váchová L, Palková Z (2015) New biosensor for detection of copper ions in water based on immobilized genetically modified yeast cells. Biosens Bioelectron 72:160–167. https://doi.org/10.1016/j.bios.2015.05.006

Wadl M, Pölzler T, Flekna G et al (2009) Easy-to-use rapid test for direct detection of *Campylobacter* spp. in chicken feces. J Food Prot 72:2483–2488. https://doi.org/10.4315/0362-028X-72.12.2483

Wei D, Oyarzabal OA, Huang TS et al (2007) Development of a surface plasmon resonance biosensor for the identification of *Campylobacter jejuni*. J Microbiol Methods 69:78–85. https://doi.org/10.1016/j.mimet.2006.12.002

Wei Q, Zhao Y, Du B et al (2012) Ultrasensitive detection of kanamycin in animal derived foods by label-free electrochemical immunosensor. Food Chem 134:1601–1606. https://doi.org/10.1016/j.foodchem.2012.02.126

Weis A, Liang F, Gao J, Barnard RT CS (2015) RNA and DNA diagnostics on microspheres: current and emerging methods. In: RNA DNA diagnostics, pp 205–224

Wetmur JG, Fresco J (1991) DNA probes: applications of the principles of nucleic acid hybridization. Crit Rev Biochem Mol Biol 26:227–259. https://doi.org/10.3109/10409239109114069

Wilson WC, Daniels P, Ostlund EN et al (2015) Diagnostic tools for bluetongue and epizootic hemorrhagic disease viruses applicable to North American veterinary diagnosticians. Vector-Borne Zoonotic Dis. 15:364–373

Wu F, Yuan H, Zhou C et al (2016) Multiplexed detection of influenza A virus subtype H5 and H9 via quantum dot-based immunoassay. Biosens Bioelectron 77:464–470. https://doi.org/10.1016/j.bios.2015.10.002

Wu H, Aoki A, Arimoto T et al (2015) Fish stress become visible: a new attempt to use biosensor for real-time monitoring fish stress. Biosens Bioelectron 67:503–510. https://doi.org/10.1016/j.bios.2014.09.015

Wu S, Zhu Y, Cai Q et al (2007) A wireless magnetoelastic α-amylase sensor. Sens Actuators, B Chem 121:476–481. https://doi.org/10.1016/j.snb.2006.04.095

Yamaguchi M, Matsuda Y, Sasaki S et al (2013) Immunosensor with fluid control mechanism for salivary cortisol analysis. Biosens Bioelectron 41:186–191. https://doi.org/10.1016/j.bios.2012.08.016

Yang G, Zhao F (2015) Electrochemical sensor for chloramphenicol based on novel multiwalled carbon nanotubes@molecularly imprinted polymer. Biosens Bioelectron 64:416–422. https://doi.org/10.1016/j.bios.2014.09.041

Yazgan I, Noah NM, Toure O et al (2014) Biosensor for selective detection of *E. coli* in spinach using the strong affinity of derivatized mannose with fimbrial lectin. Biosens Bioelectron 61:266–273. https://doi.org/10.1016/j.bios.2014.05.008

Yin Y, Qin X, Wang Q, Yin Y (2016) A novel electrochemical aptasensor for sensitive detection of streptomycin based on gold nanoparticle-functionalized magnetic multi-walled carbon nanotubes and nanoporous PtTi alloy. RSC Adv 6:39401–39408. https://doi.org/10.1039/c6ra02029a

Yoo SM, Lee SY (2016) Optical biosensors for the detection of pathogenic microorganisms. Trends Biotechnol 34:7–25

Yu S, Wei Q, Du B et al (2013) Label-free immunosensor for the detection of kanamycin using Ag@Fe_3O_4 nanoparticles and thionine mixed graphene sheet. Biosens Bioelectron 48:224–229. https://doi.org/10.1016/j.bios.2013.04.025

Zeidan E, Shivaji R, Henrich VC, Sandros MG (2016) Nano-SPRi aptasensor for the detection of progesterone in buffer. Sci Rep 6(1):1–8. https://doi.org/10.1038/srep26714

Zhang L, Wang O, Cheng ZY, Huang T (2015) Silica-coated magnetostrictive biosensors for real-time detection of *Campylobacter jejuni* in washing water. In: Meeting abstracts, The Electrochemical Society, Pennington 2015, p 1832. https://doi.org/10.1149/ma2015-02/46/1832

Zhang N, Xiao F, Bai J et al (2011) Label-free immunoassay for chloramphenicol based on hollow gold nanospheres/chitosan composite. Talanta 87:100–105. https://doi.org/10.1016/j.talanta.2011.07.108

Zhang X, Zhang YC, Zhang JW (2016) A highly selective electrochemical sensor for chloramphenicol based on three-dimensional reduced graphene oxide architectures. Talanta 161:567–573. https://doi.org/10.1016/j.talanta.2016.09.013

Zhang Y, Tan C, Fei R et al (2014) Sensitive chemiluminescence immunoassay for *E. coli* O157:H7 detection with signal dual-amplification using glucose oxidase and laccase. Anal Chem 86:1115–1122. https://doi.org/10.1021/ac4028774

Chapter 19
Nano-Biosensing Devices Detecting Biomarkers of Communicable and Non-communicable Diseases of Animals

Utkarsh Jain, Saurabh Shakya, and Kirti Saxena

Abstract Diseases are either communicable or non-communicable. Communicable diseases or infectious diseases are caused by interactions between two living organisms whereas non-communicable diseases are either hereditary or due to the living environment of animals. Animals are very susceptible to various types of infections and pathogens may live outside or inside their body very conveniently. Epidemic diseases affect animal herd, livestock and the leading cause of animal mortality which may result in significant losses to the agricultural economy. Animal diseases are a matter of concern due to the direct effect on the economy as well as the possibility of pathogen transmission in humans. The accurate detection of animal diseases is an important factor to control diseases in livestock and in wild animals. Disease diagnosis in animals is very challenging and based on the variety of examinations and tests. In this context, nano-biosensors exhibit better disease detection capacity than conventional methods. Nano-biosensors show high sensitivity, stability, reproducibility with a lower limit of detection. This chapter emphasized the insight of communicable and non-communicable animal diseases and their detection through nano-biosensing devices. Some important animal diseases are discussed along with a description of developed nano-biosensors for them. Nano-biosensor functionalized with disease-related biomarkers performed better than any other conventional methods in various aspects.

Keywords Biosensor · Communicable diseases · Non-communicable diseases · Animal diseases · Nanotechnology

---

U. Jain (✉) · K. Saxena
Amity Institute of Nanotechnology, Amity University Uttar Pradesh, Noida, UP, India
e-mail: ujain@amity.edu

S. Shakya
Viral Recombination Section, HIV Dynamics and Replication Programme, National Cancer Institute, Frederick, MD, USA

R. N. Pudake et al. (eds.), *Biosensors in Agriculture: Recent Trends and Future Perspectives*, Concepts and Strategies in Plant Sciences,
https://doi.org/10.1007/978-3-030-66165-6_19

19.1 Introduction

19.1.1 Overview (History of Animal Diseases)

Animals are an important part of our ecosystem and their existence either directly or indirectly affects the life of humans in various aspects and vice versa. Like humans, animals also suffer with several types of communicable and non-communicable diseases. Sometimes there are some common diseases also found in humans and animals as well as they can play a role of a disease carrier for each other's, for example, anthrax, brucellosis, giardiasis, and swine flu. Health risk of animal population may have an impact on food supply chains which is a threat to human health with long lasting effects and to economy as well. For the animal disease control which are communicable to man from livestock as well as wildlife animals, several organization including WHO conduct various programmes. These diseases known as zoonoses, comprised of existing infectious disease along with newly identified infectious disease to monitor the impact on human health.

History related to animal diseases describes that human and animal diseases developed parallelly. Aristotle, the father of Biology and Hippocrates, the father of medicines, demonstrate animals and man's diseases symptomatology as well as therapy in their writing. Ancient literature are evident to animal and human diseases which include Egyptian, Roman, and Greeks monographs. In history, agriculture, domesticated animals were focused particularly. After the discovery of microorganisms by Pasteur, animal diseases became a matter of concern for veterinarians because of their direct effect on human health by consuming animal origin foods like meat and dairy products. After that basic research and zoonoses have been started for the detection and treatment of animal diseases (Charles et al. 2018).

19.1.2 Communicable and Non-Communicable Diseases

Any disease either in animals or in humans may be communicable and non-communicable. The communicable term implies for interaction or communication between two living things which are known as host and parasite. The parasite exists in host cells and lives at the expense of their cellular machinery. Parasites are certain microorganisms such as virus, bacteria, protozoans, arthropods, and fungi, which are unable to survive on their own, so they depend on host organism for food, multiplication, or some other requirements. The disease-causing ability of a parasite is known as pathogenicity whereas ability to cause infection is known as virulence. The virulence property of a parasite is differed from one host to other, that it can be virulent for one but not to other hosts. Most of the parasite occurred in the atmosphere as an inactive form until they find appropriate conditions to establish themselves inside host cells. For example, *Clostridium* and *Bacillus* bacteria occurred as an inactive form known as spores for several years even in harsh environmental conditions.

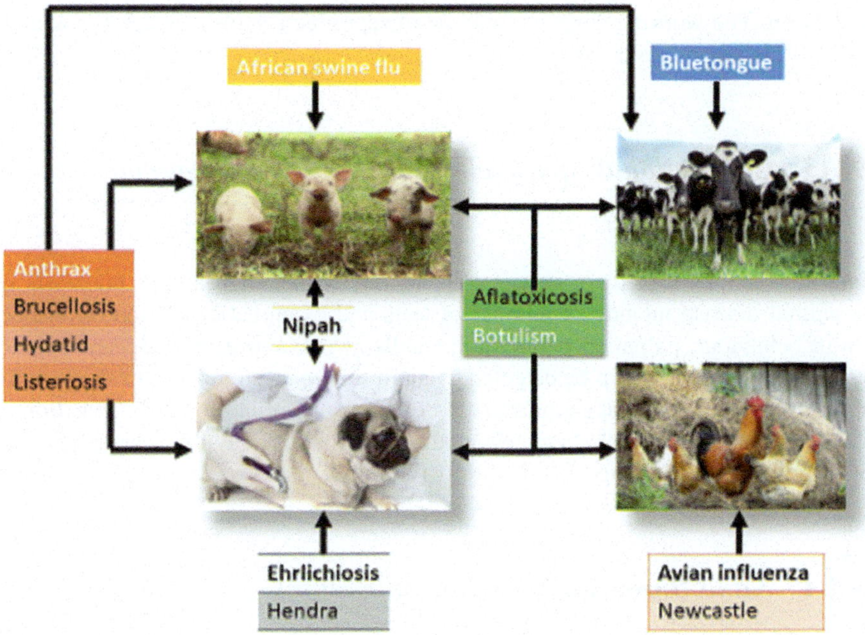

Fig. 19.1 Representation of communicable diseases and their host animals

The host cell defense system i.e. immune system provides a variety of mechanical and chemical barriers against infection, and to establish disease in the host body, it is essential to cross all these barriers. In animals, communicable diseases can be spread through direct contact of animals, waste of sick animals, unhygienic housing of animals, and presence of microorganisms in feed and water. Certain types of lice, mites, ticks, and fleas present inside or outside of body, also play a role in spreading of diseases, for example, plague (Heymann 2008). Some common types of communicable diseases and their host animals are represented in Fig. 19.1.

Non-communicable diseases, as the name suggested, not communicate from one animal to another and do not involve any virulent pathogens. Non-communicable diseases are caused by hereditary factors, poor feed, lack of nutrients, and environmental factors in surroundings. Sometimes it also caused by cuts, burns, certain toxic agents (insecticides, pesticides, fungicides, fluorine, and herbicides), or poisonous plants, and broken bones. Sometimes humans are responsible for some metabolic diseases in animals by injecting chemicals which causes the unsuitable alteration in body environment such as oxytocin is very commonly used in dairy animals to get more milk due to which contraction of uterus takes place and causes immense pain to the animal. Any disturbance in normal physiological conditions to maintain homeostasis and stability, leads to metabolic diseases. Non-communicable diseases are chronic and can be acute as well which causes a huge loss in milk, meat, and wool production. The working capacity and rate of reproduction of livestock can also be affected (Habib and Saha 2010).

19.1.3 Conventional Approaches for Detection of Animal Diseases

Since it is impossible for a veterinarian to interrogate the animal, diagnosis of the animal disease includes a number of tests and examinations based on inspection of behaviour, palpation percussion, smells, eye examination, appearance, etc. Diseases like pullorum can be detected by agglutination test, tuberculin skin test for tuberculosis, faeces and eggs examination can be done in lab for early detection of diseases. Determination of chemicals, toxins in blood, or body fluids are performed for several years for a diagnosis of variety of animal diseases.

19.1.4 Nano Biosensors and Its Applications

To control certain chronic communicable and non-communicable diseases, early and fast detection is essential in a population of animals such as a herd of cattle. In recent times, every year newly emerging diseases threaten animal health are a matter of concern. For the early stage detection of diseases in the case of animals, there is currently a lack of authentication and cost-effective diagnostics. Considering biosensing technologies, as they have the potential to overcome these problems by developing novel diagnostic tools because of their high sensitivity to detect disease-related biomarkers with a low limit of detection. The available conventional sensors are used in livestock monitoring and also as a tool for the assessment and monitoring in health monitoring and disease detection. These sensors when integrated with wireless data transfer through a server or with cloud-based systems, enabling access to the analyzed data from any internet enabled device. Applications of nanobiosensor not only help in the reduction of the current costs for reagents, handling samples, time of analysis, and cost of transportation but also helps in adapting and promoting the handling of livestock. Biosensor is a simple, miniaturized analytical device which is utilized for rapid and efficient diagnosis of infectious diseases and has the ability for real time analysis (Sagadevan and Periasamy 2014).

19.1.5 Biomarkers and Their Utilization in Biosensors

Biomarkers are typical indicators for biological, pathogenic, or pharmacologic processes which are used to identify an event or physiologic process of interest. For animal medicine, these biomarkers help in the early detection of several cardiopulmonary diseases, confirmation of doubtful processes (normal or abnormal), and contribute to the establishment of prognosis to decide a specific treatment (Fry and Dunbar 2007; Carretón et al. 2013). Also, they are used to detect the course of therapy or to predict safety or drug response. Biosensor consists of analytes, bio-recognition element (a biological element to detect analyte), and transducer surface (physio-chemical detector component) to convert any biological signal to a measurable electric signal. Biomarkers play a key role in the fabrication of biosensors to detect several important diseases like cancer, Alzheimer, organ health monitoring after a transplant, diabetes, neurological disorders, etc. Biomarkers exploit in biosensors could be antigen-antibody, dopamine, uric acid, hydrogen peroxide, ascorbic acid, hormones, microRNA, enzymes, or whole cell (bacteria/viruses) (Zetterberg et al. 2019; Dahal 2019; Viitanen et al. 2014; Scherr et al. 2016; Pereira et al. 2016; Çolak et al. 2013; Alcaraz et al. 2016; Fei et al. 2016; Peng et al. 2017). Some major diseases of animals along with their biomarkers are given in Table 19.1.

19.1.6 How Can Animal Diseases Affect Human Health?

In most of the cases, human infectious diseases which include disease agents and pathogens are acquired from animal species. Several diseases such as SARS, Nipah virus, avian influenza, cat scratch disease, West Nile disease, mad cow disease etc. are associated with animals. Recent pandemic Covid 19 is also acquired from bat species. So, humans and animal's diseases are directly proportional to each other in about 60–75% cases (Wang and Anderson 2019).

19.2 Communicable and Non-Communicable Diseases of Animals

19.2.1 Communicable Diseases of Animals

Aflatoxicosis: Aflatoxicosis is a widespread disease in animals and affects several animal species including cattle, chicks, ducklings, turkeys, horses, and pigs. Disease is caused by a mycotoxin which is produced by *Aspergillus flavus,* a most popular form of fungus. The feed items like millets, maize, soybeans, and peanuts, contaminated with fungi are source of disease. Aflatoxin is a highly toxic metabolite and mainly affects the liver through ingestion of contaminated feeds. The disease is characterized by drop in egg production particularly in poultry birds, loss of weight, and sometimes hepatocellular tumors appeared in severe conditions. Aflatoxicosis is associated with great economic loss as it reduced productivity by many folds, for

Table 19.1 The major diseases of animals and their biomarkers

S. No	Animal disease	Causing pathogen	Host	Symptoms	Associated Biomarkers
1.	Aflatoxicosis	*Aspergillus flavus*	cattle, chicks, ducklings, turkeys, horses, pigs	drop in egg production, weight loss, hepatocellular tumors	aflatoxin B_1, B_2, G_1, G_2, M_1, P_1, Q_1, and aflatoxin B_1-N^7-guanine (AFB_1-N^7-Gua)
2.	African swine fever	DNA virus of *Asfarviridae* family	domestic and wild pigs	high fever, loss of appetite, vomiting, diarrhoea, haemorrhages, loss of weight, respiratory signs, chronic skin ulcers	sCD163 & E75CV (strain of virus) antibodies, ASFV DNA
3.	Anthrax	*Bacillus anthracis*	Cattles	fever, difficulty in breathing, loss of appetite, Crepitation swelling	Calcium dipicolinate, dipicolinic acid,
4.	Avian influenza	Avian influenza A viruses	Birds, poultry	cough, cold, fever, headache, shortness of breath, and sore throat	AIV H5-hemagglutinin, IL-6, microRNA
5.	Bluetongue	bluetongue virus (BTV)	Ruminants	eye nasal discharge, high fever, drooling, inflammation	VP-7
6.	Botulism	*Clostridium botulinum*	cattle and avian species	hindlimb weakness, paralysis, death	botulinum neurotoxin serotype A
7.	Brucellosis	*Brucella*	Goats, sheep, cows, ruminants like deer, bison and elk	stillbirth or abortion	S19 and RB51

(continued)

Table 19.1 (continued)

S. No	Animal disease	Causing pathogen	Host	Symptoms	Associated Biomarkers
8.	Ehrlichiosis	*Ehrlichia*, *Anaplasma*, and *Neorickettsia* of Anaplasmataceae family	Cannines	fever, Reticuloendothelial hyperplasia, lymphadenopathy, Splenomegaly, loss of stamina, stiffness, and edema of limbs	IgM and IgG antibodies, antigenic proteins P28-19, and *Ehrlichia* Hsp60
9.	Hendra virus disease	*Hendra virus*	fruit bats, horses, humans	fever, cough, headache, tiredness, meningitis, coma	microRNA, physiological biomarkers like glucose, triglycerides, urea, protein etc.
10.	Hydatid	*Echinococcus granulosus* tapeworm	Dogs	weight loss, weakness and loss of appetite, allergy	interferon-gamma, IL-4, IL-10, specific IgE, total IgG, Glutathione S-Transferase, microRNA
11.	Listeriosis	Listeria monocytogenes	Animals, fishes, crustacean, birds and humans	encephalitis, meningitis, meningoencephalitis, stillbirth, abortion septicaemia, perinatal infections	invasion-associated protein, hemolysin, BE-LisAll
12.	Newcastle	Newcastle disease virus	domestic poultry and other species of birds	Respiratory disease, diarrohea, depression, or nervous manifestations	Nucleoprotein, glycoprotein, antibodies
13.	Nipah	Nipah virus	Fruit bat, Pigs, humans	barking pig syndrome, encephalitic syndrome	Nipah virus G-protein

example, milk production can be reduced by 15% or more (Dalvi 1986; Solis-Cruz et al. 2019).

African swine fever: African swine fever (ASF) is a severely contagious viral disease affecting domestic and wild pigs. It is caused by a DNA virus of *Asfarviridae* family which also responsible for infection in ticks of *Ornithodoros* genus. ASF associated with serious loss in production (pork, lard, leather, glue, fertilizer, and a variety of medicines produces through pigs) and economy. The virus is transmitted through a contaminated feed to animals and fomites such as shoes, knives, cloth, equipment, and vehicles. i.e. non-living things. The virus shows high resistant capacity for harsh environmental conditions. ASF has been reported in Europe, South America, Asia, Africa, and Caribbean. Symptoms of ASF include acute and chronic forms. Acute forms indicated by high fever, loss of appetite, vomiting, diarrhoea, haemorrhages in the skin, and death within few days (6–20 days). Chronic forms are characterized by loss of weight, respiratory signs, chronic skin ulcers, intermittent fever, and arthritis. Chronic forms are less intense clinically and can persist for longer period with low mortality rate (Dixon et al. 2019).

Anthrax: A large spore forming *Bacillus anthracis* bacteria is responsible for anthrax disease. It is an extremely infectious and fatal disease that causes acute mortality in cattles. The bacterial toxin is highly potent which immediately shows ill effects and a high mortality rate. After exposure to this bacterial infection, symptoms usually appear within 3–7 days and animal death occurred usually within 2 days. Cattles usually get infections by grazing and inhaling the spores of bacteria which are colourless, odourless, and tasteless. The infected animal usually expresses sudden symptoms such as fever, difficulty in breathing, loss of appetite, Crepitation swelling over the shoulder hip, and back, which is followed by death. This is a highly dangerous disease and exhibit potential loss of economy (Doganay and Demiraslan 2015).

Avian influenza: Avian influenza or bird flu is caused by avian influenza A viruses which naturally occur among wild water birds. The virus generally infects birds, poultry, animals, and sometimes human beings may also get infected through contaminated poultry. Although humans rarely get infected with this virus, however virus can change itself and gain the ability to infect humans and spread among them.

Influenza virus have the quality to change themselves time to time and certain strains spread easily in human populations. Symptoms arise within 2–8 days and include cough, cold, fever, headache, shortness of breath, and sore throat. The disease refers to high mortality rate in humans also (Li et al. 2019). In recent times poultry business is one of the major areas of interest because of high demand in food industries. But avian influenza is one of the big challenges as it may cause major economic loss. Recent advancement makes possible to detect and continuously monitoring of disease through biosensors, wearable sensors, real time monitoring, and data management (Raj and Jayanthi 2018; Vidic et al. 2017) (Fig. 19.2).

Bluetongue: Bluetongue disease refers mainly the disease of ruminants including both domestic and wild animals, however Sheep are badly infected with disease. Cattle although frequently get infection than sheep but they lack symptoms most of the times. It is normally an insect-borne disease caused by bluetongue virus (BTV) and transmitted by different species of midges Culicoides. The infected animals

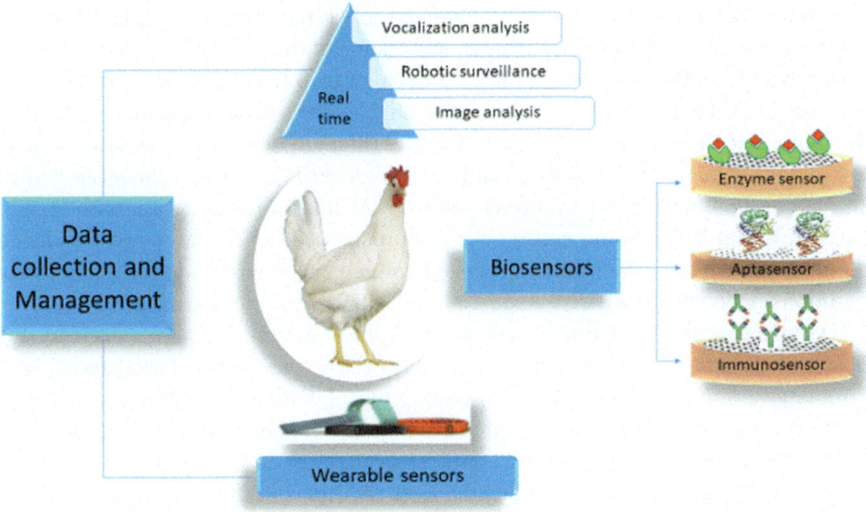

Fig. 19.2 Different methods for monitoring and detection of disease in poultry farms

usually suffer with eye, nasal discharge, high fever, drooling, and inflammation of the coronary band, haemorrhages into or under skin, lethargy, swollen teats, respiratory problems, and swelling in mouth, head, and neck (Van den Bergh et al. 2018).

Botulism: Botulism is a food borne disease caused by *Clostridium botulinum* bacteria which produce highly potential neurotoxin commonly known as botulinum. It is a rapid onset fatal disease of cattle and avian species. Botulism typically represents by hindlimb weakness followed by collapse, paralysis, and death. Animal carcasses, poorly prepared silage and rotting organic material are the main sources of bacterial infection. Although the vaccine is available for this disease but still it is not completely eradicated. The disease outbreak can be seen in poultry and waterfowl (Mariano et al. 2019).

Brucellosis: Brucellosis is a bacterium causing the disease which mainly affects livestock including goats, sheep, cows, and some wild ruminants like deer, bison, and elk. It is caused by bacteria of genus *Brucella*. The disease mainly characterized by stillbirth or abortion in animals. The disease is spread through contaminated milk, birthing fluids, the infected placenta of animals, or direct contact with other animals or human beings. Canines can also infect with bacteria, but they are rarely spread infection to human beings. Complications associated with brucellosis consist of inflammation of the brain, infection of the heart's inner lining, meningitis, and lesions on the bones and joints (Shoukat et al. 2017).

Ehrlichiosis: Ticks are the carrier of this disease which commonly developed in canine bitten by infected ticks. It is caused by intracellular obligate organisms that belong to genus *Ehrlichia*, *Anaplasma*, and *Neorickettsia* of Anaplasmataceae family. The parasite is primarily infecting the immune cells of canines and humans. Infected animals are suffered from fever, Reticuloendothelial hyperplasia, lymphadenopathy, Splenomegaly, loss of stamina, stiffness, and edema of limbs (Mylonakis et al. 2019).

Hendra virus Disease: It is mainly a disease of fruit bats which can be infectious to horses also. Hendra virus infection has infrequently been infectious to humans who came into contact with an infected horse. Infected horses may possess a range of symptoms like fever, respiratory, and neurological disorders and show rapid onset illness. The host develops symptoms within 5–21 days after infection with Hendra virus. At initial stage, some common symptoms appear as fever, cough, headache, and tiredness whereas later, meningitis and coma have been seen in severe infection conditions. Hendra virus disease can be fatal sometimes (Field et al. 2011).

Hydatid disease: Hydatidosis, echinococcosis, or hydatid disease, caused by cyst consist of *Echinococcus granulosus* tapeworm (Dog Tapeworm) larval stage, is an extremely serious condition and sometimes may be fatal. The disease is mainly infecting dogs and other canines. Infected animals shed eggs in their faeces which further infect other host by swallowing these eggs or by direct contact with infected dogs. The cyst can be remained inside the liver and lungs for months to several years. The disease show signs of weight loss, weakness, and loss of appetite. The rupture of cyst may cause serious allergic reactions even death of animals. Although Hydatid infection is potentially severe but may be treatable. The disease occurs worldwide, commonly in grazing areas (Qingling et al. 2014).

Listeriosis: Animals, fishes, crustacean, birds, and even humans can be infected with Listeriosis; which is a serious infection and become fatal sometimes. *Listeria monocytogenes* is the causative agent of this food borne disease. It has the potential to transmit from one cell to another and can cross intestinal, placental, and blood brain barriers. The disease is characterized by encephalitis, meningitis, meningoencephalitis, stillbirth, abortion septicaemia, perinatal infections. Listeriosis is a sporadic disease which spread through soil, water, milk, meat, seafood, and vegetables. There is no vaccine available for this disease so it can be prevented only through hygienic practices (Oevermann et al. 2010).

Newcastle disease: Newcastle disease virus (NDV) is a causative agent of Newcastle disease of domestic poultry and other species of birds. It is primarily represented by acute respiratory diseases, but diarrhoea, depression, or nervous manifestations can also be seen predominantly as a clinical form. Infected animals shed infection through respiratory discharge, eggs, and faeces. In poultry disease, may spread by ingesting water or food contaminated with virus. Movement of infected poultry birds, contaminated equipment, or litter plays an important role for spreading viruses. The signs of disease in an infected organism are appeared in 2–12 days of exposure of virus (Brown and Bevins 2017).

Nipah virus disease: Nipah virus or NiV belongs to genus henipavirus and paramyxovirus family causes a highly fatal disease Nipah. Natural host of nipah virus is fruit bat and currently, it affects pigs and humans very severely with high mortality rate. The virus is spread through bat urine, faeces, birthing fluids, and saliva. It is assumed that bat excreta initiate infection in pigs. The virus mainly affects the nervous system and respiratory system. The virus is highly contagious to pigs and associated with barking pig syndrome (BPS) and encephalitic syndrome

(PRES). Nipah virus may also infect dogs naturally and causes a distemper-like syndrome which is a highly fatal condition (Ang et al. 2018).

19.2.2 Non-Communicable Diseases

Ruminal tympany: It is also known as bloat. The disease mainly found in ruminant animals and characterized by high amount of gas in rumen. It can be divided into two phase, primary frothy bloat, and secondary free gas bloat. In a natural process, the rumen is dedicated for fermentation of eaten food by gut microbes and the fermentation process produces excessive amount of gas which expelled out from rumen by burping continuously but sometimes gas becomes trapped inside the rumen. This condition is remarked as bloat or ruminal tympany. In primary bloat, produced gas trapped within fermenting material which causes a raised amount of foam and unable to remove by burping. The disease is characterized by distension of abdomen on left side, distress, and difficulty in breathing. This disease can be fatal if gas accumulation process continues which also distended right side of abdomen. After visualization of such symptoms, the cattle death may occur within 3–4 h. In the secondary bloat, oesophagus becomes blocked so accumulated gas cannot be pass through it (Priyanka and Dey 2018).

Nutritional deficiencies in animals: In the case of animals, this is very difficult to find out the nutritional deficiencies. Most of the animals suffer from malnutrition. There is a lack of signs related to most of the nutritional deficiencies because signs are common and not specific such as loss of appetite, reduction in growth, and thriftiness. Multiple types of nutritional deficiencies may be caused in many animals due to starvation. Deficiency of some essential amino acids or suboptimal feed intake may result into protein deficiencies which causes weight loss and fatter carcasses in pigs. In lactating animals for example cows, protein deficiency directly affects milk production. Some fatty acids like linoleic acid are found very essential in the diet and utilized in production of another essential form i.e. longer chain fatty acids. The lack of linoleic acid in the diet leads to scaly dermatitis, necrosis in skin, unthrifty appearance, and hair loss in growing pigs. Deficiency of minerals like calcium and phosphorous may outcome as a disease known as rickets in pigs, dogs, and others. The disease is characterized by deformity, lameness, and bending of bones specifically long bones. Consequences of severe conditions are fracture and paralysis at posterior parts, generally in old ones. Salt deficiency also affects the growth of animals.

Iodine deficiency may result into hairless and week pig's new-born. New-born pigs may show enlarged thyroids and histological abnormalities. Enlarged thyroid can also be seen in many animals and known as Goitre. Iodized salt may get rid of this disease.

Iron and copper deficiencies efficiently develop anaemia in animals which is caused by a reduction in haemoglobin synthesis. Anaemia may be recognized by low RBC and haemoglobin count, enlarged heart, edema, and thumps. Deficiency of zinc, selenium, and vitamin E also have an adverse effect on growing or younger ones.

Vitamins are very essential for the development and metabolic processes in humans as well as in animals. In livestock and pets, if they provided with some supplemented commercially available diets then chances are very low to face the vitamin deficiency challenges because these are fortified with vitamins. Vitamin A deficiency affects eyes, epithelial tissues of important organ systems like the nervous, urinary, digestive, respiratory, and reproductive. Vitamin K deficiency is related with lengthened blood clotting and sometimes animals may die due to haemorrhage. Riboflavin deficiency in pigs may result into impaired reproduction. Vitamin B deficiency also affects severely to neonatal pigs in terms of voice failure, hyperirritability, and incoordination like symptoms. Biotin deficiency also characterized by excessive hair loss, dermatitis, inflammation of mouth, and mucous membrane (Innis 2000; Vieyra-Reyes et al. 2017; Shukla et al. 2016).

19.3 Development of Nano-Biosensing Devices for Detection of Animal Diseases

Nano biosensing devices exploit a biorecognition element which can be DNA/RNA, antigen/antibodies, enzymes, aptamers, and other polymers, etc. There are several studies done to develop a nano-biosensing device by using various biomarkers. Here, some types of biosensors description are given that are developed to detect several serious diseases (Table 19.2).

19.3.1 DNA Sensors

DNA based sensors are developed for the detection of several biomarkers depend upon the specific nucleic acid sequences. Various forms of nanomaterial are used for signal amplification which generated through the hybridization of two complementary DNA. Anthrax disease can be detected through a label free nanopore real time sensing device, in which a single-stranded DNA (ssDNA) was used as a molecular probe for anthrax lethal factor (Wang et al. 2014). A unique nanostructure based geno-sensor was fabricated to detect Brucella-specific probe. A gold nanoribbon surrounded by gold nano blooms were further deposited on polycrystalline gold surface by sonoelectrodeposition method. In this approach, methylene blue was used as redox marker to detect complementary sequence, sequences with one-, two-, and three bases mismatches (Rahi et al. 2015). Another DNA biosensor also studied for Brucella species detection in animals which consists of a specific oligonucleotide probe taken from S711 gene region and gold nanoparticle. This probe hybridized with complementary DNA biomarker present in a clinical sample. The device was capable to detect unamplified Brucella genomic DNA up to 7 pg μL^{-1}. The DNA biosensor

Table 19.2 List of different types of fabricated Nano-biosensors in past few years for some common animal diseases

Nano-biosensor	Disease	Nanomaterial	Detection limit	Reference
DNA biosensor	Brucellosis	Gold (nanoribbons/nanoblooms)	1.71 zmol dm^{-3}	Rahi et al. 2015
		Gold	7 pg μL^{-1}	Sattarahmady et al. 2016
	ehrlichiosis	QCM based	4.1x10^9 molecules μL^{-1}	Bunroddith et al. 2018
	Newcastle Disease	Magnetic nanoparticles (MNPs)	75 fM	Tian et al. 2016
Aptasensor	Avian influenza	Gold	0.25 HAU	Karash et al. 2016
	Anthrax	Multiwalled carbon nanotubes	20 ng m L^{-1}	Karimi and Dabbagh. 2019
	bovine viral diarrhea	Gold	800 copies mL^{-1}	Park et al. 2014
Immunosensor	Bluetongue	Gold	10 – 2 fg mL^{-1}	Yin et al. 2012
	Hydatoid	Silicon	300 fg mL^{-1}	Lv et al. 2017
	Enzootic bovine leukemia	Magnetic gold nanoparticles	0.95 mg mL^{-1}	Baniukevic et al. 2013

exhibited high sensitivity and selectivity to amplified clinical samples (Sattarahmady et al. 2016). In a study, gold nanorods are functionalized with specific ssDNA probe of 'cadF gene' to detect campylobacter species. In this study about 283 stool samples were analysed and showed 44 positive cases, that indicated sensitivity of device, rapid in comparison to PCR, culture, and real time PCR (Shams et al. 2019). Another study demonstrates the utilization of quartz crystal microbalance (QCM) in combination with PCR assay for successful detection of *Ehrlichia canis*. The QCM based DNA nano-biosensor device was reusable, simple and do not need any critical equipment. The basic principle of this device was DNA-DNA hybridization without any cross hybridization. The limit of detection of QCM was calculated as low as 4.1×10^9 molecules/μl of 289 bp *E. canis* PCR product that represented to 22 copy numbers/μl of *E. canis* (Bunroddith et al. 2018). Newcastle disease virus RNA detection was carried out by Tian and co-workers by combination of two highly efficient techniques that are loop-mediated isothermal amplification (LAMP) and opto magnetic nanoparticle-based readout system. The detection limit was 10 aM of target sequence in just 30 min (Tian et al. 2016).

19.3.2 Aptasensors

Aptasensors are fabricated by employing aptamers. Aptamers are naturally occurring or synthesized short random oligonucleotides or peptides sequences. Generally, they are developed through SELEX method. In the animal disease detection aptasensors were also utilized for rapid and accurate detection of disease biomarkers. An electrochemical graphene field-effect transistor based aptasensor has been developed for detection of anthrax toxin (Kim et al. 2013). The avian influenza detection can be possible through aptamer-based nano-biosensor devices. Several aptasensors work has been carried out for detection of avian influenza virus. For example, an impedance aptasensor was developed for the detection of highly pathogenic avian influenza (H5N1). To enhance impedance signal, a gold nanoparticle-based amplifier was designed and functionalized with H5N1-aptamer and thiocyanuric acid. The device exhibited a robust, specific, and rapid detection method for the H5N1 virus with the detection limit of 0.25 haemagglutination unit (HAU) (Karash et al. 2016). Aptasensors were also used for detection of Anthrax disease in animals. Recently, a nano-biosensor device which was based on multi-walled carbon nanotubes and fluorescence aptasensors was fabricated. In this study, researchers detected 'recombinant protective antigen domain 4 (rPAD4)' of *Bacillus anthracis* which plays an important role in the development of disease. The aptamers were labelled with gel green fluorescent dye and immobilized on carbon nanotubes to give amplified signals after hybridization of complementary sequences. The device had the potential to be used for diagnostic tests for anthrax in a fast, accurate, sensitive, and cost-effective manners (Karimi and Dabbagh 2019). Another frequently occurred disease in animals is bovine viral diarrhoea that could be also detected by aptasensors. In 2013 a biosensor based on two different aptamers was fabricated for successful screening of binding to bovine viral diarrhoea virus type 1 (BVDV type 1) in a clinical sample. The aptamer had a high affinity and specificity to this virus biomarker. In this study, the work was based upon basic principle of aptamer-aptamer sandwich type sensing platform in which one aptamer immobilized with gold nanoparticle and other hybridized with viral DNA BVDV type 1. The device was capable to detect 800 copies/ml (Park et al. 2014).

19.3.3 Immunosensors

Immunosensors are basically worked on the principle of antigen and antibody reaction. Several types of biosensors have been developed to detect animal diseases in minimum time and cost-effective way. Immunosensors utilized different type of transducer surface such as electrochemical, and optical (colorimetric, surface plasmon resonance (SPR), fluorescent, chemiluminescence), etc. Bluetongue virus associated biomarker VP7outer-core protein was detected by a nanobiosensor fabricated by exploitation of gold nanoparticle probe-based assay and bio-barcode assay technique. In this device anti-VP7 monoclonal antibodies and ssDNA are immobilized on magnetic microparticle probes. The developed biosensor exhibited the detection limit of 10^{-2}fg mL^{-1} (Yin et al. 2012). Electrochemical immunosensor consists of single gold nanowire that has been developed in 2015 for the detection of bovine viral diarrhoea virus. In this device researchers immobilized captured biomarker of virus on gold nanowire covalently by electrodeposition method. Immunosensing device was applied for detection of serum antibodies in the serum of infected animals. Team specified their device as a very sensitive with very low detection limit so could be applied for on-farm diagnosis for animal health monitoring (Montrose et al. 2015). BVDV monitoring could make possible through another microfluidic immunosensor by Heinze and co-workers in 2009. In this study they demonstrated a microfluidic based platform which consist of anti-BVDV monoclonal antibodies immobilized on carboxylated polystyrene microparticles. The performance of device had been compared with real time PCR where they found the very low detection limit in microfluidic device (Heinze et al. 2009). Enzootic bovine leucosis is a blood borne disease which affect mainly cattle and caused by BLV or bovine leukosis virus. The virus resides in white blood cells and asymptomatic most of the time. SPR based immunosensor was fabricated recently in 2019 for detection of BLV in blood serum of sick animals (Klestova et al. 2019).

An optical immunosensor for the detection of BLV has been developed by exploiting the UV and visible photoluminescence property of zinc oxide nanoparticles. The device was suitable for field use. The device showed high sensitivity and a wide range of leucosis antibody detection i.e. 0.001 mg mL^{-1} to 1 mg ml^{-1} (Ruban et al. 2017). Another surface-enhanced Raman scattering (SERS) based immunosensor has been also developed for the detection of this disease in 2013. In this study magnetic gold nanoparticles (MNP-Au) functionalized with fragmented antibodies specific for gp51 biomarker were used. They were resulted out the limit of detection as 0.95 μg mL^{-1} and limit of quantification as 3.14 μg mL^{-1} for gp51 (Baniukevic et al. 2013). A method for the diagnosis of foot and mouth disease was also developed with the help of an electrochemical immunosensor. A non-structural protein i.e. 3ABC of associated disease was detected through specific antibodies by utilization of a screen-printed electrode (Longinotti et al. 2010). Hydatid disease is associated with Echinococcus granulosis. Several studies have been carried out for the development of immunosensor to detect the associated biomarker. In a study, researcher developed silicon microcavities based nanosensor to perform antigen antibody interaction on it. An associated protein named protein kinase P38 was targeted for detection of hydatid diseases (Lv et al. 2017). Quantum dots based immuno biosensor also demonstrated in a 2017 study. In this device they immobilized Egp38 antigen on silicon pores which further functionalized with anti-p38 labelled CdSe/ZnS QDs. This was a fluorescence-based detection. They achieved detection limit 300 fg mL^{-1} for Egp38 antigen (Li et al. 2017). Very recently, an immunosensor has been developed for this disease using square wave voltammetry technique. Cysteamine/phenylene isothiocyanate linkers have been coated on gold electrodes followed by immobilization of either purified 'rabbit polyclonal antibody' or 'recombinant antigen B' (AgB) to detect hydatid antigen. Detection limits for echinococcus antigen and antibody were 0.4 pg mL^{-1} and 0.3 pg mL^{-1}, respectively. The device is low of cost, sensitive and can be used as a point of care device (Eissa et al. 2020). Newcastle disease is also a fatal disease which affect several animals worldwide. A very simple electrochemical immunosensor was developed recently for detection of NDV. In this device chicken egg yolk antibodies (IgY) was utilized as a biorecognition element and were specific for NDV antigen. The sensor had exhibited a sensitivity range of 10^6 to 10^2 EID_{50} mL^{-1} (Thinh et al. 2018). Listeriosis is also a very harmful disease and can be detected through immunosensors. A 2013 work reported a Screen-printed carbon electrode modified with gold nanoparticle exhibited great sensitivity for detection of Listeria species. *Listeria monocytogenes* presence was monitored through conjugating secondary enzyme-labelled antibodies (Davis et al. 2013). The biosensor found to be effective in detection of other enteric bacterial species such as *E. coli* and *Salmonella Typhimurium.* Another 2017 study revealed an immunosensor fabricated with a pair of monoclonal antibodies. These pair of mAbs used to recognize P60 protein associated with Listeria species (Wang et al. 2017).

19.3.4 Miscellaneous

CRISPR-Cas assay with a fluorescence-based point-of-care (POC) system has been developed for the detection of African swine fever virus detection. The LOD was found as 1 pM in 2 h (He et al. 2020). Recently, a metal oxide semiconductor (MOS) based sensor was developed for detection of *Campylobacter jejuni* in a milk sample. This microbial agent is responsible for several gastrointestinal related diseases in humans. A MOS sensor has the ability to differentiate between contaminated and control samples (Núñez-Carmona et al. 2020). Another advancement in nanobiosensors can be represented by wearable biosensors. For continuous real time monitoring of herd of animals and livestock; wearable biosensors show a great significance in the health management of animals. They provide physiological information through non-invasive, dynamic measurement of biomarkers in biofluids like saliva, sweat, tears, and interstitial fluids (Neethirajan 2017).

19.4 Future Aspects

Nano-biosensing devices are known for their high sensitivity, simplicity, and selectivity to detect animals or human diseases. They are able to provide rapid results with low cost in comparison to conventional methods (Malik et al. 2013). Animals are extremely susceptible to infection because of unhygienic living environment therefore suffer with number of communicable diseases and their direct contact with human beings make the situation more problematic. Animal communicable diseases cause significant economic loss in various ways like loss in food production system, and increased veterinary costs (Kivaria 2006; Ott et al. 1999; Dufour 1999; Neethirajan 2019). We consider that nano-biosensors have capability to advance conventional technologies for detection of disease-causing microorganisms and can improve the surveillance system to prevent any disease outbreak. Wearable devices can play an important role in future to manage animal diseases. Real time field monitoring, data networking, sampling methods, and mobile analysis will provide remarkable support through highly sensitive and selective nano-biosensors. The data through device will help in disease control and prevention of outbreak in poultry, livestock as well as wild animals.

19.5 Conclusion

Prevention of communicable diseases and non-communicable diseases is very essential for the health management of animals. Human beings are also affected with animal disease directly or indirectly and have various common diseases. The first step required for prevention of any disease is accurate diagnosis. It is important to develop some reliable, cost effective, and less time-consuming methods. Fast diagnosis is must in some disease for example anthrax which causes death within 2 days. Sometimes some diseases are remained asymptomatic like bovine leucosis, so they also need accurate detection in very low concentration. Nanotechnology exhibit several new opportunities to fabricate new devices for the successful determination of diseases. With the help of biomarkers, these nano-biosensors have given promising results in the diagnosis of animal diseases. Communicable and non-communicable diseases affecting animals at a different level can now detect at an early stage, monitored, and controlled with the help of nano-biosensor.

References

Alcaraz R, Gerber S, Weber C, Mueller E (2016) Symmetric dimethylarginine (SDMA) as a biomarker for early diagnosis of renal dysfunction in cat. Argos-Inf Vet 181:64

Ang BSP, Lim TCC, Wang L (2018) Nipah Virus infection. J Clin Microbiol 56(6):e01875–17. https://doi.org/10.1128/JCM.01875-17

Baniukevic J, Hakki Boyaci I, Goktug BA, Tamer U, Ramanavicius A, Ramanaviciene A (2013) Magnetic gold nanoparticles in SERS-based sandwich immunoassay for antigen detection by well oriented antibodies. Biosens Bioelectron 43:281–288

Brown VR, Bevins SN (2017) A review of virulent Newcastle disease viruses in the United States and the role of wild birds in viral persistence and spread. Vet Res 48(1):1–15

Bunroddith K, Viseshakul N, Chansiri K, Lieberzeit P (2018) QCM-based rapid detection of PCR amplification products of Ehrlichia canis. Anal Chim Acta 1001:106–111. https://doi.org/10.1016/j.aca.2017.10.037

Carretón E, Morchón R, González-Miguel J, Juste MC, Simón F, Montoya-Alonso JA (2013) Utility of cardiac biomarkers during adulticide treatment of heartworm disease (Dirofilaria immitis) in dogs. Vet Parasitol 197(1–2):244–250

Charles E, Cornelius, Dante G, Scarpelli, Others (2018) Animal disease, Encyclopædia Britannica, Encyclopædia Britannica, inc. August 09, https://www.britannica.com/science/animal-disease
Çolak A, Toprak B, Yalçın H, Altun Z, Diniz G, Olgun A, Yılmaz O, Aktaş S (2013) Biochemical biomarkers for detection of nephrotoxicity and cardiotoxicity in cisplatin administered rats. İzmir Dr. Behçet Uz Çocuk Hastanesi Dergisi 3(2):105–109
Dahal S (2019) Analysis of host microRNAs as potential biomarkers of Hendra virus disease in horses. Deakin University
Dalvi RR (1986) An overview of aflatoxicosis of poultry: its characteristics, prevention and reduction. Vet Res Commun 10(6):429–443. https://doi.org/10.1007/BF02214006
Davis D, Guo X, Musavi L, Lin CS, Chen SH, Wu VC (2013) Gold nanoparticle-modified carbon electrode biosensor for the detection of Listeria monocytogenes. Industrial biotechnology 9(1):31–36
Dixon LK, Sun H, Roberts H (2019) African swine fever. Antiviral Res 165:34–41. https://doi.org/10.1016/j.antiviral.2019.02.018
Doganay M, Demiraslan H (2015) Human anthrax as a re-emerging disease. Recent Pat Antiinfect Drug Discov 10(1):10–29. https://doi.org/10.2174/1574891x10666150408162354
Dufour B (1999) Technical and economic evaluation method for use in improving infectious animal disease surveillance networks. Vet Res. 30(1):27-37
Eissa S, Noordin R, Zourob M (2020) Voltammetric label free-immunosensors for the diagnosis of cystic echinococcosis. Electroanal 32(6):1170–1177
Fei J, Dou W, Zhao G (2016) Amperometric immunoassay for the detection of Salmonella pullorum using a screen-printed carbon electrode modified with gold nanoparticle-coated reduced graphene oxide and immunomagnetic beads. Microchim Acta 183(2):757–764
Field H, de Jong C, Melville D et al (2011) Hendra virus infection dynamics in Australian fruit bats. PLoS ONE 6(12):e28678. https://doi.org/10.1371/journal.pone.0028678
Fry TL, Dunbar MR (2007) A review of biomarkers used for wildlife damage and disease management. Proceedings of the 12th Wildlife DamageManagement Conference (eds: D.L. Nolte, W.M. Arjo, D.H. Stalman)
Habib SH, Saha S (2010) Burden of non-communicable disease: global overview. Diab Metabol Syndr Clin Res Rev 4(1):41–47
He Q, Yu D, Bao M et al (2020) High-throughput and all-solution phase African Swine Fever Virus (ASFV) detection using CRISPR-Cas12a and fluorescence based point-of-care system. Biosens Bioelectron 154:112068. https://doi.org/10.1016/j.bios.2020.112068
Heinze BC, Song JY, Lee CH, Najam A, Yoon JY (2009) Microfluidic immunosensor for rapid and sensitive detection of bovine viral diarrhea virus. Sens Actuators B: Chem 138(2):491–496
Heymann DL (2008). Control of communicable diseases manual (no. ed. 19). American Public Health Association
Innis SM (2000) Essential fatty acids in infant nutrition: lessons and limitations from animal studies in relation to studies on infant fatty acid requirements. Am J Clin Nutr 71(1):238S–244S
Karash S, Wang R, Kelso L, Lu H, Huang TJ, Li Y (2016) Rapid detection of avian influenza virus H5N1 in chicken tracheal samples using an impedance aptasensor with gold nanoparticles for signal amplification. J Virol Methods 236:147–156
Karimi F, Dabbagh S (2019) Gel green fluorescence ssDNA aptasensor based on carbon nanotubes for detection of anthrax protective antigen. Int J Biol Macromol 140:842–850. https://doi.org/10.1016/j.ijbiomac.2019.08.219
Kim DJ, Park HC, Sohn IY et al (2013) Electrical graphene aptasensor for ultra-sensitive detection of anthrax toxin with amplified signal transduction. Small 9(19):3352–3360. https://doi.org/10.1002/smll.201203245
Kivaria FM (2006) Estimated direct economic costs associated with tick-borne diseases on cattle in Tanzania. Trop Anim Health Prod 38(4):291–299
Klestova ZS, Yuschenko AY, Dremukh Y, Blotska OF, Venger EF, Dorozinsky GV, …, Maslov VP (2019) Diagnostics of cattle leucosis by using a biosensor based on surface plasmon resonance phenomenon. Semicond Phys Quantum Electron Optoelectron 22(1):111–118

Li Y, Jia Z, Lv G, et al (2017) Detection of Echinococcus granulosus antigen by a quantum dot/porous silicon optical biosensor.Biomed Opt Express 8(7):3458–3469. Published 2017 Jun 26. https://doi.org/10.1364/boe.8.003458

Li YT, Linster M, Mendenhall IH, Su YCF, Smith GJD (2019) Avian influenza viruses in humans: lessons from past outbreaks. Br Med Bull 132(1):81–95. https://doi.org/10.1093/bmb/ldz036

Longinotti G, Ybarra G, Lloret P et al (2010) Diagnosis of foot-and-mouth disease by electrochemical enzyme-linked immunoassay. Conf Proc IEEE Eng Med Biol Soc. 2010:674–676. https://doi.org/10.1109/IEMBS.2010.5626230

Lv X, Xin L, Lv G, Mo J, Gao Z, Jia Z, Wen H (2017) Preparation of a photoluminescent film on a silicon-on-insulator device for the simple, rapid, and quantitative detection of a hydatid disease diagnostic protein marker. IEEE Photonics J 9(4):1–7

Malik P, Katyal V, Malik V, Asatkar A, Inwati G, Mukherjee T (2013). Nanobiosensonsor: concepts and variations. Int Sch Res Noti. doi.org/10.1155/2013/327435

Mariano V, Nardi A, Gradassi S, et al (2019) A severe outbreak of botulism in cattle in Central Italy. Vet Ital. 55(1):57–62. Published 2019 Mar 31. https://doi.org/10.12834/vetit.768.3714.2

Montrose A, Creedon N, Sayers R, Barry ST, O'Riordan A (2015) Novel single gold nanowire-based electrochemical immunosensor for rapid detection of bovine viral diarrhoea antibodies in serum. J Biosens Bioelectron 2015:1–7

Mylonakis ME, Harrus S, Breitschwerdt EB (2019) An update on the treatment of canine monocytic ehrlichiosis (Ehrlichia canis). Vet J 246:45–53. https://doi.org/10.1016/j.tvjl.2019.01.015

Neethirajan S (2017) Recent advances in wearable sensors for animal health management. Sens Bio-Sens Res 12:15–29

Neethirajan S (2019) Biosensors-new frontiers in animal welfare. J Anim Sci 97:2–2

Núñez-Carmona E, Abbatangelo M, Zappa D, Comini E, Sberveglieri G, Sberveglieri V (2020) Nanostructured mos sensor for the detection, follow up, and threshold pursuing of campylobacter jejuni development in milk samples. Sens (Basel) 20(7):2009. Published 2020 Apr 3. https://doi.org/10.3390/s20072009

Oevermann A, Zurbriggen A, Vandevelde M (2010) Rhombencephalitis caused by *Listeria monocytogenes* in humans and ruminants: a zoonosis on the rise? Interdiscip Perspect Infect Dis. 2010:632513. https://doi.org/10.1155/2010/632513

Ott SL, Wells SJ, Wagner BA (1999) Herd-level economic losses associated with Johne's disease on US dairy operations. Prev Vet Med 40(3–4):179–192

Park JW, Jin Lee S, Choi EJ, Kim J, Song JY, Bock GuM (2014) An ultra-sensitive detection of a whole virus using dual aptamers developed by immobilization-free screening. Biosens Bioelectron 51:324–329. https://doi.org/10.1016/j.bios.2013.07.052

Peng F, Loo JF, Kong SK, Li B, Gu D (2017) Identification of serum MicroRNAs as diagnostic biomarkers for influenza H7N9 infection. Virol Rep 7:1–8

Pereira CS, Muzzi RA, Figueiredo VC, Muzzi LA, Oberlender G, Junior AC, Neto ML, Oliveira MM (2016) Troponin I as a biomarker of cardiac injury in dogs with sepsis. Arq Bras de Medi Vet e Zootec 68(4):919–926

Priyanka M, Dey S (2018) Ruminal impaction due to plastic materials—An increasing threat to ruminants and its impact on human health in developing countries. Vet World 11(9):1307–1315. https://doi.org/10.14202/vetworld.2018.1307-1315

Qingling M, Guanglei W, Jun Q et al (2014) Prevalence of hydatid cysts in livestock animals in Xinjiang. China Korean J Parasitol 52(3):331–334. https://doi.org/10.3347/kjp.2014.52.3.331

Rahi A, Sattarahmady N, Heli H (2015) Zepto-molar electrochemical detection of Brucella genome based on gold nanoribbons covered by gold nanoblooms. Sci Rep. 5:18060. Published 2015 Dec 14. https://doi.org/10.1038/srep18060

Raj AAG, Jayanthi JG (2018) IoT-based real-time poultry monitoring and health status identification. 2018 11th International Symposium on Mechatronics and its Applications (ISMA), Sharjah, pp 1–7. https://doi.org/10.1109/isma.2018.8330139

Ruban Y, Shpyrka N, Pareniuk O, Galat M, Taran MS, Ishchenko L, Shavanova K (2017). BLV leucosis biosensor based on ZnO nanorods photoluminescence. In 2017 IEEE 37th International Conference on Electronics and Nanotechnology (ELNANO) (pp 333–337), IEEE, April

Sagadevan S, Periasamy M (2014) Recent trends in nanobiosensors and their applications-a review. Rev Adv Mater Sci 36:62–69

Sattarahmady N, Movahedpour A, Heli H, Hatam GR (2016) Gold nanoparticles-based biosensing of Leishmania major kDNA genome: visual and spectrophotometric detections. Sens Actuators B: Chem 235:723–731

Scherr SM, Daaboul GG, Trueb J, Sevenler D, Fawcett H, Goldberg B, Connor JH, Ünlü MS (2016) Real-time capture and visualization of individual viruses in complex media. ACS nano 23;10(2):2827–2833, Feburary

Shams S, Bakhshi B, Tohidi Moghadam T, Behmanesh M (2019) A sensitive gold-nanorods-based nanobiosensor for specific detection of Campylobacter jejuni and Campylobacter coli. J Nanobiotechnol17(1):43. Published 2019 Mar 26. https://doi.org/10.1186/s12951-019-0476-0

Shoukat S, Wani H, Ali U, Para PA, Ara S, Ganguly S (2017) Brucellosis: a current review update on zoonosis. J Immunol Immunopathol 19(2):61–69

Shukla AK, Tiwari PK, Pakhare A, Prakash C (2016) Zinc and iron in soil, plant, animal and human health. Indian J Fertil 12(11): 133–149

Solis-Cruz B, Hernandez-Patlan D, Petrone VM, et al (2019) Evaluation of cellulosic polymers and curcumin to reduce aflatoxin b1 toxic effects on performance, biochemical, and immunological parameters of broiler chickens. Toxins (Basel) 11(2):121. Published 2019 Feb 16. https://doi.org/10.3390/toxins11020121

Thinh TQ, Quan TVV, Thuy TH, Xuan CT, Tuan MA. (2018). A label-free electrochemical immunosensor for detection of Newcastle disease virus. In International Conference on the Development of Biomedical Engineering in Vietnam (pp. 699–703), Springer, Singapore, June

Tian B, Ma J, Zardán Gómez de la Torre T, Bálint A, Donolato M, Hansen MF, …, Strömberg M (2016) Rapid newcastle disease virus detection based on Loop-mediated isothermal amplification and optomagnetic readout. Acs Sensors 1(10):1228–1234

Van den Bergh C, Coetzee P, Venter EH (2018) Reassortment of bluetongue virus vaccine serotypes in cattle. J S Afr Vet Assoc 89(0):e1–e7. Published 2018 Dec 5. https://doi.org/10.4102/jsava.v89i0.1649

Vidic J, Manzano M, Chang C et al (2017) Advanced biosensors for detection of pathogens related to livestock and poultry. Vet Res 48:11. https://doi.org/10.1186/s13567-017-0418-5

Vieyra-Reyes P, Millán-Aldaco D, Palomero-Rivero M, Jiménez-Garcés C, Hernández-González M, Caballero-Villarraso J (2017) An iron-deficient diet during development induces oxidative stress in relation to age and gender in Wistar rats. J Physiol Biochem 73(1):99–110

Viitanen SJ, Laurila HP, Lilja-Maula LI, Melamies MA, Rantala M, Rajamäki MM (2014) Serum C-reactive protein as a diagnostic biomarker in dogs with bacterial respiratory diseases. J Vet Intern Med 28(1):84–91

Wang L, Han Y, Zhou S, Wang G, Guan X (2014) Nanopore biosensor for label-free and real-time detection of anthrax lethal factor. ACS Appl Mater Interfaces 6(10):7334–7339. https://doi.org/10.1021/am500749p

Wang LF, Anderson DE (2019) Viruses in bats and potential spillover to animals and humans. Curr Opin Virol 34:79–89. https://doi.org/10.1016/j.coviro.2018.12.007

Wang W, Liu L, Song S, Xu L, Zhu J, Kuang H (2017) Gold nanoparticle-based paper sensor for multiple detection of 12 Listeria spp. by P60-mediated monoclonal antibody. Food Agric Immunol 28(2): 274–287

Yin HQ, Jia MX, Yang S, Jing PP, Wang R, Zhang JG (2012) Development of a highly sensitive gold nanoparticle probe-based assay for bluetongue virus detection. J Virol Methods 183(1):45–48. https://doi.org/10.1016/j.jviromet.2012.03.027

Zetterberg H, Bozzetta E, Favole A, Corona C, Cavarretta MC, Ingravalle F, Blennow K, Pocchiari M, Meloni D (2019) Neurofilaments in blood is a new promising preclinical biomarker for the screening of natural scrapie in sheep. PloS ONE 14(12):e0226697. https://doi.org/10.1371/journal.pone.0226697

Chapter 20
Recent Advances in Biosensor Development for Poultry Industry

Nidhi Chauhan, Ramesh Namdeo Pudake, Utkarsh Jain, and Sapna Balayan

Abstract Poultry sector is becoming important for the sustainable food chain. It is now rapidly growing across the world. Various biosensors for antibiotics and pathogens identification have been developed in comparison to traditional methods in poultry industry. Poultry market has already occupied a major share of worldwide farmer's income. To prevent bacterial growth in farm birds many antibiotics are injected. Ultimately these antibiotic residues are found in them which results in a huge life-threatening scenario for the person who is consuming these animal products. The antibiotics that are widely used in poultry animals create antibiotic resistance in humans. Several biosensors are developed for detecting antibiotics residue in poultry birds. Biosensors are also used for the detection of infection-causing pathogens in farm birds and they have proven to be very useful when compared to other traditional screenings. Because of the increasing demand for eggs and meat due to the increasing population, it is expected that the poultry sector will expand in the future. Emerging nanotechnology is a key factor development of biosensors. Therefore, the chapter briefly explains the nanotechnology-based sensors for antibiotics and pathogens detection in poultry birds.

Keywords Poultry · Antibiotics · Antibiotic detection · Pathogens · Biosensor

20.1 Introduction

Presently, the poultry sector is becoming one of the rapidly growing components of the agriculture-based industry across the world for higher economic growth. But the growing field of poultry farming in large-scale is resulting in the deterioration and, higher rate destruction of environment, it may cause economic loss (Kabir 2009). It has been predicted that till 2025 poultry industry will be the largest producer of meat across the world (King et al. 2018). According to "The United States Department of Agriculture" (USDA) globally the poultry production was around 111,000 thousand

N. Chauhan (✉) · R. N. Pudake · U. Jain · S. Balayan
Amity Institute of Nanotechnology, Amity University Uttar Pradesh, Noida, UP, India
e-mail: nchauhan1@amity.edu

R. N. Pudake et al. (eds.), *Biosensors in Agriculture: Recent Trends and Future Perspectives*, Concepts and Strategies in Plant Sciences,
https://doi.org/10.1007/978-3-030-66165-6_20

metric tons during 2015. It is assumed that poultry production worldwide over the next few decades will be projected to be raised by 24 percent that will surely become around 131,255 thousand metric tons in 2025. Till 2050 the estimation for the global population is awaited to be increased to 9 billion and also increase in the income growth, which will be leading to the higher demand for animal-based protein diet. This explains the important part played by the poultry industry for a sustainable food supply (King et al. 2018).

Different factors play a role in the continuous growth of this sector including rising population, industrialization, rising income, geographical distribution. All these mentioned factors are believed to be responsible for the rapid expansion of poultry industry worldwide. There are several small poultry dressing plants everywhere in the developed and developing countries for meat and egg processing. Around every locality, we can find these poultry products in close proximity.

Also, more and more employment will be generated for the poor people in this industry. Modernization too is an important factor in the rise of poultry sector. Poultry meat especially chicken is highly consumed, due to its low-fat content, higher protein level, and producing less greenhouse gas. Also, when compared to other birds' chickens have a higher feed conversion ratio (FCR) and more meat is produced with lesser feed input (King et al. 2018).

In both rural and urban areas poultry has proven to be a valuable thing especially for small holders or shopkeepers. This industry may help them to earn their livelihood and the easiest way to enter the business world. With the growing demand for poultry products, there are many drawbacks associated with this sector too. The poultry sector is having many disastrous challenges such as bird flu and other contagious diseases. Pathogen detection in poultry animals has remained a big challenge for the poultry industry in the last decades. Various pathogen detection methods have already been developed which is discussed further in this chapter (Mottet and Tempio 2017).

Due to the increase in demand for poultry meat, the industry will explore different methods and incorporate new technologies to increase their production, to continue the sustainable food chain, and meet the demands for safe food at an affordable prices (King et al. 2018). And the challenges faced by the poultry industry are briefly discussed further.

20.2 Challenges Faced by the Poultry Sector

Along with the growing market of poultry sector, it has been indicated that this industry is facing many challenges during pathogen attack on the animals. Apart from spoiling human health, it has a significant effect on our environment and natural resources too. Any change in the environmental factor directly results in climate change through pollution of air and water and production of feeding materials (Hu et al. 2017). The most important risk factor in poultry industries is represented in Fig. 20.1 and discussed in brief.

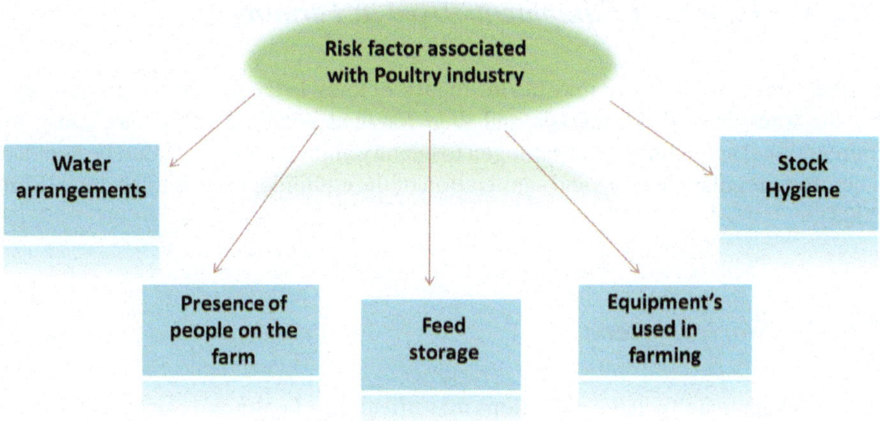

Fig. 20.1 Important risk factor in poultry industry

20.2.1 Water

Water source which is used for various purposes in poultry farm must be clean enough to ensure animal's good health, as most of the contagious diseases are due to dirty or unhygienic water (Sims 2008).

20.2.2 Work Force on the Farm

Not only visitors but caretakers on the farm can also be a disease carrier through their dirty shoes or clothes. So, to avoid this proper cleaning of the shoes and wearing protective gear is advisable. This personal protective equipment will also help in preventing the disease transfer to the operator (Sims 2008).

20.2.3 Feed Storage

Storage of the feeding material or may be transporting it under unhygienic condition can worsen the situation. The feed can be a breeding ground for vectors or may become a favorable place for the infectious pathogen to grow. So, the farmers should take care while storing the feed (Sims 2008).

20.2.4 *Hygiene of Equipments Used in Farming*

Farmers unintentionally or may be due to lack of knowledge use the same equipment for the transfer of dead animals and then for feed handling. This may cause the unintentional spreading of the pathogen to healthy animals, so this should be avoided. Therefore, regular cleaning and sanitization of the equipment needs to be done (Sims 2008).

20.2.5 *Stock Hygiene*

Animals kept in unhygienic conditions may affect their health and can result in heavy losses. Therefore, the maintenance of hygiene is very much important in maintaining bird stocks. The growers must take utmost care while disposing of the dead birds. This needs to be done as early as possible to prevent the transmission of infectious disease to other birds (Sims 2008).

20.3 Concern Regarding the Use of Antibiotics and Transmissible Diseases from Poultry

Antibiotics are being in use very commonly as veterinary drugs. The farm animals receive antibiotic treatment for microbial resistance. But it is a big and serious concern for public health as after treatment of animals with antibiotics, antibiotic residue remains in the animals that have proven to be very harmful to the person who is consuming it. The maximum acceptable amount of any medicinal drug or substance and their intermediates in the edible products in food extracted from animals is decided based on various safety studies and can be termed as acceptable daily intake (ADI). Scientists have shown that medicine residues in the food that originated from animals ultimately go through the human immune system. Several immunological reactions occur in humans after exposure of a certain level of these antibiotic residues. These residues may also interfere with our intestinal bacterium that helps to maintain our immune system (Nel 2006; Kriebel et al. 2001). Some commonly used antibiotics for the antimicrobial treatment in animals are: Sulfonamides (Mubito et al. 2014), Tetracyclines (Doyle 2006), Quinolones (Er et al. 2013), and Streptomycin (Pyun et al. 2008). There is a high association between the residues of antibiotics given to poultry animals and health issues in people such as digestive disturbance, neurological disease, hypersensitivity, and many more after consumption.

On the other hand, poultry industry displays an intimidate effect on the human population, as poultry animals are a vector of many infectious diseases especially bird flu that has already shown its mortal effects in the past recent years (Mangsi et al. 2014; Babapour et al. 2012). Therefore, measures must be taken to reduce the adverse effect of poultry infection on human beings. Nanotechnology can be used as an essential tool for the development of a platform for pathogen detection in poultry birds.

20.4 Nanoparticles in Poultry Sector

Recently, due to the potential advantages of nanoparticles, interest in utilizing them in poultry production has significantly increased. In the future, it will play a vital role in poultry industry with the integration of biotechnology. For maintaining hygienic and more nutritious condition, nanomaterials are introduced in food produced with chicken eggs that help in the blocking of cholesterol in the bloodstream. Also, in the future for hygienic and nutritious meat, nanotechnology can be incorporated in the poultry sector. Also, it can be used in the challenges like feed efficiency, drug delivery, modification of content in the egg (Fig. 20.2). These and other potential applications and various aspects have been reviewed earlier by some of the authors (Anwar et al. 2019; Hassan et al. 2020; Kannaki and Verma 2006), and here we will be focusing on the use of sensors in poultry industry.

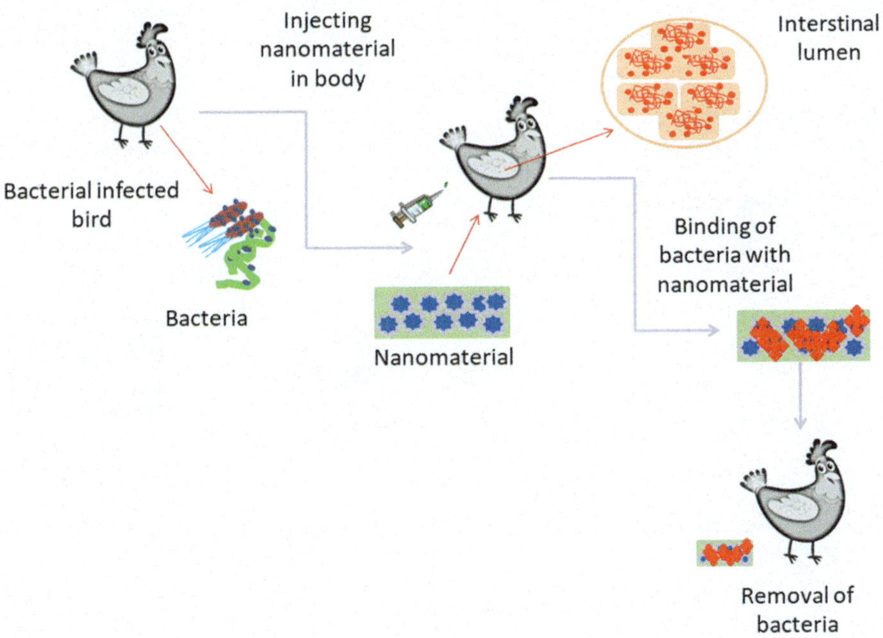

Fig. 20.2 Nanomaterial for binding pathogen in the poultry birds (Anwar et al. 2019)

20.5 Biosensors for Poultry Industry

Integration of two-stream biology and physics results into biosensors and are the platform for the detection of biological compounds. The biosensor consists of a recognition element and a transducing device. On recognition element, biological (enzyme/protein, antigen/antibody) components are immobilized, and the analyte binds to them (Fig. 20.3). The transducer is used for the detection of electrons/ions transmitted or generated on the surface (biological signals are converted into an electrical signal for the detection) (Mungroo and Neethirajan 2014). Biosensors provide higher sensitivity, selectivity, low-cost fabrication, durable, less time-consuming, and are easy to handle. Biosensors are being developed for the identification of different antibiotics and diseases causing pathogens in poultry (Astill et al. 2020; Kannaki and Verma 2006). Many biosensors with different platforms that are recently developed are further discussed in this chapter.

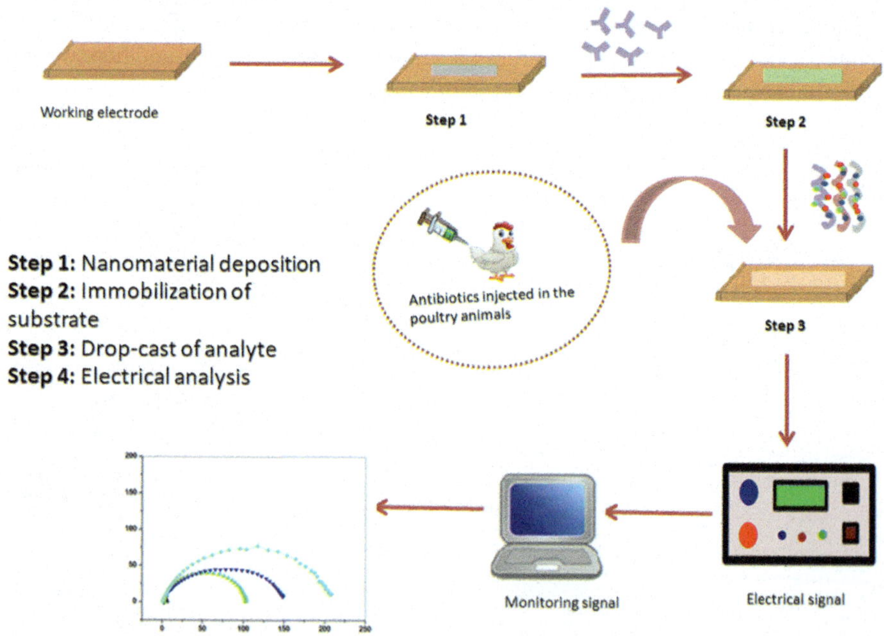

Fig. 20.3 Schematic diagram showing the working of antibiotic detection in poultry animals through biosensor

20.5.1 *Biosensors for Antibiotic Detection in Poultry*

Various detection methods have been developed for the detection of antibiotics in poultry animals such as LC-MS (liquid chromatography-mass spectroscopy), ELISA (enzyme-linked immunosorbent assay), High-performance liquid chromatography (HPLC), Thin-layer chromatography (TLC) (Okerman et al. 2001; Wang et al. 2009) and tube or plate test. All the above screening procedures are proven to be useful for their small volume sample preparation property, detection of multiple samples at a time, and broad specificity. Apart from these advantages, it has many drawbacks too such as it requires long hours of duration to be processed, no real-time detection and it lacks quick and sensitive detection practically (Pikkemaat 2009). Due to the various drawbacks in the above used analytical methods, there is a need for quick, sensitive, and specific method. In this regard, a scientists have designed several biosensors that fulfill almost all the properties in relation to an ideal detection like sensitive, accurate, user-friendly, and low cost. A quick semi-quantitative mechanism-based approach is the basic principle of a biosensor that made them a good and efficient idea toward the detection of antibiotic residual volume present in poultry animals (Cháfer et al. 2010). Some biosensors for the detection of antibiotics in poultry field are listed in Table 20.1 and are discussed below.

20.5.1.1 Molecular Imprinted Polymer (MIP) Based Sensor

The molecular imprinting polymer (MIP) has been a popular tool for drug and other molecular detection since the beginning of molecularly imprinting technology (MIT) in the 1970s. Many scientists have been focusing on using MIPs for the detection of molecules in traces and achieved significant success in the quantification of drugs at trace levels. Recently, the combination of MIP with magnetic particles have provided additional benefits, like elevated adsorption capacity to template molecule, unique selective recognition, and magnetic adsorption. The use of these materials for drug analysis with their advantages and limitations has been reviewed recently (Ansari and Karimi 2017). In the case of poultry, a magnetic electrochemical sensor fabricated was based on the molecular imprinting technology with multiwalled carbon nanotubes (MWCNTs) to carry out the quick determination of kanamycin traces in a complex sample matrix. Also, PMMA (polymethyl methacrylate) is used as a polymeric film for the grafting of MWCNTs that enhances compatibility for the compactness of the imprinted sensor structure. This MIP sensor has enhanced conductivity of electrode that initiates transfer of electrons for the kanamycin detection in complex samples (Long et al. 2015). The results had indicated that, under the test condition the linear relationship between the kanamycin concentration (from 1.0×10^{-10} M L^{-1} to 1.0×10^{-6} M L^{-1}) and response current from the sensor. The sensor had a detection limit of 2.3×10^{-11} M L^{-1} in the real samples.

Table 20.1 Antibiotic detection in poultry industry using various biosensors

S. No.	Transducer	Analyte	Limit of detection	References
1	Luminescent	Tetracycline	10 mg kg^{-1}	(Pikkemaat et al. 2010)
2	Colorimetric	Tetracycline	0.1 ng mL^{-1}	(Weber et al. 2005)
		Streptogramin	2.7 ng mL^{-1}	(Weber et al. 2005)
		Macrolide	1.7 ng mL^{-1}	(Weber et al. 2005)
3	Electrochemical biosensors	Penicillin	41.2 μA μM^{-1} cm^{-1}	(Li et al. 2019)
		Tetracycline	26.4 μA μM^{-1} cm^{-1}	(Li et al. 2019)
		Ceftazidime	0.01 μM	(Sun et al. 2018)
		Sulfadiazine	0.16 μM	(Sun et al. 2019)
		Acetaminophen	0.032 μM	(Sun et al. 2019)
		Ceftiofur	0.01 ng mL^{-1}	(Stevenson et al. 2019)
		Sulfadimethoxine	3.7×10^{-16} M	(Mohammad-Razdari et al. 2019a)
4	Drop-Type Chemiluminescence	Enrofloxacin	0.039 pM mL^{-1}	(Ahn and Lim 2015)
		Ciprofloxacin	0.022 pM mL^{-1}	(Ahn and Lim 2015)
5	Magnetic-nanobead-based biosensor	Oxytetracycline	0.88 ng mL^{-1}	(Lu et al. 2015)
6	Montmorillonite film-based sensor	Tetracycline	0.10 μM	(Gan et al. 2014)
7	Screen-printed gold electrode-based biosensor	Tetracycline	0.96 μM L^{-1}	(Masawat and Slater 2007)
		Chlortetracycline	0.58 μM L^{-1}	(Masawat and Slater 2007)
		Oxytetracycline	0.35 μM L^{-1}	(Masawat and Slater 2007)
8	Surface plasmon resonance (SPR)	Chloramphenicol	0.005 μg kg^{-1}	(Ferguson et al. 2005)
9	Optical-surface plasmon resonance (SPR)	Quinolones	0.13 ng gm^{-1}	(Huet et al. 2009)
10	Impedimetric biosensor	Tetracycline	3×10^{-17} M	(Mohammad-Razdari et al. 2019b)
11	Electrochemical aptasensor	Tetracycline	3.7×10^{-17}	(Benvidi et al. 2018)
12	Label-Free immunosensor	Kanamycin	15 pg mL^{-1}	(Yu et al. 2013)
13	Amperometric immunosensor	chloramphenicol	45 pg mL^{-1}	(Kim et al. 2010)

20.5.1.2 Electrochemical Biosensor

Electrochemical biosensors are composed of the 3-electrode system with counter, working, and reference electrodes. The electrodes are placed in an electrochemical cell containing solution and different studies are performed for the measurements like amperometric and potentiometry. The electrochemical biosensor works on the principle of change in voltage and current during the reaction in an electrochemical cell. Researchers fabricated an electrochemical biosensor with a multisegmented nanowire/nanoparticle hybrid array, for the detection of penicillin and tetracycline antibiotics simultaneously in meat samples (Li et al. 2019). The limit of detection (LOD) of the sensor observed for penicillin was 41.2 $\mu A \mu M^{-1}$ cm^{-2} and for tetracycline was 26.4 μA μM^{-1} cm^{-2}. In another study, the fabrication of an electrochemical biosensor for the identification of ceftiofur residue in the meat samples within 15 minutes has been reported (Stevenson et al. 2019), with the detection limit of 0.01 ng mL^{-1} in test buffer and 10 ng mL^{-1} in spiked samples.

20.5.1.3 Piezoelectric Sensors

Piezoelectric sensors are the sensors that operate on piezoelectric effect to measure quantities like temperature, pressure, etc. In a reported study, researchers have fabricated a piezoelectric sensor based on the imprinting polymer technique (Karaseva et al. 2016) for diagnosis of two different antibiotics penicillin and ampicillin in a meat sample. The LOD of the sensor was reported as 0.04 μg mL^{-1} for penicillin and 0.09 μg mL^{-1} ampicillin.

20.5.1.4 Aptamer Based Electrochemical Biosensor

An aptamer-based electrochemical biosensing platform for the identification of tetracycline in milk, meat, chicken/egg, drinking water, and pharmaceutical was fabricated by Kim and colleagues. The screen-printed gold electrode (SPE) was modified with streptavidin and the surface of streptavidin was immobilized with the biotinylated ssDNA aptamer. For analyzing the binding of tetracycline with ssDNA aptamer electrochemical measurements like cyclic voltammetry (CV) and square wave voltammetry (SWV) were performed. The limit of detection for the sensor was recorded in a range from 10 nM to micromolar (Kim et al. 2010). In another study, Benvidi and coworkers described the fabrication of an electrochemical aptasensor for identifying the concentration of tetracycline using multiwalled carbon nanotubes and electropolymerized poly (LL-glutamic acid) for the immobilization of anti-tetracycline aptamer. For analysis, electrochemical CV, differential pulse voltammetry (DPV), and Electrochemical impedance spectroscopy (EIS) measurements were performed (Benvidi et al. 2018). Also, an electrochemical biosensing platform for sulfadimethoxine (SDM) antibiotic identification in meat samples was fabricated on a pencil graphite electrode improved with Au-RGo NPs by Mohammad-Razdari and colleagues. The developed sensing platform shows excellent stability, selectivity, and high reproducibility (Mohammad-Razdari et al. 2019a).

20.5.1.5 Surface Plasmon Resonance (SPR)

SPR biosensors based on optical biosensing technology, the working principle of SPR is associated with modification in the refractive index with a binding analyte and bio recognize molecule on the sensing platform (Piliarik et al. 2009). For detection of antibiotics in poultry, SPR immunoassay was developed by Heut et al. (2008) for antibiotics 13 (fluoro) quinolones and flumequine in animals which are used in food production. The platform is based on the antigen-antibody combination. The SPR biosensor was validated for screening (fluoro) quinolones in poultry, egg, and fish (Heut et al. 2009)

20.5.1.6 Impedimetric Biosensor

Impedimetric aptasensor biosensor based on AuNPs/RGO NPs improved with pencil graphite electrode was fabricated for the identification of tetracycline in meat, milk, egg, shrimp, fish, chicken. The sensor shows that the combination of aptamer and EIS helps in the quantitatively and qualitatively detection of tetracycline (Mohammad-Razdari et al. 2019b).

20.5.1.7 Label-Free Immunosensor

A label-free immunosensing platform for the identification of kanamycin antibiotic was developed by using thionine mixed graphene sheet (TH-GS) modified with silver hybridized mesoporous ferroferric oxide NPs ($Ag@Fe_3O_4$ NPs). For the electrochemical characterization, cyclic voltammetry and square wave voltammetry measurements were applied. The sensor provides fast detection within 3 min and the observed LOD was 15 pg mL^{-1} (Yu et al. 2013).

20.5.1.8 Amperometric Immunosensor

For the chloramphenicol antibiotic identification in meat sample, an amperometric immunosensing platform was fabricated using CdS NPs modified-dendrimer which were bonded with the conducting polymer (poly $5, 2' : 5', 2''$-terthiophene-$3'$-carboxyl acid [poly-TTCA]) layer. To increase the sensitivity of the sensor the polymer was modified with Au NPs, dendrimers, and CdS NPs. The LOD of the sensor was recorded as 45 pg mL^{-1} (Kim et al. 2010).

20.5.2 Biosensors for Pathogen Detection

For pathogens identification in the poultry products various types of biosensing platforms are developed and are discussed below (Table 20.2).

20.5.2.1 Wave Fiber Optic Biosensor

Salmonella enterica is a prime pathogen present in food and is most common in poultry products. Therefore, for identification of *S. enterica* in chicken and shell egg evanescent wave optical fiber assay was developed that provide results in 8 hours with higher sensitivity and specificity. The LOD of the sensor was recorded as 104 CFU mL^{-1} (Valadez et al. 2009).

20.5.2.2 Microfluidic Device-Based Nano Biosensor

In this assay, *S. typhimurium* was identified with the help of a microfluidic nano biosensor based on magnetic beads and quantum dots (QDs) as a fluorescent label. Magnetic beads were used for coating anti-*Salmonella* to separate and concentrate a small volume of samples of the bacterium. A sandwich complex was made using antibody-conjugated QDs that label the cells. Fluorometer was used to measure the signals for quantitative detection of *S. typhimurium* (Kim et al. 2015).

20.5.2.3 Impedance Biosensor

Different types of impedance-based biosensing platforms were fabricated for the detection of pathogens in poultry animals. Jasim and his colleagues. developed impedance-based MEMS (micro-electromechanical system) biosensing platform for the screening of *Salmonella* serogroups (Jasim et al. 2017, 2019). In another study, Liu and coworkers fabricated an integrated impedance biosensor for the identification of *Salmonella* serotypes (B, D, and E) (Liu et al. 2018). Wang and the team fabricated an impedance biosensor based on immunomagnetic separation to identify the concentration of *Salmonella* (Wang et al. 2020).

20.5.2.4 Colorimetric Biosensor

Surface plasma resonance (SPR), Quartz crystal microbalance (QCM), and optical waveguide biosensors were used to identify glycan based Avian Influenza virus (AIV) (Hidari et al. 2007; Suenaga et al. 2012). HA (Influenza hemagglutinin surface protein) binding to gold NPs attached sialic acid helped in the detection of influenza virus without any amplification process or prior treatment in solution. The reaction

Table 20.2 Disease causing pathogen detection in poultry animals using biosensors

S. No.	Transducer	Pathogen detected	Detection limit	References
1	Microfluidic biosensor	*Salmonella typhimurium*	10^3 CFU mL^{-1}	(Kim et al. 2015)
2	QCM biosensor	Avian Influenza virus	1×10^4 PFU mL^{-1}	(Hewa et al. 2009)
3	Surface plasmon biosensor	Influenza virus	50 PFU mL^{-1}	(Lee et al. 2015)
		S. typhimurium	1×10^6 CFU mL^{-1}	(Lan et al. 2008)
4	Electrochemical biosensor	Influenza virus	80–100 virus particles μL^{-1}	(Nidzworski et al. 2014)
		S. pullorum	89 CFU mL^{-1}	(Fei et al. 2016)
		Avian influenza virus H7	1.6 pg mL^{-1}	(Huang et al. 2016)
5	Electrochemical impedance biosensor	*S. serogroups*	7 Cell mL^{-1}	(Jasim et al. 2019)
6	Optical fiber and light scattering sensors	*Salmonella*	10^3 CFU mL^{-1}	(Abdelhaseib et al. 2016)
7	Multiplex optical fiber biosensor	*Listeria monocytogenes*, *E. coli* and *Salmonella*	~10^3 CFU mL^{-1}	(Ohk and Bhunia 2013)
8	Immunosensors	*E. coli*	7.1×10^2 cells mL^{-1}	(Tokarskyy and Marshall 2008)
9	Wave Fiber Optic Biosensor	*S. enterica*	104 CFU mL^{-1}	(Valadez et al. 2009)
10	Microfluidic colorimetric biosensor	*E. coli*	50 CFU mL^{-1}	(Zheng et al. 2019)
11	Enzyme-Free biosensor	*S. typhimurium*	50 CFU mL^{-1}	(Huang et al. 2018)
12	Phage-based capacitive biosensor	*Salmonella* spp	200 CFU mL^{-1}	(Niyomdecha et al. 2018)
13	nano-bead based colorimetric immune-biosensors	*S. typhimurium*	10 CFU mL^{-1}	(Alamer et al. 2017)
14	Impedance biosensor with immunomagnetic	*Salmonella*	10^2 CFU mL^{-1}	(Wang et al. 2020)
16	Impedance biosensor	*Salmonella serogroups*	7 cells mL^{-1}	(Jasim et al. 2019)
17	Integrated impedance biosensor	*Salmonella* serotypes B and D	8 cells mL^{-1}	(Liu et al. 2018)
18	Impedance-based MEMS (micro-electromechanical system) biosensor	*Salmonella* serogroups	7 cells mL^{-1}	(Jasim et al. 2017)

(continued)

Table 20.2 (continued)

S. No.	Transducer	Pathogen detected	Detection limit	References
19	Colorimetric immunoassay	*S. enteritidis, S. typhimurium Staphylococcus aureus,* and *Campylobacter jejuni*	10 CFU mL^{-1}	(Alamer et al. 2018)

produced after binding of HA with the sialic acid released a signal in this colorimetric assay which shows linear relation with the viral strain (Gopinath et al. 2013; Takahashi et al. 2013; Lee et al. 2013). Alamar and group developed a colorimetric immunoassay for the detection of *S. enteritidis, S. Typhimurium, Campylobacter jejuni,* and *Staph aureus*, using a cotton swab and nanobead conjugated antibody (Alamer et al. 2018)

20.5.2.5 Piezoelectric Biosensor

A quartz crystal microbalance assay was utilized for the identification of avian influenza virus-based coupling of anti-M1 monoclonal antibody attached to Au NPs and recorded LOD was 1×10^4 PFU mL^{-1} with a linear range of 10^3–10^8 PFU mL^{-1} (Hewa et al. 2009).

20.5.2.6 Surface Plasmon Biosensor

An immunosensor based on plasmon assisted fluorescence was designed to detect influenza virus through anti-M1 antibody attached to Au NPs coated with carbon nanotubes. When virus combines with the NPs, a fluorescent signal was released following the addition of cadmium telluride quantum dots, which shows the lower detection limit of the sensor at 50 PFU mL^{-1} (Lee et al. 2015).

20.5.2.7 Electrochemical Biosensor

Several other biosensors were also made for the detection of pathogens such as an electrochemical sensor was employed based on an anti-M1 antibody specific to the virus which detected all the influenzas serotypes. The detection limit was observed as 80–100 virus particles μL^{-1} (Nidzworski et al. 2014).

Salmonella pullorum is a deadly organism that causes high mortality in chickens. For the detection of *S. pullorum,* an immunomagnetic assay based on magnetic beads and an electrochemical label comprising of gold-coated graphene oxide (rGO/AuNPs) were designed. The developed electrochemical immunosensor was based on the conjugation of antibody and magnetic beads, the process was carried out on the surface of screen-printed electrode (SPCE), by immobilizing rGO/antibody complex and using differential pulse voltammetry. The studies show that gold-coated rGO initiates signal amplification more strongly. The LOD of the system was 89 CFU mL^{-1} and the linear range for the experiment was 10^2–10^6 CFU mL^{-1} (Fei et al. 2016). For *E. coli* K12 identification in frozen chicken, Helali and the team fabricated a platform based on the antigen and antibody immobilization. The physisorption technique was used for the immobilization of anti-*E.coli* on the gold surface. For determining *E. coli*, the EIS and surface plasmon resonance studies were performed. The lower limit of detection for the sensor was 10^3 CFU mL^{-1} (Helali et al. 2018).

20.5.2.8 Microfluidic Colorimetric Biosensor

A novel microfluidic colorimetric biosensor using gold nanoparticles (AuNPs) was developed for the identification of *E.coli* pathogen in chicken samples. The sensor is based on the capture and detection of antibodies, where capture antibodies were used to modify magnetic nanoparticles and detection antibodies were used to modify polystyrene microspheres (PSs). The lower detection limit of the biosensor was 50 CFU mL^{-1} (Zheng et al. 2019).

20.5.2.9 Enzyme-Free Biosensor

A study by Huang and colleagues describes the fabrication of a biosensor for identification of *Salmonella Typhimurium (S. Typhimurium)*, the developed platform was an enzyme-free biosensor. The sensor was developed using signal reporter as curcumin (CUR) and for signal amplification 1,2,4,5-tetrazine (Tz)-trans-cyclooctene (TCO) was used. The LOD for the developed platform was calculated as 50 CFU mL^{-1} (Huang et al. 2018).

20.5.2.10 Phage-Based Capacitive Biosensor

For the identification of Salmonella *spp.*, the biosensor fabricated was based on a capacitive flow injection system. In this system, the gold surface was modified with *Salmonella*-specific M13 bacteriophage using crosslinker (glutaraldehyde). The sensor can produce the results after 40 min of the sample injection, the observed LOD for the sensor was 200 CFU mL^{-1} (Niyomdecha et al. 2018).

20.6 Conclusion

For the rapidly growing poultry sector, it is necessary to provide hygienic food, so that the effect of antibiotics may not be suffered by humans. Antibiotic resistance in poultry industry which has become a crisis worldwide for the human population has come out to the life-threatening event in recent past years. Mortality rates have been increasing day by day due to the consumption of poultry animals or animal products. Drug-resistance in humans arises when the drug-resistant bacteria enters our immune system on the consumption of poultry products as they receive a high amount of antibiotics for the prevention of bacterial diseases. Research study has already shown that antibiotic residue in animals when transferred to the human body there occurs an imbalance between our good microbial bacteria and harmful disease-causing bacteria.

There should be some alternative ways to prevent diseases and control antibiotic use in farm animals. Today many platforms are already available in the market for the diagnosis of the antibiotics and pathogens present in the food. Hygiene practices must be employed in the farms which are neglected generally, and strict rules should be implemented to check the maximum residual limit in farm animals. Some safeguards must be utilized for controlling the rate of infection in both animals and humans. All the antibiotics are not dangerous but their use beyond Maximum Residue Limits (MRL) is a frightening event. Along with these management practices, the use of nanosensors can be of great help, and more research in this area is needed to reduce the cost and to improve the reliability of these sensors. We firmly believe that, as the field of sensors developments is supported by more and more research, there will be many opportunities for the poultry industry to benefit from it.

References

Abdelhaseib MU, Singh AK, Bailey M, Singh M, El-Khateib T, Bhunia AK (2016) Fiber optic and light scattering sensors: complimentary approaches to rapid detection of *Salmonella enterica* in food samples. Food Control 61:135–145

Ahn J, Lim HB (2015) Drop-type chemiluminescence (DCL) system and sample treatment platform using magnetic nanoparticles to determine enrofloxacin and its metabolite in a chicken meat. Food Anal Methods 8(1):79–85

Alamer S, Chinnappan R, Zourob M (2017) Development of rapid immuno-based nanosensors for the detection of pathogenic bacteria in poultry processing plants. Procedia Technology 27:23–26

Alamer S, Eissa S, Chinnappan R, Zourob M (2018) A rapid colorimetric immunoassay for the detection of pathogenic bacteria on poultry processing plants using cotton swabs and nanobeads. Microchim Acta 185(3):164

Ansari S, Karimi M (2017) Recent configurations and progressive uses of magnetic molecularly imprinted polymers for drug analysis. Talanta 167:470–485

Anwar MI, Awais MM, Akhtar M, Navid MT, Muhammad F (2019) Nutritional and immunological effects of nano-particles in commercial poultry birds. World's Poultry Science Journal 75(2):261–272

Astill J, Dara RA, Fraser ED, Roberts B, Sharif S (2020) Smart poultry management: smart sensors, big data, and the internet of things. Comput Electron Agric 170:105291

Babapour A, Azami L, Fartashmehr J (2012) Overview of antibiotic residues in beef and mutton in Ardebil, North West of Iran. World Appl Sci J 19(10):1417–1422

Benvidi A, Yazdanparast S, Rezaeinasab M, Tezerjani MD, Abbasi S (2018) Designing and fabrication of a novel sensitive electrochemical aptasensor based on poly (L-glutamic acid)/MWCNTs modified glassy carbon electrode for determination of tetracycline. J Electroanal Chem 808:311–320

Cháfer-Pericás C, Maquieira A, Puchades R (2010) Fast screening methods to detect antibiotic residues in food samples. TrAC Trends Anal Chem 29(9):1038–1049

Doyle E (2006) Veterinary drug residues in processed meats-potential health risk. A review of the scientific literature. Fri Briefings. Food Research Institute, University of Wisconsin. http://fri.wisc.edu/docs/pdf/FRIBrief_VetDrgRes.pdf

Er B, Onurdağ FK, Demirhan B, Özgacar SÖ, Öktem AB, Abbasoğlu U (2013) Screening of quinolone antibiotic residues in chicken meat and beef sold in the markets of Ankara. Turkey Poultry science 92(8):2212–2215

Fei J, Dou W, Zhao G (2016) Amperometric immunoassay for the detection of Salmonella pullorum using a screen-printed carbon electrode modified with gold nanoparticle-coated reduced graphene oxide and immunomagnetic beads. Microchim Acta 183(2):757–764

Ferguson J, Baxter A, Young P, Kennedy G, Elliott C, Weigel S, Gatermann R, Ashwin H, Stead S, Sharman M (2005) Detection of chloramphenicol and chloramphenicol glucuronide residues in poultry muscle, honey, prawn and milk using a surface plasmon resonance biosensor and Qflex® kit chloramphenicol. Anal Chim Acta 529(1–2):109–113

Gan T, Shi Z, Sun J, Liu Y (2014) Simple and novel electrochemical sensor for the determination of tetracycline based on iron/zinc cations–exchanged montmorillonite catalyst. Talanta 121:187–193

Gopinath SC, Awazu K, Fujimaki M, Shimizu K (2013) Neu5Acα2, 6Gal and Neu5Acα2, 3Gal receptor specificities on influenza viruses determined by a waveguide-mode sensor. Acta Biomater 9(2):5080–5087

Hassan S, Hassan FU, Rehman MS (2020) Nano-particles of trace minerals in poultry nutrition: potential applications and future prospects. Biol Trace Elem Res 195(2):591–612

Helali S, Sawelem Eid Alatawi A, Abdelghani A (2018) Pathogenic Escherichia coli biosensor detection on chicken food samples. J Food Saf 38(5):e12510

Hewa TM, Tannock GA, Mainwaring DE, Harrison S, Fecondo JV (2009) The detection of influenza A and B viruses in clinical specimens using a quartz crystal microbalance. J Virol Methods 162(1–2):14–21

Hidari KI, Shimada S, Suzuki Y, Suzuki T (2007) Binding kinetics of influenza viruses to sialic acid-containing carbohydrates. Glycoconj J 24(9):583–590

Hu Y, Cheng H, Tao S (2017) Environmental and human health challenges of industrial livestock and poultry farming in China and their mitigation. Environ Int 107:111–130

Huang J, Xie Z, Xie Z, Luo S, Xie L, Huang L, Fan Q, Zhang Y, Wang S, Zeng T (2016) Silver nanoparticles coated graphene electrochemical sensor for the ultrasensitive analysis of avian influenza virus H7. Anal Chim Acta 913:121–127

Huang F, Xue L, Zhang H, Guo R, Li Y, Liao M, Wang M, Lin J (2018) An enzyme-free biosensor for sensitive detection of Salmonella using curcumin as signal reporter and click chemistry for signal amplification. Theranostics 8(22):6263

Huet AC, Charlier C, Singh G, Godefroy SB, Leivo J, Vehniäinen M, Nielen MW, Weigel S, Delahaut P (2008) Development of an optical surface plasmon resonance biosensor assay for (fluoro) quinolones in egg, fish, and poultry meat. Anal Chim Acta 623(2):195–203

Huet AC, Charlier C, Weigel S, Godefroy SB, Delahaut P (2009) Validation of an optical surface plasmon resonance biosensor assay for screening (fluoro) quinolones in egg, fish and poultry. Food Addit Contam 26(10):1341–1347

Jasim I, Abdullha A, Shen Z, Zhang S, Alalem M, Dewik M, Almasri M (2017) An impedance biosensor for simultaneous detection of low concentration of Salmonella serogroups in poultry

samples. In 2017 19th International conference on solid-state sensors, actuators and microsystems (TRANSDUCERS) 2017 Jun 18, pp 726–729. IEEE
Jasim I, Shen Z, Mlaji Z, Abdullah A, Liu J, Ghosch S, El-Dweik M, Zhang S, Almasri M (2019) An impedance biosensor for simultaneous detection of low concentration of Salmonella serogroups in poultry and fresh produce samples. Biosens Bioelectron 126:292–300
Kabir SM (2009) The role of probiotics in the poultry industry. Int J Mol Sci 10(8):3531–3546
Kannaki TR, Verma PC (2006) The challenges of 2020 and the role of nanotechnology in poultry research. Poultry Research Priorities to 2020 122:273
Karaseva N, Ermolaeva T, Mizaikoff B (2016) Piezoelectric sensors using molecularly imprinted nanospheres for the detection of antibiotics. Sensors and Actuators B: Chemical 225:199–208
Kim DM, Rahman MA, Do MH, Ban C, Shim YB (2010a) An amperometric chloramphenicol immunosensor based on cadmium sulfide nanoparticles modified-dendrimer bonded conducting polymer. Biosens Bioelectron 25(7):1781–1788
Kim YJ, Kim YS, Niazi JH, Gu MB (2010b) Electrochemical aptasensor for tetracycline detection. Bioprocess Biosyst Eng 33(1):31
Kim G, Moon JH, Moh CY, Lim JG (2015) A microfluidic nano-biosensor for the detection of pathogenic Salmonella. Biosens Bioelectron 67:243–247
King T, Osmond-McLeod MJ, Duffy LL (2018) Nanotechnology in the food sector and potential applications for the poultry industry. Trends Food Sci Technol 72:62–73
Kriebel D, Tickner J, Epstein P, Lemons J, Levins R, Loechler EL, Quinn M, Rudel R, Schettler T, Stoto M (2001) The precautionary principle in environmental science. Environ Health Perspect 109(9):871–876
Lan YB, Wang SZ, Yin YG, Hoffmann WC, Zheng XZ (2008) Using a surface plasmon resonance biosensor for rapid detection of Salmonella typhimurium in chicken carcass. J Bionic Eng 5(3):239–246
Lee C, Gaston MA, Weiss AA, Zhang P (2013) Colorimetric viral detection based on sialic acid stabilized goldnanoparticles. Biosens Bioelectron 42:236–241
Lee J, Ahmed SR, Oh S, Kim J, Suzuki T, Parmar K, Park SS, Lee J, Park EY (2015) A plasmon-assisted fluoro-immunoassay using gold nanoparticle-decorated carbon nanotubes for monitoring the influenza virus. Biosens Bioelectron 64:311–317
Li Z, Liu C, Sarpong V, Gu Z (2019) Multisegment nanowire/nanoparticle hybrid arrays as electrochemical biosensors for simultaneous detection of antibiotics. Biosens Bioelectron 126:632–639
Liu J, Jasim I, Abdullah A, Shen Z, Zhao L, El-Dweik M, Zhang S, Almasri M (2018) An integrated impedance biosensor platform for detection of pathogens in poultry products. Scientific reports 8:16109
Long F, Zhang Z, Yang Z, Zeng J, Jiang Y (2015) Imprinted electrochemical sensor based on magnetic multi-walled carbon nanotube for sensitive determination of kanamycin. J Electroanal Chem 755:7–14. https://doi.org/10.1016/j.jelechem.2015.07.018
Lu C, Tang Z, Liu C, Kang L, Sun F (2015) Magnetic-nanobead-based competitive enzyme-linked aptamer assay for the analysis of oxytetracycline in food. Anal Bioanal Chem 407(14):4155–4163
Mangsi AS, Khaskheli MU, Soomro AH, Shah MG (2014) Antibiotic residues detection in raw beef meat sold for human consumption in sindh. Pakistan. Int. J. Res. Appl. Nat. Soc. Sci. 2(7):15–20
Masawat P, Slater JM (2007) The determination of tetracycline residues in food using a disposable screen-printed gold electrode (SPGE). Sensors and Actuators B: Chemical 124(1):127–132
Mohammad-Razdari A, Ghasemi-Varnamkhasti M, Izadi Z, Rostami S, Ensafi AA, Siadat M et al (2019a) Detection of sulfadimethoxine in meat samples using a novel electrochemical biosensor as a rapid analysis method. J. Food Composition Anal. 82:e103252. https://doi.org/10.1016/j.jfca.2019.103252
Mohammad-Razdari A, Ghasemi-Varnamkhasti M, Izadi Z, Ensafi AA, Rostami S, Siadat M (2019b) An impedimetric aptasensor for ultrasensitive detection of Penicillin G based on the use of reduced graphene oxide and gold nanoparticles. Mikrochim Acta 186:372. https://doi.org/10.1007/s00604-019-3510-x

Mottet A, Tempio G (2017) Global poultry production: current state and future outlook and challenges. World's Poultry Science Journal 73(2):245–256
Mubito EP, Shahada F, Kimanya ME, Buza J (2014) Sulfonamide residues in commercial layer chicken eggs in Dar-es-Salaam. Tanzania American Journal of Research Communication 2(4):124–132
Mungroo NA, Neethirajan S (2014) Biosensors for the detection of antibiotics in poultry industry—a review. Biosensors 4(4):472–493
Nel H (2006) The establishment and standardization of a veterinary antimicrobial resistance surveillance programme in South Africa. Dissertation (MSc in Veterinary Tropical Diseases, University of Pretoria). http://hdl.handle.net/2263/23033
Nidzworski D, Pranszke P, Grudniewska M, Król E, Gromadzka B (2014) Universal biosensor for detection of influenza virus. Biosens Bioelectron 59:239–242
Niyomdecha S, Limbut W, Numnuam A, Kanatharana P, Charlermroj R, Karoonuthaisiri N, Thavarungkul P (2018) Phage-based capacitive biosensor for Salmonella detection. Talanta 188:658–664
Ohk SH, Bhunia AK (2013) Multiplex fiber optic biosensor for detection of Listeria monocytogenes, Escherichia coli O157: H7 and Salmonella enterica from ready-to-eat meat samples. Food Microbiol 33(2):166–171
Okerman L, Croubels S, De Baere S, Hoof JV, De Backer P, De Brabander H (2001) Inhibition tests for detection and presumptive identification of tetracyclines, beta-lactam antibiotics and quinolones in poultry meat. Food Addit Contam 18(5):385–393
Pikkemaat MG (2009) Microbial screening methods for detection of antibiotic residues in slaughter animals. Anal Bioanal Chem 395(4):893–905
Pikkemaat MG, Rapallini ML, Karp MT, Elferink JA (2010) Application of a luminescent bacterial biosensor for the detection of tetracyclines in routine analysis of poultry muscle samples. Food Addit Contam 27(8):1112–1117
Piliarik M, Vaisocherová H, Homola J (2009) Surface plasmon resonance biosensing. In: Biosensors and Biodetection, pp. 65–88. Humana Press
Pyun CW, Abd El-Aty AM, Hashim MM, Shim JH, Lee SK, Choi KD, Park KH, Shin HC, Lee C (2008) Monitoring of streptomycin and dihydrostreptomycin residual levels in porcine meat press juice and muscle via solid-phase fluorescence immunoassay and confirmatory analysis by liquid chromatography after post-column derivatization. Biomed Chromatogr 22(3):254–259
Sims LD (2008) Risks associated with poultry production systems. In: International conference poultry in the twenty-first century (Vol. 1, p. 24)
Stevenson HS, Shetty SS, Thomas NJ, Dhamu VN, Bhide A, Prasad S (2019) ultrasensitive and rapid-response sensor for the electrochemical detection of antibiotic residues within meat samples. ACS Omega 4(4):6324–6330
Suenaga E, Mizuno H, Kumar PK (2012) Influenza virus surveillance using surface plasmon resonance. Virulence 3(5):464–470
Sun Y, He J, Waterhouse GI, Xu L, Zhang H, Qiao X, Xu Z (2019a) A selective molecularly imprinted electrochemical sensor with GO@ COF signal amplification for the simultaneous determination of sulfadiazine and acetaminophen. Sensors and Actuators B: Chemical 300:126993
Sun Y, Li X, Wang X, Yu Z, Wang T (2018) An electrochemical sensor for determination of Ceftazidime based on poly (crystal violet) doped platinum nanoparticles modified electrode. Int J Electrochem Sci 13:1859–1868
Sun Y, Xu L, Waterhouse GI, Wang M, Qiao X, Xu Z (2019b) Novel three-dimensional electrochemical sensor with dual signal amplification based on MoS2 nanosheets and high-conductive NH2-MWCNT@ COF for sulfamerazine determination. Sensors and Actuators B: Chemical 281:107–114
Takahashi T, Kawagishi S, Masuda M, Suzuki T (2013) Binding kinetics of sulfatide with influenza A virus hemagglutinin. Glycoconj J 30(7):709–716
Tokarskyy O, Marshall DL (2008) Immunosensors for rapid detection of Escherichia coli O157: H7—perspectives for use in the meat processing industry. Food Microbiol 25(1):1–2

Valadez AM, Lana CA, Tu SI, Morgan MT, Bhunia AK (2009) Evanescent wave fiber optic biosensor for Salmonella detection in food. Sensors 9(7):5810–5824

Wang S, Xu B, Zhang Y, He JX (2009) Development of enzyme-linked immunosorbent assay (ELISA) for the detection of neomycin residues in pig muscle, chicken muscle, egg, fish, milk and kidney. Meat Sci 82(1):53–58

Wang L, Xue L, Guo R, Zheng L, Wang S, Yao L, Huo X, Liu N, Liao M, Li Y, Lin J (2020) Combining impedance biosensor with immunomagnetic separation for rapid screening of Salmonella in poultry supply chains. Poult Sci 99(3):1606–1614

Weber C, Link N, Fux C, Zisch A, Weber W, Fussenegger M (2005) Broad-spectrum protein biosensors for class-specific detection of antibiotics. Biotechnol Bioeng 89:9–17. https://doi.org/10.1002/bit.20224

Yu S, Wei Q, Du B, Wu D, Li H, Yan L, Ma H, Zhang Y (2013) Label-free immunosensor for the detection of kanamycin using Ag@ Fe3O4 nanoparticles and thionine mixed graphene sheet. Biosens Bioelectron 48:224–229

Zheng L, Cai G, Wang S, Liao M, Li Y, Lin J (2019) A microfluidic colorimetric biosensor for rapid detection of Escherichia coli O157: H7 using gold nanoparticle aggregation and smart phone imaging. Biosens Bioelectron 124:143–149

Chapter 21
Smart Aquaculture: Integration of Sensors, Biosensors, and Artificial Intelligence

Dolly Sharma and Ranjit Kumar

Abstract Aquaculture involves cultivating aquatic plants and a wide variety of aquatic animals used for human consumption, medicines, and supplements. This industry is consistently growing due to growing demands from people preferring organic seafood and marine supplements, such as fish oil, marine collagen, seaweed, algae, etc. This extremely complex task of setting up and maintaining the aquaculture industry requires fine-tuning a large number of variables to produce an optimal outcome. Technology integrated with efficient algorithms helps create a framework for the intelligent aquaculture industry. Sensing and machine learning algorithms play a key role here, as they enable automated real-time monitoring of the environment along with quick decision-making to mitigate risks. This chapter presents a review of major tasks involved, namely technology used for collecting real-time data such as physical sensors, biosensors, optical sensors, etc., the role of automation and automated decision-making.

Keywords Machine learning · Artificial intelligence · Sensors · Aquaculture

21.1 Introduction

Aquaculture is synonymous with aquatic farming and has been in existence for a very long time. The FAO (Food and Agriculture Organization) of United Nations Organization (UNO) defined aquaculture as rearing or farming of aquatic organisms such as fish, molluscs, crustaceans, aquatic plants, crocodiles, alligators, turtles, and amphibians (FAO 2008). There are three kinds of culture environments: freshwater,

D. Sharma
Computer Science and Engineering, Shiv Nadar University, Gautam Buddha Nagar, Noida, Uttar Pradesh, India
e-mail: dolly.sharma@snu.edu.in

R. Kumar (✉)
Department of Chemistry, School of Engineering, University of Petroleum and Energy Studies (UPES), Bidholi, Dehradun, Uttarakhand, India
e-mail: ranjit.kumar@ddn.upes.ac.in

R. N. Pudake et al. (eds.), *Biosensors in Agriculture: Recent Trends and Future Perspectives*, Concepts and Strategies in Plant Sciences,
https://doi.org/10.1007/978-3-030-66165-6_21

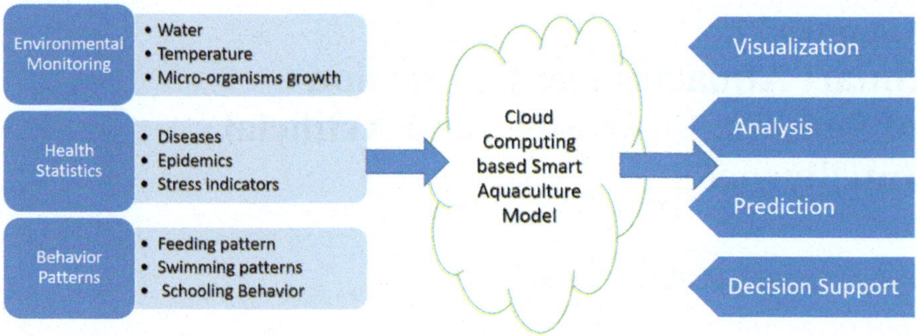

Fig. 21.1 Smart aquaculture model

brackishwater, and mariculture. The aquaculture units could be ponds, tanks, cages, raceways, barrages, rice-cum-fish paddies, rafts, hatcheries, nurseries, etc. Setting up the right environment is key to succeeding in this industry. Some of the components that must be carefully thought through during the initial phase of setup are: a selection of the species that you are planning to grow/breed, procurement, and transportation, selecting an appropriate location and type of medium to set up the farm, and choosing the right culture type.

Once the big picture is ready, every small aspect needs to be carefully planned and executed. Technology is great support for designing a successful aquaculture model, as an integrated smart environment makes real-time monitoring and decision-making possible. There are many such models presented in the past, where data from the sensors is collected and fed into a machine learning model for predictions and then further into a decision support system. These models work well in small experimental settings. However, with larger setups and the requirement of constant monitoring, there is a need for a scalable big data and cloud computing-based smart aquaculture system. This setup will also bring down the hardware cost. Some such efforts have been initiated by (Rao et al. 2018; Dzulqornain et al. 2018), where data analytics is integrated with cloud computing.

In this chapter, we discuss the current scenario of the aquaculture industry. We aim to present the need to design a functional smart aquaculture setup, where operations can be standardized for better management. A high-level model for smart aquaculture is shown in Fig. 21.1.

21.2 Monitoring the Quality of Water

The monitoring of the quality of water in aquaculture tanks is essential in order to monitor the health of fish. Various parameters such as temperature, pH, ammonia, nitrite/nitrate, hardness, alkalinity, turbidity, gas concentration, and dissolved oxygen level need to be monitored in aquaculture. Earlier traditional ways like potentiometry, titration, and conductivity measurement methods were used to monitor these parameters (Amrita and Babiyola 2018). Modern methods use sensory data to analyze the water conditions. Different types of sensors, including optical sensors, thermal sensors, biosensors, electrochemical sensors, etc., are being used (Parra et al. 2018).

The temperature of the water has a direct impact on feeding patterns, growth of fish, stress, and disease breakout. Warm water has less oxygen solubility than cold water. The level of dissolved oxygen (DO) in water is related to the amount of oxygen consumed, which in turn affects the size of fish, feeding rate, and activity level. The optimal level of DO is essential for fish respiration, as well as for the survival of phytoplankton. Phytoplankton breaks down the toxic ammonia into harmless nitrogen. Through this process, phytoplankton indirectly maintains the pH level within the acceptable range of 6.5–9.0 for fish culture. When pH rises above 9, ammonium ion (NH_4^+) in the water is converted to toxic ammonia (NH_3), which is lethal for fish. Acidic water with pH less than 5 leads to leaching of metals from rocks and sediments. These metals concentration also needs to be monitored as it adversely affects fishes' metabolism rates and their ability to take in water (Su et al. 2020).

Li and coworkers proposed a hybrid model to predict dissolved oxygen level, based on multi-scale features using ensemble empirical mode decomposition (EEMD) (Li et al. 2018). Wireless sensor networks (WSN) have been used to sense and measure pH, NH_4^+, and temperature for a fish farm. Sensors based on a specially designed Ion Sensitive Field Effect Transistor (ISFET) can be used to monitor different parameters in aquaculture. The sensor module can collect data and transmit the data to the wireless module by means of a 9600 b/s asynchronous wired communication. The wireless node retransmits the data to the central unit via radio frequency (RF) (López et al. 2009). Liu and coworkers proposed a real-value genetic algorithm support vector regression (RGA-SVR), based on genetic algorithms and Support Vector Regression, that predicts water quality (Liu et al. 2013).

21.3 Health Monitoring of Aquatic Animals

Customized food options and optimal feeding pattern is crucial to aquatic animal's survival. One such example is cottonseed, which is an inexpensive and very widely available plant-based protein source and is liked by most aquatic animals in the diet (Li and Robinson 2006). However, it is only used in limited quantities as it is not easily digested by aquatic animals and is potentially toxic. Thus, depending upon the species, a specific percentage of cottonseed can be incorporated into aquatic food in combination with other components.

Other than taking care of deciding the quality and quantity of food, it is also essential to observe the feeding behavior of aquatic animals. Biosensors and other optical sensors have been very helpful in monitoring the feeding patterns of aquatic animals. This task cannot be accomplished manually because of low visibility underwater. Image processing techniques have been used to analyze the uneaten food pellets that pollute the water and cause sickness to animals. Continuously learning the feeding patterns also leads to a cost-effective, sustainable, and environment-friendly aquaculture industry setup. Liu et al. (2015) used an underwater camera with LED lights for this task. They used a linear-time component-labeling algorithm to figure out the number of pellets left. Their algorithm has a counting error of less than 8% and is suitable for practical applications. The complex task of identifying animal's food preferences and identifying their hunger is very important in maintaining good animal health. Razman et al. (2019) designed a method to identify hunger state for *Lates calcarifer*. Machine learning algorithms were used to analyze the 16 features collected through time-series data using window size of 0.5, 1, 1.5, and 2 min from the videos. The study demonstrated that they could identify the stage of hunger and use of automated feeding resulted in growing heavier and longer fish. Nanobiosensors are also used for monitoring and controlling microbes in water (Dar et al. 2019).

Other than maintaining healthy feeding protocols, it is also essential to monitor the health of aquatic animals for diseases, epidemics, and stress. Biosensors are very useful in health monitoring. Surface Plasmon Resonance (SPR)-based biosensors detect poisoning in Paralytic Shellfish (Fonfría et al. 2007), and are essential for the safety of consumers. Based on molecular interaction detection by SPR biosensors, anti-GTX2/3 (Gonyautoxin 2,3) antibody (GT13-A), and saxitoxin-CM5 chip; an inhibition assay was developed. Most countries set the Paralytic Shellfish Poison (PSP) toxin limit of 80 μg per 100 g of meat and this process is capable of detecting 15–400 μg per 100 g of shellfish meat. This method detects decarbamoyl saxitoxin-dcSTX, gonyautoxin 2,3 (GTX2/3), decarbamoyl gonyautoxin 2,3 (dcGTX2/3), gonyautoxin 5 (GTX5), C1/2, and saxitoxin (STX). The detection of STX can be improved by improving the cross-reactivity among the analogues.

When a fish is exposed to an internal or external stressful situation, it activates the hypothalamic-pituitary-interrenal (HPI) axis as a primary response followed by secondary and tertiary responses (Sopinka et al. 2016). The authors presented a study of various stress indicators and factors that are responsible. Understanding how stress effects to fish and the ways by which it can be identified will be beneficial for fishes.

Biosensors also play a vital role in evaluating the freshness of fish. Table 21.1 enlists different types of biosensors used in aquaculture. Amperometric electrodes based on oxygen or hydrogen peroxide detection, together with an oxidase system or an aerobic microorganism are commonly used in food and drink industries because of their simplicity and high selectivity (Luong et al. 1991). The important areas of applications of biosensors are proximate analysis, nutritional labeling, determination of pesticide residues, naturally occurring toxins and antinutrients, processing changes, microbial contamination, enzymatic inactivation, and Biochemical Oxygen Demand (BOD) of wastes (Rajee and Alicia 2019; Suleiman and Guilbault 1994).

Biochemical Oxygen Demand (BOD) is an important parameter for the well-being of aquatic animals. It can be monitored using biosensors based on bacteria like *Pseudomonas* sp., *Bacillus* sp., and thermophilic bacteria. These are very fast biosensors and are capable of determining BOD values as low as 1 ppm within 30 min (Venugopal 2002; Rajee and Alicia 2019). Various biosensors have been developed to detect pathogenic bacteria such as *Salmonella* sp., *E. coli*, etc., in the contaminated water discharged from the aquaculture (Monzó et al. 2015).

Table 21.1 Types of biosensors used in aquaculture

Biosensor	Description	Application	References
Optical biosensors	Measures the change in amplitude, phase, frequency, or polarization of light	Detection of pathogens, pesticide and drug residues, hygiene monitoring, heavy metals, and other toxic substances in the food to check whether the food is safe for the consumption Detection of antibiotics	(Narsaiah et al. 2012) (Zhao et al. 2020)
Aptamer-based biosensors	High sensitivity and selectivity of transducer-linked molecular recognition element for biosensing that are used in electrochemical, optical, or mass-sensitive analytical techniques	Disease diagnosis, prophylactic, and therapeutic	(Chong and Low 2019)
Enzyme-based biosensor	Uses a combination of an enzyme with a transducer to generate a signal relative to concentration of target analyte	Detection of Trimethylamine in fish Antibiotics and antimicrobials detection, BOD, insecticides detection	(Mitsubayashi et al. 2004) (Rajee and Alicia 2019)
Fluorometric biosensor	Works on chemiluminescence and fluorometric detection	Nitrite detection in aquaculture water	(Pires et al. 2019)
Nanomaterial-based biosensor	Carbon nanotubes-based biosensor	Detects microbes such as bacteria, viruses, and parasites and also heavy metals in food and water	(Dar et al. 2019)
Electrochemical biosensors	Based on catalysis reaction by enzymes that generates or consumes electrons	Detection of ostreid herpesvirus 1, which poses threat to shellfish aquaculture	(Toldrà et al. 2020)

Biosensors are also used for the detection of pesticides found in the food chain of fish as reported. Campàs et al. (2007) reported an array of biosensors to detect marine toxins. Biosensors based on the inhibition of urease can be developed for the determination of Hg, Ag, and Cu in the ppb range (Majlesi et al. 2019). Detection of antibiotics like Penicillin in the range 5–30 mmol was achieved using immobilized *E. coli* with a response time of 8 min (Galindo et al. 1990). A highly sensitive colorimetric biosensor for antibiotics like chloramphenicol (CAP) was reported by incorporating ELISA onto surfaces of microporous and nanofibrous membranes with a detection limit up to 0.3 ng/mL (Zhao et al. 2020).

Optical biosensors measure polarization as well as frequency, phase, and amplitude of the light wave and are used to identify food borne pathogens. Some of such optical biosensors are microarrays, optical fibers, and SPR (Narsaiah et al. 2012).

21.4 Role of Technology in Aquaculture Industry

Information and Communication Technology plays a huge role in the success of the aquaculture industry, especially the large-scale setups where physical monitoring at all times is almost impossible. The role of ICT starts from the very beginning, such as the use of GIS-based methods for selecting the right location for setup (Kapetsky et al. 1990; Gimpel et al. 2018; Falconer et al. 2019). The authors presented an automated system that used a dataset composed of many features, including spatial features; and analyzed the potential of development of aquaculture setup. They proposed that some other features that could be helpful are the location of feed mills, availability of workforce, land use, water use, level of groundwater, frequency of floods, other factors to optimize logistics. Their experiment was set up in Louisiana, where they studied double cropping for catfish and crawfish. This area has been of interest to many researchers in the last three decades. GIS-based planning tools such as AquaSpace, enable mapping of 30 indicators based on opportunities (Gimpel et al. 2018).

21.4.1 Study Growth Patterns

Other parameters that help keep a tab on fish health are growth pattern, fish count, fish density, movement patterns, etc. Keeping an accurate tally of fish count majorly involves tagging and vision-based methods. A fish counting method (Le and Xu 2017) was proposed and they identified that underwater images are not very high quality and have a lot of noise. Thus, they designed a threshold-based algorithm that was able to identify the fishes using only the contour information and removing the noise. The situations where two or more fishes were overlapping in the image were also resolved using the skeletal endpoints. The experiments were carried out in a small experimental setup, however, the performance in large complex real-world setups would be more challenging. Another effective method to keep track of fish count is Tagging (Pine et al. 2003). The studies integrate different types of tags such as fin clips, telemetry tags, etc., with computer software so that the population size and the mortality-related data can be analyzed. This data when integrated with data from other sources gives significant insight into factors leading to an ideal breeding scenario. The way multiple sensors integrated with computer vision techniques have been incorporated into computational models, has been studied by many researchers and in-depth reviews of the methods have also been studied (Saberioon et al. 2017; Hassan et al. 2016). A latest trend in aquaculture is to create the feed using a lower percentage of fish meal and higher percentage of several plant ingredients and nutrients. The experiments involved creating a multi-choice feeding system in an aquaculture setup (da Silva et al. 2016).

21.4.2 Real-Time Monitoring

Another important aspect where technology can support is real-time automated monitoring. An automated fish farm monitoring system (Chen et al. 2016) was proposed where environmental monitoring could be done by a mobile device. They tested the performance of the wireless sensor networks in a simulation setup using MSP430 chip and ZigBee wireless transmission, in addition to the feeder, inflator to manage oxygen levels, heaters to maintain temperature, and RGB light modulation system. The proposed system proved to be power efficient as well as cost efficient according to the simulation settings. A system where real-time monitoring can be valuable, if the corrective actions can also be taken on a timely basis without any human intervention. An integrated model using environmental, biological, economical, and technical data was simulated to implement a decision support system (DSS) using Particle Swarm Optimization (Cobo et al. 2019). This DSS provided support in both operational and strategic decision-making, and at both levels: day-to-day basis as well as long-term planning. With the availability of high computational power with lower cost, it is now possible to train deep learning models for smart fish farming and can be used for fish counting, classification of species, analyze behavior patterns, feeding patterns, size, analyze water (Yang et al. 2020).

Another DSS was proposed (Shahriar and McCulluch 2014), where two methods were explored for decision-making in aquaculture closure. They used data such as rain, river flow, salinity and temperature for time-series prediction and expert rules and classification. The experiment was conducted on the dataset from a shellfish farm from Tasmania, where they collected one-year data for training purpose and the next two months data for testing.

Thus, we see that technology plays a huge role in optimal functioning of aquaculture industry. Machine learning algorithms (Kang et al. 2017) built and trained specifically for this industry, can reshape it and help in standardization across the globe. These algorithms are capable of parameter tuning for optimal results, prediction, and decision-making; thus, enabling the stakeholders to maximize profits.

21.5 Conclusion and Future Directions

Aquaculture industry is growing consistently as the demand is growing. It is important that the industry brings profit, thus there is a lot of interest in the area of optimizing the operations leading to low cost and low energy. There is an additional requirement that all stakeholders must also address, that is, sustainability. The leading cause of pollution caused by this industry is the leftover food pellets. Thus, the setups which have recirculating aquacultures (Sönmez et al. 2016), need to make sure that the water supply is not contaminated. Also, methods such as double cropping and integration of aquaculture systems to irrigation water should be encouraged. Futuristic planning of aquaculture industry has to involve an efficient model that incorporates all relevant physical sensors, biosensors, audio/video monitoring, and IoT devices, which provides input to a big data and cloud computing environment, for real-time monitoring of this large-scale data.

References

Amrita CM, Babiyola D (2018) Analysing the water quality parameters from traditional to modern methods in Aquaculture. Int J Sci Environ Technol 7:1954–1961

Campàs M, Prieto-Simón B, Marty JL (2007) Biosensors to detect marine toxins: assessing seafood safety. Talanta 72:884–895. https://doi.org/10.1016/j.talanta.2006.12.036

Chen JH, Sung WT, Lin GY (2016) Automated monitoring system for the Fish Farm aquaculture environment. IEEE International Conference on Systems, Man, and Cybernetics, Hong Kong, China, 2015, pp. 1161–1166. https://doi.org/10.1109/SMC.2015.208

Chong CM, Low CF (2019) Synthetic antibody: prospects in aquaculture biosecurity. Fish Shellfish Immunol 86:361–367. https://doi.org/10.1016/j.fsi.2018.11.060

Cobo Á, Llorente I, Luna L, Luna M (2019) A decision support system for fish farming using particle swarm optimization. Comput Electron Agric 161:121–130. https://doi.org/10.1016/j.compag.2018.03.036

da Silva RF, Kitagawa A, Sánchez Vázquez FJ (2016) Dietary self-selection in fish: a new approach to studying fish nutrition and feeding behavior. Rev Fish Biol Fish 26:39–51. https://doi.org/10.1007/s11160-015-9410-1

Dar AH, Rashid N, Majid I et al (2019) Nanotechnology interventions in aquaculture and seafood preservation. Crit Rev Food Sci Nutr 0:1–10. https://doi.org/10.1080/10408398.2019.1617232

Dzulqornain MI, Harun Al Rasyid MU, Sukaridhoto S (2018) Design and Development of smart aquaculture system based on IFTTT model and cloud integration. MATEC Web Conf 164. https://doi.org/10.1051/matecconf/201816401030

Falconer L, Middelboe AL, Kaas H et al (2019) Use of geographic information systems for aquaculture and recommendations for development of spatial tools. Rev Aquac 664–677. https://doi.org/10.1111/raq.12345

FAO (2008) Coordinating working party on fishery statistics. J Chem Inf Model 53:287. https://doi.org/10.1017/CBO9781107415324.004

Fonfría ES, Vilariño N, Campbell K et al (2007) Paralytic shellfish poisoning detection by surface plasmon resonance-based biosensors in shellfish matrixes. Anal Chem 79:6303–6311. https://doi.org/10.1021/ac070362q

Galindo E, Bautista D, García JL, Quintero R (1990) Microbial sensor for penicillins using a recombinant strain of *Escherichia coli*. Enzyme Microb Technol 12:642–646. https://doi.org/10.1016/0141-0229(90)90001-7

Gimpel A, Stelzenmüller V, Töpsch S et al (2018) A GIS-based tool for an integrated assessment of spatial planning trade-offs with aquaculture. Sci Total Environ 627:1644–1655

Hassan SG, Hasan M, Li D (2016) Information fusion in aquaculture: a state-of the art review. Front Agric Sci Eng 3:206–221. https://doi.org/10.15302/J-FASE-2016111

Monzó J, Insua I, Fernandez-Trillo F, Rodriguez P (2015) Fundamentals, achievements and challenges in the electrochemical sensing of pathogens. Analyst 140:7116–7128

Kang M-S, Jung Y-G, Jang D-H (2017) A study on the search of optimal aquaculture farm condition based on machine learning. J Inst Internet Broadcast Commun 17:135–140. https://doi.org/10.7236/jiibc.2017.17.2.135

Kapetsky JM, Hill JM, Worthy LD, Evans DL (1990) Assessing potential for aquaculture development with a Geographic Information System. J World Aquac Soc 21:241–249. https://doi.org/10.1111/j.1749-7345.1990.tb00535.x

Le J, Xu L (2017) An automated fish counting algorithm in aquaculture based on image processing. Proceedings of the 2016 International Forum on Mechanical, Control and Automation (IFMCA 2016), Shenzhen, China, 30–31 December 2017; 113, pp 358–366. https://doi.org/10.2991/ifmca-16.2017.56

Li C, Li Z, Wu J et al (2018) A hybrid model for dissolved oxygen prediction in aquaculture based on multi-scale features. Inf Process Agric 5:11–20. https://doi.org/10.1016/j.inpa.2017.11.002

Li MH, Robinson EH (2006) Use of cottonseed meal in aquatic animal diets: a review. N Am J Aquac 68:14–22. https://doi.org/10.1577/a05-028.1

Liu H, Xu L, Li D (2015) Detection and recognition of uneaten fish food pellets in aquaculture using image processing. Sixth Int Conf Graph Image Process (ICGIP 2014) 9443:94430G. https://doi.org/10.1117/12.2179138

Liu S, Tai H, Ding Q et al (2013) A hybrid approach of support vector regression with genetic algorithm optimization for aquaculture water quality prediction. Math Comput Model 58:458–465. https://doi.org/10.1016/j.mcm.2011.11.021

López M, Martínez S, Gómez JM et al (2009) Wireless monitoring of the pH, $NH_4{}^+$ and temperature in a fish farm. Procedia Chem 1:445–448. https://doi.org/10.1016/j.proche.2009.07.111

Luong JHT, Groom CA, Male KB (1991) The potential role of biosensors in the food and drink industries. Biosens Bioelectron 6(7):547–554. https://doi.org/10.1016/0956-5663(91)80018-S

Majlesi M, Malekzadeh J, Berizi E, Toori MA (2019) Heavy metal content in farmed rainbow trout in relation to aquaculture area and feed pellets. Foods Raw Mater 72:884–895. https://doi.org/10.21603/2308-4057-2019-2-329-338

Mitsubayashi K, Kubotera Y, Yano K et al (2004) Trimethylamine biosensor with flavin-containing monooxygenase type 3 (FMO3) for fish-freshness analysis. Sens Actuators, B Chem 103:463–467. https://doi.org/10.1016/j.snb.2004.05.006

Narsaiah K, Jha SN, Bhardwaj R et al (2012) Optical biosensors for food quality and safety assurance-A review. J Food Sci Technol 49:383–406. https://doi.org/10.1007/s13197-011-0437-6

Parra L, Sendra S, García L, Lloret J (2018) Design and deployment of low-cost sensors for monitoring the water quality and fish behavior in aquaculture tanks during the feeding process. Sens (Switzerland) 18:750. https://doi.org/10.3390/s18030750

Pine William E, Pollock Kenneth H, Hightower Joseph E, Kwak Thomas J, Rice JA (2003) A review of tagging methods for estimating fish population size and components of mortality. Fish 28:10–23

Pires NMM, Dong T, Yang Z (2019) A fluorimetric nitrite biosensor with polythienothiophene-fullerene thin film detectors for on-site water monitoring. Analyst 144:4342–4350. https://doi.org/10.1039/c8an02441c

Rajee O, Alicia TK (2019) Biotechnological application in aquaculture and its sustainability constraints. Int J Adv Biotechnol Res 10:1–15

Rao PV, Reddy AR, Sucharita V (2018) Big data analytics in aquaculture using hive and hadoop platform. Explor Converg Big Data Internet Things IGI Glob 29–35. https://doi.org/10.4018/978-1-5225-2947-7.ch002

Razman MAM, Susto GA, Cenedese A et al (2019) Hunger classification of Lates calcarifer by means of an automated feeder and image processing. Comput Electron Agric 163:104883. https://doi.org/10.1016/j.compag.2019.104883

Saberioon M, Gholizadeh A, Cisar P et al (2017) Application of machine vision systems in aquaculture with emphasis on fish: state-of-the-art and key issues. Rev Aquac 9:369–387. https://doi.org/10.1111/raq.12143

Shahriar MS, McCulluch J (2014) A dynamic data-driven decision support for aquaculture farm closure. Procedia Comput Sci 29:1236–1245. https://doi.org/10.1016/j.procs.2014.05.111

Sönmez AY, Sazykina M, Bilen S et al (2016) Assessing contamination in sturgeons grown in recirculating a uaculture system by Lu—Biosensors and metal accumulation. Fresenius Environ Bull 25:1028–1037

Sopinka NM, Donaldson MR, O'Connor CM et al (2016) Stress indicators in fish. Fish Physiol 35:405–462. https://doi.org/10.1016/B978-0-12-802728-8.00011-4

Su X, Sutarlie L, Loh XJ (2020) Sensors, biosensors, and analytical technologies for aquaculture water quality. Research 2020:8272705. https://doi.org/10.34133/2020/8272705

Suleiman AA, Guilbault GG (1994) Recent developments in piezoelectric immunosensors. A review. Analyst 119(11):2279–2282. https://doi.org/10.1039/AN9941902279

Toldrà A, Furones MD, O'Sullivan CK, Campàs M (2020) Detection of isothermally amplified ostreid herpesvirus 1 DNA in Pacific oyster (Crassostrea gigas) using a miniaturised electrochemical biosensor. Talanta 207:120308. https://doi.org/10.1016/j.talanta.2019.120308

Venugopal V (2002) Biosensors in fish production and quality control. Biosens Bioelectron 17:147–157. https://doi.org/10.1016/S0956-5663(01)00180-4

Yang X, Zhang S, Liu J et al (2020) Deep learning for smart fish farming : applications, opportunities and challenges. arXiv Prepr arXi 2004

Zhao C, Si Y, Pan B et al (2020) Design and fabrication of a highly sensitive and naked-eye distinguishable colorimetric biosensor for chloramphenicol detection by using ELISA on nanofibrous membranes. Talanta 217:121054. https://doi.org/10.1016/j.talanta.2020.121054

Chapter 22
Biosensor as a Potential Tool for On-Site Detection of Insect Pathogens

Mudasir Gani, Taskeena Hassan, Pawan Saini, Khalid Hussain Bhat, Rakesh Kumar Gupta, and Kamlesh Bali

Abstract Insects constitute the largest and most diverse group of animals on the earth and are potentially the largest pathogen reservoir. Several pathogens are known to infect insects and cause lethal or sub-lethal effects. These entomopathogens are widespread in the natural environment and invade and reproduce in an insect and spread to infect other ones through horizontal and vertical transmission. Historically, insect pathogens have been of interest due to their economic impact on beneficial insects in terms of morbidity and mortality or their potential as specific biological control agents of insect pests. The present methodologies for insect pathogen detection viz. microscopic examination, the culture of pathogens, biochemical, immunological, and molecular techniques are entirely laboratory-based and cannot be performed under field conditions. Hence, there is a great scope and need for the development of sensitive and specific techniques for early, rapid, and on-site detection of insect pathogens. One of the useful emerging technology is the use of a biosensor for on-site and point-of-care detection of pathogens of beneficial insects for efficient disease management and for on-site detection and distribution of native insect pathogens for commercial production of insect-specific biopesticides. The biosensor technology has the potential to surpass conventional or present

M. Gani (✉)
Division of Entomology, Faculty of Agriculture, Sher-e-Kashmir University of Agricultural Sciences and Technology, Kashmir 193201, J&K, India
e-mail: mudasir32@gmail.com

T. Hassan
Department of Zoology, Aligarh Muslim University, Aligarh 202002, Uttar Pradesh, India

P. Saini
Central Sericultural Research and Training Institute, Central Silk Board, Ministry of Textiles, Govt. of India, Pampore 192121, J&K, India

K. H. Bhat
Division of Basic Sciences and Humanities, Faculty of Agriculture, Sher-e-Kashmir University of Agricultural Sciences and Technology, Kashmir 193201, J&K, India

R. K. Gupta · K. Bali
Division of Entomology, Sher-e-Kashmir University of Agricultural Sciences and Technology, Chatha 180009, J&K, India

R. N. Pudake et al. (eds.), *Biosensors in Agriculture: Recent Trends and Future Perspectives*, Concepts and Strategies in Plant Sciences,
https://doi.org/10.1007/978-3-030-66165-6_22

detection techniques in time, accuracy and cost owing to recent developments in antibody production, nanotechnology, and microfluidics along with the availability of improved antibody immobilization strategies. This chapter provides a comprehensive review of the insect pathogen detection techniques developed and used so far, as well as their limitations and the potential of biosensors for on-site detection of insect pathogens.

Keywords Insect · Pathogen · Diagnosis · Biopesticide · Nanotechnology · Transducer · Immunosensor

22.1 Introduction

Insects represent the largest and most diverse group of animals on the Earth with an estimated 5.5 million different species (Stork 2018). Like other animals, insects are sensitive to environmental, nutritional, and microbial factors which result in various types of diseases caused by noncellular agents (viruses), prokaryotes (bacteria), eukaryotes (fungi and protists), and multicellular animals (nematodes) (Gani et al. 2019). The microbes exist everywhere in nature and have the capability to invade and reproduce in an insect, spread, and infect other insects, and cause lethal or sub-lethal effects. Historically, insect pathogens have been of interest due to their economic impact on beneficial insects in terms of morbidity and mortality or their potential as biological control agents of insect pests. Accordingly, the insect diseases are broadly divided into two categories which include: (a) diseases of beneficial insects (productive insects, i.e. honey bee, silkworm, lac insect; pollinators such as honey bees, syrphid flies, bumblebees, etc.; biological control agents, i.e. coccinellids, praying mantis, predatory stink bugs, etc.) and (b) diseases of injurious or harmful insects (insect pests of cultivated crop plants like cotton bollworm, tobacco caterpillar, etc.; storage insect pests like rice weevil, pulse beetle, etc.). The protection of beneficial insects from diseases is an important aspect of insect pathology as beneficial insects provide useful products such as honey, silk, lac, etc., and act as pollinators and biological control agents. Much of the research in this area concerns the causal agents of insect diseases, prevalence, virulence, transmission, genetics, genomics, and prevention and management of diseases (Chen et al. 2014; Gani et al. 2018). On the other hand, insect pathogens can regulate insect pest populations (Anderson and May 1980; Cory and Myers 2003; Elderd 2013) and have been developed as biological insecticides for the management of the insect pests of agricultural and forest crops (Moscardi 1999; Gupta et al. 2016). Here focus is on the exploitation of insect pathogens for biological pest control, their mass production, formulation, and application to insect pest populations in a manner analogous to chemical pesticides (Del Rincón-Castro and Ibarra 2011; Gani et al. 2019).

The insect disease epidemics are determined by the quantity of initial inoculum and the rate of transmission. Therefore, diagnostic techniques which allow early, and rapid detection of initial inoculum are needed. The detection of insect pathogens is being routinely conducted by traditional techniques based on symptomology, in vitro culturing, and microscopy (Evans and Shapiro 1997; Gani et al. 2017a). However, it takes a long time for bacteria and fungus to grow on a culture media and produce diagnostic spores. Such a delay is not acceptable when trying to decide on a management strategy for beneficial insects or when rapid detection is required to eradicate the spread of a newly introduced pathogen. Further, there are pathogens such as viruses, rickettsia, and microsporidia (obligate pathogens) which cannot be cultured on artificial media. The alternative methods for insect pathogen detection viz. immunoassays and nucleic-acid amplification {polymerase chain reaction (PCR)} are also entirely laboratory-based and require well-trained staff and equipments (Gani et al. 2017a). These techniques also lack the convenience of "on-site" testing and require a complex workflow from sample collection, sample labeling, sample storage, and transport to appropriate facility and further followed by sample processing, assays, analysis, and interpretation of results.

Normally, the disease is recognized by the presence of clear symptoms at later stages of infection. Many diseases remain undetected (latent, sub-lethal) and are recognized only once the disease has clear symptoms or causes lethal effects (Vogel et al. 2019). Many insect pathogens are present in asymptomatic form and hence the absence of pathogens from insect populations cannot be ruled out. Moreover, the pathogen load or the initial rate of infection may often be low in insect populations which demonstrates that highly sensitive and specific detection techniques are also needed. Therefore, it is imperative for insect pathologists to develop sensitive, specific, and portable assays capable of early, rapid, and on-site diagnosis. In recent years, major advances has been made in the field of nanotechnology which have several potential applications in agriculture (Fig. 22.1). The biosensor for disease

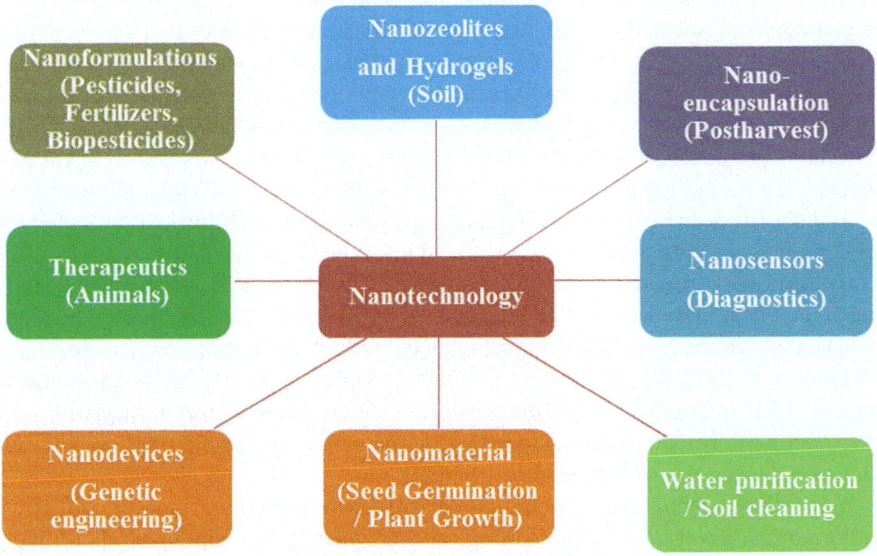

Fig. 22.1 Potential applications of nanotechnology in agriculture

diagnosis was developed by the integration of a powerful nanotechnology approach with molecular biology. The biosensor is one of the useful emerging technology for on-site detection of pathogens of beneficial insects for effective disease management and for on-site detection and geographical distribution of native insect pathogens for commercial production of insect-specific biopesticides as a component of biological control. The native strains are always preferred in biological control owing to their adaptability, sustainability, and efficacy under a given set of agro-ecosystem. The identification and utilization of native strains hold ample scope for their widespread multiplication and commercial use as novel biopesticides. The biosensors were developed and used for the detection of diverse pathogens (Ahmed et al. 2014), however, this technology has not been widely applied to insect pathogen detection.

22.2 Current Approaches for Insect Pathogen Detection

The microscopic examination of insect pathogens has been well known as one of the most available, easy to perform, and inexpensive techniques. However, microscopy has inherent problems such as late detection, presence of secondary pathogens, and inaccuracy as some pathogens can be overlooked under a microscope being very small in size (Iwano and Ishihara 1981; Kawarabata and Ishihara 1984; Poláchová et al. 2019). The culture of pathogens is done by the aseptic transfer of an inoculum from a source (insect, soil, food, etc.) to a suitable growth media or host resulting in amplification of cell numbers for quantitative determination. This propagation of pathogens may be performed using antibiotics to suppress the growth of other strains of pathogens that may also reside in the inoculum. The subsequent transfer of pathogens to selective or differential media generates colonies that can be distinguished based on colony morphology by ocular inspection and identified by rigorous biochemical, microscopic or nucleic acid-based assays. Koch's postulates are used to identify unknown pathogens in diseased samples (Koch 1884; Rivers 1937). Koch's postulates are based on four criteria to establish the identity the causal agent of a disease: (1) the pathogen is always associated with a disease; (2) the pathogen must be isolated from the diseased sample and grown in pure culture; (3) the pathogen from the pure culture must be re-inoculated into a healthy host and should cause the same disease symptoms; and (4) the pathogen must be re-isolated from the newly inoculated host and proven to be the same as the original one. Electron microscopy (EM) has also been used for insect pathogen identification, description, and discovery (Adams et al. 1977; Van Iddekinge et al. 1983; Kumar et al. 2011). The problems in the use of EM in insect pathogen diagnosis include sample transportation, processing, testing conditions, variations induced by the personnel in sample fixation and time of exposure, and use of high-end instruments.

Serological techniques including complement fixation, immunodiffusion, radioimmune assays, western blotting, and enzyme-linked immunosorbent assay (ELISA) have been used to detect and quantify insect pathogens (McCarthy and Gettig 1986; Harlow and Lane's 1988; Liu et al. 2011). These methods employ antibodies for the detection of a given insect pathogen from analytical samples isolated from infected insects. The latex agglutination test for the detection of flacherie (Shimizu et al. 1983), cytoplasmic polyhedrosis (Shimizu and Arakawa 1986), and densonucleosis (Shimizu and Tauchi 1991) viruses in *Bombyx mori* L. (Lepidoptera: Bombycidae) have been used. Arakawa (1989) introduced a latex agglutination test for the detection of nucleopolyhedrovirus (NPV) infection in *B. mori*. Arakawa and Shimizu (1990) developed a protein-A linked latex agglutination test using polyclonal antibodies for the detection of NPV infection and the test was more sensitive than the direct latex agglutination test. A colloidal textile dye-based dipstick immunoassay for the detection of *B. mori* nucleopolyhedrovirus (BmNPV) was developed by Nataraju et al. (1994), with comparable sensitivity and cost-effectiveness. Shamim et al. (1994) produced murine monoclonal antibodies (MCAs) against BmNPV which reacted well with polyhedra of BmNPV and to a variable extent with the NPV of other species, namely *Amsacta albistriga*, *Helicoverpa armigera,* and *Spodoptera litura.* These anti-BmNPV MCAs showed no cross-reactivity with *Nosema bombycis*, *Serratia marcescens*, group A *Streptococci*, *Staphylococcus aureus*, *Escherichia coli*, and normal hemolymph proteins, except *Bacillus thuringiensis.* The MCAs specifically recognized BmNPV polyhedra and were found effective for the detection of BmNPV infection in *B. mori* at early stages. Shamim et al. (1995) developed a protein-A linked monoclonal antibody latex agglutination test (PALMAL) for BmNPV detection in *B. mori.* The latex beads were precoated with protein-A and then sensitized with monoclonal antibody (MA-231) (125pgml^{-1}) directed against polyhedrin protein of BmNPV. PALMAL test has the capability to detect 1×10^5 polyhedra/test and is more sensitive than the direct agglutination test. The use of specific monoclonal antibody and the presence of free-floating antigen-binding sites (Fab) of the immunoglobulin to react with antigen have improved the sensitivity of the PALMAL test (Torrance 1980).

ELISA, Southern hybridization, and PCR assay were used to detect *Orgyia pseudotsugata* multiple nucleopolyhedrovirus (OpMNPV) in Douglas-fir tussock moth larvae (Thorne et al. 2007). Kumar et al. (2007) developed a direct antigen coating enzyme-linked immunosorbent assay (DAC-ELISA) to detect NPV infection in *Helicoverpa armigera* using polyhedrin-specific polyclonal antibodies of *Helicoverpa armigera* nucleopolyhedrovirus (HaNPV). The polyclonal antibodies were highly specific and useful to detect HaNPV at the early stages of infection. Monoclonal antibodies were also generated against *Autographa californica* nucleopolyhedrovirus (AcNPV) which reacted well with polyhedrin protein of AcNPV and with a wide range of baculoviruses (Roberts and Naser 1982). The western blot was developed and described as an extremely sensitive technique for the detection of traces of AcNPV in very crude materials (impure polyhedra, infectious haemolymph, and larval homogenates) (Naser and Miltenburger 1983). Poláchová et al. (2019) developed a sandwich upconversion-linked immunosorbent assay (ULISA) for *M. plutonius* detection using rabbit polyclonal anti-Melissococcus antibody and successfully demonstrated its practical applicability. But these assays are time-consuming and are not suitable for rapid on-site diagnosis. Further, the selectivity and sensitivity of

ELISA and western blotting techniques depend on the antigen–antibody interaction and are well suited for samples without any contaminants or interfering molecules such as proteins, DNA, and non-target cells.

Another method for insect pathogen detection is the amplification and subsequent analysis of pathogen-specific nucleic acid by PCR and sequencing. The PCR technology is commonly used for the identification and detection of insect pathogens owing to its sensitivity to detect specific DNA templates at nanogram and picogram concentrations (Gani et al. 2017b; Dieffenbach and Dveksler 2003). An important step in the PCR procedure is the extraction and purification of clean nucleic acid as a template for the reactions. While it is relatively easy and simple to isolate the pathogen from the insect cadaver or living infected insect, it is technically more cumbersome to extract quality DNA from the stock solution of the pathogen for PCR. There are also chances of getting false results due to the presence of contaminants in the sample. Further, simultaneous processing of large numbers of samples increases the risk of contamination and false positive results. These methodologies are now emphasized by real-time PCR and reverse transcriptase PCR. However, the implementation of these techniques for insect pathogens detection can be limited by external factors. For example, the original stock inoculum often contains high levels of fats, carbohydrates, and tissue debris which requires a sample purification step prior to analysis. Further, there are chances of non-specific DNA amplification due to the presence of "naked" DNA in analytical samples that may act as a template for the amplification of these superfluous products. Ravikumar et al. (2011) developed a novel PCR-based assay for individual and simultaneous detection of three major pathogens {*N. bombycis*, NPV, and densovirus (DNV)} infecting the silkworm, *B. mori*. A multiplex real-time PCR assay for the simultaneous detection of three major pathogens of *B. mori* (*N. bombycis,* NPV, and densovirus) was developed by Wu et al. (2017). The *Helicoverpa armigera* Nucleopolyhedrovirus (HaNPV) isolate from *Heliothis peltigera* was confirmed by tissue polymerase chain reaction and sequence analysis by Eroglu et al. (2019).

The Next Generation Sequencing (NGS) technology has greatly helped in insect pathogen discovery and detection (Liu et al. 2011; Cox-Foster et al. 2007). The most common NGS platforms are Roche 454 pyrosequencing (454 Life Science), Illumina (Solexa) sequencing, and SOLiD sequencing (ABI Biosystems). Huang et al. (2019) determined the nucleotide sequence of the whole *Troides aeacus* NPV genome for identification using NGS technology. However, this technology involves a series of wet (sample processing, library preparation, and sequencing) and dry (bioinformatics analysis and interpretation of data) laboratory steps. Besides, NGS-based approaches are highly sophisticated and financially non-viable options. Another key disadvantage is that microbial nucleic acids from infected samples are dominated by insect host background which limits the overall analytical sensitivity of the genome sequencing approach for pathogen detection (Liu et al. 2011). The current techniques for insect pathogen detection are listed in Table 22.1. Due to the limitations of the above-discussed techniques the development of new diagnostic techniques with high sensitivity and specificity for on-site detection of insect pathogens is highly desirable.

Table 22.1 Summary of current techniques used for detection of the insect pathogens

Technique	Detection limit	Constraints	Reference
Microscopy	Neither specific, sensitive nor accurate	Late detection, Misdiagnosis	Wu et al. (2017)
Immunoassays (Complement fixation, Immunodiffusion, Radioimmune assays, ELISA, Western blotting, Latex agglutination test)	850 baculovirus occlusion bodies	Cross-reactivity, possibility of false positive/negative results	Ndao (2009), Thorne et al. (2007)
LAMP	50 viral copies ng^{-1} genomic DNA (equivalent to 150 viral copies per reaction)	Designing of primers, sensitivity	Chaivisuthangkura et al. (2009), Sahoo et al. (2016)
Lateral Flow Dipstick immunoassay	100 viral copies	Suitable for primary screening only, good antibody preparation	Jaroenram et al. (2009), Koczula and Gallotta (2016)
LAMP-LFD	0.2 pg DNA	Designing of primers, sensitivity	Zhou et al. (2015)
Southern hybridization	6000 baculovirus occlusion bodies	Complex and labour intensive; time-consuming	Thorne et al. (2007)
PCR	8.0 baculovirus occlusion bodies	DNA extraction; use of reagents, high-end instruments, and skilled manpower	Thorne et al. (2007)
Quantitative real-time PCR (qPCR)	0.013 pg of viral DNA	DNA extraction; use of reagents, high-end instruments, and skilled manpower	Krokene et al. (2013)
Next Generation Sequencing	Varies with the sequencing platform	Use of complicated data analysis methods, Difficult to identify novel viruses due to non-homology to known viruses	Liu et al. (2011)

22.3 Biosensors

The biosensor technology is being developed as an alternative to laboratory techniques for pathogen detection. It is one of the useful emerging techniques developed by assembling complex analytical platforms into a single miniature device for on-site and point-of-care detection of pathogens by monitoring a biological reaction at the surface of transducers (Helali et al. 2006). Biosensors are portable, easy to use, and do not require skilled personnel, laboratory equipments, or reagents, thus, capable of on-site diagnostics, cutting short the cumbersome procedures, and significantly reducing the costs involved. A biosensor is actually a compact analytical device with a ligand-specific biorecognition element or bioreceptor (antibody, enzyme, nucleic acid, receptor, aptamers, peptide or protein, lectin, cells, tissue, or whole organism) coated on a sensor surface integrated directly or indirectly with a signal conversion unit called a transducer (Tiwari et al. 2017). The specific physiological interaction between the target analyte and biorecognition element is translated, by the transducer, into a measurable electric signal, which is further interpreted by a computer-aided readout system for the user (Fig. 22.2) (Arora et al. 2010). The biorecognition element can sense the presence of an analyte and determines the selectivity or specificity of the biosensor. The transducer element has a major influence on sensitivity as it translates the selective recognition of the analyte into a quantifiable signal. The miniaturization achieved by utilization of nanostructures in biosensors for sensing or detection of analyte results in the formation of a nanobiosensor (Girigoswami and Akhtar 2019).

Biosensors are mainly classified based on the biorecognition element and the transducer type. The immunosensors employ antibody or antigen as the biorecognition element and are well suited for the development of point-of-care and on-site diagnostics (Fig. 22.3) (Conroy et al. 2009). Immunosensors represent a potential tool for simple, label-free, sensitive, and rapid qualitative and quantitative detection of pathogens. The result of the immunosensor depends on a reaction between an antigen and an antibody. The high selectivity and sensitivity of immunosensors to recognize a biomolecular component depends on the specific interactions and the equilibrium association constants (10^{10} M^{-1} and greater) between an antigen and its corresponding antibody (Farka et al. 2017). The sensing is carried out using competitive and non-competitive assay format for the development of an immunosensor and

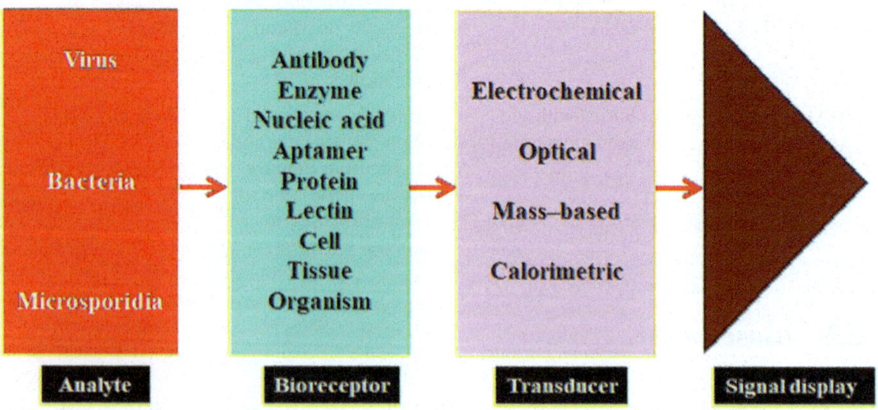

Fig. 22.2 Principle of biosensor. The bioreceptor is coated on a sensor surface and integrated directly or indirectly with the transducer. The specific physiological interaction between the analyte and bioreceptor is translated by the transducer into a measurable electric signal for interpretation

Fig. 22.3 A simple representation of an antibody-based biosensor (immunosensor) used for the detection of insect pathogens. Here antibody is immobilized on a sensor surface for the capture of an analyte and nanoparticles are used as signal amplification labels. The antigen–antibody interaction produces a specific physicochemical change which is then converted (via a transducer) to a signal which the user can interpret (*Source* Modified from Ayyar and Arora 2013; Farka et al. 2017)

the choice is made based on the molecular size and reactivity of the analyte (Farka et al. 2017). The non-competitive approach is used for analytes holding two or more epitopes and a competitive approach is generally preferred for smaller or less reactive analytes. In recent years, several reports on biosensor development for sensitive, specific, and on-site detection of human, animal, and plant pathogens were published (Du and Zhou 2018; Devarakonda et al. 2017; Khater et al. 2017) (Table 22.2). Most

Table 22.2 List of biosensors for sensitive, specific, and on-site detection of different pathogens

Virus	Antibody type	Transducer type	Detection time	Reference
Tobacco Mosaic virus	mAb	SPR	Not specified	Boltovets et al. (2004)
Autographa californica MNPV	mAb	SPR	Not specified	Baac et al. (2006)
West Nile virus	mAb	Voltametric	30 min	Nguyen et al. (2009)
Bovine viral diarrhoea virus	pAb and mAb	Conductometric	8 min	Luo et al. (2010)
Avian influenza virus (subtype H5)	mAb and pAb	Voltametric	10 min	Kamikawa et al. (2010)
Plum pox virus	pAb	Impedimetric Immunosensor	Not specified	Jarocka et al. (2011)
Rabies virus	mAb	SPR	Not specified	Xu et al. (2012)
Influenza virus H1N1	mAb	Electrochemical immunosensor	30 min	Devarakonda et al. (2017)
Newcastle disease virus	chicken egg yolk antibody (IgY)	Electrochemical immunosensor	Not specified	Tran et al. (2019)
European foulbrood, *Melissococcus Plutonius*	pAb	Amperometric Immunosensor	2 h	Mikušová et al. (2019)
Staphylococcus aureus	pAb	Electrochemical immunosensor	20 min	Wang et al. (2019)
Salmonella typhimurium	mAb and pAb	Conductive immunosensor	30 min	Wonsawat et al. (2020)

mAb: monoclonal antibody; pAb: polyclonal antibody, SPR: Surface Plasmon Resonance

of the reported applications used nanomaterials such as gold nanoparticles (Pingarron et al. 2008; Kwon et al. 2017; Lu et al. 2019; Zheng et al. 2019; Alagappan et al. 2020), catalytic nanoparticles (Ramanavicius et al. 2005; Luo et al. 2006; Farka et al. 2018), quantum dots (Annio et al. 2018; Faridbod and Sanati 2019; Zhao et al. 2020), or carbon-based materials (Sotiropoulou et al. 2003; Lin et al. 2017; Sharma et al. 2020) for enhanced assay properties.

The biosensor is of high potential interest for in-field applications and the rapid diagnostic kit suitable for field use is represented by the lateral flow assay (LFA). The LFA is a paper-based platform for the detection and quantification of analytes in complex mixtures, where the sample is placed on a test device and the results are displayed within minutes (Tomkies et al. 2009). The Lateral Flow Device (LFD) is comprised of a release pad, a nitrocellulose test membrane, and an absorbent pad and is used to detect biological antigens in infected tissue extracts (Danks and Barker 2000). Due to their affordability, easy handling LFAs are being widely used for disease detection (Koczula and Gallotta 2016). Moreover, they possess long shelf life and do not need refrigerated storage, which makes them most suited for field application. LFAs are classified into "lateral flow immunoassays" (LFIAs) when antibodies are exclusively used as recognition elements, and Nucleic acid LFA (NALFA) that detect PCR amplicons. LFDs have proved to be reliable and rapid diagnostic tools for a variety of viruses, bacteria, and fungi (Danks et al. 2003). The gold nanoparticles (Au NPs) are the most common labels for lateral flow immunoassays (Ge et al. 2014). The biosensor technology holds a lot of potential for early disease diagnosis and timely initiation of appropriate treatment against the diseases of beneficial insects and on-site detection and geographical distribution of native insect pathogens for commercial production of insect-specific biopesticides.

A point-of-care device had been reported, when an amperometric immunosensor based on a sandwich assay was used for in-field diagnosis of European foulbrood (EFB) pathogen, *Melissococcus plutonius* (Mikušová et al. 2019). They have immobilized anti-Melissococcus antibody on a gold surface of a screen-printed sensor via a self-assembled monolayer of cysteamine activated with glutaraldehyde. The better performance was obtained after the formation of the sandwich with the peroxidase-labelled antibody in the amperometric mode. The limit of detection (LOD) of the reported device was 6.6×10^4 CFU mL^{-1} with a wide linear range between 10^5 CFU mL^{-1} and 10^9 CFU mL^{-1}. The immunosensor was successfully employed in the diagnosis of honeybee samples and helped in preventing *M. plutonis* infection spread and losses of honeybee colonies.

A lateral flow immunoassay was developed as a rapid and sensitive test to diagnose EFB in the field using a highly specific anti-*Melissococcus plutonius* monoclonal antibody (Tomkies et al. 2009). The immunoassay was found to be very effective in detecting *M. plutonius* in 96–100% of EFB-infected samples with no cross-reactivity with other bee brood pathogens in laboratory trials. The field validation revealed correct diagnoses in 96% of samples and the EFB LFDs are now used as a diagnostic tool for routine confirmation of *M. plutonius* infection in the field, thus helping in disease detection and control.

Surface Plasmon Resonance (SPR)-based sensor chip with three bio-functional layers (amine-reactive crosslinker with a disulfide bond that chemisorbs to gold film, protein-A, and a mouse IgG monoclonal antibody raised against a surface protein of the target viral pathogen) was used to directly detect the intact *Autographa californica* multiple nucleopolyhedrovirus (AcMNPV) (Baac et al. 2006). Here AcMNPV was introduced as a target pathogen to couple with a substrate-immobilized antibody while tobacco mosaic virus (TMV) was used as a non-binding control. The consecutive exposure of the sensor chip to both viruses, i.e. AcMNPV and TMV demonstrated the selective response to the AcMNPV with an SPR angular shift of 0.037° for 10^7 pfu mL^{-1} (plaque forming units per ml) virus concentration. The improvement in electronic components and the use of nanoparticles as labels allow for a significant reduction in the detection limit of SPR-based sensors.

22.4 Transducers

A transducer is a device that converts the biorecognition event, i.e. the specific physiological interaction between the antibodies immobilized on the sensor surface and their target analyte into a measurable electric signal. The biosensors are classified into electrochemical, optical, mass-based, and calorimetric based on the detection method and transducer system (Velusamy et al. 2010; Monosik et al. 2012). The most commonly used techniques for pathogen detection include electrochemical and optical transducers.

22.5 Types of Biosensors

22.5.1 Electrochemical Biosensors

Electrochemical biosensors are those that measure the change in electrical properties following biorecognition due to the consumption or production of the ions or electrons (Mohanty and Kougianos 2006). They operate by reacting with the analyte of interest and produce an electrical signal that is proportional to the analyte concentration. The electrochemical sensors are the first scientifically proposed as well as successfully commercialized biosensors. Electrochemical detection has been used in conjunction with paper-based biosensors because of its high sensitivity, selectivity, portability, cost-effectiveness, reproducibility, and compatibility with micromachining technology. The application of nanoparticles in electrochemical techniques leads to major developments in electrochemical immunosensing. The nanomaterials (carbon nanotubes and graphene) lower the limit of detection (LOD) in electrochemical biosensors.

Electrochemical impedance spectroscopy (EIS) has received considerable attention in the field of immunosensors, especially due to the possibility of ultrasensitive, non-destructive, and rapid electrochemical sensing. EIS embodies a powerful technique to investigate the physicochemical properties of the biorecognition events connected to a respective transducer. The electrode is a key component of an electrochemical biosensor that is employed as solid support for the immobilization of biomolecules (antibody, enzyme, and nucleic acid) and movement of electrons. Typically, the electrochemical biosensor is comprised of two or three electrodes on the same substrate. In the two-electrode configuration, one functions as the working electrode while the other works as the reference electrode. The three-electrode configuration consists of the reference, working, and counter or auxiliary electrodes. The working electrode also known as the sensing or redox electrode serves as the transduction element in the biochemical reaction. The counter electrode in connection with the electrolytic solution applies current to the working electrode. The reference electrode is located at a distance from the reaction site to maintain a known and stable potential. These electrodes should be both chemically stable and conductive. Therefore, gold, platinum, carbon (e.g. graphite), and silicon compounds are commonly used depending on the analyte. Electrochemical biosensors are further classified into potentiometric, amperometric, voltammetric, and impedimetric/conductometric (Velusamy et al. 2010). The most used is impedimetric followed by amperometric immunosensors.

22.5.1.1 Potentiometric

The potentiometric biosensors monitor the potential of a system at a working electrode with respect to the reference electrode under zero current flow. They are based on ion-selective electrodes (ISE) and ion-sensitive field effect transistors (ISFET) (Pohanka and Skladal 2008). The biorecognition event is converted into a change in potential signal which is detected by a reference electrode. Here the biosensor format typically consists of a perm-selective outer layer and a bioactive element, such as urease for enhancement of the assay performance (Leonard et al. 2003). The light-addressable potentiometric sensor (LAPS) which combines potentiometric and optical detection has been applied for pathogen detection (Gehring et al. 1998; Dill et al. 1999).

22.5.1.2 Amperometric

The enzymes in amperometric biosensors produce electroactive product due to oxidation or reduction of analytes on the working electrode (carbon, gold, etc.). The resultant current can then be detected. The amperometric sensors can be fabricated by using disposable and customized screen-printed electrochemical electrodes (screen-printed carbon electrodes) by depositing conducting inks (carbon, silver, etc.) in a pre-determined arrangement and thickness (Palchetti and Mascini 2008). These systems are robust, economical, and sensitive and can be used in conjunction with mediators such as ferrocene dicarboxylic acid (FEDC) or iodine to improve selectivity (Leonard et al. 2003). Besides, there is major potential to miniaturize these systems that lead to smaller sample volume requirements. Amperometric biosensors are more sensitive and well suited for mass production than the potentiometric ones (Pohanka and Skladal 2008; Ghindilis et al. 1998).

22.5.1.3 Impedimetric

The impedimetric biosensor set-up follows either impedance (Z) or its components resistance (R) and capacitance (C). The inverse value of resistance is known as conductance and for this reason, these types of biosensors are also named as conductometric. Many strategies have been used to design impedimetric biosensors. Electrochemical impedance spectroscopy (EIS) is a powerful tool and a non-destructive technique for investigating the interfacial properties related to biorecognition events (Bahadır and Sezginturk 2016). EIS is the most common technique used in impedimetric biosensors, where the impedance is measured over a wide range of potential frequencies (typically from 100 kHz to 1 MHz) (Hammond et al. 2016). Impedimetric immunosensors have shown promising applications owing to sensitive, simultaneous, and label-free detection of pathogens. In impedimetric immunosensors the antibodies are immobilized on the electrodes, semiconductor chips, or optical fibre (Bahadır and Sezginturk 2016). The impedimetric biosensors are less frequently developed and used as compared to potentiometric and amperometric biosensors (Pohanka and Skladal 2008).

22.5.2 Optical Biosensor

The transducers that measure changes in the intensity of light are known as optical biosensors. Here the biorecognition sensing element is integrated with an optical transducer system with the basic objective to produce a signal which is proportionate to the concentration of a measured substance (analyte). The optical biosensors are broadly divided into two general modes: label-free and label-based. The detected signal is generated directly by the interaction of the analysed material with the transducer in a label-free mode. The label-based sensing involves the use of a label and the optical signal is then generated by a colorimetric, fluorescent, or luminescent method. The optical biosensors have high sensitivity and compatibility with several well-established optical phenomena such as luminescence, fluorescence, phosphorescence, reflectance, colorimetry, light polarization and rotation, spectroscopy, interference, ellipsometry, and surface plasmon resonance (SPR) (Luong et al. 2008; Velusamy et al. 2010; Huang et al. 2011). However, fluorescence and SPR-based biosensors are most common (Velusamy et al. 2010).

Fluorescence is defined as "a short-lived type of luminescence created by electromagnetic excitation which is generated when a substance absorbs light energy at a short (higher energy) wavelength and then emits light energy at a longer (lower energy) wavelength" (Behlke et al. 2005). The time interval between absorption and emission is very short, usually at the order of 10^{-9} to 10^{-8} s in fluorescence (Serrano-Andres and Serrano-Perez 2012). The fluorescent biosensors have high selectivity and sensitivity and profound applications in disease diagnosis. The brightness of the fluorescent labels, packing density of the fluorophores, and the detection optics parameters determine the sensitivity of a fluorescence-based biosensor. The polymeric materials have made remarkable improvements in the design and functioning of fluorescent-based biosensors (Liu and Tang 2013). The conjugated polymers having light harvesting and electron delocalizing properties are also used in fluorescent sensors (Girigoswami and Akhtar 2019). Recently, the Fluorescent Resonance Energy Transfer (FRET) based ratiometric sensors have been designed with the advantage that the ratio between two fluorescence intensities is independent of the external factors such as fluctuation of the source of excitation and sensor concentration (Zhang et al. 2008; Jun et al. 2010). The different designs of fluorescent biosensors can be developed depending on the nature of the sensing element. In a single fluorophore-based biosensor, the change in fluorescence intensity of a single fluorescent reporter is monitored. This change in fluorescence intensity is due to various mechanisms such as accessibility of the solvent or surrounding environment polarity or fluorophore protein interactions are changed. The fluorophores-based biosensors are designed based on the principle of FRET. In such biosensors, fluorophore pairs, particularly a FRET donor and an acceptor, are fused with the recognition element.

SPR is the excitation of an electromagnetic wave that propagates along with the interface of two different media such as metal and sample buffer at a specific angle of incident light beam (Damborsky et al. 2016; Guo 2012). The signal generated is based on total internal reflection. The angle at which the resonance occurs is sensitive to change in refractive index or formation of a nanoscale film thickness. These changes can be quantified by monitoring the light intensity minimum shift over time. SPR-based biosensors have demonstrated to be user friendly, versatile, and are very powerful tools for biomolecular interaction studies in immunoassays. SPR measurements mostly consist of direct detection (Nakamura et al. 2003), sandwich (Wei et al. 2003), and competitive inhibition (Gobi et al. 2004) assays. Direct detection is appropriate where the direct binding of the analyte yields a sufficient response and its detection limits can be enhanced using a sandwich or inhibition assay. In a sandwich assay, the measurement is carried out in two steps viz. the analyte is bound to the antibodies on the sensor surface and then the sensor surface is incubated with a solution containing secondary antibodies (which can be labelled by nanoparticles or enzymes), which binds to the previously captured analyte and increase the sensor response. The sample is first mixed with labelled antibodies and then the mixture is brought into contact with the sensor surface exhibiting immobilized analyte molecules in an inhibition assay. The free antibodies remaining in the solution can subsequently bind to the surface.

A practical SPR instrument consists of an optical detector part which measures intensity shift, a sensor chip with a gold surface, and a layer enabling ligand immobilization that is integrated with a fluidics system enabling a flow-through operation. The interaction component (ligand) is permanently attached to the chip surface and another interacting component (analyte) flows over the surface and binds to the ligand in a practical experiment. There are three types of SPR analyses: kinetic, equilibrium, and concentration analysis for practical applications (Damborsky et al. 2016). Kinetic and equilibrium analyses are used to characterize any molecular interaction viz. ligand–analyte binding, receptor characterization, antibody–antigen interaction, etc. The SPR technique has multiple applications in the concentration analysis of any analyte where the availability of a specific ligand and its immobilization on the SPR chip is important. The concentration is obtained by measuring direct binding or from the rate of binding in a mass transport limited mode.

22.5.3 Mass-Based Biosensor

Mass-based biosensors (also known as gravimetric biosensors) detect a change in mass by using piezoelectric materials that change their resonant frequency thus, generating acoustic waves (Holford et al. 2012). These waves propagate either along (surface acoustic wave, SAW) or through (bulk acoustic wave, BAW) the substrate surface (Rocha–Gaso et al. 2009). The piezoelectric quartz crystals can either be in the form of resonating crystals (quartz crystal microbalance, QCM), or as surface acoustic wave (SAW) devices. A QCM biosensor consists of a quartz crystal wafer in the middle of two metal electrodes, and the resonant frequency of QCM changes according to the mass change at the crystal's surface. The change in frequency is proportional to the mass of the sensor. The combining of QCM device with highly specific antigen–antibody interaction enables the direct detection of micro-organisms (Ngo et al. 2014). QCM based biosensors have been used in the rapid detection of pathogens (Ozalp et al. 2015) and toxins (Neethirajan et al. 2018) due to their ease of use, cost-effectiveness, shorter analysis time, and the ability to produce label-free measurement.

SAW-based biosensors detect acoustic waves generated by mass loading on the surface of the piezoelectric crystal via the interdigital transducers (IDT) (White and Voltmer 1965). The IDT allows the acoustic energy to be strongly confined to the surface irrespective of the thickness of the substrate. The analyte recognition by the immobilized receptors changes the velocity of SAW and produces signal by the driving electronics (Lange et al. 2008). The piezoelectric materials produce an electric signal in response to mechanical forces and the commonly employed piezoelectric materials in sensors are the quartz crystals (Holford et al. 2012). The crystals are made to vibrate at a specific frequency by the application of an electric signal and the oscillation frequency of the crystal depends on the applied frequency. A shear wave is generated at a thin (5 μm) guiding layer deposited at the sensor surface by applying an electrical field to quartz crystal (Gronewold 2007). The binding events are detected by changes in the physical properties of the shear wave via the angular phase shift, and changes in the oscillation amplitude. SAW sensors have been developed as a powerful and promising system for detecting various biological recognition events (Gronewold 2007; Lange et al. 2008). The bio-capture layer on the surface of the crystal exhibits specific binding with the analyte. The mass-based sensors have antibodies as the most common bio-capture molecules and such sensors are known as piezoelectric immunosensors.

22.5.4 Calorimetric Biosensor

The calorimetric biosensors or Thermal biosensors measure the heat generated in an enzymatic reaction. The calorimetric sensors have many advantages over amperometric and optical sensors such as relatively easy fabrication, label-free detection, and less complex detection (Kopparthy et al. 2015). These types of biosensors are developed using either thermistors or thermopiles. The thermistors transform the heat (generated or lost) during a reaction into an electrical signal. One thermistor is placed at the inlet of the reactor containing the immobilized enzyme and the second thermistor is placed at the exit of the reactor to measure the temperature after conversion of the substrate to product. The thermistor records temperature changes by altering its resistance. High temperature causes a reduction in resistance of the thermistor. The difference in resistance between thermistors located at the entry and exit points to the reactor is recorded as a measure of the temperature change produced as a result of the passage of the sample. Thermopiles are widely used in calorimetric biosensors due to their compatibility with miniaturized devices and high common mode thermal noise rejection ratio. Reference junctions of the thermopiles are controlled either by a constant heat source, heat sink, or by vacuum encapsulation.

The entire calorimetric sensing set-up is surrounded by an insulated jacket to prevent the loss of heat generated during the reaction. The analyte acts as the substrate for the enzyme. The enzyme is immobilized in a packed bed column which is placed in the centre of the insulated chamber. The sample is passed through a coiled column of aluminium that serves as a heat exchanger to maintain a uniform temperature of the sample in the reaction chamber. When the analyte is exposed to such immobilized enzymes or antibodies on the sensor active area, the biochemical reaction begins and its evolution in terms of the total amount of heat generated and kinetics is proportional to the concentration of the reactants and rate constants of the reaction. Enzyme-catalysed reactions involve enthalpy change and show high selectivity and are highly suitable as thermal biosensors for a broad range of applications.

The most common calorimetric technique, differential scanning calorimetry (DSC) is used to measure the heat generated by the differential thermal activity of the biomolecular sample and reference material in two calorimetric chambers whose temperatures are scanned over time (Lerchner et al. 2008). However, the conventional DSC systems are bulky, require large sample volume, and are not well suited for rapid detection of analytes (MicroCal 2017). Therefore, microcalorimetric biosensing has been developed using microsystem and microfabrication technologies to overcome the weaknesses of the conventional calorimeters (Hohne et al. 2004). The lower background signal noise is a key performance parameter of microcalorimetric biosensors. The technological advancement in microsystem and microfabrication technologies has the potential to show great promise in the utilization of calorimetric biosensors in disease diagnosis.

22.6 Antibodies as Biorecognition Elements

Biorecognition is an important aspect in biosensor design and the biorecognition probe needs to be chosen carefully. The antibodies (immunoglobulins—Ig's) are the most popular class of biorecognition elements owing to their high binding affinity and specificity towards the target (Connelly and Baeumner 2012). The Ig is comprised

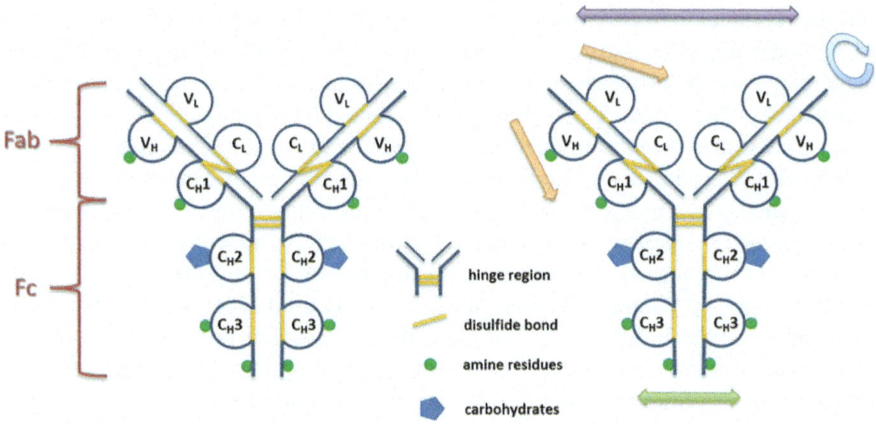

Fig. 22.4 Structure of antibodies comprised of four polypeptide chains, two identical shorter light chains (V_L) and two identical larger heavy chains (V_H). A light chain is coupled to a heavy chain via a disulfide bond and the heavy chains are connected with each other by covalent disulfide bridges and noncovalent bonds which correspondingly form the typical Y-shaped antibody structure. The two Fabs are antigen-binding fragments, and Fc is a crystallizable fragment. In the right-hand structure, flexibility is outlined: Fab arm waving (violet), Fab rotation (blue), Fab elbow bend (orange), and Fc wagging (green) (*Source* Farka et al. 2017)

of four polypeptide chains with two identical shorter light chains (VL) and two identical larger heavy chains (VH). A light chain is connected to a heavy chain by a disulfide bond and the heavy chains are connected with each other by covalent disulfide bridges and noncovalent bonds which correspondingly form the typical Y-shaped antibody structure (Fig. 22.4). The light and heavy chains consist of conserved protein structures called domains corresponding to approximately 110 amino acids. The light chains have variable (VL) and constant (CL) domains. The heavy chains also have one variable domain (VH) and three (IgG, IgA, and IgD) or four (IgM and IgE) constant domains (CH_1, CH_2, CH_3, CH_4) depending on the type of antibody. The arms of the "Y" consist of the N-termini of each heavy chain associated with one of the light chains to create two antigen-binding domains termed as the "antigen-binding fragment" (Fab fragment or domain). The tail of the "Y" is formed by a combination of the C-termini of the two heavy chains and is termed the "crystallizable fragment" (Fc fragment or domain). The hinge region located between the CH_1 and CH_2 domains gives flexibility (arm rotation and waving) to the Fab fragment. Due to the variability of the V_L and V_H domains, the Fab fragment determines the specificity and ability to interact and bind the immunogen. Each antibody molecule has two antigen-binding sites formed by the regions, located at the tip of each arm of the "Y" shaped structure.

Antibodies are classified as polyclonal (PAb), monoclonal (MAb), and recombinant (RAb) antibodies. The predominant antibody form used in biosensors is monoclonal and polyclonal. However, recombinant antibodies are now gaining increasing significance for biosensor applications. MAbs are produced by hybridoma technology and are monospecific and more homogeneous than PAbs. For the use of monoclonal antibody the important considerations include biochemical/physicochemical (class, subclass, molecular weight, light-chain composition, either N- and C-terminal amino acid sequence, secondary and tertiary structure) and biological/immunological properties (antigenic specificity, the ability for complement binding and activation, binding capacity, cytotoxic properties, ability to modify relevant antigens, capacity to stimulate immunocompetent cells and to induce secretion of cytokines or other mediators) of the antibody. Besides specificity (determination of unintentional reactivity with tissues distinct from the intended target) and cross-reactivity (reactivity with a range of tissues by immunohistochemical procedures) of the monoclonal antibody are also important. PAbs are produced by animal immunization and are extracted from the living animal serum. The PAbs are polyspecific in nature and recognize different epitopes on the same antigen. RAbs are monoclonal antibodies produced in bacterial cells using phage display technology.

The nanobodies (Nbs) found in camelids (llamas, dromedaries, alpacas) or in non-mammals (ratfish and sharks) are well suited for the development of immunosensor as they can be easily modified by site-specific functional groups with a negligible loss of specificity and affinity (Flajnik et al. 2011). The Nbs are single-domain antibodies with molecular weights of 12–14 kDa and constitute the smallest domains of natural antibodies with antigen-binding capacity. Nbs are encoded by a single gene segment and can be produced in bacteria and yeast with less cost than conventional monoclonal antibodies.

The immobilization procedure is crucial in the development of an immunosensor and glass, gold, metal oxides, and carbon (i.e. graphite, graphene, glassy carbon, etc.) are the most common materials of optical and electrochemical transducer surfaces. The different procedures used for the immobilization of antibody or antigen on the transducer surface include physical adsorption, covalent chemical binding, and the use of ordered layers obtained by deposition of self-assembled layers (SAM). The most common practice is the use of ordered layers on the transducer surface for oriented immobilization and better presentation of the antibody to the target analyte.

22.6.1 Antibodies: Production, Purification, and Selection

The in vitro offers a high degree of control and standardization and is the preferred method of antibody production. It is advantageous over in vivo production method with respect to consistency of production, absence of contaminant immunoglobulins in the crude harvest, and viral safety. Reduction of animal usage is another advantage. However, for each in vitro production run, the microbial contamination in the culture vessels needs to be thoroughly examined prior to harvesting. The bulk culture fluid should be free from bacteria, mycoplasmal, and mycotic contamination and should be tested for the presence of viruses using a general test involving inoculation into suitable cell substrates. Simultaneously only one cell line should be cultivated in a production area and records must be kept of the cell lines cultivated in parallel with an evidence of the absence of cross contamination between them. In vivo production involves the exploitation of the immune system of leporine, murine, ovine, or avian hosts. The strain, origin, genotype, and age of the animal used for production should be specified. The animal used should be from a closed, specific pathogen-free (SPF) colony which is routinely monitored for the diseases. During in vivo production, the maximum permissible number of serial passages should be defined and restricted.

The purification is done to remove tissue debris, unwanted host cell derived proteins, nucleic acids, carbohydrates, viruses, and other impurities. The purification should be effective and not compromise the biological activity of the product. Further purification procedures should not impair relevant immunobiological features of the immunoglobulin. The choice of the nucleotide probe to detect DNA contamination should be relevant to the system used. A reduction factor at each stage of purification, and overall, should be established by using concentrations of viruses, host cell proteins, other impurities, and DNA in excess of that expected during normal production. Validation of the purification process should also include justification of the working conditions such as column loading capacity, length of use of the columns, column regeneration, and sanitisation.

When selecting antibodies for the detection of pathogens, the key parameters include sensitivity, selectivity, stability, immobilization, labelling, antibody size, ease of production, cost, commercial availability, and capacity for engineering affinity. Many bacterial strains share homologues of surface-presented proteins leading to the detection of multiple cell-types by a single antibody as such the selection of a highly specific epitope on the pathogen is a key consideration. Therefore, a constitutively expressed antigen which is species-specific is targeted. Where possible, the expression of this target antigen should not be highly dependent on the growth matrix of the concerned pathogen. Finally, the antibody should bind with its cognate target with sufficient strength to permit interrogation (high affinity). The identification of an antibody that satisfies these requirements can be facilitated by ELISA to reduce the number of potential antibodies to a smaller number followed by sensor-based analysis to identify the candidate with the best affinity for the target epitope. This antibody can further be used for incorporation on an immunosensor-based platform. These factors are optimized by recombinant antibodies as genetic modification facilitates improvements in selectivity, stability, size, and in addition aid effective immobilization (Townsend et al. 2006). Display libraries and high throughput screening facilitate improvements in sensitivity due to the ability to screen much larger recombinant libraries. The use of genetic insertion of tags for immobilization and coupling chemistries are useful methods to help in the orientation and immobilization of novel recombinant antibodies. Besides antibody selection, the immobilization of antibodies to the sensor surface is an important factor that influences biosensing mechanism. The immobilization depends on the properties of the sensor interface. For this purpose, gold, silver, glass, platinum, and silica type of surfaces have been studied. The sensor surface should possess less signal-to-noise ratio, excellent electrical and thermal conductivity, low diffusion rates, stability, and large surface area for high density probe immobilization (Holford et al. 2012). It is difficult to get all these characteristics on one sensor surface and further to keep the device small to be portable. However, the development of new generation sensor platforms compatible with the aforementioned sensor needs is possible with the improvement in matrices such as carbon nanotubes, self-assembled monolayers (SAMs), fabricated nanoparticles, and quantum dots (Holford et al. 2012). The most commonly used methods for coupling the antibodies to the sensor surface include passive absorption, matrix entrapment, encapsulation, covalent coupling, and affinity tags with their own pros and cons.

22.7 Importance of Nanoparticles in Biosensors

The nanoparticles (NPs) have been used in biosensors for improvement of pathogen detection efficiency, expeditiously, and cost effectively. The diagnostic nanoparticles have the ability to amplify the signal, provide a more effective environment for anchoring of the biological element and in the miniaturization. NPs improve the reproducibility of nanoimmunosensors because a high and targeted surface area allows faster and controlled charge transfer at lower overpotentials. The most important properties of nanoparticles include particle size and charge, core and surface properties, shape and flexibility, as well as multivalency and controlled synthesis. Besides, the enormous surface to volume proportion, higher reactivity, stability, and controlled particle size are important physicochemical characteristics superior to the mass material. These properties determine the targeting ability, toxicity, stability, and persistence of nanoparticles. The nanoparticles are classified into organic (proteins and peptide nanoparticles), inorganic (gold, silver, iron oxide, and calcium phosphate), and hybrid nanoparticles (composed of more than one set of nanoparticles, e.g. mixture of polymeric nanoparticles and liposomes) on the basis of their origin. The nanoparticles that are used for diagnostic applications include metallic nanoparticles (gold, silver), magnetic nanoparticles, carbon-based nanomaterials (carbon nanotubes, graphene, fullerene, carbon nanohorns, and carbon dots), and luminescent nanocrystals (quantum dots, photon-upconverting nanoparticles).

The gold (Au) NPs are the most common signal amplification labels for various kinds of immunoassays and immunosensors due to their unique optical, chemical, electrical, and catalytic properties (Fenzl et al. 2016). Besides, easy preparation and modification, superior biocompatibility, and intensive colour of Au NPs have drawn the focus of scientists on their use in the bioanalytical field (Jans and Huo 2012). The excellent stability compared to that of other metals, pronounced colours and sensitive plasmon changes makes Au NPs the first nanomaterial to be considered for improving the performance of classic immunoassays. The silver (Ag) NPs are the most common commercially produced and excellent candidates for detection tags in electrochemical sensing as they offer improved electrochemical activity and can be oxidized more easily as compared to Au NPs. However, the significantly lower stability, limited biocompatibility compared to those of Au NPs and difficulty to functionalize restricts the practical applications of Ag NPs in immunosensing (Li and Xu 2014). The magnetic NPs-based analytical applications are represented mainly by three strategies viz. magnetic preconcentration of an analyte due to its interaction with biofunctionalized nanoparticles, use of magnetic NPs as tags for the visualization and sensitive detection of immunocomplexes with a specific analyte and the integration of magnetic NPs into the transducer material or the modification of the sensor surface to improve the analytical signal. The functionalized magnetic NPs are referred as magnetic nanosensors by some authors.

Carbon nanotubes (CNTs) are divided into two main categories: single-walled carbon nanotubes (SWCNTs) and multiwalled carbon nanotubes (MWCNTs) and are often used in immunosensing due to their unique physical and chemical properties (Oliveira et al. 2014; Jiang et al. 2015). The graphene is a good candidate for the preparation of biosensors due to its outstanding mechanical, electrochemical, and thermal properties, optical transparency, and flexibility (Ramnani et al. 2016). Fullerene C60 is the smallest stable and most abundant member of the fullerene group that is well suited for the development of electrochemical sensors (Pilehvar and De Wael 2015). The fullerene C60 modified electrodes have an increased electroactive surface area and high electric conductivity, however, insolubility in water and hydrophobicity complicate the conjugation of biologically active molecules with fullerene (Yanez-Sedeno et al. 2016). Carbon nanohorns (CNHs) are convenient for amplification of the electrochemical signal and as a scaffold for immobilization of the antibodies or antigens (Liu et al. 2014). The electrochemical properties of carbon dots (carbon quantum dots and graphene quantum dots) have also been exploited to develop electrochemical biosensors (Wang and Dai 2015). The graphene variants and carbon nanotubes play an important role in the enhancement of the electron transfer rate in electrochemical immunosensors.

Therefore, the sensitive and specific immunosensor will allow farmers or extension workers to detect pathogen at early stages resulting in more efficient disease management in beneficial insects. The immunosensor will also be used for the on-site detection and distribution of native isolates of insect pathogens for the commercial production of insect-specific biopesticides. Moreover, long shelf life and no refrigeration for storage make the immuno-diagnostic kits very well suited for field use.

22.8 Conclusion

The present microbiological and molecular methods for pathogen detection are time-consuming, expensive, relatively insensitive, and entirely laboratory-based. Hence, cost-effective, sensitive, and specific detection techniques suitable for field diagnosis are highly desired. We believe that the biosensor technology is of high potential interest for in-field applications and exhibits great advantages over other techniques in terms of time, simplicity, and quantitative analysis. Besides, the biosensor technology exhibits high sensitivity and specificity and thus, can serve as good detector of pathogens with unprocessed or crude samples. The use of nanoparticles as signal amplification labels has the ability to improve the performance of biosensor for real-time monitoring of pathogens. A significant progress has been made in biosensor technology for the detection of human and veterinary pathogens in recent years. This technology is still in its infancy with respect to insect pathogen detection. However, the commercialization of biosensors has not yet maintained pace with the research activities carried out throughout the world due to cost and technical issues. It is important to continue working on these technologies for a successful transition from laboratory to field diagnosis. Therefore, to improve the development of nanodiagnostics, an increased interdisciplinary collaboration and knowledge exchange between scientists from various disciplines is necessary.

References

Adams JR, Goodwin RH, Wilcox TA (1977) Electron microscopic investigations on invasion and replication of insect baculoviruses *in vivo* and *in vitro*. Biol Cell 28(3):261–268

Ahmed A, Rushworth JV, Hirst NA, Millner PA (2014) Biosensors for whole-cell bacterial detection. Clinl Microbiol Rev 27(3):631–646

Alagappan M, Immanuel S, Sivasubramanian R, Kandaswamy A (2020) Development of cholesterol biosensor using Au nanoparticles decorated f-MWCNT covered with polypyrrole network. Arabian J Chem 13(1):2001–2010

Anderson RM, May RM (1980) Infectious diseases and population cycles of forest insects. Science 210(4470):658–661

Annio G, Jennings TL, Tagit O, Hildebrandt N (2018) Sensitivity enhancement of forster resonance energy transfer immunoassays by multiple antibody conjugation on quantum dots. Bioconjugate Chem 29(6):2082–2089

Arakawa A (1989) Diagnosis of nuclear polyhedrosis by latex agglutination test in the silkworm, *Bombyx mori*. J Seric Sci Japan 58:257–258

Arakawa A, Shimizu S (1990) Agglutination test using protein-A coated latex particles for the detection of nuclear polyhedrosis virus of silkworm, *Bombyx mori* (Lepidoptera: Bombycidae). Appl Ent 25:519–521

Arora S, Pastorella G, Byrne B, Marsili E, O'Kennedy R (2010) Microbial cells and biosensing: a dual approach–exploiting antibodies and microbial cells as analytical/power systems. In: Zacharis CK, Tzanavaras PD (eds) Reviews in pharmaceutical and biomedical analysis. Bentham Science Publishers, Sharjah, UAE, pp 63–75

Ayyar BV, Arora S (2013) Antibody-based biosensors for detection of veterinary viral pathogens. Adv Anim Vet Sci 1(4S):37–44

Baac H, Hajo´s JP, Lee J, Kim D, Kim SJ, Shuler ML (2006) Antibody-based surface plasmon resonance detection of intact viral pathogen. Biotech Bioeng 94:815–819

Bahadir EB, Sezginturk MK (2016) A review on impedimetric biosensors. Artif Cells Nanomedicine Biotechnol 44(1):248–262

Behlke MA, Huang L, Bogh L, Rose S, Devor EJ (2005) Fluorescence and fluorescence applications. Integrated DNA Technologies:1–13

Boltovets PM, Snopok BA, Boyko VR, Shevchenko TP, Dyachenko NS, Shirshov YM (2004) Detection of plant viruses using a surface plasmon resonance via complexing with specific antibodies. J Virol Methods 121:101–106

Chaivisuthangkura P, Srisuk C, Rukpratanporn S, Longyant S, Sridulyakul P, Sithigorngul P (2009) Rapid and sensitive detection of*Penaeus monodon* nucleopolyhedrovirus by loop-mediated isothermal amplification. J Virol Methods 162(1–2):188–193

Chen YP, Pettis JS, Corona M, Chen WP, Li CJ et al (2014) Israeli acute paralysis virus: epidemiology, pathogenesis and implications for honey bee health. PLoS Pathog 10(7): e1004261. https://doi.org/10.1371/journal.ppat.1004261

Connelly J, Baeumner A (2012) Biosensors for the detection of waterborne pathogens. Anal Bioanal Chem 402:117–127

Conroy PJ, Hearty S, Leonard P, O'Kennedy RJ (2009) Antibody production, design and use for biosensor–based applications. Semin Cell Dev Biol 20:10–26

Cory JS, Myers JH (2003) The ecology and evolution of insect baculoviruses. Ann Rev Ecol Evol Syst 34:239–272

Cox-Foster DL, Conlan S, Holmes EC, Palacios G, Evans JD, Moran NA, Quan PL, Briese T, Hornig M, Geiser DM, Martinson V (2007) A metagenomic survey of microbes in honeybee colony collapse disorder. Science 318(5848):283–287

Damborsky P, Svitel J, Katrlik J (2016) Optical biosensors. Essays Biochem 60:91–100

Danks C, Barker I (2000) On-site detection of plant pathogens using lateral-flow devices. Bull. EPPO/OEPP 30:421–426

Danks C, Ostoja-Starzewska S, Flint J, Banks J (2003) The development of a lateral flow device for the discrimination of OTA producing and non-producing fungi. Aspects Appl Biol 68:21–27

Del Rincón-Castro MC and Ibarra JE (2011) Entomopathogenic viruses. In: Biological control of insect pests, 1st edn. Studium Press, India, pp 29–64

Devarakonda S, Singh R, Bhardwaj J, Jang J (2017) Cost-effective and handmade paper-based immunosensing device for electrochemical detection of influenza virus. Sensors 17(11):2597

Dieffenbach CW, Dveksler GS (2003) PCR primer: a laboratory manual (No. Ed. 2). Cold Spring Harbor Laboratory Press

Dill K, Stanker LH, Young CR (1999) Detection of *Salmonella* in poultry using a silicon chip-based biosensor. J Biochem Biophys Methods 41:61–67

Du X, Zhou J (2018) Application of biosensors to detection of epidemic diseases in animals. Res Vet Sci 118:444–448

Elderd BD (2013) Developing models of disease transmission: insights from ecological studies of insects and their baculoviruses. PLoSP Pathog 9(6):e1003372

Eroglu GB, Nalcacioglu R, Demirbag Z (2019) A new *Helicoverpa armigera* nucleopolyhedrovirus isolate from *Heliothis peltigera* (Denis & Schiffermuller) (Lepidoptera: Noctuidae) in Turkey. Turk J Biol, 43(5):340–348

Evans H, Shapiro M (1997) Viruses. In: Lacey LA (ed) Manual of techniques in insect pathology. Academic Press, San Diego, CA, pp 17–53

Faridbod F, Sanati AL (2019) Graphene quantum dots in electrochemical sensors/biosensors. Curr Anal Chem 15(2):103–123

Farka Z, Čunderlová V, Horáčková V, Pastucha M, Mikušová Z, Hlaváček A, Skládal P (2018) Prussian blue nanoparticles as a catalytic label in a sandwich nanozyme-linked immunosorbent assay. Analyt Chem 90(3):2348–2354

Farka Z, Juřík T, Kovar D, Trnkova L, Skladal P (2017) Nanoparticle-based immunochemical biosensors and assays: recent advances and challenges. Chem Rev 117:9973–10042

Fenzl C, Hirsch T, Baeumner AJ (2016) Nanomaterials as versatile tools for signal amplification in (bio)analytical applications. TrAC Trends Anal Chem 79:306–316

Flajnik MF, Deschacht N, Muyldermans SA (2011) Case of convergence: why did a simple alternative to canonical antibodies arise in sharks and camels? PLoS Biol 9:e1001120

Gani M, Chouhan S, Babulal Gupta RK, Khan G, Kumar NB, Saini P, Ghosh MK (2017a) *Bombyx mori* nucleopolyhedrovirus (BmNPV): its impact on silkworm rearing and management strategies. J Biol Control 31(4):189–193

Gani M, Gupta RK, Zargar SM, Kour G, Monobrullah M, Kandasamy T, Mohanasundaram A (2017b). Molecular identification and phylogenetic analyses of multiple nucleopolyhedrovirus isolated from *Lymantria obfuscata* (Lepidoptera: Lymantriidae) from India. Appl Entomol Zool 52(3):389–399. https://doi.org/10.1007/s13355-017-0490-3

Gani M, Mary Josepha AV, Satish L, Chouhan S, Gadwala M, Ahmad MN, Sivaprasad V (2018) Prevalence and molecular characterization of viruses causing diseases in *Bombyx mori* L. from different climatic regions of India. J Biol Control 32(4):189–193

Gani M, Taskeena Hassan, Saini P, Gupta RK, Bali K (2019) Molecular phylogeny of entomopathogens. In: Khan MS, Ahmad W (eds) Microbes and their metabolites for sustainable insect pest and vector management. Springer Nature Switzerland. https://doi.org/10.1007/978-3-030-23045-6_3

Ge X, Asiri AM, Du D, Wen W, Wang S, Lin Y (2014) Nanomaterial-enhanced paper-based biosensors. TrAC Trends Anal Chem 58:31–39

Gehring AG, Patterson DL, Tu SI (1998) Use of a light-addressable potentiometric sensor for the detection of *Escherichia coli* O157:H7. Anal Biochem 258:293–298

Ghindilis AL, Atanasov P, Wilkins M, Wilkins E (1998) Immunosensors: electrochemical sensing and other engineering approaches. Biosens Bioelectron 13:113–131

Girigoswami K, Akhtar N (2019) Nanobiosensors and fluorescence based biosensors: an overview. Int J Nano Dimens 10(1):1–17

Gobi KV, Tanaka H, Shoyama Y, Miura N (2004) Continuous flow immunosensor for highly selective and real-time detection of sub-Ppb levels of 2-hydroxybiphenyl by using surface plasmon resonance imaging. Biosens Bioelectron 20:350–357

Gronewold TMA (2007) Surface acoustic wave sensors in the bioanalytical field: recent trends and challenges. Anal Chim Acta 603:119–128

Guo X (2012) Surface plasmon resonance based biosensor technique: a review. J Biophoton 5:483–501

Gupta RK, Gani M, Kaul V, Bhagat RM, Bali K, Samnotra RK (2016) Field evaluation of *Lymantria obfuscata* multiple nucleopolyhedrovirus for the management of Indian gypsy moth in Jammu & Kashmir, India. Crop Prot 80:149–158. https://doi.org/10.1016/j.cropro.2015.11.008

Hammond JL, Formisano N, Estrela P, Carrara S, Tkac J (2016) Electrochemical biosensors and nanobiosensors. Essays Biochem 60:69–80

Harlow E, Lane D (1988) Antibodies: a laboratory manual. Cold Spring Harbor Laboratory Press, Plainview, New York, p 726

Helali S, Martelet C, Abdelghani A, Maaref MA, Jaffrezic-Renault N (2006) A disposable immunomagnetic electrochemical sensor based on functionalised magnetic beads on gold surface for the detection of atrazine. Electrochim Acta 51:5182–5186

Hohne GWH, Hemminger WF, Flammersheim HJ (2004) Differential scanning calorimetry, 2nd edn. Springer, Berlin

Holford TR, Davis F, Higson SP (2012) Recent trends in antibody based sensors. Biosens Bioelectron 34:12–24

Huang C, Li J, Tang Y, Chen Y, Jin G (2011) Detection of duck hepatitis virus serotype 1 by biosensor based on imaging ellipsometry. Curr Appl Phys 11:353–357

Huang YF, Chen TH, Chang ZT, Wang TC, Lee SJ, Kim JC, Kim JS, Chiu KP, Nai YS (2019) Genomic sequencing of *Troides aeacus* nucleopolyhedrovirus (TraeNPV) from golden birdwing larvae (*Troides aeacus formosanus*) to reveal defective Autographa californica NPV genomic features. BMC genomics, 20(1):1–16

Iwano H, Ishihara R (1981) Inhibitory effect of several chemicals against the hatch of *Nosema bombycis* spores [pebrine of silkworm]. J Seric Sci Japan 50(4):276–281

Jans H, Huo Q (2012) Gold nanoparticle-enabled biological and chemical detection and analysis. Chem Soc Rev 41:2849–2866

Jarocka U, Wąsowicz M, Radecka H, Malinowski T, Michalczuk L, Radecki J (2011) Impedimetric immunosensor for detection of plum pox virus in plant extracts. Electroanalysis 23(9):2197–2204

Jaroenram W, Kiatpathomchai W, Flegel TW (2009) Rapid and sensitive detection of white spot syndrome virus by loop-mediated isothermal amplification combined with a lateral flow dipstick. Mol Cell Probes 23(2):65–70

Jiang L, Han J, Li F, Gao J, Li Y, Dong Y, Wei Q (2015) A sandwich-type electrochemical immunosensor based on multiple signal amplification for α-fetoprotein labeled by platinum hybrid multiwalled carbon nanotubes adhered copper oxide. Electrochim Acta 160:7–14

Jun Y, Li C, Wang D, Lui S (2010) FRET-derived ratiometric fluorescent K+ sensors fabricated from thermoresponsivepoly (n-isopropylacrylamide) microgels labeled with crown ether moieties. J Phys Chem B 114:12213–12220

Kamikawa TL, Mikolajczyk MG, Kennedy M, Zhang P, Wang W, Scott DE, Alocilja EC (2010) Nanoparticle–based biosensor for the detection of emerging pandemic influenza strains. Biosens Bioelectron 26:1346–1352

Kawarabata T, Ishihara R (1984) Infection and development of Nosema bombycis (Microsporida: Protozoa) in a cell line of *Antheraea eucalypti*. J invertebrate Pathol 44(1):52–62

Khater M, de la Escosura-Muñiz A, Merkoçi A (2017) Biosensors for plant pathogen detection. Biosens Bioelectron 93:72–86

Koch R (1884) Die Aettiologie der Tuberkulose. Mitt Kaiser Gesundh, pp 1–88

Koczula KM, Gallotta A (2016) Lateral flow assays. Essays Biochem 60(1):111–120

Kopparthy VL, Tangutooru SM, Guilbeau EJ (2015) Label free detection of L-Glutamate using microfluidic based thermal biosensor. Bioengineering 2:2–14

Krokene P, Heldal I, Fossdal CG (2013) Quantifying Neodiprion sertifer nucleopolyhedrovirus DNA from insects, foliage and forest litter using the quantitative real-time polymerase chain reaction. Agric For Entomol 15(2):120–125

Kumar CS, Rao GR, Sireesha K, Kumar PL (2011) Isolation and characterization of baculoviruses from three major lepidopteran pests in the semi-arid tropics of India. Indian J Virol 22(1):29–36

Kumar CS, Sireesha K, RangaRao GV, Reddy AS, Waliyar F, Rambabu C, Kumar PL (2007) Production of polyclonal antibodies for detection of nucleopolyhedrovirus infecting *Helicoverpa armigera*. Indian J Plant Prot 35(2):210–212

Kwon JH, Kim HT, Lee JH, Kim R, Heo M, Shin J, Lee HY, Cha YJ, Lee J (2017) Signal self-enhancement by coordinated assembly of gold nanoparticles enables accurate one-step-immunoassays. Nanoscale 9(42):16476–16484

Lange K, Rapp BE, Rapp M (2008) Surface acoustic wave biosensors: a review. Anal Bioanal Chem 391:1509–1519

Leonard P, Hearty S, Brennan J, Dunne L, Quinn J, Chakraborty T, O'Kennedy R (2003) Advances in biosensors for detection of pathogens in food and water. Enzyme Microbiol Tech 32:3–13

Lerchner J, Maskow T, Wolf G (2008) Chip calorimetry and its use for biochemical and cell biological investigations. Chem Eng Process 47:991–999

Li H, Xu D (2014) Silver nanoparticles as labels for applications in bioassays. TrAC Trends Anal Chem 61:67–73

Lin Y, Zhou Q, Tang D, Niessner R, Knopp D (2017) Signal-on photoelectrochemical immunoassay for aflatoxin B1 based on enzymatic product-etching MnO_2 nanosheets for dissociation of carbon dots. Analyt Chem 89(10):5637–5645

Liu B, Tang BZ (2013) Fluorescent sensors. Macromol Rapid Commun 34:704–709

Liu F, Xiang G, Yuan R, Chen X, Luo F, Jiang D, Huang S, Li Y, Pu X (2014) Procalcitonin sensitive detection based on graphene-gold nanocomposite film sensor platform and single-walled carbon nanohorns/hollow Pt chains complex as signal tags. Biosens Bioelectron 60:210–217

Liu S, Vijayendran D, Bonning BC (2011) Next generation sequencing technologies for insect virus discovery. Viruses 3(10):1849–1869

Lu M, Zhu H, Bazuin CG, Peng W, Masson JF (2019) Polymer-templated gold nanoparticles on optical fibers for enhanced-sensitivity localized surface plasmon resonance biosensors. ACS Sens 4(3):613–622

Luo X, Morrin A, Killard AJ, Smyth MR (2006) Application of nanoparticles in electrochemical sensors and biosensors. Electroanalysis 18(4):319–326

Luo Y, Nartker S, Miller H, Hochhalter D, Wiederoder M, Wiederoder S, Setterington E, Drzal LT, Alocilja EC (2010) Surface functionalization of electrospun nanofibers for detecting *E. coli*

O157:H7 and BVDV cells in a direct–charge transfer biosensor. Biosens Bioelectron 26:1612–1617
Luong JH, Male KB, Glennon JD (2008) Biosensor technology: technology push versus market pull. Biotechnol Adv 26:492–500
McCarthy WJ, Gettig RR (1986) Current developments in baculovirus serology. In: Granados RR, Federici BA (eds) The biology of baculoviruses, vol I. CRC Press, Boca Raton, FL, pp 147–148
MicroCal LLC (2017) Use of isothermal titration calorimetry to measure enzyme kinetic constants. Available [Online]: www.microcalorimetry.com
Mikušová Z, Farka Z, Pastucha M, Poláchová V, Obořilová R, Skládal P (2019) Amperometric immunosensor for rapid detection of honeybee pathogen melissococcus plutonius. Electroanalysis 31(10):1969–1976
Mohanty SP, Kougianos E (2006) Biosensors: a tutorial review. IEEE Potent 25:35–40
Monosik R, Stredansky M, Sturdik E (2012) Biosensors—classification, characterization and new trends. Acta Chim Slov 5:109–120
Moscardi F (1999) Assessment of the application of baculoviruses for control of Lepidoptera. Ann Rev Entomol 44:257–289
Nakamura C, Hasegawa M, Nakamura N, Miyake J (2003) Rapid and specific detection of herbicides using a self-assembled photosynthetic reaction center from purple bacterium on an SPR chip. Biosens Bioelectron 18:599–603
Naser WL, Miltenburger HG (1983) A rapid, sensitive, and specific assay system for the detection of the *Autographa californica* nuclear polyhedrosis virus using a monoclonal antibody. Mitteilungen der Deutschen Gesellschaft fuer Allgemeine und Angewandte Entomologie, Germany, FR
Nataraju B, Sivaprasad V, Datta RK, Gupta SK, Shamim M (1994) Colloidal textile dye-based dipstick immunoassay for the detection of nuclear polyhedrosis virus (BmNPV) of silkworm. Bombyx mori L. J Invertebrate Pathol 63(2):135–139
Ndao M (2009) Diagnosis of parasitic diseases: old and new approaches. Interdiscip Perspect Infect Dis 68(69):70–74
Neethirajan S, Ragavan V, Weng X, Chand R (2018) Biosensors for sustainable food engineering: challenges and perspectives. Biosensors (Basel) 8:E23
Ngo VKT, Nguyen DG, Nguyen HPU, Tran VM, Nguyen TKM, Huynh TP et al (2014) Quartz crystal microbalance (QCM) as biosensor for the detecting of Escherichia coli O157:H7. Adv Nat Sci Nanosci 5:045004
Nguyen BT, Koh G, Lim HS, Chua AJ, Ng MM, Toh CS (2009) Membrane–based electrochemical nanobiosensor for the detection of virus. Anal Chem 81:7226–7234
Oliveira ON, Iost RM Siqueira JR, Crespilho FN, Caseli L (2014) Nanomaterials for diagnosis: challenges and applications in smart devices based on molecular recognition. ACS ApplMater Interfaces 6:14745−14766
Ozalp VC, Bayramoglu G, Erdem Z, Arica MY (2015) Pathogen detection in complex samples by quartz crystal microbalance sensor coupled to aptamer functionalized core-shell type magnetic separation. Anal Chim Acta 853:533–540
Palchetti I, Mascini M (2008) Electrochemical biosensors and their potential for food pathogen and toxin detection. Anal Bioanal Chem 391:455–471
Pilehvar S, De Wael K (2015) Recent advances in electrochemical biosensors based on fullerene-C60 nano-structured platforms. Biosensors 5:712–735
Pingarron JM, Yanez-Sedeno P, González-Cortes A (2008) Gold nanoparticle-based electrochemical biosensors. Electrochim Acta 53(19):5848–5866
Pohanka M, Skladal P (2008) Electrochemical biosensors—principles and applications. J Appl Biomed 6:57–64
Poláchová V, Pastucha M, Mikušová Z, Mickert MJ, Hlaváček A, Gorris HH, Skládal P, Farka Z (2019) Click-conjugated photon-upconversion nanoparticles in an immunoassay for honeybee pathogen Melissococcus plutonius. Nanoscale 11(17):8343–8351
Ramanavicius A, Kaušaitė A, Ramanavičienė A (2005) Polypyrrole-coated glucose oxidase nanoparticles for biosensor design. Sens Actuators B Chem 111:532–539

Ramnani P, Saucedo NM, Mulchandani A (2016) Carbon nanomaterial-based electrochemical biosensors for label-free sensing of environmental pollutants. Chemosphere 143:85–98
Rivers TM (1937) Viruses and Koch's postulates. J Bacteriol 33(1):1
Roberts PL, Naser W (1982) Characterization of monoclonal antibodies to the Autographa californica nuclear polyhedrosis virus. Virology 122(2):424–430
Rocha-Gaso MI, March-Iborra C, Montoya-Baides A, Arnau-Vives A (2009) Surface generated acoustic wave biosensors for the detection of pathogens: a review. Sensors 9:5740–5769
Sahoo PR, Sethy K, Mohapatra S, Panda D (2016) Loop mediated isothermal amplification: an innovative gene amplification technique for animal diseases. Vet World 9(5):465
Serrano-Andres L, Serrano-Perez JJ (2012) Calculation of excited states: molecular photophysics and photochemistry on display. Netherlands; Springer Sci Busin Media 14:483–560
Shamim M, Baig M, Datta RK, Gupta SK (1994) Development of monoclonal antibody-based sandwich ELISA for the detection of nuclear polyhedra of nuclear polyhedrosis virus infection in *Bombyx mori* L. J Invertebr Pathol 63:151–156
Shamim M, Baig M, Nataraju B, Datta RK, Gupta SK (1995) Evaluation of protein-a linked monoclonal antibody latex agglutination test for diagnosis of nuclear polyhedrosis virus (BmNPV) of silkworm *Bombyx mori* L. J Immunoass 16(2):155–166
Sharma A, Sharma N, Kumari A, Lee HJ, Kim T, Tripathi KM (2020) Nano-carbon based sensors for bacterial detection and discrimination in clinical diagnosis: a junction between material science and biology. Appl Mater Today 18:100467
Shimizu S, Arakawa A (1986) Latex agglutination test for the detection of the cytoplasmic polyhedrosis virus and the densonucleosis virus of the silkworm, *Bombyx mori* L. J Seric Sci Japan 56:169–170
Shimizu S, Ohaba M, Kanda K, Aizawa X (1983) Latex agglutination test for the detection of flacherie virus of the silkworm, *Bombyx mori* L. J Invertebr Pathol 42:151–155
Shimizu S, Tauchi S (1991) Protein-A coated latex linked antiserum test for the detection of the densonucleosis virus of the silkworm, *Bombyx mori* L. J Invertebr Pathol 57:124–125
Sotiropoulou S, Gavalas V, Vamvakaki V, Chaniotakis NA (2003) Novel carbon materials in biosensor systems. Biosens Bioelectron 18(2–3):211–215
Stork NE (2018) How many species of insects and other terrestrial arthropods are there on Earth? Annl Rev Entomol 63:31–45
Thorne CM, Otvos IS, Conder N, Levin DB (2007) Development and evaluation of methods to detect nucleopolyhedroviruses in larvae of the Douglas-fir tussock moth, *Orgyia pseudotsugata* (McDunnough). App Environ Microbiol 73(4):1101–1106
Tiwari S, Shukla A, Tiwari DK, Tiwari V (2017) Biosensors: a novel diagnostic approach. Int J Sci Environ Technol 6(1):908–915
Tomkies V, Flint J, Johnson G, Waite R, Wilkins S, Danks C, Watkins M, Cuthbertson AG, Carpana E, Marris G, Budge G (2009) Development and validation of a novel field test kit for European foulbrood. Apidologie 40(1):63–72
Torrance L (1980) Use of protein-A to improve sensitization of latex particles with antibodies to plant virus. Ann Appl Biol 96:45–50
Townsend S, Finlay WJ, Hearty S, O'Kennedy R (2006) Optimising recombinant antibody function in SPR immunosensing: the influence of antibody structural format and chip surface chemistry on assay sensitivity. Biosens Bioelectron 22:268–274
Tran LT, Tran TQ, Ho HP, Chu XT, Mai TA (2019) Simple label-free electrochemical immunosensor in a microchamber for detecting newcastle disease virus. J Nanomater 2019:1–9
Van Iddekinge BH, Smith GE, Summers MD (1983) Nucleotide sequence of the polyhedrin gene of Autographa californica nuclear polyhedrosis virus. Virology 131(2):561–565
Velusamy V, Arshak K, Korostynska O, Oliwa K, Adley C (2010) An overview of food borne pathogen detection: in the perspective of biosensors. Biotechnol Adv 28:232–254
Vogel E, Santos D, Mingels L, Verdonckt TW, Broeck JV (2019) RNA interference in insects: protecting beneficials and controlling pests. Front Physiol 9:1912

Wang H, Xiu Y, Chen Y, Sun L, Yang L, Chen H, Niu X (2019) Electrochemical immunosensor based on an antibody-hierarchical mesoporous SiO_2 for the detection of Staphylococcus aureus. RSC Adv 9(28):16278–16287

Wang Z, Dai Z (2015) Carbon nanomaterial-based electrochemical biosensors: an overview. Nanoscale 7:6420–6431

Wei J, Mu Y, Song D, Fang X, Liu X, Bu L, Zhang H, Zhang G, Ding J, Wang W, Jin Q, Luo G (2003) A novel sandwich immunosensing method for measuring cardiac troponin I in sera. Anal Biochem 321:209–216

White RM, Voltmer FW (1965) Direct piezoelectric coupling to surface elastic waves. Appl Phys Lett 7:314

Wonsawat W, Limvongjaroen S, Supromma S, Panphut W, Ruecha N, Ratnarathorn N, Dungchai W (2020) A paper-based conductive immunosensor for the determination of *Salmonella typhimurium*. Analyst 13:1–26

Wu S, He YQ, Lu XM, Zhang XF, Shuai JB, Yu HZ, Jin CC, Mo HF (2017) Early and simultaneous detection of Nosema bombycis (Microsporidia: Nosematidae), nucleopolyhedrovirus (Baculoviridae), and densovirus (Parvoviridae) by multiplex real-time polymerase chain reaction in Bombyx mori (Lepidoptera: Bombycidae). Can Entomol 149(2):265–275

Xu J, Wan JW, Yang ST, Zhang SF, Xu N, Li N, Li JP, Wang HY, Bai X, Liu WS (2012) A surface plasmon resonance biosensor for direct detection of the rabies virus. Acta Vet Brno 81:107–111

Yanez-Sedeno P, Gonzalez-Cortes A, Aguí L, Pingarron JM (2016) Uncommon carbon nanostructures for the preparation of electrochemical immunosensors. Electroanalysis 28:1679–1691

Zhang XL, Xiao Y, Qian XH (2008) Aratiometric fluorescent probe based on FRET for imaging Hg^{2+} ions in living cells. Angew Chem Int Ed 47:8025–8029

Zhao X, Gao W, Zhang H, Qiu X, Luo Y (2020) Graphene quantum dots in biomedical applications: recent advances and future challenges. In: Handbook of nanomaterials in analytical chemistry. Elsevier, Oxford, UK, pp 493–505

Zheng L, Cai G, Wang S, Liao M, Li Y, Lin J (2019) A microfluidic colorimetric biosensor for rapid detection of *Escherichia coli* O157: H7 using gold nanoparticle aggregation and smart phone imaging. Biosens Bioelectron 124:143–149

Zhou Y, Wu J, Lin F, Chen N, Yuan S, Ding L, Gao L, Hang B (2015) Rapid detection of *Bombyx mori* nucleopolyhedrovirus (BmNPV) by loop-mediated isothermal amplification assay combined with a lateral flow dipstick method. Mol Cell Probes 29(6):389–395

The manufacturer's authorised representative in the EU is Springer Nature Customer Service Centre GmbH, Europaplatz 3, 69115 Heidelberg, Germany. If you have any concerns regarding our products, please contact ProductSafety@springernature.com

Printed and bound by CPI Group (UK) Ltd, Croydon, CR0 4YY

22/09/2026

02220955-0002